T0281073

PLANT PATHOGEN DETECTION AND DISEASE DIAGNOSIS

BOOKS IN SOILS, PLANTS, AND THE ENVIRONMENT

Soil Biochemistry, Volume 1, edited by A. D. McLaren and G. H. Peterson
Soil Biochemistry, Volume 2, edited by A. D. McLaren and J. Skujiņš
Soil Biochemistry, Volume 3, edited by E. A. Paul and A. D. McLaren
Soil Biochemistry, Volume 4, edited by E. A. Paul and A. D. McLaren
Soil Biochemistry, Volume 5, edited by E. A. Paul and J. N. Ladd
Soil Biochemistry, Volume 6, edited by Jean-Marc Bollag and G. Stotzky
Soil Biochemistry, Volume 7, edited by G. Stotzky and Jean-Marc Bollag
Soil Biochemistry, Volume 8, edited by Jean-Marc Bollag and G. Stotzky
Soil Biochemistry, Volume 9, edited by G. Stotzky and Jean-Marc Bollag
Soil Biochemistry, Volume 10, edited by Jean-Marc Bollag and G. Stotzky

Organic Chemicals in the Soil Environment, Volumes 1 and 2, edited by C. A. I. Goring and J. W. Hamaker
Humic Substances in the Environment, M. Schnitzer and S. U. Khan
Microbial Life in the Soil: An Introduction, T. Hattori
Principles of Soil Chemistry, Kim H. Tan
Soil Analysis: Instrumental Techniques and Related Procedures, edited by Keith A. Smith
Soil Reclamation Processes: Microbiological Analyses and Applications, edited by Robert L. Tate III and Donald A. Klein
Symbiotic Nitrogen Fixation Technology, edited by Gerald H. Elkan
Soil–Water Interactions: Mechanisms and Applications, Shingo Iwata and Toshio Tabuchi with Benno P. Warkentin

The Rhizosphere: Biochemistry and Organic Substances at the Soil–Plant Interface, Roberto Pinton, Zeno Varanini, and Paolo Nannipieri

Woody Plants and Woody Plant Management: Ecology, Safety, and Environmental Impact, Rodney W. Bovey

Metals in the Environment: Analysis by Biodiversity, M. N. V. Prasad

Plant Pathogen Detection and Disease Diagnosis: Second Edition, Revised and Expanded, P. Narayanasamy

Additional Volumes in Preparation

Handbook of Plant and Crop Physiology: Second Edition, Revised and Expanded, edited by Mohammad Pessarakli

Plant Roots: The Hidden Half, Third Edition, Revised and Expanded, edited by Yoav Waisel, Amram Eshel, and Uzi Kafkafi

Handbook of Postharvest Technology, edited by A. Chakraverty, Arun S. Mujumdar, G. S. V. Raghavan, and H. S. Ramaswamy

Enzymes in the Environment: Activity, Ecology, and Applications, edited by Richard G. Burns and Richard Dick

Environmental Chemistry of Arsenic, edited by William T. Frankenberger, Jr.

PLANT PATHOGEN DETECTION AND DISEASE DIAGNOSIS

Second Edition
Revised and Expanded

P. Narayanasamy

Centre for Plant Protection Studies
Tamil Nadu Agricultural University
Coimbatore, India

CRC Press
Taylor & Francis Group
Boca Raton London New York

CRC Press is an imprint of the
Taylor & Francis Group, an **informa** business

CRC Press
Taylor & Francis Group
6000 Broken Sound Parkway NW, Suite 300
Boca Raton, FL 33487-2742

First issued in paperback 2019

© 2001 by Taylor & Francis Group, LLC
CRC Press is an imprint of Taylor & Francis Group, an Informa business

No claim to original U.S. Government works

ISBN-13: 978-0-8247-0591-6 (hbk)
ISBN-13: 978-0-367-39702-9 (pbk)

Visit the Taylor & Francis Web site at
http://www.taylorandfrancis.com

and the CRC Press Web site at
http://www.crcpress.com

To my parents
for their love and affection

Foreword

Plant Diseases take a heavy toll on yield and quality. Major famines in the past such as the Irish famine of the 1840s and the Bengal famine of the 1940s, were caused by disease epidemics that damaged potato and rice crops, respectively. The occurrence of physiological specialization among many disease-causing organisms further compounds problems relating to disease control and management. Many new diseases caused by different kinds of microbial plant pathogens that inflict appreciable losses have been reported. The need for rapid and reliable identification of the causes of such diseases to plan effective management strategies and to prevent the entry of new pathogens into the country is being increasingly realized. In light of the current situation this book is extremely timely. The coverage is comprehensive and the subject has been dealt with in an authoritative manner.

I am confident that the updated second edition of this book will be found very useful by research scientists and scholars as well as extension workers and policy makers. We owe a deep debt of gratitude to Dr. Narayanasamy for this labor of love.

M. S. Swaminathan
Chairman, M.S. Swaminathan Research Foundation,
and Holder of the UNESCO Chair of Eco-technology
Chennai, India

Preface to the Second Edition

The first edition (1997) of this book was well received by all concerned with crop disease diagnosis and management. The importance of rapid detection and precise identification of microbial pathogens for achieving effective crop disease management has been increasingly realized, as reflected by numerous research papers published in the past few years. Several new microbial plant pathogens have been identified, and the phylogenetic relationships of many pathogens, have been established based on new information. The inadequacy of morphological and biochemical characteristics alone to establish the identity of microbial pathogens has been demonstrated. Several studies have demonstrated the need to make wider use of modern molecular methods that provide more reliable information for the detection, identification, and differentiation of microbial plant pathogens.

Many national and international seminars and symposia have focused the attention of researchers and personnel connected with crop disease management on the imperative need to alert crop growers to techniques that can yield rapid and reliable results. The primary objective is to encourage the application of techniques that are simple, cost-effective, and adaptable for different field conditions. This second edition has been revised and expanded to information that has been compiled since the publication of the first edition. Newer techniques that are more sensitive and rapid are presented in Appendices to appropriate chapters. It is hoped that researchers, teachers, extension specialists, and students will find this second edition to be a useful source of information for planning their strategies to contain microbial pathogens that cause economically significant crop diseases.

It is with a sense of great respect and admiration that I profusely thank Prof. M. S. Swaminathan, Holder of the UNESCO Chair of Eco-technology, who has played a leading role in the Green Revolution and ecological preservation in India, for writing the Foreword for this book. I also offer my appreciation to all my colleagues for their useful suggestions, especially to Dr. T. Ganapathy for help in preparing the illustrations and to Mrs. K. Mangayarkarasi for assistance in preparing the manuscript. Permission given by editors and publishers to reproduce photographs is acknowledged individually in the text.

Finally, I feel much gratified for the immense affection of my wife Mrs. N. Rajakumari, son Mr. N. Kumaraperumal, and daughter, Mrs. Nirmala Suresh, who provided enormous encouragement and an enviable environment during the preparation of the second edition.

P. Narayanasamy

Preface to the First Edition

There are many kinds of plant pathogens ranging from ultramicroscopic entities to well-defined multicellular organisms, with wide variations in pathogenic potential. They are known to be the principal causes of destructive diseases of many economically important crops cultivated all over the world. Many of them are widely distributed and survive in varied habitats. It is well recognized that development of effective crop disease management depends on the rapid detection and precise identification of the pathogen(s) causing the disease in question. In this context, knowledge of the different methods available for the detection and identification of pathogens is a basic requirement for the successful management of disease(s) affecting the various crops in any location. This volume provides this vital information on currently applied methods of detection and diagnosis.

During the past decade, many sensitive methods of detecting microbial plant pathogens have been developed as a result of intensive research efforts undertaken in different laboratories. In this book, both conventional and modern molecular methods of detecting plant pathogens have been described in an easily understandable manner to enable researchers both in the laboratory and the field, who may not be very familiar with molecular techniques, to understand and follow these methods in their investigations. Many examples for the detection, identification, differentiation, and quantification of different kinds of microbial pathogens are included, with protocols in appendices of the appropriate chapters. The usefulness and limitations of the different methods are indicated. Researchers and personnel of disease diagnostic centers, plant protection, plant quarantine, and seed certification services will find a wide choice of techniques to suit the requirements and facilities available. It is hoped that this book will serve as a valuable reference in this aspect of plant pathology.

I am very happy to express my sincere thanks to all my colleagues and students who helped me in preparing the manuscript and illustrations for this book. Dr. K. Umamaheswaran and Dr..T. Ganapathy need special mention. Photographs and figures provided by scientists are individually acknowledged in the text. I wish to thank the editors and publishers for permission to reproduce figures and photographs included in this volume.

I thank Mr. M. Anifa and Mrs. K. Mangayarkarasi for typing the manuscript.

P. Narayanasamy

Contents

Contents

PLANT PATHOGEN DETECTION AND DISEASE DIAGNOSIS

1
Introduction

The incidence of plant diseases even in the prehistoric period can be inferred from fossils about 250,000 years old, and it has been estimated that about 80,000 diseases may affect various crops, resulting in losses as high as U.S. $60 billion throughout the world annually (Klausner, 1987; Agrios, 1969; Chu et al., 1989). Crop diseases are caused by distinct groups of organisms, predominantly fungi, bacteria, mollicutes, viruses, and viroids. Nematodes and protozoa are able to cause disease in some crops. Effective management of crop diseases depends essentially on rapid detection and accurate identification of the pathogens causing them.

Pathogen detection and disease diagnosis may be required for several purposes: a) to determine the presence and quantity of the pathogen(s) in a crop in order to take plant protection measures; b) to assess the effectiveness of application of cultural, physical, chemical, or biological methods of containing the pathogens; c) to certify seeds and planting materials for plant quarantine and certification programs; d) to determine the extent of disease incidence and consequent yield loss; e) to assess pathogen infection in plant materials in breeding programs; f) to detect and identify new pathogens rapidly to prevent further spread; g) to study taxonomic and evolutionary relationships of plant pathogens; h) to resolve the components of complex diseases incited by two or more pathogens; and i) to study pathogenesis and gene functions.

In this book discussion is confined to microbial plant pathogens; the information useful for the diagnosis of crop diseases incited by microbial pathogens is presented in the following chapters.

1.1 NATURE AND CAUSES OF PLANT DISEASES

All plant species are affected by one or more diseases. Cultivated plants are infected by thousands of diseases all over the world. Each kind of crop may be af-

1

fected by 100 or more diseases, causing different magnitudes of losses. Crops may suffer as a result of diseases induced by pathogenic (biotic) or physiogenic (abiotic) causes. Identification of the cause of the disease is the basic and primary requirement of disease diagnosis. Many plant pathogens induce characteristic symptoms in susceptible plant species, whereas some pathogens, especially viruses, may produce general systemic symptoms similar to those caused by environmental factors such as nutritional disorders.

Pathogenic causes of plant diseases may be microbes, such as fungi, bacteria, viruses, viroids, and phytoplasmas and others, such as nematodes and parasitic phanerogams. The physiogenic (nonpathogenic) causes may be nutrient deficiencies, toxicity due to excess of minerals, lack or excess of soil moisture, temperature extremes, light, oxygen, air pollution, variations in soil pH, etc. Pathogenic diseases can be transmitted from infected plants to healthy plants by using appropriate inoculation methods. On the other hand, physiogenic diseases are nontransmissible, and the affected plant may recover from the disease if the adverse condition is removed. The pathogenic diseases are far more numerous and varied than the physiogenic diseases, and hence the diagnosis of such diseases becomes more difficult. Such a complex situation is revealed by the tomato, which is reported to be infected by 80 species of fungi, 11 species of bacteria, 16 viruses, and several nematodes, whereas the apple and potato are infected by about 200 pathogenic diseases (Agrios, 1969).

The nature of the pathogen can be established by isolating it from infected plants and inducing the disease by inoculating healthy plants with the organism isolated. Different steps in Koch's postulates have to be followed; then the organism isolated can be considered the pathogen causing the disease. Frequently such a straightforward approach has not been found to be possible, and the use of reliable methods for correct identification of the pathogen has become necessary in the case of microbial plant pathogens. Apart from classification of the pathogens based on taxonomic characteristics, development of serological techniques and nucleic acid hybridization methods have been employed for rapid detection, identification, and assay of microbial pathogens in the recent years.

1.2 DISEASE DIAGNOSIS AND MANAGEMENT STRATEGIES

Rapid detection, identification, and quantification of pathogens are essential prerequisites for the development of effective disease management systems based on strategies such as cultural methods, biological control, use of chemicals, and breeding or engineering resistant cultivars. The conventional methods of isolating the pathogens from infected plants and identifying them based on taxonomical criteria are time-consuming, labor-intensive, and thus become expensive, although no sophisticated equipment is required. The study of molecular genetics of pathogenicity (ability to infect susceptible plant and cause disease) has provided basic information on the nucleotide sequences of the genes responsible for infection (Chapter 3). Such information is essential for nucleotide sequence analysis and designing specific primers based on pathogenicity genes for detection and differentiation of microbial pathogens by nucleic acid-based techniques.

Characterization of pathogens by biochemical methods requires relatively less time and labor. Analysis of gel electrophoretic protein profiles, fatty acid, and nutritional profiling are useful for the rapid identification of microbial pathogens especially the bacterial pathogens (Chapter 6). Electron microscopy (Chapter 7) and serological techniques such as enzyme-linked immunosorbent assay (ELISA) have been demonstrated to be rapid, sensitive, and reliable for the detection of pathogens in diverse sources such as plants, soil, water, and air (Chapter 8). Nucleic acid–based techniques, especially polymerase chain reaction (PCR)-based assays have wide application, since they are the most sensitive, reliable, and accurate tools for the detection, identification, differentiation, and establishing genetic relatedness between microbial pathogens (Chapter 9).

The various methods developed for microbial pathogen detection and disease diagnosis have the ultimate aim of providing essential information for crop disease management. The detection techniques have been employed for the assessment of disease intensity and crop losses caused by different pathogens. Production of disease-free seeds and propagative plant materials, such as cuttings, bulbs, and tubers, has been successful because of the effectiveness of detection methods. Cultural practices, chemical use, and prevention of the development of resistance in pathogens to chemicals and determination of relative resistance/susceptibility of cultivars are the management strategies relying on the effectiveness of detection techniques. Thus the selection and application of appropriate methods of disease diagnosis may be helpful not only to prevent the diseases reaching destructive proportion, but also to prevent excessive use of chemicals leading to environmental pollution and hazards to human beings and animals. The Disease Diagnostic Centers (DDCs) have the vital role in diagnosing crop diseases and providing technical advice to the growers who have to make necessary management decisions to prevent incidence and/or spread of crop diseases (Chapter 11) and to realize maximum monetary gain.

SUMMARY

The nature and causes of different diseases affecting various crops are described briefly to produce a basic understanding of the pathogens inducing such diseases. The need for the detection of pathogens in plants and other habitats is emphasized. Application of diagnostic methods for studying different aspects of host plant-pathogen interaction and disease management is indicated.

2
Characteristics of Pathogenic Microbes

Plant pathogenic microbes such as fungi, bacteria, and viruses belong to the same groups as the pathogens infecting human beings and animals (Fig. 2.1). None of the plant pathogens can infect humans or animals. However, the insect-transmitted propagative plant viruses can cause diseases both in plants and in their vectors. The plant pathogens are characterized by their ability to grow and multiply in the affected plants, resulting in characteristic symptoms, and to spread from diseased plants to healthy ones, causing new infections.

The plant pathogens become intimately associated with the host plant, drawing nutrients and water from the host, their parasitic nature leads to reduced efficiency in normal growth and reproduction of the host. Parasitism and pathogenicity are intimately associated, as the ability of the parasite to invade and become established in the host plant results in different types of disease symptoms, depending on the host-microbe combination. Some pathogens, including downy mildews, powdery mildews, rusts, and the entire group of viruses, require the presence of living hosts; they are known as obligate parasites. Others can live on living or dead hosts or host tissues; they are called nonobligate parasites. Nonobligate parasites include facultative saprophytes, which live most of the time or most of their life cycles as parasites but are capable of existing as saprophytes if required, and facultative parasites, which live most of the time on dead organic matter but can become parasitic under certain conditions. There frequently appears to be no correlation between the degree of parasitism of a pathogen and the severity of disease, since some weak pathogens may cause more damage to plants than more virulent pathogens. Pathogens may produce different kinds of enzymes, toxins, and other metabolites which may be responsible for structural and physiological changes which lead to specific symptoms of diagnostic value.

The characteristics of plant pathogenic microbes can be studied by using different traditional methods involving light microscopy in the case of fungi and bacteria, whereas electron microscopy is required for studying the characteristics

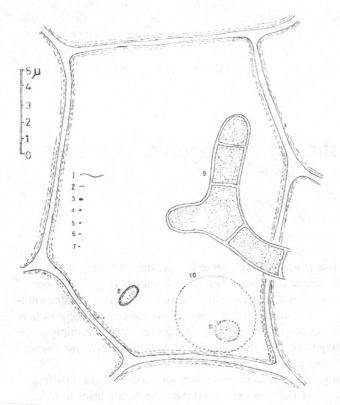

Figure 2.1   Microbial plant pathogens in a plant cell to indicate their relative sizes and shapes: 1, Beet yellows virus; 2, tobacco mosaic virus; 3, wheat striate mosaic virus; 4, wound tumor virus; 5, apple mosaic virus; 6, tobacco necrosis virus; 7, hemoglobin molecule; 8, bacterial cell; 9, fungus; 10, nucleus; 11, nucleolus. Academic Press Inc., USA. (Adapted from Agrios, 1969.)

of viruses and phytoplasmas (Chapter 7). For bacterial pathogen identification, in addition to determining morphological characteristics by light microscopy, it is essential to carry out many biochemical and physiological tests (Chapter 6).

2.1   LIGHT MICROSCOPY

Light microscopes can be used to examine the different vegetative/sexual reproductive structures formed by the fungi either in infected plant tissues or in cultures. Many fungal pathogens may be identified up to generic level and in some cases even up to species level by studying characters by means of light micro-scopes, leading to rapid diagnosis of the diseases caused by them.

2.1.1 Rapid Examination of Pathogens

The fungal pathogens present on the seeds and other plant parts can be directly ex-amined by a stereomicroscope, which is useful for observing fungal spore-bearing structures such as acervuli, pycnidia, or sclerotia produced on infected plant or-gans and seeds and for rapidly assessing the extent (or percentage) of infection of seeds. The fungal structures produced outside the plant tissues can be scraped by a sharp scalpel or razor blade and examined under a compound microscope. Dif-ferent models of compound microscopes are manufactured by various companies. The resolving power of these microscopes varies widely to suit different purposes. The compound microscope, irrespective of model, essentially consists of objec-tive and ocular lenses, light source, base, condenser, stage, arm, and body (Fig. 2.2) and of other accessories that allow it to be used as a phase contrast/fluores-cent microscope for certain specific studies. The description of various features of the microscope is provided by the manufacturer.

The magnification of a microscope depends on the magnification of the ob-jective and ocular lenses used, and the total magnification of an object can be de-termined by multiplying the magnification of the objective by that of the ocular lens. The resolving power of the microscope depends on the wavelength of the light source and the optical quality of the lenses. By substituting an ultraviolet light source for white (normal) light, the resolving power of the microscope can be increased. Proper illumination is required for achieving optimal resolution. The iris diaphragm regulates the amount of light passing through the condenser, and excessive light may obscure the specimen as a result of lack of contrast.

The upper end of the body tube supports the ocular lens (eyepiece) with a magnification of $10\times$-$12.5\times$ for viewing the image of the object (specimen). The monocular microscope has only one ocular lens, whereas the binocular micro-scope has two ocular lenses, with a provision for adjusting the distance between the two lenses for better viewing. The lower end of the body is fitted with a rotat-ing nosepiece to which objective lenses of varying magnification, usually desig-nated as low-power, high-power, and oil immersion lenses, are attached. More powerful microscopes are provided with additional ocular and objective lenses. The microscope may have a fixed stage with two clips to hold the slides or a me-chanical stage with a slide holder, and the assembly can be moved by means of ad-justment knobs. The substage attached to the stage consists of a condenser, an iris diaphragm, and a mirror. The condenser, consisting of several lenses, causes light to converge on the object, whereas the iris diaphragm regulates the angle and amount of light required. The light from the source is reflected by the concave side of the mirror via the condenser. Improved models have a built-in light source,

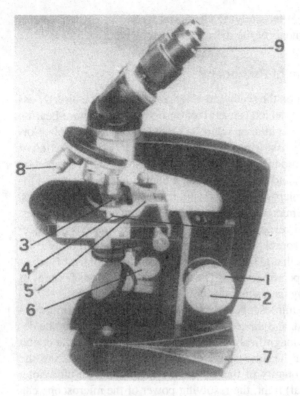

Figure 2.2 Light microscope: 1, coarse adjustment; 2, fine adjustment; 3, microscope slide; 4, slide holder; 5, mechanical stage; 6, condenser adjustment knob; 7, base; 8, objective lens; 9, ocular lens.

which is easily focused. The images of objects are viewed directly when low- and high-power objectives are used, whereas a drop of immersion oil is placed over the slide and then the oil immersion objective is put into position and then focused for viewing. All objective and ocular lenses should be thoroughly cleaned with lens paper moistened with xylol.

2.1.1.1   Induction of Sporulation

Fungal spores with characteristic features may not be present on the infected plant tissues. Incubation of such tissues in moist chambers maintaining high humidity may induce production of spores which can be examined under the microscope to allow the pathogen to be identified. Seed lots are incubated on moist blotters in a humid atmosphere to induce the production of identifiable structures or to produce symptoms of infection. Usually seeds of temperate crops are incubated at 14°-

20°C, whereas seeds of tropical crops are incubated at 28°C. In cultures of certain pathogenic fungi sporulating structures are not formed, unless they are exposed to certain treatments. Exposure of fungal cultures to wavelengths of near ultraviolet (NUV) region light is the most effective means of inducing sporulation in pathogens such as *Alternaria solani* and *Septoria lycopersici*. Cool white daylight fluorescent lamps emitting an appreciable amount of NUV radiation are suitable for this purpose. A cycle of 12 h of ultraviolet radiation alternating with 12 h of darkness is effective in the case of many fungi. However, the requirement of the individual fungal pathogen has to be determined. Production of sexual spores by many *Ascomycotina* pathogens is favored by very low temperatures. By storing plant tissues infected by powdery mildew pathogens at 0°–5°C for several weeks, formation of cleistothecia may be induced.

When seeds are incubated in several layers of filter paper, they usually germinate and symptoms of infection may appear or intensify. *Leptosphaeria maculans* forms pycnidia on the seed coats of *Brassica* spp., and dark brown or black streaks appear on the coleoptile. Using a stereomicroscope, such infected seeds can be easily recognised. *Alternaria brassicae* and *A. brassicola* can be differentiated by the presence of conidia and absence of pycnidia, though *L. maculans* and *Alternaria* spp. produce similar symptoms of infection. Some pathogens produce some specific identifiable product(s), when incubated under NUV light. *Leptosphaeria nodorum* infection of wheat seeds leaves fluorescent products in the blotter (Kietreiber, 1980). The roll-tower or paper doily method of germination testing may also be useful to detect pathogen infection in seeds of cereals, legumes, and peanuts during the long period of incubation (Neergaard, 1977).

2.1.2 Preparation of Temporary and Semipermanent Mounts

The spores or sporulating structures formed on plant tissues as such or after incubation may be removed by using a scalpel or needles and placed into a drop of mounting fluid kept on a glass slide and stained. The mounting fluids or media and stains commonly used are presented in Appendix 2(i). After applying enough stain, a cover slip is placed over the mounting medium, carefully preventing entry of an air bubble, and is examined under the microscope for the presence of spores and sporulating structures. For taking free-hand sections, rectangular pieces of infected leaf tissues are usually placed in the partially split end of the pith cylinder taken from cassava stem or carrot. Then, using sharp razor blades, thin sections are cut either to observe the pathogenic structures in deep-seated tissues or to study the histopathological characteristics of the infected tissues. The sections are transferred to water kept in watch glasses or petri plates. Thin sections are selected and transferred to drops of mounting medium placed on glass slides by needles and stained and examined under the microscope. These slides can be preserved for some time by sealing the edges of the cover slip with a sealant (such as euparal or glyceel).

Using a strip of transparent cellophane tape, a temporary mount of fungal pathogen can be prepared. A drop of lactophenol cotton blue or aniline blue stain is placed in the center of a clean glass slide. A strip of transparent cellophane tape, about 10 cm in length, is held between the thumb and forefinger, and the sticky side of the tape is firmly pressed onto the surface of a sporulating fungal colony grown over a suitable medium in the petri dish. The tape is then gently removed and the sticky surface is placed over the drop of stain solution kept on the glass slide lengthwise and pressed onto the slide. The extended ends of the tape, if it is longer than the slide, may be folded over the ends of the slide. The spores and sporulating structures produced by the fungal pathogens adhering to the tape can be examined by this simple and rapid technique.

2.1.3   Preparation of Permanent Slides

The host-parasite relationships may be studied in detail by fixing the plant tissues infected by pathogens in different fixatives and embedding them in paraffin then cutting sections by using the microtome. Details of various steps have been excellently described by Johansen (1940). The different steps in the preparation of permanent slides are described briefly. The fixatives and the staining procedures used widely are presented in Appendices 2(ii) and (iii), respectively. The usefulness of the permanent preparations for rapid identification of plant pathogens is, however, limited.

2.1.3.1   Microtome Sectioning

a. Fixation of plant tissues.   Preparation of permanent slides involves killing and fixation of infected plant tissues in appropriate fixatives. Fixation aims at preservation of all cellular and structural elements in the natural living condition as far as possible. It may also allow ready recognition of structures which are obscured or become invisible when observed in the living conditions and help harden soft structures which may otherwise be damaged during subsequent treatments. The optimal time required for fixation depends on the plant materials and fixation fluids. Formalin-acetic acid-alcohol (FAA) mixture, Carnoy's fluid, chrom-acetic acid at different concentrations, and Navashin fluid have generally been used.

b. Washing.   It is necessary to wash the plant materials thoroughly, after allowing enough time for fixation, prior to dehydration and infiltration. When formalin-acetic acid-alcohol (FAA) mixture is used for fixation and tertiary butyl alcohol for dehydration, washing out killing fluid is not necessary. But washing in two changes of 50% ethyl alcohol may be needed in other methods. Woody materials have to be washed for 2 days in running tap water and then softened for a period of 3–6 weeks in a 50% aqueous solution of hydrofluoric acid. If Carnoy's fluid is used for fixation, washing in two changes of 95% ethyl alcohol, followed immediately by paraffin infiltration, is required. Materials fixed with chrom-acetic acid or Navashin fluid should be washed free of killing fluids in either running tap water or several changes of water.

c. Dehydration. All traces of water should be removed by using absolute alcohol before substituting it with the clearing agent or solvent of paraffin. Dehydration with tertiary butyl alcohol is satisfactory in most cases. The dehydration process begins with immersion in ethyl alcohol, the concentration of which increases from 5% to 11%, 18%, and 30%. The material is left in each solution for 2 hours. A series of solutions of water, ethyl, and tertiary butyl alcohols is used for further dehydration as recommended by Johansen (1940). Finally dry erythrosin dye is added to the last solution to give a red tinge to the material. This is useful for easy orientation of the material during embedding and microtoming. The material is placed into three changes of pure tertiary butyl alcohol; in one change it has to be immersed overnight.

d. Infiltration. The transfer of the dehydrated material in butyl alcohol to paraffin has to be done gradually. The material is transferred to a mixture of equal parts of paraffin oil and tertiary butyl alcohol and allowed to remain for 1 h or more. Then it is placed on the top of the solidified but not cold paraffin (Parowax) kept in a vial and covered with a butyl alcohol-paraffin mixture. The vial, with its contents, is immediately placed into the paraffin oven. As the paraffin melts and butyl alcohol is evaporated, the material will slowly sink through the melting paraffin (Parowax), until it settles on the bottom of the vial. The paraffin oil prevents damage to the tissues due to heat and also permits the wax to diffuse in gradually. After about an hour, the contents of the vial are poured off and replaced with melted pure Parowax. The material is transferred to two changes of Parowax during the next 6 h. Finally the material is transferred to a good-quality paraffin that will melt at temperatures around 56°C. It can be embedded after about 30 min.

e. Embedding. Embedding plant tissues in paraffin is the most common method. The paraffin infiltration procedures used by early workers required the use of clearing agents such as bergamot oil, cedar oil, and xylol (xylene). Later on, tertiary butyl alcohol was found to be a good substitute for the clearing agents mentioned and helped to reduce the time needed, select suitable materials, and eliminate excessive hardening of tissues.

The embedding process involves pouring of the contents of the vial into suitable receptacles, aligning the material in proper order, and rapidly cooling the entire mass. Simple folded paper trays or ready-made trays are used. The vial containing the material is placed over a gas flame, and, as the wax melts, the contents are quickly transferred to the tray. If necessary, melted paraffin may be poured just to cover the material. By using a needle heated slightly in the flame, pieces of material are arranged in the desired order, leaving enough space between specimens. After arranging the material, the paraffin should be cooled as quickly as possible. In warm weather conditions, ice water may be used for this purpose. The blocks of paraffin in which the plant materials are embedded are stored in small boxes.

f. Microtoming.   Microtomes are of two types, the sliding and rotary. The rotary microtome is generally used for sectioning materials embedded in paraffin, whereas the sliding microtome is employed for cutting sections of materials processed by other methods. In some models, the forward movement is directly related to the up-and-down movement (Minot microtome), whereas in later models (Spencer No. 820) the horizontal and vertical movements are entirely independent, providing greater stability and precision. Microtome knives are interchangeable, and use of different knives by different individuals is desirable.

With a sharp scalpel a straight furrow is cut across the paraffin block in which the plant material is embedded. A piece of tissue is then separated by holding the block firmly in the hands and breaking it apart along the cut. Paraffin is trimmed down the piece of tissue until it is enclosed in a thin shell of paraffin measuring about 3 mm at the top and 5 mm at the bottom. Square or rectangular hardwood blocks are dipped into melted paraffin to a depth of about 2 mm, and a mound of paraffin is formed on top of the wooden block when it is cooled. A scalpel or a similar flat spatula is heated and touched first to the paraffin on the wooden block and then to the bottom of the embedded piece, placing the two together while the paraffin is still more or less melted. Gently and carefully touching the two paraffins with the hot scalpel fuses them to form a single mass on the wooden block. The wooden block with the embedded plant tissue is placed into cold water to cool the paraffin. Excess paraffin is trimmed away using a sharp straight scalpel till the face of the block is either a square or a rectangle so that a straight ribbon of paraffin is formed when the block is cut by the knife of the microtome.

The wooden holder is inserted into the clamp and fixed firmly, taking care to see that the part of the wooden block projecting from the clamp is not more than 3 mm in length. The micrometer scale in the microtome is set to create sections of the desired thickness. The wheel is moved with a steady and even stroke. A ribbon is formed as sections are cut one by one. A straight ribbon is the most desirable.

The ribbon is placed on a piece of black cardboard or thick paper and cut into small sections of the required length (about 5 cm) by a sharp scalpel. The glass slides are smeared with Haupt's adhesive to form a very thin film and flooded with 3% formalin immediately. A section of ribbon is transferred to the slide, which is then placed on a warm plate at 43°C to flatten out the wrinkles in the paraffin. Then the slides are cooled and sections are positioned properly before the excess water is drained off. Absorbent paper can be used to remove as much water as possible. The slides are put aside in a dust-free place to dry completely and then stored until they are stained.

The paraffin has to be removed before staining with xylol (xylene). The slides are placed either individually or in groups in a jar containing xylol for 5 min or more. They are then slowly withdrawn from xylol and transferred to a jar containing a mixture of absolute alcohol and xylol (1:1). After 5 min, the slides are exposed to a mixture of absolute alcohol and ether (anesthetic ether) (1:1) and about 1% celloidin for 5–10 min. The slides are air dried and immersed in ethyl alcohol at 95% or 70% for 5 min and then in 35% ethyl alcohol for another 5 min. If an alcohol solution of stain is to be used, immersion in 35% alcohol can be omitted. Various stains used to stain the sections of tissues infected by fungal pathogens are indicated in Section 2.1.3.2. After staining of the sections, they are dehydrated before mounting them in Canada balsam. If aqueous stains are used, the sections are immersed in 35% alcohol and then in 70% and 95% alcohol. Immersion in 35% and 70% alcohol is not required if alcoholic stains are used, and they can be placed in 95% alcohol directly for dehydration.

The slides are then placed in a differentiator consisting of USP clove oil (1 part) and a 1:1 mixture of absolute alcohol and xylol (1 part) and moved slowly back and forth for about 10 sec. They are transferred to a jar of xylol to which a trace of absolute alcohol is already added to remove the moisture, if any. The slides are moved back and forth for a few seconds and then placed into a jar of pure xylol.

Before mounting, the slides are taken out and the lower surfaces are wiped with a clean cloth. A small drop of thin balsam is placed on top of the sections. A clean cover slip is dipped into a mixture of equal parts of 95% alcohol and xylol, excess fluid is removed by touching to a paper towel, and, after quickly passing through a clean alcohol flame, the cover slip is carefully placed over the balsam, taking care not to allow any air bubble into the balsam. The slides are then dried on a warming table or in an incubator at 60°C for 1–2 days. They are then sealed with a cellulose-based sealant (glyceel) or a mixture of camsal, sandarac, eucalyptol, and paraldehyde (euparal).

2.1.3.2 Stains

Various stains have been used to stain plant tissues to study the characteristics of cellular organization and organelles. Some of the more commonly used stains are described. For staining procedures refer to Appendix 2(ii) A–C.

a. Hematoxylin. Hematoxylin is a natural dye derived from the logwood *Hematoxylin campechianum.* The stain is always prepared in combination with different metallic salts, such as iron (always the ferric form), aluminum, and copper, since the dye as such has little affinity for tissues. Iron or aluminum acts as a mordant. Harris's or Delafield's hematoxylin solution is frequently used, with good results.

1. Harris's hematoxylin.

Hematoxylin crystals	5.0 g
Aluminum ammonium sulfate	3.0 g
50% Ethyl alcohol	1000 ml

The dye is dissolved along with the salt by heating. Mercuric oxide (6 g) is then added, boiled for 30 min, and filtered. Alcohol (50%) is added to return the material to the original volume and acidified by adding hydrochloric acid at the rate of 1 ml/100 ml solution.

2. Delafield's hematoxylin.

Hematoxylin crystals	4.0 g
95% Ethyl alcohol	25.0 ml
Ammonium aluminium sulfate (saturated solution)	400.0 ml

To the saturated ammonium aluminium sulfate, hematoxylin dissolved in alcohol is added drop by drop, and the mixture is exposed to light and air for 4 days. Then glycerol (10 ml) and methyl alcohol (100 ml) are added and allowed to stand for about 2 months for ripening.

3. Heidenhain's Iron Hematoxylin. A 10% solution of iron hematoxylin in absolute alcohol is prepared. This solution is diluted with distilled water to a 0.5% concentration, when the stain is to be used.

b. Fast green FCF. The dye fast green FCF belongs to the acidic diaminotriphenyl methane group. It stains tissues rapidly and does not fade even after long periods. It is prepared as 1% aqueous or 0.1% alcohol solution. Better results are obtained by using stain prepared by adding sufficient dry dye to a mixture of methyl cellosolve, absolute alcohol, and clove oil (1:1:1) to produce a dark green 0.5% solution.

c. Safranin. The stain safranin is frequently used in morphological and cytological investigations. Safranin solution is prepared by dissolving the dye (4.0 g) in methyl cellosolve (200 ml), adding alcohol (100 ml) and distilled water (100 ml) to this solution, then adding sodium acetate (4.0 g) and formalin (8.0 ml). Sodium acetate intensifies the color, while formalin acts as a mordant. As safranin generally overstains the sections, differentiation is necessary. This is achieved by adding picric acid to the dehydrating alcohol (95%). Excess stain has to be washed with distilled water.

2.1.3.3 Processing Plant Tissues Infected by Fungal Pathogens

The plant tissues infected by fungal pathogens are fixed in various fixatives and stained to study their morphological characteristics and the effect of infection on host tissues/organelles. Portions of potato tubers infected by *Spongospora subterranea,* which causes powdery scab disease, may be fixed in a chromacetic acid medium and stained in iron hematoxylin and fast green. Roots infected by *Plasmodiophora brassicae,* which causes cabbage club root disease, may be fixed in formalin-acetic acid-alcohol, followed by tertiary butyl alcohol dehydration. Medium chrom-osmo-acetic fluid or Navashin's fluid is suitable for fixing the tissues. Iron hematoxylin and fast green stains are useful.

The sporangiophores and sporangia in the pustules formed by white rust pathogen *Albugo* spp. can be studied by fixing the tissues in a chrom-acetic fluid or formalin-aceto-alcohol medium followed by staining with iron hematoxylin and counterstaining with orange G.

Leaves infected by powdery mildew fungi such as *Erysiphe cichoracearum* and *Sphaerotheca pannosa* may be fixed either in FAA or Navashin's fluid, then stained with iron hematoxylin. Leaves infected by *Taphrina deformans,* which causes peach leafcurl disease, are fixed in formalin-propiono-alcohol and then stained with either safranin and fast green or iron hematoxylin. Wheat leaves or stem infected by rust pathogens are fixed in formalin-aceto-alcohol and stained with a quadruple combination.

Bright-field microscopy was used for accurate detection of infection sites and fluorescence microscopy for examination of the initial histological responses associated with the host-pathogen interaction. Wheat leaves were spot-inoculated with the conidial suspension of *Pyrenophora tritici–repentis,* causing tan spot disease. The leaf pieces were cleared and stained with aniline blue (0.06%), cotton blue (0.035%), or Calcofluor White M2R (0.2%) as well as with aniline blue fluorochrome (0.001%) or acid fuchsin (1.0%) followed by fast green (0.5%). Resin-embedded sections were stained with acid fuchsin (1.0%) followed by toluidine blue O (0.05%) or Calcofluor White M2R (0.2%) (Dushnicky et al., 1998). Kuo Kerchung (1999) reported that cotton blue may be used along with safranin and Calcofluor White M2R can be used for staining paraffin sections for the detection and differentiation of fungal structures produced in the infected host plants.

2.1.4 Preparation of Ultrathin Sections for Light Microscopy

Excellent preparations for light microscopy can be obtained by glutaraldehyde-osmium fixation, though this procedure is generally followed for electron microscopy (Chapter 7).

2.1.4.1 Fixation

The plant tissue is cut into small pieces by using a sharp razor blade in a pool of 3% glutaraldehyde in phosphate buffer, pH 7.0–7.4. The tissue is then transferred to a vial containing 3% buffered glutaraldehyde at 4°C. The fixative is changed after 2–3 hr and incubated for 12–16 hr in a refrigerator. The tissue should be washed free of glutaraldehyde by using several changes of cold buffer at 20 min intervals. The tissue is postfixed in 1% buffered osmium tetroxide for 1–2 hr at 4°C and washed with buffer thrice at 10 to 30 minute intervals.

2.1.4.2 Embedding

Epoxy resins are used as embedding matrix. Epon 812 is commonly employed, after dehydration of fixed plant tissues in a graded series of alcohol, as in paraffin embedding. Embedding methods that employ nonsolvents and solvents of plastic have been used. As a nosolvent method, propylene oxide, which is miscible with alcohol and plastic in different proportions, can be used. The plastic mixture with the following components gives good results:

Epon 812	54.90 g
DDSA (dodecenyl succinic anhydride)	29.10 g
NMA (nadic methyl anhydride)	30.75 g
DMP-30 [2, 4, 6-tri (dimethylaminomethyl) phenol]	0.90 g

The following steps may be used:

Propylene oxide (PO)	3 changes at 5 min intervals or 5, 10, and 15 min intervals
3 parts PO + 1 part plastic mixture	15–30 min
1 part PO + 1 part plastic mixture	30–60 min
1 part PO + 3 parts plastic mixture	60 min to overnight
Plastic mixture alone	12–24 hr

The tissue pieces may be embedded in plastic bottle covers, gelatin capsules, or aluminium foil boats and hardened for about 48 hr at 60°C.

2.1.4.3 Staining

After trimming the block to a pyramid under a binocular stereoscopic microscope, the block is mounted in an ultrmicrotome. Sections that display green interference colors are cut. The floating sections from the trough are collected and mounted on clean glass slides with drops of water. A hair loop or eyelash whisker can be used to transfer the thin sections. The slides are placed in an oven at 60°C or on a warming plate to permit evaporation of water and settling of sections on the slide. Generally no adhesive may be necessary. If required, egg albumin or gelatin adhesive may be used for coating the slides before mounting the sections.

The slides are then removed from the oven or warming plate and cooled to room temperature. The sections are treated with 1% periodic acid for 5 min at room temperature to remove osmium and to prevent rapid fading of the stain. The sections are rinsed in distilled water and slides are wiped by suitable absorbent paper. The required amount of stain is added in drops to the sections, and the slide is placed on either the warming plate or an alcohol (sprit) lamp for different periods, depending on the nature of the plant tissue. The sections are washed with distilled water to remove the excess stain and dried on a warming plate. The slides are then immersed in two changes of xylene and dried by draining. Suitable mounting medium is added to the section and covered. For further details refer to Berlyn and Miksche (1976).

2.2 METHODS OF ISOLATION OF PATHOGENS

Plant pathogens that cause different diseases on various crop plants have to be isolated in pure culture, and their morphological characters and physiological and biochemical activities are studied with a view to using them as bases for their identification and differentiation from closely related species. Facultative parasites and facultative saprophytes which have the ability to grow in cell-free media can be isolated and maintained on appropriate media for characterization of these pathogens. But obligate fungal pathogens, viruses, viroids, and most of the mycoplasma-like organisms (phytoplasmas) have to be maintained on live host plant species by adopting different methods of transmission (Chapter 4).

2.2.1 Preparation of Media

The growth and sporulation of microbial pathogens are favored to different extents on the basis of the nature of the media used and cultural conditions provided. Both solid and liquid media have been used to cultivate pathogens, depending on the objectives of the experiments. Agar is the basic component of all media used in all solid media. The usefulness of granulated cassava as a substitute for agar has been reported by Nene and Sheila (1994). Media rich in carbohydrates and slightly acidic (pH 6-6.5) in nature favor the growth of fungi, whereas bacteria prefer media with neutral or slightly alkaline pH. The media have to be sterilized at the required temperature and pressure. Many proprietory media are available for ready use, and addition of water alone is required for such media. The compositions of various media used for cultivation of fungal and bacterial pathogens are presented in Appendix 2(iv).

2.2.2 Isolation and Identification of Fungal Pathogens

The fungal pathogens can infect different plant parts, causing visible symptoms of the disease after the completion of the incubation period. The pathogens may be isolated from infected plant tissues such as leaves, stems, fruits, and roots by following different techniques for studying characters through which they can be identified.

2.2.2.1   Isolation from Leaves and Other Plant Parts

The infected leaves are thoroughly washed in sterile water. Then the infected tissues along with adjacent small unaffected tissues are cut into small pieces of 2–5 mm squares and transferred by using flame-sterilized forceps to sterile petri dishes containing 0.1% mercuric chloride solution. The tissue pieces are surface-sterilized in this solution for 30–60 sec and washed in sterile water two or three times. Clorox (10%), sodium hypochlorite (1%), or hydrogen peroxide (50%) may also be used to sterilize the surface. The tissue pieces are aseptically transferred to petri dishes containing a nutrient medium (such as potato dextrose medium) supplemented with streptomycin sulfate at the rate of three to five pieces per plate. The plates are incubated at room temperature (25°–27°C). The fungal mycelium growing on the nutrient medium is then transferred to agar slants kept in tubes.

The fungal pathogens from stems, roots, or fruits in which they may be present in deep-seated tissues have to be isolated by culturing pieces of internal tissues. The infected tissues are thoroughly washed in sterile water and then swabbed with cotton wool dipped into 80% ethanol, followed by exposure to an alcohol flame for a few seconds. The outer layer of tissues are quickly removed by a flame-sterilized scalpel. Small pieces from the central core of tissues in the area of the advancing margin of infection are removed by a sterilized scalpel or scissors and sterilized by dipping into 90% alcohol then flaming for a few seconds. The tissues, thus sterilized, are transferred to nutrient agar kept in petri dishes and incubated. The fungal mycelium growing from the infected tissues is transferred to agar slants kept in tubes.

2.2.2.2   Purification of Fungal Cultures

The cultures of fungal pathogens growing in agar slants have to be purified by either the single hyphal tip method or single spore isolation for precise identification. A small bit of agar medium containing fungal growth is transferred to the center of petri dishes containing nutrient medium, using a flame-sterilized inoculation needle, and incubated at room temperature for a few days. As the fungus grows, the advancing edge of the fungal growth will have well separated hyphal tips which are marked by a glass marking pencil by observing the bottom petri dish under the low power of the microscope. The bits of agar bearing a single hyphal tip marked earlier are carefully removed by a flame-sterilized inoculation needle and individually transferred to agar slants in tubes in which the hyphal tips will grow into a pure colony.

The spores of the fungal culture growing in the agar slant are suspended in sterile water by transferring the fungal growth to sterile water kept in a sterilized test tube then vigorously shaking the tube for a few minutes. This spore suspension is serially diluted by transferring 1 ml aliquots to a series of tubes containing 9 ml of sterile water. After attaining optimal dilution, 1 ml aliquots of the spore suspensions are mixed with melted nutrient agar at about 45°C and poured into sterile petri dishes, and the medium is spread to cover the entire surface by tilting the dishes suitably. The petri dishes are incubated at room temperature and examined at intervals of a few hours under the low power of the microscope. Individual germinating spores are marked by using a glass marking pencil as in the single hyphal tip method. The germinating spores along with the medium in the marked area are individually transferred to agar slants. The spores will grow into pure colonies.

The characteristics of asexual and sexual spores and spore-bearing structures of fungal pathogens are studied and used as bases of identification and differentiation of various species of fungal pathogens. The general outline for the classification of fungi up to the level of subclass is presented in Chapter 3.

2.2.3 Isolation and Identification of Bacterial Pathogens

The bacterial pathogens infecting leaves and other plant parts are generally isolated by preparing a bacterial suspension. Infected leaves showing clear symptoms of infection are cut into small bits 3 mm in diameter, surface-sterilized with 70% ethyl alcohol or 0.1% mercuric chloride solution for 30–60 sec and washed repeatedly in several changes of sterile water. The bacterial suspension may be prepared in two ways: The surface-sterilized infected tissue bits may be immersed in sterile water in a test tube and incubated for 6 h at room temperature, as in the case of *Xanthomonas oryzae* pv. *oryzae,* which causes bacterial leaf blight in rice. This suspension is then streaked on potato-peptone-glucose agar (PPGA) medium, using a sterilized inoculation needle.

In the second method, the surface-sterilized infected tissues are crushed in sterile water kept in an aspetic mortar with a pestle to get a suspension of bacteria, as in the case of *Xanthomonas campestris* pv. *malvacearum,* which causes bacterial leaf blight in cotton. This suspension is then streaked on nutrient agar medium kept in petri dishes by a sterile inoculation needle. The plates are incubated at room temperature for 24–48 h. The bacteria in individual colonies are then transferred to agar slants and purified by the serial dilution plate technique.

The serial dilution plate method is based on the principle that when appropriately diluted each viable bacterial cell will develop into a colony, ensuring purity and homogeneity of cells in the bacterial culture. A known volume of bacterial suspension (1 ml) is added to known volume (9 ml) of sterile water blank and agitated to produce a uniform suspension of bacterial cells. Serial dilutions of this suspension 10^{-2}, 10^{-3}, 10^{-4}, 10^{-5}, 10^{-6}, and above are prepared by pipetting 1 ml aliquots into dilution blanks of 9 ml each. Then 1 ml aliquots of different dilutions are added to separate petri dishes to which sterilized melted nutrient agar medium is added at the rate of 15 ml/dish. The plates are incubated in an inverted position for a few days at 25°C. The bacteria from well-isolated colonies are streaked on nutrient agar medium or special selective media suitable for the bacterial species. Tetrazolium medium (Kelman, 1954) is used for the isolation of *Pseudomonas solanacearum,* which causes bacterial wilt of banana, whereas 523 medium (Kado, 1971) is used for the isolation of *Xanthomonas campestris* pv. *malvacearum.* In the case of the xylem-limited bacterium *Xylella fastidiosa,* causing oleander leaf scorch disease, isolation of the pathogen on a growth medium more effectively detected the bacterial infection in oleander plants than enzyme-linked immunosorbent assay (ELISA) technique (Purcell et al., 1999) indicating that conventional techniques are still useful for the identification of some bacterial pathogens, although the molecular techniques have been shown to be generally more sensitive and rapid.

The bacterial pathogens are identified on the basis of colony characters, Gram's staining, and the properties determined by various biochemical and physiological tests (Chapter 6). The general outline for the classification of bacteria is presented in Chapter 3.

2.2.4 Isolation and Identification of Viruses

Plant viruses that infect different plant species are obligate parasites that require the presence of living cells/plants. They cannot be isolated and grown on cell-free media as fungi or bacteria, but they have to be maintained on susceptible host plants by artificial inoculation at regular intervals. Sap inoculation and/ inoculation employing appropriate vectors are done, and a plant species that favors rapid multiplication of the given virus is used as the source plant. If a virus is found along with another virus in the same plant, it has to be separated by inoculating on a diagnostic host which either forms local lesions or characteristic systemic symptoms which are distinct from those caused by the contaminating virus. An inoculum from a single local lesion can be prepared and inoculated on healthy plants, and the culture of virus can be obtained in pure form by such repeated inoculations. Insect vectors also can be used to separate the virus from contaminants on the basis of the specific nature of the virus-vector relationship.

The viruses can be purified, after extraction of the sap from infected source plants, by following appropriate purification procedures, which depend on the host-virus combinations (Matthews, 1991; Narayanasamy and Doraiswamy, 1996). The purified virus preparations are required for studying various properties of the viruses, such as virus particle morphological characteristics, the nature of

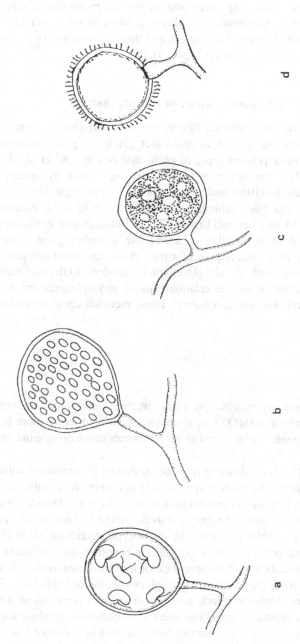

Figure 2.3 Asexual and sexual spores produced by members of Phycomycotina: a, Sporangium with zoospores; b, sporangium with aplanospores, c, oospore; d, zygospore.

the viral genome, the molecular weights of capsid protein and viral nucleic acid, amino acid and nucleotide sequences, and serological properties. The host range of viruses, reaction on diagnostic host plants, stability in vitro, and nature of the virus-vector relationship are also studied, and these properties are considered in the classification of plant viruses (Chapter 3).

2.2.5   Isolation and Identification of Phytoplasmas

The plant-infecting mycoplasma-like organisms (MLOs), currently termed phytoplasmas, were recognized as a distinct group of plant pathogens after the studies on certain yellows types of plant diseases by Doi et al. (1971) and Ishiie et al. (1971). Though intensive efforts were made by researchers for over two decades, isolation and cultivation of these organisms in cell-free media have been not been achieved, except in the case of *Spiroplasma citri* (Markham et al., 1974; Chen and Liao, 1975; Williamson and Whitcomb, 1975). Hence it has not been possible to study the morphological features and other biochemical properties required for their characterization and precise iden- tification. However, with the development of modern molecular methods of detection and identification, the relationships of phytoplasmas can be studied with reasonable reliability and certainty. These methods are described in Chap- ters 8 and 9.

2.3   FUNGI

The fungi are generally microscopic plants without chlorophyll and conductive tissues. Among about 100,000 fungal species known to exist, more than 8000 species are pathogenic to plants, and about 50 species cause disease in humans and animals.

The fungal body (thallus) consists of numerous filamentous hyphae weav- ing into a mycelium. The hyphae may or may not have cross-walls (septa). The hyphal walls may be uniform in thickness or irregularly thickened, tapering into thinner or broader portions. The fungal cells delimited by well-defined cell walls contain one or two nuclei per cell. The coenocytic mycelium, as in Phycomy- cotina, contains many nuclei, and the mycelium becomes one continuous tubular, branched, or unbranched multinucleate cell or multinucleate hyphae, if septa are formed later. Mycelial growth occurs through elongation of hyphal tips. Fungi be- longing to Plasmodiophorales lack true mycelium and form naked, amoeboid, multinucleate plasmodia, whereas members of Chytridiales produce a system of strands which are dissimilar and continuously varying in diameter, known as rhi- zomycelium.

The fungi reproduce efficiently by both asexual and sexual means. Specialized propagative bodies called spores or conidia are formed in the asexual reproductive cycle, whereas the sexual spores, called oospores, zygospores, ascospores, and basidiospores, are formed after a sexual process. In the Phycomycotina, the asexual spores are produced in a saclike sporangium and are later released through an opening of the sporangium or when the sporangial wall ruptures. The spores may be either motile by means of flagella (zoospores) or nonmotile (aplanospores) (Fig. 2.3). Conidia are formed by Ascomycotina and Deuteromycotina fungi by cutting off of terminal or later cells from the specialized hyphae known as conidiophores (Fig. 2.4). In some fungi, individual hyphal cells may be enlarged and develop thick walls. They are separated at maturity

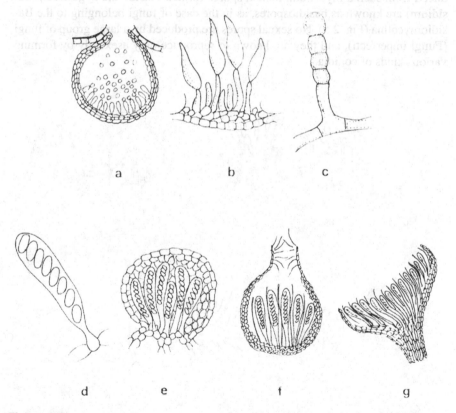

Figure 2.4 Asexual and sexual spores produced by Ascomycotina: a, Pycnidium with conidia; b, conidiophore bearing single conidium; c, conidiophore bearing chain of conidia; d, ascus with ascospores; e, cleistothecium with asci; f, perithecium with asci; g, apothecium with asci.

and can germinate to form a new colony. These structures are chlamydospores. Sclerotia are formed when a group of vegetative cells are enclosed in thick walls. Asexual spores may be formed in specialized reproductive structures, such as pycnidia, acervuli, and sporodochia.

In Phycomycotina, during sexual reproduction, union of two cells (gametes) of equal size and similar appearance results in the production of a zygote called the zygospore in some fungi; in other fungi the oospore is produced when gamates of unequal size unite. The sexual spores in Ascomycotina are formed in the zygote cell (ascus) produced by the union of two unequal-sized gamets. Usually eight ascospores are produced in each ascus, which may or may not be enclosed in the fruiting body called the ascocarp. In some fungi, any one cell of a mycelium may fuse with any cell of another compatible mycelium. Spores (teliospores) are produced from such a mycelium. Sexual spores formed outside the zygote cell (basidium) are known as basidiospores, as in the case of fungi belonging to the Basidiomycotina (Fig. 2.5). No sexual spores are produced by a large group of fungi (Fungi imperfecti), and they are known to reproduce only asexually by forming various kinds of conidia.

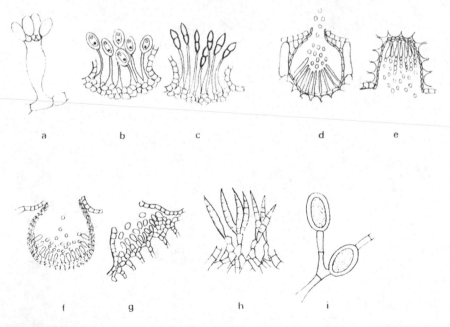

a   b   c   d   e

f   g   h   i

Figure 2.5 Spore-forming structures produced by Basidiomycotina and Deuteromycotina: a, Basidium with basidiospores; b, uredium with uredospores; c, telium with teliospores; d, pycnium with pycniospores; e, aecium with aeciospores; f, pycnidium with conidia; g, acervulus with conidia; h, sporodochium with conidia; i, chlamydospores.

2.4 BACTERIA

About 200 of 1600 bacterial species are known to infect various kinds of plants. Most of the bacteria are strictly saprophytic and help decompose large quantities of organic wastes produced by industrial establishments or by dead plants and animals. Many bacterial species are beneficial to humans because of their role in building up the nutritional level of the soil. Bacteria are simple in structure; they may be rod-shaped, spherical, ellipsoidal, spiral, comma-shaped, or filamentous. Some are motile by means of flagella; many of them are nonmotile. Some bacterial species transform themselves into spores, whereas spores are formed at the tips of some filamentous forms.

All plant pathogenic bacteria except two species of *Streptomyces* are rod-shaped. *Streptomyces* sp. is filamentous. The rod-shaped bacterial cells are short and cylindrical, measuring 0.6–3.5 μm × 0.3–1.0 μm. Deviations from this rod shape in the form of a club, a Y or V shape, and branched forms may also be observed. Pairs of cells or short chains of cells may also be formed in certain cases. The bacterial cells may be enveloped by a thin or thick slime layer made of viscous, gummy materials. The slime layer may be found as a larger mass around the cell called capsule. Most of the plant pathogenic bacteria are motile, and the flagella may be present either singly or in groups or distributed over the entire cell surface. The cells of *Streptomyces* spp. consist of nonseptate branched threads which usually have a spiral formation and produce conidia in chains on aerial hyphae. The bacterial colonies may vary in size, shape, color, elevation, form of edges, etc., depending on the species, and these properties may be characteristic of some species. The size of colonies may vary from 1 mm to several centimeters in diameter, and they may be circular, oval, or irregular with smooth, wavy, or angular edges. The colonies may have flat, raised, dome-shaped, or wrinkled elevation and are of different colors: white, yellow, red, or gray.

Rod-shaped plant pathogenic bacteria reproduce asexually by the process referred to as fission or binary fission. During this process, the cytoplasmic membrane grows toward the center of the cell, forming a transverse membranous partition dividing the cytoplasm into two approximately equal parts. Between the two layers of cytoplasmic layers, two layers of cell wall material, continuous with the outer wall, are laid down. As the formation of these cell walls is completed, they separate, splitting the two cells apart. The nuclear material is duplicated, as the layers of cell wall are formed in the center and become distributed equally between the two daughter cells formed from the dividing single cell. The bacteria can divide at an astonishingly rapid rate, dividing once in every 20 min under favorable conditions.

The sexual process known as conjugation occurs in some species. During this process, two compatible cells come into contact side by side and a small portion of the deoxyribonucleic acid (DNA) of the male cell (donor) is transferred to the female (receptor) cell. The female cell then multiplies by fission, resulting in daughter cells that contain the characteristics of both donor and receptor cells.

Variations in the genetic constitution of bacteria may be brought about by three different phenomena, viz. mutation, transformation, and transduction, in addition to conjugation. Mutation may occur in a very small percentage of cells, resulting in changes in genetic material leading to permanent changes in certain characteristics of the bacteria. In some bacteria, the genetic material is liberated from one bacterial cell by either secretion or rupture of the cell wall. A portion of this free genetic material (DNA) gains entry into a genetically compatible bacterium of the same or closely related species, making the recipient cell become genetically different. Transduction requires a bacteriophage as a vector for the transfer of genetic material (DNA) from one bacterium to another. The bacteriophage acquires a portion of genetic material of the infected bacterium. When it is liberated and infects another bacterial cell, the genetic material of the first cell becomes integrated with the DNA of the second bacterium. Thus different characters may be transferred from one bacterial cell to another, which shows new or different characters.

The ability of bacteria to multiply very rapidly and reach high populations within a short period makes them an important factor to be considered in any ecosystem in general and as plant pathogens in particular.

2.5 PHYTOPLASMAS

Mycoplasmas are known to exist as saprophytes in soil and sewage, and some have been found to be pathogens of human beings and other animals, such as cattle. The possibility of mycoplasmas being plant pathogens was indicated by Doi et al. (1967) and Ishiie et al. (1967). The presence of pleomorphic bodies in the phloem cells of infected mulberry plants was observed; these bodies degenerated, and the infected plants showed remission of the disease after application of tetracycline antibiotics. These observations suggested that some yellows-type plant diseases may be due to phytoplasmas rather than viruses, as believed earlier. Subsequently several other plant diseases were reported to be due to this new group of pathogens.

Most of the phytoplasmas have not been isolated in pure culture, and hence information on their cultural characters is lacking. Because of their morphological similarities to other mycoplasmas as observed in ultra-thin sections using electron microscope, they were named mycoplasma-like organisms.

Phytoplasmas and animal mycoplasmas are included in the class Mollicutes. Phytoplasmas differ from the bacteria in the absence of cell wall and penicillin-binding sites. They are, therefore, resistant to penicillin to which bacteria are sensitive. Phytoplasmas have a triple-layered plasma membrane. The name Mollicutes is derived from the property of lack of rigidity of cells of phytoplasmas.

They are able to pass through pores of even 220 nm diameter, even though the diameter of a viable cell may be more than 300 nm. The phytoplasmas are pleomorphic (varying in shape and size) generally and form characteristic fried egg colonies on solid media. They have an absolute dependence on sterols for growth; bacteria do not require sterols for their growth.

Among the phytoplasmas *Sprioplasma citri,* the pathogen that causes citrus stubborn and corn stunt diseases, has been isolated in pure culture, and its pathogenicity has been proved by following Koch's postulates (Davis and Lee, 1982). All attempts to prove the pathogenicity of other phytoplasmas have been unsuccessful.

2.6 VIRUSES

Viruses are submicroscopic pathogens that cause diseases in all known organisms from bacteria to highly evolved human beings. They are obligate parasites, requiring the presence of living cells for their replication and development. They cause diseases such as the common cold, influenza, polio, rabies, and the recently recognized acquired immunodeficiency syndrome (AIDS) in human beings, and cattle suffer from foot and mouth disease, which is of viral origin. Plants are affected by one or more viruses, and frequently losses caused by them are appreciable.

Plant viruses differ distinctly from other pathogens not only in their size, but also in their shape, which varies from rod-shape or bacilliform to spherical or polyhedral. They are very simple in their constitution, primarily possessing a protein coat which encloses either a ribonucleic acid (RNA) or DNA molecule(s) as the genome. Viruses have a unique method of replication which is not seen in any other known organisms. The protein coat and genomic nucleic acid are synthesized separately in different sites in susceptible cells and assembled together at an appropriate time to form progeny virus particles.

Plant viruses require an agency for transmission, though they can spread through leaf contact, infected seeds, and seed materials in certain cases. The viruses have to be introduced into the susceptible cells through the aid of insects, mites, nematodes, fungi, or parasitic dodders, since they cannot enter the plants through intact epidermis or natural openings. Mechanical inoculation or grafting is done for experimental transmission of viruses to healthy plants.

Plants respond differently to different viruses, depending on the level of their resistance to a given virus. Susceptible plants react to viruses by producing either local lesions, when the virus is confined to the initially infected tissues, or characteristic symptoms, in tissues/organs away from the site of infection/inoculation, when the virus becomes systemic. In addition to macroscopic external symptoms, the viruses induce, in certain host plants, the formation of inclusion bodies with characteristics of diagnostic value. No other plant pathogenic organism is known to induce the production of inclusion bodies in plants. As the plant viruses neither produce any structure outside the infected tissue nor are liberated from the infected cell, different methods have been used to detect their presence in infected plants.

Some viruses, including tobacco mosaic virus and cucumber mosaic virus, are found in the form of many strains which differ in their host range, type of symptoms induced, method of transmission, species of vectors, and physical and chemical properties. Related strains usually interfere with the development of another; this phenomenon is known as cross-protection. A mild strain of a virus can be used to protect cultivars against infection with severe strains of the same virus (Chapter 5).

Studies on the molecular biological characteristics of viruses have helped researchers to understand the functions of different cistrons (genes) in the viral genomes and the various strategies adopted by viruses for replication that result in increase in the viral population under optimal conditions. Biotechnological approaches have been made to develop transgenic crop plants with built-in resistance to several virus diseases.

2.7 VIROIDS

Viroids are the simplest among the plant pathogens capable of causing diseases, for example, potato spindle tuber disease, which is the earliest recognized viroid disease (Diener, 1971). Viroids are subviral pathogens that replicate independently and reach sufficient concentrations, when introduced into cells/tissues of susceptible plants, to produce characteristic symptoms of disease.

The viroids are small covalently closed circular RNA molecules with a highly base-paired, rather stiff rodlike native conformation. Potato spindle tuber viroid (PSTVd), which has been studied in detail, possesses a serial arrangement of 26 double-stranded segments interrupted by bulge loops of varying sizes. The extended secondary structure is folded into a more globular tertiary conformation, as the single-stranded loops do not interact. The viroid structure shows both stability and flexibility. The native rodlike structure of viroids is converted, through thermal denaturation, into a hairpin-containing circle in a highly cooperative fashion (Henco et al., 1979).

The viroid nucleic acids have only a few hundred nucleotides varying from 247 in avocado sunblotch viroid (ASBVd) to 373 in columnea latent viroid (CLVd) (Table 2.1). The molecular weights of viroid RNAs thus vary to an extent. The sizes of cucumber pale fruit viroid (CPFVd) and hop stunt viroid (HSVd) are quite similar and these viroids are biologically indistinguishable (Shikata, 1985).

Table 2.1 Plant Pathogenic Viroids and Their Principal Hosts

Viroid	Genome-size*	Principal/additional host (s)	References
Apple dimple fruit viroid (ADFVd)	306	Apples	Serio et al., 1996
Apple scar skin viroid (ASSVd)	330	Apples	Zhu et al., 1995b
Avocado sun blotch viroid (ASBVd)	247	Avocado	Dale and Allen 1979; Thomas and Mohamed, 1979; Symons, 1981
Blueberry mosaic viroid (BBMVd)	351	Carnations *Corymbosum vaccinium*	Zhu et al., 1995a
Burdock stunt viroid (BSVd)	RNA 1 RNA 2	Burdock	Chen et al., 1983; Tien Po, 1985
Chrysanthemum chlorotic mottle viroid (CCMVd)	—	Chrysanthemum	Romaine and Horst, 1975
Chrysanthemum stunt viroid (CSVd)	354 356	Chrysanthemum	Diener and Lawson, 1973; Hollings and Stone 1973; Gross et al., 1982; Hooftman et al., 1996
Citrus cahexia viroid	—	Oranges	Terranova et al., 1995
Citrus excortis viroid (CEVd)	359 371	Citrus, tomato, faba beans	Semancik and Weathers, 1972; Sanger 1972; Gross et al., 1982
Coconut cadang cadang viroid (CCCVd)	ccRNA1-246-297 ccRNA2-492-574	Coconut, Oilpalms	Randles 1975; Randles 1985; Randles et al., 1976; Haseloff et al., 1982; Mohamed et al., 1982
Coconut tinangaja viroid (CTiVd)	254	Coconut	Hodgson et al., 1998
Coleus blumei viroid (CBVd)	364	Coleus	Spieker et al., 1996

Table 2.1 Continued

Viroid	Genome-size*	Principal/additional host (s)	References
Columnea latent viroid (CLVd)	373	Columnea	Owens et al., 1978; Spieker, 1996
Cucumber pale fruit viroid (CPFVd)	330 320	Cucumber	Van Dorst and Peters, 1974; Sanger et al., 1976; Sano et al., 1983
Daple apple viroid (DAVd)	320 331	Apple	Zhu et al., 1995b
Grapevine yellow speckle viroid (GYSVd)		Grapevine	Szychowski, et al., 1995
Hop latent viroid (HLVd)		Hops	Adams et al., 1995
Hop stunt viroid (HSVd)	297 315	Hops, citrus, grapes	Sasaki and Shikata, 1977; Shikata, 1985
Japanese pear fruit dimple viroid (JPFDVd)	330	Japanese pear	Osaki et al., 1996
Mexican papita viroid (MPVd)		*Solanum cardiophyllum*	Martinez-Soriano, 1996
Peach latent mosaic viroid (PLMVd)	336	Peach	Shamloul et al., 1995
Pear blister canker viroid (PBCVd)		Pear	Ambros et al., 1995
Pear rusty skin viroid (PRSVd)	334	Pear	Zhu et al., 1995a
Potato spindle tuber viroid (PSTVd)	359, 356	Potato, tomato	Diener, 1971; Gross et al., 1978; Behjatnia et al., 1996
Tomato apical stunt viroid (TPSVd)		Tomato	Walter, 1981
Tomato bunchy top viroid (TBTVd)		Tomato	McLean, 1931
Tomato planta macho viroid (TPMVd)		Tomato	Galino et al., 1982

* Number of nucleotides.

SUMMARY

The general characteristics of fungi, bacteria, phytoplasmas, viruses, and viroids
which cause destructive diseases on economically important crops are described
to give the reader basic knowledge of these disease-causing microbes. These microorganisms can be differentiated on the basis of the differences in characteristics of the morphological and structural features.

APPENDIX 2(i): MOUNTING MEDIA AND STAINS FOR FUNGAL PATHOGENS

A. Mounting Media
    i. Lactophenol

Phenol (pure crystals)	20.0 g
Lactic acid (SG 1.21)	20.0 g
Glycerol	40.0 g
Water	20.0 ml

    ii. Anhydrous lactophenol

Phenol	20.0 g
Lactic acid	20.0 g (16.0 ml)
Glycerol	40.0 g (31.0 ml)

    iii. Glycerine jelly

Gelatine	1.0 g
Glycerol	7.0 g
Water	6.0 ml
Phenol to give 1% concentration	

B. Stains
    i. Cotton blue (or trypan blue)

Anhydrous lactophenol	67.0 ml
Distilled water	20.0 ml
Cotton blue or trypan blue	0.1 g

    ii. Erythrosin

Erythrosin	1.0 g
Ammonia (10%)	100.0 ml

    iii. Lacto-fuchsin

Acid fuchsin	0.1 g
Lactic acid	100.0 ml

This solution and Gurr's water mounting medium are mixed in a 1:1 ratio
(Carmichael, 1955).

APPENDIX 2(ii): FIXATIVES USED FOR PREPARATION OF PERMANENT SLIDES

A. Formalin-acetic acid-alcohol (FAA) mixture
    Ethyl alcohol (50% or 70%)   90.0 ml
    Glacial acetic acid   5.0 ml
    Formalin   5.0 ml

Alcohol at lower concentration is used for fixing delicate tissues, whereas higher concentration may be required for woody tissues. Fixation time is 18 h or more.

B. Carnoy's fluids
    i.   Ethyl alcohol (100%)   15.0 ml
         Glacial acetic acid   5.0 ml
    ii. Ethyl alcohol (100%)   30.0 ml
         Glacial acetic acid   5.0 ml
         Chloroform   15.0 ml
         Fixation time varies from 15 to 60 min.
C. Chamberlain's chrom-osmo-acetic acid mixture
    Chromic acid   1.0 g
    Glacial acetic acid   3.0 ml
    Osmic acid (1% aqueous solution)   1.0 ml
    Distilled water   100 ml

This mixture is suitable for filamentous fungi. Chrom-acetic acid fluid has been used at different concentrations also.

D1. Weak chrom-acetic acid
     Chromic acid   2.5 ml
     Acetic acid   5.0 ml
     Distilled water added to make up 100 ml.
D2. Medium chrom-acetic acid
     Chromic acid   7.0 ml
     Acetic acid   10.0 ml
     Distilled water added to make up to 100 ml.
     Chromic acid and acetic acid are used as 10% aqueous solution.
     Fixation time is 24 h or more.
E.   Randolph's modified Navashin fluid
     Solution A: Chromic acid   1.0 g
                 Glacial acetic acid   7.0 ml
                 Distilled water   92.0 ml
     Solution B: Neutral formalin   30.0 ml
                 Distilled water   70.0 ml

Mix solutions A and B in equal proportions before use. Fixation time varies from 12 to 24 h.

APPENDIX 2(iii): STAINING PROCEDURES (JOHANSEN, 1940)

A. **Heidenhain's iron hematoxylin**
   i. Use xylol to remove paraffin; pass through a mixture of xylol and absolute alcohol (1:1) for 10 min, then through a mixture of alcohol and ether (1:1) + 1% celloidin for 3 min; air dry the slides till they become opaque; immerse successively in 70% alcohol for 5 min, 35% alcohol, and finally water, and rinse in distilled water.
   ii. Prepare the mordant solution containing ferric ammonium sulfate crystals (15.0 g), glacial acetic acid (5.0 ml), conc. H_2So_4 (0.6 ml), and distilled water (500 ml); place the slides into this solution for 1–2 h; wash thoroughly in running water for 5 min and then rinse in distilled water.
   iii. Place the slides into aqueous hematoxylin solution (0.5%) for 1–2 h or more; wash the excess stain with water.
   iv. Destain by immersing the slides in ferric ammonium sulfate (2%) or ferric chloride as long as required; wash in running water for 30–60 min.
   v. Dehydrate slides by passing them successively through 50%, 70%, and 95% alcohol for 5 min in each concentration.
   vi. Pass the slides through a mixture of absolute alcohol and xylol (1:1) for 5 min and then through two changes of xylol for 5 min each and mount in balsam.

B. **Iron hematoxylin and safranin**
   i. Follow steps as in A(i).
   ii. Place the slides in 3% aqueous ferric ammonium sulfate solution (used as mordant) for 2 to 3 hr; wash in running water for 5 min.
   iii. Stain in hematoxylin for 2 to 3 hr, followed by differentiation in 3% aqueous ferric ammonium sulfate; transfer the slides to water when sections turn colorless and wash in running water for 1 hr or more.
   iv. Stain in safranin for 12–15 hr, followed by differentiation using either 70% alcohol acidified with a few drops of HCl or 95% picro alcohol for not more than 10 sec.
   v. Follow steps in A(v) and A(vi).

C. **Conant's quadruple stain**
    i.   Follow steps as in A(i) up to 70% alcohol step.
    ii. Place the slides in 1% safranin in 50% alcohol for 2–24 hr; rinse thoroughly in distilled water.
    iii. Transfer the slides to a saturated aqueous solution of crystal violet for about 1 min and rinse in distilled water, followed by dehydration through two changes of absolute alcohol.
    iv. Stain the slides by rapidly dipping 5 to 10 times in 1% fast green in absolute alcohol; transfer to a saturated solution of gold orange (or orange G) in clove oil and agitate the slides till the alcohol is completely diffused into the clove oil.
    v.   Place the slides successively in a series of three more jars of orange clove oil solution for several minutes in each for further differentiation and clearing of the background.
    vi. Rinse the slides in xylol and mount the sections in balsam.

APPENDIX 2(iv): MEDIA USED FOR CULTIVATION OF FUNGAL AND BACTERIAL PATHOGENS

A. Media for fungal pathogens
    i.   Potato dextrose agar

Potato (peeled)	200.0 g
Dextrose	20.0 g
Agar	20.0 g
Water	1000.0 ml

Sterilize for 20 min at 1.06 Kg/cm^2

    ii. Potato sucrose agar

Potato extract	500.0 ml
Sucrose	20.0 g
Water	500.0 ml

Potato extract is prepared by placing peeled pieces of potato (1800 g) in muslim cloth, suspending in water (4500 ml), and boiling for 10 min.

    iii. Oatmeal agar

Oats	100.0 g
Agar	15.0 g
Water	1000 ml

    iv. Yeast extract glucose agar

Yeast extract	10.0 g
Glucose	10.0 g
Agar	15.0 g
Tap water	1000 ml

    v.   Water agar (plain agar)
         Agar   20.0 g
         Water   1000 ml
    vi. Czapek (Dox) agar
         Sodium nitrate   2.0 g
         Potassium dihydrogen Phosphate (KH_2PO_4)   1.0 g
         Magnesium sulfate ($MgSO_4 \cdot 7H_2O$)   0.5 g
         Potassium chloride (KCl)   0.5 g
         Ferrous sulfate ($FeSO_4 \cdot 7H_2O$)   0.01 g
         Sucrose   30.0 g
         Agar   20.0 g
         Water   1000 ml
B. Media for bacterial pathogens
    i.   Nutrient broth
         Bactopeptone   5.0 g
         Beef extract   3.0 g
         Water   1000 ml
    ii. Nutrient glucose agar
         Beef extract   3.0 g
         Bactopeptone   5.0 g
         Glucose   5.0 g
         Sodium chloride   5.0 g
         Agar   15.0 g
         Tap water   1000 ml
    iii. Potato-peptone-glucose-agar medium (PPGA)
         Potato extract   500.0 ml
         Peptone   5.0 g
         Glucose   5.0 g
         Sodium chloride   3.0 g
         Sodium monohydrogen phosphate (Na_2HPO_4)   3.0 g
         Potassium monohydrogen phosphate (K_2HPO_4)   0.5 g
         Agar   18.0 g
         Water   500 ml

Potato extract is prepared by placing peeled pieces of potato (200 g) in muslin cloth, suspending in water (500 ml), and boiling for 10 min.

    iv. Tetrazolium medium (TTC) (Kelman, 1954)
         Dextrose   10.0 g
         Peptone   10.0 g
         Cis amino acids   1.0 g
         Agar   18.0 g
         Water   1000 ml

The basal medium in 200 ml aliquots is sterilized at 1.06 Kg/cm^2 for 20 min. Prepare tetrazolium chloride solution by dissolving 1.0 g of 2,3,5-triphenyl tetra-zolium chloride in 100 ml of distilled water and sterilize at 121°C for 8 min and store in darkness. Add 1 ml of this solution to 200 ml of basal medium to yield 0.05% concentration before pouring the melted medium into the petri dishes.

v. Medium 523 (Kado, 1971)

Sucrose	10.0 g
Casein acid hydrolysate	8.0 g
Yeast extract	4.0 g
Potassium monohydrogen phosphate (K_2HPO_4)	2.0 g
Magnesium sulfate	0.3 g
Agar	15.0 g
Water	1000 ml

vi. Brinkerhoff medium (Brinkerhoff, 1960)

Dextrose	20.0 g
Potassium monohydrogen phosphate (K_2HPO_4)	50.0 g
Calcium carbonate	10.0 g
Agar	15.0 g
Water	1000 ml

vii. Wakimoto's medium (Wakimoto, 1960)

Potato	200.0 g
Sucrose	15.0 g
Peptone	5.0 g
$Na_2HPO_4 \cdot 12\ H_2O$	2.0 g
$Ca(NO_3)_2$	0.5 g
Water	1000 ml

viii. Dye's medium (Dye, 1962)

Glucose	10.0 g
K_2HPO_4	2.0 g
Ammonium phosphate	1.0 g
$MgSO_4$	0.2 g
NaCl or KCl	0.2 g
Water	1000 ml
pH	7.0

ix. Semiselective agar medium (T-5) (Gitaitis et al., 1997)

NaCl	5.0 g
$NH_4H_2PO_4$	1.0 g
K_2HPO_4	1.0 g
$Mg\ SO_4\ H_2O$	0.2 g
D-Tartaric acid	3.0 g
Phenol red	0.01 g

Agar	20.0 g
Water	1000 ml

After autoclaving add:

Bacitracin	10 mg
Vaniomycin	6 mg
Cycloheximide	75 mg
Novobiocin	45 mg
Penicillin G	5 mg

Adjust pH to 7.4

x. Semiselective medium (PCCG) (Hara et al., 1995; Ito et al., 1998)

Prepare potato semisynthetic agar (PSA) medium containing:

Potato decoction from	300.0 g
Peptone	5.0 g
Sucrose	15.0 g
$Ca(NO_3)_2$	0.5 g
$Na_2 HPO_4 \cdot 12H_2O$	2.0 g

PCCG medium contains:

PSA	1000 ml
Gella gum	18.0 g
Crystal violet	5 mg
Polymyxin B	4×10^5 units
Chloramphenicol	7.5 mg
Cycloheximide	50 mg
Tetrazolium chloride	2.5 mg

xi. Semiselective medium—Cefazolin trehalose agar (CTA) medium (Fessehaie et al., 1999)

$K_2 HPO4$	3.0 g
$NaH_2 PO4$	1.0 g
$MgSo4 \cdot 7H_2O$	0.3 g
$NH_4 Cl$	1.0 g
D (+)-trehalose	9.0 g
D (+)-glucose	1.0 g
Yeast extract	1.0 g
Cefazolin	0.025 g
Lincomycin	0.0012 g
Phosphomycin	0.0025 g
Cycloheximide	0.25 g
Agar	14.0 g
Distilled water	1 litre

3
Symptoms of Plant Diseases

3.1 EXTERNAL SYMPTOMS

Plants are attacked by different groups of pathogens individually or sometimes by more than one pathogen-producing complex and more severe disease. The type of external symptoms can, in most cases, indicate the nature of the pathogen responsible for the disease. Fungi, bacteria, viruses, and phytoplasmas cause distinct types of symptoms in most host-pathogen interactions. However, there are some diseases which show similarity in symptoms, though they are induced by different groups of pathogens. Further careful examination under microscope or by other methods may be necessary to establish the nature of the pathogen(s). Internal symptoms, such as histological and cytological changes caused by viruses, have diagnostic value, facilitating the identification of viruses in certain host-virus combinations.

3.1.1 Symptoms Induced by Fungal Pathogens

Fungal pathogens may cause local or systemic symptoms. Fungal infections are generally found to be restricted to infected organs, such as leaves, stem, or flowers. In some cases they cause symptoms on different plant parts, when they reach the vascular tissues. Generally the affected tissues show necrosis or rotting, and in some cases hypoplasia or stunting of plant organs or whole plants or hyperplasia or excessive growth resulting in malformation or modification of organs or the entire plant.

The following are the common symptoms induced by fungal pathogens: root rot, collar rot, stem rot, stem canker, gummosis, club root, galls, warts, blight, blast, leaf spot, shot holes, anthracnose, rusts, powdery mildews, fruit rot, capsule rot, and head rot. These symptoms are generally restricted to the tissues or organs in which infection is initiated. Damping off, wilt, smut, and downy mildew may be due to the systemic infection of plants by pathogens. These symptoms are distinct and indicate that involvement of fungal pathogens (Figs. 3.1–3.4). However, in some cases, either the incubation period required by the fungal pathogen may be very long or the symptoms induced may be indistinct. In such cases the use of indicator plants may be useful. For example, pear abnormal leaf spot disease caused by *Alternaria kikuchiana* can be detected rapidly by using the progeny (PS-95) from the cross between Niitaka and Waseaka, as an indicator plant (Nam KiWoong et al., 1996).

    The fungi produce specialized asexual spore forms such as sporangia or conidia either on the surface of the infected tissues or within such tissues in spore-forming structures, such as sorus, acervulus, pycnidia, or sporodochia. By using a compound microscope the characteristics of the spores may be studied in detail. At later stages of disease development or in pure cultures of the pathogen, sexual spores may be formed. The characters of sexual and asexual spores are primarily used in the identification and classification of fungal pathogens. However, some cultural characteristics may be useful to differentiate the fungal pathogens as in

A

Figure 3.1   A, Chickpea root rot (*Rhizoctonia solani*) (ICRISAT); B, rice stem rot (*Sclerotium oryzae*) (IRRI); C, pepper (chili) stem rot (*Sclerotium rolfsii*) (AVRDC).

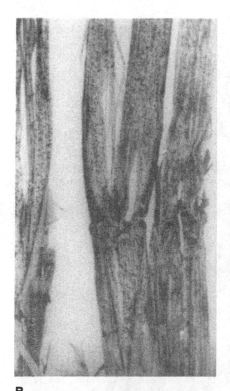

B

C

Figure 3.1 Continued.

the case of *Colletotrichum* spp. pathogenic to rubber. The slower growth rate at temperatures ranging from 15°C to 32.5°C and higher level of tolerance to fungicides, benomyl, carbendazim, and thiophanate methyl of *Colletotrichum actuatum* were found to be useful characteristics to differentiate this pathogen from *C. gloeosporioides (Glomerella cingulata)* (Jayasinghe and Fernando, 1998). For more details on taxonomy and classification of fungi, refer to Bold et al. (1980).

The general outline for the classification of fungi including slime molds up to the level of subclass is as follows (after Bold et al., 1980):

Kingdom Mycetate
Division 1 Gymnomycota
  Subdivision 1 Acrasiogymnomycotina
    Class 1 Acrasiomycetes
  Subdivision 2 Plasmodiogymnomycotina
    Class 1 Protosteliomycetes
    Class 2 Myxomycetes

A

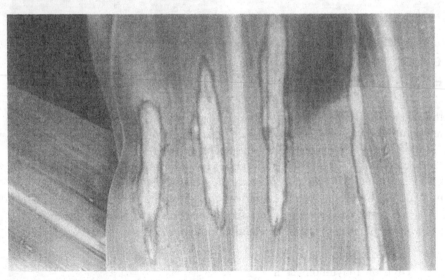

B

Figure 3.2 A, sorghum downy mildew (*Sclerospora sorghi*) (ICRISAT); B, sorghum leaf blight (*Exserohilum sorghi*) (ICRISAT).

Division 2 Mastigomycota
  Subdivision 1 Haplomastiogmycotina
    Class 1 Chytridiomycetes
    Class 2 Hyphochytridiomycetes
    Class 3 Plasmodiophoromycetes
  Subdivision 2 Diplomastigomycotina
    Class 1 Oomycetes
Division 3 Amastigomycota
  Subdivision 1 Zygomycotina
    Class 1 Zygomycetes
    Class 2 Trichomycetes
  Subdivision 2 Ascomycotina
    Class 1 Ascomycetes
    Subclass 1 Hemiascomycetidae
    Subclass 2 Plectomycetidae

A   B

Figure 3.3 A, Rice sheath rot (*Sarocladium oryzae*) (IRRI) B, sorghum kernel smut
(*Sphacelotheca sorghi*) (ICRISAT); C, sorghum grain molds (*Curvularia lunata, Fusarium
moniliforme, F. semitectum*) (ICRISAT).

C

Figure 3.3   Continued.

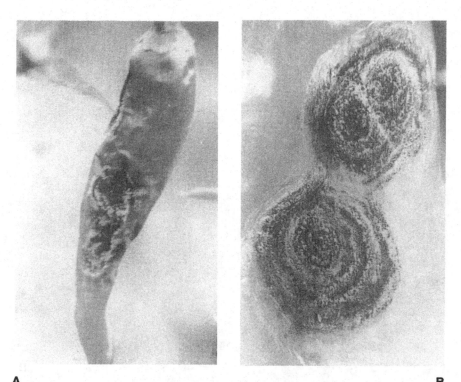

A   B

Figure 3.4 A, Pepper (chili) fruit rot (*Colletotrichum gloeosporioides, C. capsici, C. acutatum, C. coccodes*); (AVRDC); B, fruiting structures (acervuli) of *Colletotrichum* spp. in concentric rings (AVRDC).

3.1.2 Symptoms Induced by Bacterial Pathogens

Pathogenic bacteria characteristically induce water-soaked lesions in the infected tissues at the initial stages, and these lesions turn necrotic later. Formation of encrustations or bacterial ooze from infected tissues is another distinguishing feature associated with bacterial diseases. As the infection progresses, leaf spots, blights, scabs, cankers, tumors, wilts, and soft rots of fruits, tubers, and roots may be the prominent types of symptoms caused by bacterial pathogens (Figs. 3.5 and 3.6). Although some symptoms, such as leaf spot and blight, may have similarities with those due to fungal pathogens, microscopic examination will provide a definite indication of the nature of the pathogen. The absence of spores and fungal structures and presence of characteristic bacterial ooze from the cut ends of the tissue under examination may indicate that the disease is likely to be due to bacteria. Isolation of the bacterium using appropriate medium and study of other properties determined by biochemical tests (Chapter 6) are necessary for the identification of the

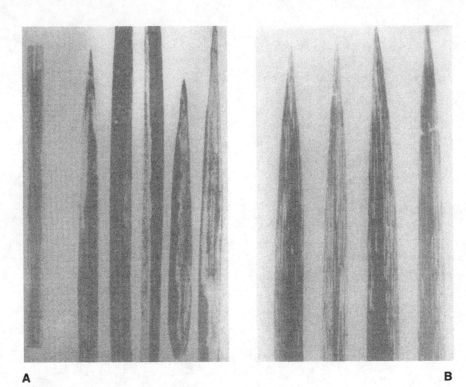

A   B

Figure 3.5   A, Rice bacterial leaf blight (*Xanthomonas campestris* pv. *oryzae* (IRRI); B, rice bacterial leaf streak (*Xanthomonas campestris* pv. *oryzae transluscens*) (IRRI).

Figure 3.6   Cassava bacterial stem gall (*Agrobacterium tumefaciens*).

causative bacterium. Detailed information regarding the cultural characteristics and different tests to be done is provided by Schaad (1988).

The general outline for classification of bacteria is included in *Bergey's Manual of Systematic Bacteriology* (Krieg and Holt, 1984). Naming a new species requires the results of polyphasic tests which include nucleic acid analyses such as DNA-DNA and DNA-rDNA hybridization, chemotaxonomic comparisons such as cell-wall composition, lipid composition, isoprenoid quinones, soluble and total proteins, fatty acid profiles, enzyme characterizations, in addition to biochemical and nutritional tests, so that determinative keys are formulated (Young et al., 1992). The International Society of Plant Pathology established the criteria for use of the term "pathovar" for those plant pathogenic bacteria that did not satisfy the criteria for species designation. The nomenclature of "pathovar" is applied at the infrasub-specific level for bacteria distinguished chiefly based on the differences in pathogenicity on a particular host plant species or a set of plant species.

The plant pathogenic bacterial genera are classified as follows (Young et al., 1992):

Division: Firmicutes
    Family: No family classification
    Genus: *Arthrobacter*
               Bacillus
               Clavibacter
               Curtobacterium
               Nocardia
               Rhodococcus
Division: Gracilicutes
    Alpha subclass Class: Proteobacteria
    Family: Acetobacteriaceae
    Genus: *Acetobacter*
    Family: Rhizobiaceae
    Genus: *Agrobacterium*
    Family: Not classified
    Genus: *Rhizomonas*
    Beta subclass
    Family: Comamonadaceae
    Genus: *Acidovorax (Pseudomonas)*-rRNA
                    homology group III
               Xylophilus
    Family: Not named
    Genus: *Burkholderia (Pseudomonas)*—rRNA
                    homology group II

Gamma subclass
Family:	Enterobacteriaceae
Genus:	*Enterobacter*
	Erwinia
Family:	Pseudomonadaceae (needs emendation)
Genus:	*Pseudomonas* (rRNA homology group I)
Family:	Hypomicrobiaceae
Genus:	*Xanthomonas*
Family:	Not classified
Genus:	*Xylella*
Genera of uncertain affinity	
	Rhizobacter
	Streptomyces

Economically important pathogenic bacterial species and their principal hosts are presented in Table 3.1.

3.1.3 Symptoms Induced by Phytoplasmas

Although the symptoms induced by phytoplasmas have some similarity to those caused by viruses, there are certain characteristic symptoms that may be useful to differentiate the diseases due to phytoplasmas and those due to viruses. The phytoplasmas cause general stunting or dwarfing of affected plant parts or whole plants. Chlorosis and smalling of leaves are also frequently observed in affected plants. Antholysis of floral parts is the most characteristic feature of phytoplasmal diseases, it results from virescence, phyllody, and proliferation of floral tissues. Floral parts are transformed into green leaflike structures. Partial or total sterility of infected plants may be commonly noted. These symptoms are observed in plants infected by diseases such as aster yellows, eggplant little leaf, sesamum phyllody, and witches' broom disease of potato, peanut (groundnut), and grain legumes. Proliferation of axillary buds and formation of a large number of thin shoots are observed prominently in eggplant (brinjal) little leaf, rice yellow dwarf, and sugarcane grassy shoot diseases (Fig. 3.7). Reduction in leaf size and internodal length and a tendency for the leaves to stand out stiffly, giving a spikelike appearance to the infected branches, are the distinguishing symptoms of sandal spike disease. Tomato big bud disease is characterized by hypertrophy of floral parts, leading to upright disposition of flower buds, which remain swollen and unopened. The general outline for the classification of mycoplasmas is as follows:

Taxonomy of Mycoplasmas (Archer and Daniels, 1982)
Class: Mollicutes   Order: Mycoplasmatales

Table 3.1 Plant Pathogenic Bacteria

Bacterial pathogen	Principal crop host(s)
Acidovorax avenae sub sp. *avenae*	Maize, rice
A. avenae subsp. *utrulli*	Tomato, watermelon
Agrobacterium tumefaciens	Capsicum, castor, peaches, roses, stonefruits, sunflower
A. vitis	Grapes
Bacillus megaterium pv. *cerealis*	Wheat
Burkholderia andropogonis	*Amaranthus* sp., tulips
B. caryophylli	*Dianthus caryophyllus*
B. cepacia	Onions
B. gladioli pv. *gladioli*	Tulips
B. glumae	Rice
B. plantarii	Rice
B. solanacearum	Banana, casava, cotton and sweet potato
Clavibacter michiganensis subsp. *insidiosus*	Tomato
C. michiganensis subsp. *michiganensis*	Tomato
C. michiganensis subsp. *nebraskensis*	Maize
C. michiganensis subsp. *sepedonicus*	Potato
C. xyli subsp. *cyanodontis*	Bermuda grass, maize
C. xyli subsp. *xyli*	Sugarcane
Curtobacterium (=*Corynebacterium*) *flaccumfaciens* pv. *betae*	Sugarbeet
C. flaccumfaciens pv. *flaccumfaciens*	French bean
C. flaccumfaciens pv. *poinsettiae*	Poinsettia
Erwinia alni	Black alder (*Alnus glutinosa*), Italian alder (*A. cordata*)
E. amylovora	Apples, pears
E. corotovora subsp. *atroseptica*	Potato
E. corotovora subsp. *betavasculorum*	Sugarbeet
E. carotovora subsp. *carotovora*	Cabbage, potato
E. chrysanthemi pv. *chrysanthemi*	Chrysanthemum
E. chrysanthemi pv. *dianthicola*	Carnations
E. chrysanthemi pv. *dieffenbachiae*	Dieffenbachia
E. chrysanthemi pv. *zeae*	Maize
E. nigrifluens	Walnuts
E. salicis	Willow trees (*Salix* spp.)
E. stewartii	Maize
E. tracheiphila	Melons
Pantoea ananas	Rice
Pantoea (=*Erwinia*) *stewartii* subsp. *stewartii*	Maize

Table 3.1 Continued

Bacterial pathogen	Principal crop host(s)
Pseudomonas caricapapayae	Papaya
P. cichorii	*Mentha* sp., *Ocimum basilicum*
P. fuscovaginae	Rice
P. marginalis	Pear
P. marginalis pv. *alfalfae*	Alflafa
P. marginalis pv. *marginalis*	Garlic
P. savastanoi pv. *glycinea*	Soybeans
P. savastanoi pv. *phaseolicola*	French bean
P. syringae pv. *actinidiae*	Kiwifruits
P. syringae pv. *atrofaciens*	Cereals
P. syringae pv. *atropurpurea*	*Lolium* sp.
P. syringae pv. *avellanae*	Hazelnuts
P. syringae pv. *helianthi*	Sunflower
P. syringae pv. *lachrymans*	Cucumber
P. syringae pv. *maculicola*	Crucifers
P. syringae pv. *mori*	Mulberry
P. syringae pv. *morsprunorum*	Cherries, French bean, tobacco
P. syringae pv. *oryzae*	Rice
P. syringae pv. *persicae*	Peaches
P. syringae pv. *pisi*	Peas
P. syringae pv. *savastanoi*	Olives
P. syringae pv. *sesami*	Sesamum
P. syringae pv. *syringae*	Apricots, taba beans, maize, mangoes
P. syringae pv. *tabaci*	Tobacco
P. syringae pv. *tagetis*	Jerusalem artichoke, sunflower
P. syringae pv. *theae*	Cocoa
P. syringae pv. *tomato*	Soybeans, tomato
P. syzygii	Cloves
P. viridiflava	Cauliflower, onion
P. zingiberi	Ginger
Ralstonia (=Pseudomonas) solanacearum	Capsium, egg-plant (brinjal), ginger, peanut, sesamum, sweet potato, tobacco
Rathayibacter (=Calvibacter) tritici	Wheat
Streptomyces ipomoeae	Ipomoea
S. scabies	Potato
Xanthomonas albilineans	Sugarcane
X. arboricola pv. *juglandis*	Walnuts
X. arboricola pv. *pruni*	Plums
X. axonopodis pv. *alfalfae*	Alfalfa
X. axonopodis pv. *cajani*	Pigeonpea
X. axonopodis pv. *citri*	Citrus

Table 3.1 Continued

Bacterial pathogen	Principal crop host(s)
X. axonopodis pv. *coracanae*	Finger millet
X. axonopodis pv. *cyamopsidis*	Cluster beans
X. axonopodis pv. *dieffenbachiae*	Anthurium, dieffenbachia
X. axonopodis pv. *glycines*	Soybeans
X. axonopodis pv. *malvaceaum*	Cotton
X. axonopodis pv. *manihotis*	Cassava
X. axonopodis pv. *phaseoli*	Beans
X. axonopodis pv. *poinsetticola*	Poinsettia
X. axonopodis pv. *ricini*	Castor
X. axonopodis pv. *sesbaniae*	*Sesbania* sp.
X. axonopodis pv. *sojense*	Soybeans
X. axonopodis pv. *tamarindi*	Tamarind
X. axonopodis pv. *vasculoram*	Surgarcane
X. axonopodis pv. *vignicola*	Cowpea
X. axonopodis pv. *vitians*	Cabbage, Lettuce
X. bromi	*Bromus catharticus*
X. campestris pv. *arecae*	Arecanuts
X. campestris pv. *azadirachtae*	Neem
X. campestris pv. *beticola*	*Piper betel*
X. campestris pv. *campestris*	Capsicum, crucifers, tomato
X. campestris pv. *carotae*	Carrots
X. campestris pv. *coriandri*	Coriander
X. campestris pv. *cucurbitae*	Squashes
X. campestris pv. *eucalypti*	Eucalyptus
X. campestris pv. *holicola*	Sorghum
X. campestris pv. *incancanae*	*Matthiola incana*
X. campestris pv. *mangiferae-indicae*	Mangoes
X. campestris pv. *musacearum*	Musa spp
X. campestris pv. *passiflorae*	Passion fruits
X. campestris pv. *populi*	Populus
X. campestris pv. *sesami*	Sesamum
X. campestris pv. *undulosa*	Wheat
X. campestris pv. *viticola*	Grapevine
X. campestris pv. *gingibericola*	Ginger
X. campestris pv. *zinniae*	Zinnia
X. cassavae	Cassava
X. cucurbitae	Cucurbits
X. fragariae	Strawberry
X. hortorum pv. *carotae*	Apiaceae
X. hortorum pv. *pelargonii*	Pelargonium
X. melonis	Melons

Table 3.1 Continued

Bacterial pathogen	Principal crop host(s)
X. oryzae pv. oryzae	Rice
X. oryzae pv. oryzicola	Rice
X. pisi	Peas
X. populi	Poplar tree
X. sacchari	Sugarcane
X. translucens pv. cerealis	Wheat
X. translucens pv. graminis	Grasses
X. translucens pv. phlei	Grasses
X. translucens pv. poa	Poa annua
X. translucens pv. secalis	Oat
X. translucens pv. translucens	Barley, wheat
Xanthomonas vesicatoria	Capsicum, tomato
Xylella fastidiosa	Almonds, citrus, coffee, grapevine, peach, pear, plum.
Xylophilus ampelinus	Grapevine

Source: Young et al., 1996.

Family I Mycoplasmataceae
  i) Sterol required for growth
  ii) Genome size about 0.5×10^9 Da
  iii) Reduced nicotinamide-adenine dinucleotide (NADH) oxidase localized in cytoplasm
Genus I *Mycoplasma* (about 50 species)
  i) Do not hydrolyze urea
Genus II *Ureaplasma*
  i) Hydrolyzes urea (one species with many serotypes)
Family II Acholeplasmataceae
  i) Sterol not required for growth
  ii) Genome size about 1.0×10^9 Da
  iii) NADH oxidase localized in membrane
Genus *Acholeplasma* (6 species)
Family III Spiroplasmataceae
  i) Helical cells formed during some phase of growth
  ii) Sterol required for growth
  iii) Genome size about 1.0×10^9 Da
  iv) NADH oxidase localized in cytoplasm
Genus *Spiroplasma* (one species)
Genera of uncertain taxonomic position:
  I *Thermoplasma* (one species)
  II *Anaeroplasma* (two species)

A   **B**

Figure 3.7 A, Yellow dwarf MLO-infected rice plant; B, little leaf MLO-infected eggplant.

Phylogenetic studies based on analysis of 16S ribosomal RNA (rRNA) or both 16S rRNA and ribosomal protein gene operon sequences have shown the phylogenetic position of phytoplasmas as members of the class Mollicutes. The phytoplasmas have been classified into 14 groups and 42 subgroups based on the characteristics determined by restriction fragment length polymorphism (RFLP) analysis of polymerase chain reaction (PCR)-amplified 16S rRNA sequences as follows (Davis and Sinclair, 1998):

1. Group   16SrI   (Aster yellows group)
    Subgroup   I (A)   Tomato big bud (BB)
                     I (B)   Michigan Aster yellows (MIAY)
                     I (C)   Clover phyllody (Cph)
                     I (D)   Paulownia witches' broom (PaWB)
                     I (E)   Blueberry stunt (BBSt)
                     I (F)   Apricot chlorotic leaf roll (ACLR-AY)
                     I (K)   Strawberry multiplier (STRAWB2)

2. Group   16Sr II   (Peanut witches' broom group)
    Subgroup   II (A)   Peanut witches' broom (PnWB)
                II (B)   Witches' broom of lime (WBD2)
                II (C)   Fababean phyllody (FBP)
                II (D)   Sweet potato little leaf (SPLL)
3. Group   16Sr III   (X-disease group)
    Subgroup   III (A)   X-disease (CX)
                III (B)   Clover yellow edge (CYE)
                III (C)   Pecan bunch (PB)
                III (D)   Goldenrod yellows (GR1)
                III (E)   Spirea stunt (SP1)
                III (F)   Milkweed yellows (MW1)
                III (G)   Walnut witches' broom (WWB)
                III (H)   Poinsettia branch-inducing (Poi B1)
                III (I)   Virginia grapevine yellows (VGY III)
4. Group   16Sr IV   (Coconut lethal yellows group)
    Subgroup   IV (A)   Coconut lethal yellowing (LY)
                IV (B)   Tanzanian coconut lethal decline (LDT)
5. Group   16Sr V   (Elm yellows group)
    Subgroup   V (A)   Elm yellows (EY1)
                V (B)   Cherry lethal yellows (CLY)
                V (C)   Flavescence dorée (FD)
6. Group   16Sr VI   (Clover proliferation group)
    Subgroup   VI (A)   Clover proliferation (CP)
                VI (B)   "Multicipita" phytoplasma
7. Group   16Sr VII   (Ash yellows group)
                VII (A)   Ash Yellows (AshY)
8. Group   16Sr VIII   (Loofah witches'-broom group)
    Subgroup   VIII (A)   Loofah witches'-broom (LfWB)
9. Group   16Sr IX   (Pigeonpea witches'broom group)
    Subgroup   IX (A)   Pigeonpea witches'broom (PPWB)
10. Group   16Sr X   (Apple proliferation group)
    Subgroup   X (A)   Apple proliferation (AP)
                X (B)   Apricot chlorotic leafroll (ACLR)
                X (C)   Pear decline (PD)
                X (D)   Spartium witches'broom (SPAR)
                X (E)   Black alder witches' broom (BAWB)
11. Group   16Sr XI   (Rice yellow dwarf group)
    Subgroup   XI (A)   Rice yellow dwarf (RYD)
                XI (B)   Sugarcane white leaf (SCWL)
                XI (C)   Leaf hopper-borne (BVK)

12.	Group	16Sr XII	(Stolbur group)
	Subgroup	XII (A)	Stolbur (STOL)
		XII (B)	Australian grapevine yellows (AUSGY)
13.	Group	16Sr XIII	(Mexican periwinkle virescence group)
	Subgroup	XII (A)	Mexican periwinkle virescence (MPV)
14.	Group	16Sr	(Bermudagrass white leaf group)
	Subgroup	XIV (A)	Bermudagrass white leaf (BGWL)

3.1.4 Symptoms Induced by Viruses

Plant viruses cause a variety of symptoms, depending on the host plant species, and different unrelated viruses may induce similar symptoms in the same host plant species. Dependence on the symptoms alone for the identification of viruses may lead to erroneous conclusions. However, information on host range and the reactions of diagnostic or differential hosts has been used for the identification of some viruses which have not been purified and adequately characterized.

The viruses induce primary symptoms on inoculated leaves, which exhibit chlorotic or necrotic local lesions or vein clearing. Later, when the virus becomes systemic, secondary symptoms develop on other plant parts. The secondary symptoms may be grouped as color changes, teratological symptoms, death or necrosis, and abnormal growth forms. Color changes may vary from mosaic on leaves to color breaking in flowers. Various kinds of changes in size and shape of plant part may be seen as leaf roll, leaf curl, enations, leaf crinkle, galls, and tumors (Figs. 3.8–3.10). Necrosis may be localized, as in chlorotic or necrotic lesions or ringspots on leaves or streaks on stems. Extensive necrosis may occur in phloem tissue, resulting in various growth abnormalities and gum formation. The plant species that react with local lesions have been extensively used for detection, identification, and assay of plant viruses, especially those for which nonbiological assay methods are either not available or difficult to perform (Table 4.2). The virus infection may result in conspicuous changes in the general growth and appearance of infected plants. General stunting of branches for entire plants, rosette nature of leaves, and bunching or crowding of leaves at the apex leading to bunchy top are some of the symptoms associated with viruses. For more details Bos (1970a) and Narayanasamy and Sabitha Doraiswamy (1996) may be consulted.

Plant viruses may be isolated from plants either infected naturally or inoculated artificially by different methods when the symptoms are expressed. Different methods of purification are followed, and they may be grouped on the basis of structural, physical, chemical, and serological properties. The particle morphological characteristics, presence or absence of envelope, nature of viral genome,

Figure 3.8 Local Lesions induced by cowpea aphid-borne mosaic virus on *Chenopodium amaranticolor.*

and strandedness of viral nucleic acid are some of the important characters used for grouping them. Plant viruses are classified into different groups, each having a type member, definitive members, and possible members that are not fully characterized (Fig. 3.11).

The nature of viral genome may be used as the basis for grouping the plant viruses as follows (Murphy et al., 1996).

A

B

Figure 3.9 A, Vein banding caused by cowpea aphid-borne mosaic virus on cowpea CV 152; B, vein clearing caused by yellow vein mosaic virus of okra.

A

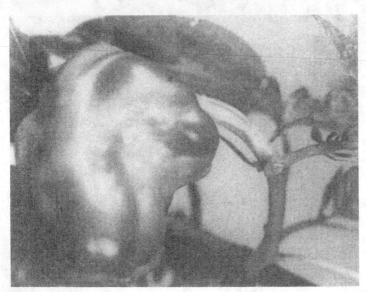

B

Figure 3.10 A, Yellow leaf curl virus-infected tomato (AVRDC); B, tomato spotted wilt virus-infected pepper fruit (AVRDC); C, rosette virus complex disease of peanut (ICRISAT).

58

C

Figure 3.10 Continued.

I. Viruses with DNA as the genome
A. Viruses with double-stranded DNA
    B.   Spherical particles
        1.   *Caulimovirus*—Caulliflower mosaic virus
    BB. Bacilliform particles
        2.   *Badnavirus*—Cocoa swollen shoot virus
AA. Viruses with single-stranded DNA
    B.   Spherical particles
        3.   Geminiviridae (Family)
            Monogeminivirus—monopartite—Maize streak
               virus
            Bigeminivirus—bipartite—Bean golden mosaic
               virus
            Hybrigemini virus—monopartite—Beet curly top
               virus

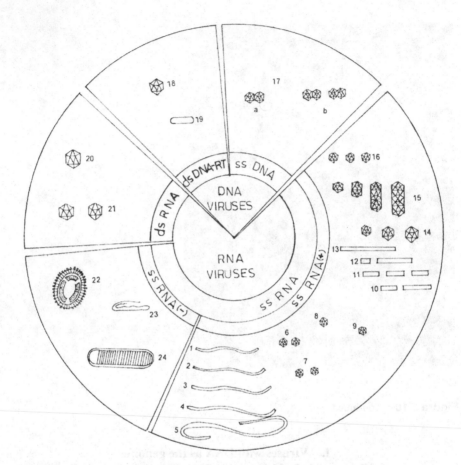

Figure 3.11   Plant virus groups/families: 1, *Potexvirus;* 2, *Capillovirus;* 3, *Carlavirus;* 4, *Potyvirus;* 5, *Closterovirus;* 6, *Enamovirus;* 7, Comoviridae; 8, Tombusviridae; 9, *Dinathovirus, Luteovirus, Machlovirus, Marafivirus, Necrovirus, Sobemovirus,* and *Tymovirus;* 10, *Furovirus;* 11, *Hordeivirus;* 12, *Tobravirus;* 13, *Tobamovirus;* 14, *Ilarvirus;* 15, *Alfamovirus;* 16, Bromoviridae—*Cucumovirus* and *Bromovirus;* 17, *Geminivirus*—a, subgroups I and II; b, subgroup III; 18, *Caulimovirus;* 19, *Badnavirus;* 20, *Reoviridae;* 21, *Alphacryptovirus, Betacryptovirus;* 22, Bunyaviridae—*Tospovirus;* 23, *Tenuivirus;* 24, Rhabdoviridae.

II.   Viruses with RNA as genome

A.   Viruses with double-stranded RNA

 B.   Spherical particles

  4.   Reoviridae (Family)

   Subgroup 1—*Phytoreovirus*—Wound tumor virus

Subgroup 2—*Fijivirus*—Fiji disease virus
Subgroup 3—*Oryzavirus*—Rice ragged stunt virus
5. Partitiviridae (Family)
Alphacryptovirus
Betacryptovirus
AA. Viruses with ss RNA (negative strand)
B. Bacilliform particles
6. Rhabdoviridae (Family)
Cytorhabdovirus—Lettuce necrotic yellows virus
Nucleorhabdovirus—Potato yellow dwarf virus
BB. Spherical particles
7. Bunyaviridae (Family)
Tospovirus—Tomato spotted wilt virus
BBB. Thread-like particles
8. *Tenuivirus*—Rice stripe virus
AAA. Viruses with single-stranded RNA (positive strand)
B. Spherical particles
C. Monopartite
9. Sequiviridae (Family)
Sequivirus—Parsnip yellow fleck virus
Waikavirus—Rice tungro spherical virus
10. Tombusviridae (Family)
Tombusvirus—Tomato bushy stunt virus
11. *Dianthovirus*—Carnation ringspot virus
12. *Luteovirus*—Barley yellow dwarf virus
13. *Machlomovirus*—Maize chlorotic mottle virus
14. *Marafivirus*—Maize *raydo fino* virus
15. *Necrovirus*—Tobacco necrosis virus
16. *Sobemovirus*—Southern bean mosaic virus
17. *Tymovirus*—Turnip yellow mosaic virus
CC. Bipartite
18. *Enamovirus*—Pea enation mosaic virus
19. *Idaeovirus*—Raspberry bushy dwarf virus
20. Comoviridae (Family)
Comovirus—Cowpea mosaic virus
Nepovirus—Tobacco ringspot virus
Fabavirus—Broadbean wilt virus
CCC. Tripartite
21. Bromoviridae (Family)
Cucumovirus—Cucumber mosaic virus
Bromovirus—Brome mosaic virus
Ilarvirus—Tobacco streak virus
Alfamovirus—Alfalfa mosaic virus

BB. Rodshaped particles (rigid)
    C. Monopartite
        22. *Tobamovirus*—Tobacco mosaic virus
    CC. Bipartite
        23. *Tobravirus*—Tobacco rattle virus
        24. *Furovirus*—Soilborne wheat mosaic virus
    CCC. Tripartite
        25. *Hordeivirus*—Barley stripe mosaic virus
BBB. Rodshaped particles (flexible)
        26. *Potexvirus*—Potato virus X
        27. *Capillovirus*—Apple stem grooving virus
        28. *Trichovirus*—Apple chlorotic spot virus
        29. *Vitivirus*—Grapevine virus A
        30. *Carlavirus*—Carnation latent virus
        31. Potyviridae (Family)
            Potyvirus—Potato virus Y
            Bymovirus—Barley yellow mosaic virus
            Rymovirus—Ryegrass mosaic virus
        32. *Closterovirus*—Beet yellows virus
BBBB. Bacilliform (positive strand)
        33. *Oleavirus*—Olive latent 2 virus

Detailed descriptions of the characteristics of plant viruses are provided in AAB descriptions of plant viruses published periodically by Association of Applied Biologists, U.K. The definitive members/possible members under each genus are presented below:

1.	Alfamovirus:	Alfalfa mosaic
2.	Alphacryptovirus:	Alfalfa 1, Beet 1, Beet 2, Carnation 1, Poinsettia cryptic, Radish yellow edge, Red pepper, Spinach temperate, White clorer 1
3.	Badnavirus:	Banana streak, Cocoa swollen shoot, Commelina yellow mottle, Rice tungro bacilliform, Schefflera ringspot, Sugarcane bacilliform
4.	Betacryptovirus:	Carnation 2, Red clover 2, White clover 2
5.	Bigeminivirus:	Abutilon mosaic, Ageratum yellow vein, Bean golden mosaic, Bhendi yellow vein mosaic, Cassava African mosaic, Cotton leaf curl, Mungbean yellow mosaic, Pepper huasteco, Potato yellow mosaic, soybean

crinckle leaf, Squash leaf curl, Tobacco leaf curl, Tomato golden mosaic, Tomato Indian leaf curl, Tomato leaf curl, Tomato mottle, Tomato yellow leaf curl.

6. Bromovirus: Broadbean mottle, Brome mosaic, Cassia yellow blotch, Cowpea chlorotic mottle, Melandrium yellow fleck

7. Bymovirus: Barley mild mosaic, Barley mild mottle, Barley yellow mosaic, Oat mosaic, Rice necrosis, Wheat spindle streak, Wheat yellow mosaic

8. Capillovirus: Apple stem grooving, Cherry A, Citrus tatter leaf, Lilac chlorotic leaf spot, Nandina stem pitting, Potato T.

9. Carlavirus: Blueberry scorch, Carnation latent, Chrysanthemum B, Cole latent, Cowpea mild mottle, Garlic latent, Hop latent, Hop mosaic, Lily symptomless, Mulberry latent, Passiflora latent, Pea streak, Potato M, Potato S, Red clover vein mosaic, Shallot latent.

10. Carmovirus: Blackgram mottle, Cardamine chlorotic fleck, Carnation mottle, Cowpea mottle, Cucumber soil-borne galinsago mosaic, Elderberry latent, Glycine mottle, Hibiscus chlorotic ringspot, Melon necrotic spot, Narcissus tip necrosis, Pelargonium flower break, Pelargonium line pattern, Saguaro cactus, Turnip crinkle.

11. Caulimovirus: Blueberry red ring ringspot, Carnation etched ring, Cassava vein mosaic, Cauliflower mosaic, Dahlia mosaic, Figwort mosaic, Mirabilis mosaic, Peanut chlorotic streak, Petunia vein clearing, Soybean chlorotic mottle, Strawberry vein banding.

12. Closterovirus: Apple chlorotic leaf spot, Beet yellows, Beet yellow stunt, Carnation necrotic fleck, Carrot yellow leaf, Citrus tristeza, Clover yellows, Grapevine cork bark-associated, Grapevine leaf roll-associated, Grapevine

stem pitting-associated, Lettuce infectious yellows, Pineapple wilt-associated, Sweet potato sunken vein, Tomato infectious chlorosis, Wheat yellow leaf.

13. Comovirus: Andean Potato mottle, Bean pod mottle, Bean rugose mosaic, Bean severe mosaic, Broad bean stain, Cowpea mosaic, Cowpea severe mosaic, Radish mosaic, Red clover mottle, Squash mosaic.

14. Cucumovirus: Cowpea ringspot, Cucumber mosaic, Peanut stunt, Robinia true mosaic, Tomato aspermy

15. Cytorhabdovirus: American wheat striate mosaic, Barley yellow striate mosaic, Cereal northern mosaic, Festuca leaf streak, Lettuce necrotic yellows, Strawberry crinkle

16. Dianthovirus: Carnation ringspot, Red clover necrotic mosaic, Sweet clover necrotic mosaic

17. Enamovirus: Pea enation mosaic

18. Faba virus: Broadbean wilt, Lamium mild mosaic

19. Fiji virus: Maize rough dwarf, Rice black-streaked dwarf, Sugarcane Fiji disease

20. Furovirus: Beet necrotic yellow vein, Beet soil-borne, Broadbean necrosis, Indian peanut clump, Oat golden stripe, Peanut clump, Potato mop-top, Rice stripe necrosis, Wheat soil-borne mosaic

21. Hordeivirus: Barley stripe mosaic, *Lychnis* ringspot, *Poa* semilatent

22. Hybrigemini virus: Beet curly top, Horseradish curly top, Tomato pseudo curlytop.

23. Idaeovirus: Rasberry bushy dwarf

24. Ilarvirus: American plum line pattern, Apple mosaic, Asparagus 2, Citrus leaf rugose, Citrus variegation, Elm mottle, Hydrangea mosaic, Lilac ring mottle, Prune dwarf, Prunus necrotic ringspot, Rose mosaic, Spinach latent, Tobacco streak, Tomato 1, Tulare apple mosaic

25. Ipomovirus: Sweet potato mild mottle

26. Luteovirus: Barley yellow dwarf, Bean leaf roll, Beet mild yellowing, Beet western yellows,

Carrot red leaf, Faba beans, Groundnut rosette assistor, Indonesian soybean dwarf, Pepper vein yellows, Potato leaf roll, Soybean dwarf, Strawberry mild yellow edge, Subterranean clover red leaf, Tobacco necrotic dwarf, Tomato yellow top, Turnip mild yellows

27. Machlomovirus: Maize chlorotic mottle

28. Maculravirus: Maculra mosaic, Narcissus latent

29. Maize white line: Maize white line mosaic
    mosaic virus (group)

30. Marafivirus: Bermuda grass etched line, Maize *rayado fino,* Oat blue dwarf

31. Monogeminivirus: Chickpea chlorotic dwarf, *Chlorosis* striate mosaic, *Digitaria* streak, Maize streak, *Panicum* streak, Sugarcane streak, Wheat dwarf.

32. Nanavirus: Banana bunchy top; Coconut foliar decay, Faba bean necrotic yellows, Subterranean clover stunt

33. Necrovirus: Tobacco necrosis

34. Nepovirus: Arabis mosaic, Artichoke Italian latent, Blueberry leaf mottle, Cherry leaf roll, Cherry rasp leaf, Grapevine chrome mosaic, Grapevine fan leaf, Mulberry ringspot, Peach rosette, Rasberry ringspot, Satsuma dwarf, Strawberry latent ringspot, Tobacco ringspot, Tomato black ring, Tomato ringspot

35. Nucleorhabdovirus: Eggplant dwarf mottled, Maize mosaic, Potato yellow dwarf, Sonchus yellow net,

36. Oleavirus: Olive latent 2

37. Oryzavirus: Rice ragged stunt

38. Ourmiavirus: Epirus cherry, Melon, Ourmia, Olive latent 2 (currently included under new genus Oleavirus)

39. Parsnip yellow fleck: Dandelion yellow mosaic, Parsnip yellow
    virus (group) fleck (currently included under genus Sesquivirus)

40. Phelum mottle virus Cocks foot mild mosaic, Panicum mosaic
    (group):

41. Phytoreovirus: Clover wound tumor, Rice dwarf

42. Potexvirus: Bamboo mosaic, Cactus X, Cassava com-
                                   monmosaic, Clover yellow mosaic, Cym-
                                   bidium mosaic, Foxtail mosaic, Hosta X,
                                   Hydrangea ringspot, Lily X, Narcissus mo-
                                   saic, Nerine X, Papaya mosaic, Pepino mo-
                                   saic, Potato aucuba, Potato X, Strawberry
                                   mild yellow edge—associated, Viola mot-
                                   tle, White clover mosaic.

43. Potyvirus: Alstroemeria mosaic, Amazon lily mosaic,
                                   Artichoke latent, Asian prunus latent, Azuki
                                   bean mosaic, Banana bract mosaic, Bean
                                   common mosaic, Bean common mosaic
                                   necrosis, Bean yellow mosaic, Beet mosaic,
                                   Bidens mosaic, Blackeye cowpea mosaic,
                                   Carnation vein mottle, Cassava brown
                                   streak, Celery mosaic, Chilli veinal mottle,
                                   Clover yellow vein, Cocksfoot streak, Cow-
                                   pea aphid-borne mosaic, Dasheen mosaic,
                                   Datura columbian, Garlic yellow streak,
                                   Johnson grass mosaic, Kalanchoe mosaic,
                                   Konjak mosaic, Leek mosaic, Lily mottle,
                                   Maize dwarf mosaic, Narcissus yellow
                                   stripe, Onion yellow dwarf, Ornithogalum
                                   mosaic, Palm mosaic, Papaya leaf-distor-
                                   tion mosaic, Papaya mosaic, Papaya
                                   ringspot, Passion fruit woodiness, Peanut
                                   mottle, Peanut stripe, Pea seed-borne mo-
                                   saic, Pepper mottle, Pepper vein mottle,
                                   Plum pox, Potato A, Potato V, Potato Y, Ra-
                                   nunculus mottle, Sorghum mosaic, Soybean
                                   mosaic, Sugarcane mosaic, Sweet potato
                                   feathery mottle, Sweet potato latent, Tamar-
                                   illo mosaic, Taro feathery mottle, Tobacco
                                   etch, Tobacco vein mottling, *Trifolium
                                   montana* mosaic, Tulip breaking (mosaic),
                                   Turnip mosaic, Vanilla necrosis, Water
                                   melon mosaic 1, Watermelon mosaic 2,
                                   Yam mosaic, Zucchini yellow fleck, Zuc-
                                   chini yellow mosaic

44. Rhabdobirus: Beet leaf curl, Citrus leprosis, Gentiana, Or-
                                   chid fleck.

45.	Rymovirus:	Agropyron mosaic, Brome streak mosaic, *Hordeum* mosaic, Ryegrass mosaic, Wheat leaf streak, Wheat streak, Wheat streak mosaic.
46.	Satellivirus:	*Panicum* mosaic, Tobacco mosaic, Tobacco necrosis
47.	Sesquivirus:	Parship yellow fleck, Dandelion yellow mosaic
48.	Sobemovirus:	Blueberry shoestring, Calopo yellow mosaic, Cocks foot mottle, Clover mottle, Olive latent 1, Rice yellow mottle, Southern bean mosaic, Sowbane mosaic, Subterranean clover mottle, Turnip rosette mosaic
49.	Tenuivirus:	*Echinochloa* hoja blanca, European wheat stripe mosaic, Maize stripe, Maize yellow stripe, Rice grassy stunt, Rice hoja blanca, Rice stripe
50.	Tobamovirus:	Cucumber green mottle, *Odontoglossum* ringspot, Paprika mild mottle, Pepper mild mottle, Ribgrass mosaic 4, Sunnhemp mosaic, Tobacco mild green mosaic, Tobacco mosaic, Tomato mosaic, Tobacco rattle
51.	Tobravirus:	Pea early browing, Tobacco rattle
52.	Tombusvirus:	Artichoke mottled crinkle, Carnation Italian ringspot, Cucumber necrosis, Cymbidium ringspot, Oat chlorotic stunt, Petunia asteroid mosaic, Tomato bushy stunt.
53.	Tospovirus:	Groundnut ringspot, Impatiens necrotic, Peanut bud necrosis, Peanut yellow spot, Tomato chlorotic spot, Tomato spotted wilt
54.	Trichovirus:	Apple chlorotic leaf spot, Cherry mottle leaf, Grape vine A, Grape vine B (currently included under new genus Vitivirus).
55.	Tymovirus:	Andean potato latent, Cocoa yellow mosaic, Egg plant mosaic, Kennedya yellow mosaic, Okra mosaic, Onion yellow mosaic, Physalis mottle, Poinsettia mosaic, Turnip yellow mosaic
56.	Umbravirus:	Carrot mottle, Groundnut rosette, Sunflower yellow blotch
57.	Varicosavirus:	Lettuce big vein, Lettuce ring necrosis

58.   Vitivirus: Grape vine A, Grape vine B
59.   Waikavirus: Maize chlorotic dwarf, Rice tungro spher-
                                          ical

3.1.5   Symptoms of Viroid Diseases

Plant viroid diseases do not exhibit any specific symptoms that can be used to differentiate them from diseases caused by viruses. This may be a possible reason
why this group of disease-causing agents could not be recognized prior to the report of Diener (1971a) on potato spindle tuber disease. However, individual viroid
diseases can be identified by the symptoms on given plant species. Infected potato
plants may show different degrees of stunting with foliage turning slate gray with
dull leaf surface. But the characteristic symptom of the disease is seen on the tubers, which are abnormally elongated, assuming spindle or cylindrical shape with
prominent eyes (Diener, 1979).

Tomatoes infected by bunchy top viroid are markedly stunted. Smalling and
distortion of leaflets, necrosis of leaves and stems, and crowding of leaves near the
apices of branches are the other symptoms induced by this viroid (McClean,
1931).

Citrus exocortis disease is characterized by scaling of the bark below the
graft union and stunting of the trees. Trifoliate orange (*Poncirus trifoliata*), and
Citrus spp. and varieties may be used to detect latent infection (Benton et al.,
1950; Olson, 1968). Chrysanthemum plants infected by stunt viroid show severe
stunting, reduction in leaf size, and paling of foliage as the chief symptoms (Dimock, 1947). Chrysanthemum chlorotic mottle viroid, on the other hand, induces
mild mottling or variegation of young leaves. The cv. 'Deep Ridge' can be used
as the diagnostic host for this viroid (Horst, 1975).

The diagnostic symptoms of cucumber pale fruit disease are observed on
fruits, which become pale, shorter, and slightly pear-shaped. The flowers are
stunted and crumpled. The leaves are malformed and become chlorotic later (van
Dorst and Peters, 1974).

Coconuts affected by cadang-cadang disease have small irregularly shaped
lamina with bright yellow or orange spots. As the discolored spots coalesce, older
leaves may show mottling or turn yellow and become brittle. The nuts become
progressively smaller in size year after year, elongated, or distorted. Nut-bearing
capacity is progressively reduced as the crown size is reduced with cessation of
flower production. Ultimately the growing bud dies and falls off, leaving bare
trunk (Price, 1971).

Hop stunt disease is recognized by shortened internodes of the main and lateral branches and curling of upper leaves (Sasaki and Shikata, 1977a).

3.2 CYTOPATHOLOGICAL CHARACTERISTICS

3.2.1 Fungal Diseases

Pathogenic fungi gain entry into susceptible host plants through natural openings or by direct penetration of the epidermal cells. Most of the obligate parasites that cause powdery mildews, downy mildews and rust absorb nutrients from the cells into which haustoria alone are formed, keeping the host cells alive for longer periods. Facultative parasites, on the other hand, produce many kinds of biologically active substances, such as enzymes, toxins, growth regulators, polysaccharides, and antibiotics. The substances may affect the cells directly or cause structural changes.

The fungi produce characteristic spore-bearing structures which arise from the internal mycelium and emerge either through stomata or by piercing of the epidermis. Most of the pathogens that cause downy mildews, powdery mildews, rusts, leafspots, and blight can be identified by examining the sporangia or conidia which are formed either on free conidiophores or in specialized structures such as sori, acervuli, or pycnidia. The pathogens that cause damping off and root rots produce macerating enzymes that produce extensive structural breakdown in affected tissues. The mycelium and spore-bearing structures are formed in these tissues. The wilt pathogens are found in the conducting vessels. They produce toxic metabolites and enzymes, which may break down the cell walls, leading to death of cells and plugging of vessels. Fungal structures, such as mycelium, microconidia, and macroconidia, may be present in the affected tissues.

Anatomical research will be useful in studying the characteristics of fungi present in the tissues of affected plants. However, there is hardly any specific change induced in infected tissue by fungal pathogen that may have diagnostic value for the identification of the pathogen that is causing the disease in question.

3.2.2 Bacterial Diseases

After the entry of bacteria through wounds or natural openings in susceptible plants, bacteria that cause diseases such as soft rots (*Erwinia* spp.), fire blight of apple (*E. amylovora*), and wild fire of tobacco (*Pseudomonas tabaci*) live and reproduce in the intercellular spaces for some time. Then they secrete enzymes capable of breaking down the middle lamella and macerating the tissues, leading to loss of turgor, cell collapse, and ultimate death of cells. The toxins produced by these pathogens hasten the loss of water and electrolytes from the cells. As the cells are killed in large numbers, bacterial ooze may appear on the leaf surface. Lysigenous cavities may be formed as a result of breakdown of cell walls, and the cavities are filled with cellular contents and masses of bacterial cells. Cankers may arise from the cavities formed in the cortical tissues.

Some pathogenic bacteria that cause wilts invade the conducting vascular tissues, in which they are able to reproduce rapidly and spread to other organs or tissues. They secrete enzymes, toxins, or slimy extracellular polysaccharides (EPSs). These metabolites may act on the cells of vascular tissue, leading to the breakdown of cells and accumulation of cellular materials. This may cause clogging of vessels, resulting in reduction in or complete blockage of translocation of water and nutrients. The infected plants show progressive stunting, wilting, and death.

Excessive cell division (hyperplasia) and cell enlargement (hypertrophy) are the characteristic features of crown gall disease caused by *Agrobacterium tumefaciens*. Infection by this pathogen results in galls or tumors on roots, stem, and other organs. In the infected plants, diversion of nutrients to the tumor or gall tissues occurs, and normal tissue development and other essential processes are hampered. As a result, the growth and consequently the yield of affected plants are appreciably reduced.

3.2.3 Virus Diseases

Plant viruses induce characteristic macroscopic or external symptoms (see Chapter 3.1.4) that may be of diagnostic value. So also some plant viruses induce distinct histological changes in infected plants, and such changes may help to differentiate the virus in question from other pathogens that may infect the same host plant species.

3.2.3.1 Anatomical changes

Viruses such as potato leaf foll virus and sugar beet curly top virus cause phloem degeneration. Necrosis is confined to primary phloem in potatoes infected by leaf roll virus and no abnormal growth of phloem tissue is seen. There is excessive deposition of callose in the phloem of stem and tubers. Presence of callose in high concentration may be observed by staining the cells with resorcin blue. Sugar beet curly top virus causes growth abnormalities in the initial stages of phloem degeneration and necrosis of primary and secondary phloem, leading to cell collapse and lesion formation.

Grapevine leaf roll virus induces a characteristic structural abnormality in vascular tissues. Formation of tuberculae, cellulose bars and rods that traverse the lumens of interfascicular and phloem parenchyma cells, is a diagnostic symptom of this disease (Bos, 1970a). Crimson clover wound tumor disease is characterized by the abnormal development of phloem cambial cells. Phloem parenchyma forms meristematic tumor cells in the phloem of leaf, stem, and root (Lee and Black, 1955).

Pitting of the wood in apple stem pitting disease is due to the failure of some cambial initials to differentiate into normal cells. A wedge of phloem tissue is formed and becomes embedded in the newly formed xylem tissue. Later the phloem tissue becomes necrotic (Hilborn et al., 1965).

3.2.3.2 Cytological changes

Virus infections result in cytological effects such as starch accumulation, inhibition of plastid development, and chloroplast destruction. But a virus-specific effect leading to the formation of intracellular inclusions is observed in certain host plants after infection by viruses. These inclusions have characteristic features useful for the identification of viruses. The inclusions may be formed in either the cytoplasm or the nucleus of infected cells and may be either amorphous or crystalline in nature. Among the 49 criteria listed for the classification of plant viruses, characteristics of inclusion bodies and their intracellular location are included (Harrison et al., 1971). Members of 20 virus groups are known to induce inclusions which form one of the main characteristics of the respective virus group. These inclusions are useful for diagnostic and taxonomic purposes and for identification and characterization of virus-specific, noncapsid proteins and possible sites of viral synthesis (Hiebert et al., 1984).

Viruses which produce characteristic inclusions may be identified by following simple techniques. The inclusions are present in large numbers in the epidermal cells of leaves showing distinct symptoms of virus infection. Epidermal strips without fixing are stained with trypan blue, mounted in water, and examined under the microscope. Trypan blue is dissolved in hot 0.9% aqueous NaCl to yield a 0.5% solution, and this stock solution may be diluted in 0.9% NaCl to produce 1/2000–1/5000 dilution at the time of examination. Amorphous inclusions are deeply stained, whereas nuclei have less intense color (McWhorter, 1941b). Phloxine (1%) is recommended by Rubio-Huertos (1950). Inclusion bodies appear as bright red structures, and nuclei stain pink. A combination of pyronin (0.2%) and methyl green (0.5%) in 0.1 M acetate buffer, pH 5.3, can be used to stain the inclusions, differentially. Nuclei turn blue and inclusions are red, because of their chemical constitution. The blue color is due to the presence of DNA and the red to the presence of RNA (Rubio-Huertos, 1972). With different combinations of calcomine orange, Luxol brilliant green, Congo rubin-methyl green, phloxine, and methylene blue, Christie (1967) and Christie and Edwardson (1977) reported that inclusion bodies produced by tobacco mosaic virus, and tobacco etch virus could be differentiated. Christie et al. (1988) showed that by using two differential stains—Azure A and Luxol brilliant green-calcomine orange—characteristic inclusions induced by several viruses could be detected, and the causal virus may be identified by this simple technique. The maize dwarf mosaic virus, maize stripe virus, maize mosaic virus, and maize rayado fino virus could be detected and identified by examining the intracellular inclusions associated with infection by these viruses (Overman et al., 1992). Using the same combinations of strains, the inclusions induced by other genera of viruses such as *Furovirus* and *Tenuivirus* were detected. A short microwave treatment for 10–15 s provided better staining intensity and reduced the staining time at room temperature by 10 min (Hoefert et al., 1992; Christie et al., 1995) [Appendix 3(1)].

Intracellular inclusions have been examined in plant materials fixed with different fixatives. By using Dalton's fixative containing mercuric chloride and a small quantity of acetic acid, followed by washing with iodine in alcohol, excellent results have been obtained. Details regarding the composition of different fixatives and protocols are described in the review by Rubio-Huertos (1972).

By using Azure A and orange-green (calcomine orange 2 RS and Luxol brilliant green BL) combinations, the inclusions associated with viruses belonging to 17 virus groups, viz. bromovirus, carlavirus, caulimovirus, closterovirus, comovirus, cucumovirus, geminivirus, luteovirus, nepovirus, phytoreovirus, plant rhabdovirus, potexvirus, potyvirus, tobacco necrosis, tobamovirus, tombus virus, and tymovirus, can be detected in leaves infected by these viruses (Hiebert et al., 1984).

Cylindrical inclusions of many potyviruses, including tobacco etch virus (Hiebert and McDonald, 1973), bean yellow mosaic virus (Nagel et al., 1983), turnip mosaic virus (Hiebert and McDonald, 1973), and tobacco vein mottling virus (Hellman et al., 1983), have been purified. Hiebert et al. (1984) have developed a procedure for the purification of nuclear inclusions induced by tobacco etch virus. Viroplasms formed by cauliflower mosaic virus can be purified by the procedure developed by Al Ani et al. (1980).

Polyclonal antisera using rabbits have been raised against inclusion bodies induced by potyviruses and caulimoviruses. These antisera prepared against nonstructural virus-associated proteins are useful in detecting their presence in different host plants. The immunodiffusion test (Purcifull et al., 1973), immunofluorescence test, ferritin-antibody electron microscopy (Breese and Hsu, 1971), liquid precipitin test (Hiebert et al., 1974), Western blotting, and radioimmunoassay (Towbin et al., 1979) and enzyme-linked immunosorbent assay (ELISA) (Falk and Tsai et al., 1987) have been employed for the detection of inclusion body proteins in various host plant species infected by different viruses.

As the cylindrical inclusion proteins, the amorphous inclusion proteins, and the TEV nuclear inclusion proteins account for about 70% of the total protein-coding capacity of the potyviral genome, the antisera raised against these inclusions may be used for more precise viral diagnosis and study of relationships between potyviruses. The serological relationships between the cylindrical inclusion proteins (CIPs) have been shown to be useful to study relationship between potyviruses. The antisera were raised against CIPs of bean yellow mosaic (BYMV), clover yellow vein (C1YVV), turnip mosaic (TuMV), sweet potato feathery mottle (SPFMV), and maize dwarf mosaic (MDMV) potyviruses. The antisera to CIPs of all these potyviruses except MDMV cross-reacted to most or all of the purified CIPs of 18 potyviruses tested in Western blots. The antiserum to MDMV-CIP showed significant cross-reaction only to the CIP for sorghum mosaic virus (Hammond, 1998). This approach of using CIPs for preparing antisera will be preferable, because the cumbersome procedures required for virus purification can be avoided. Antisera raised against virus capsid proteins represent only about 10% of potyviral protein-coding capacity, and such antisera have only limited value in establishing the relationship between potyviruses. Another advantage derived from the study of inclusion bodies, which are, in most cases, an aggregation of the virus particles inducing their formation, is the potential for purifying the viruses in a novel way as in the case of citrus tristeza virus (Lee et al., 1982).

Cytological changes induced by viruses may be studied in detail by examining ultrathin sections under an electron microscope (Chapter 7). In many host-virus combinations, specific or broad changes in ultrastructure of the cellular constituents, which represent the signature of the virus (McWhorter, 1965), may be discernible. Such cytopathological features of infected cells may be used as the basis for identification of virus groups or even individual viruses.

a. Tombus viruses.   Two cytological changes occur together commonly in plants infected by tombus viruses: a) appearance of cytoplasmic multivesicular bodies and b) formation of virus-containing bleblike evaginations of the tonoplast into the vacuole (Martelli, 1981). Additionally the intranuclear presence of virus and membraneous inclusions may be observed in certain host-virus combinations.

b. Comovirususes and nepoviruses.   The members of the comovirus and nepovirus groups do not produce any distinct cytopathological change with which they can be distinguished. However, most of them produce vesiculate-vacuolate cytoplasmic inclusions, tubules containing rows of virus particles, and cell wall outgrowths. These changes are quite characteristic of these two groups and useful as intracellular markers of diagnostic value (Martelli and Russo, 1984). Another distinct feature of infection by como- and nepoviruses is the production of tubules enclosing virus particles. They are usually single-walled, but in some cases, for example, strawberry latent ringspot virus, double-walled tubules may be formed (Roberts and Harrison, 1970).

c. Tymoviruses.   The tymoviruses form a homogeneous group and so also they induce group-specific, as well as virus-specific, cytological changes that may help in their identification. Clumping of altered chloroplasts and intranuclear accumulations of empty viral capsids are the two diagnostic markers of this group. Clumping of chloroplasts, first reported in chinese cabbage infected with turnip yellow mosaic virus (Rubio-Huertos, 1950), is observed in all hosts infected by tymoviruses. Periplastial flask-shaped vesicles bounded by a double membrane and containing finely stranded material appear as a very early change induced by these viruses (Gerola et al., 1966; Matthews, 1977). Empty virus protein shells, when produced excessively, move into the nucleus, forming large crystalline aggregates (Lesemann, 1977).

d. Luteoviruses. The luteoviruses do not appear commonly to cause any cytological changes that may be considered characteristic of this group. However, the presence of virus particles and membranous vesicles in the nuclear area and of virions in plasmodesmata has been noted consistently in infections by viruses that have been studied. Esau and Hoefert (1972) observed the accumulation of viruses around the nucleolus sometimes in crystalline array in the early stages of infection with beet western yellows virus (BWYV). The vesicles found in plants infected by BWYV and potato leaf roll virus (PLRV) are bound by a single or double membrane and have a network of fibrils resembling nuclei acid (Esau and Hoefert, 1972; Shepardson et al., 1980). On the basis of the ultrastructural differences induced, barley yellow dwarf virus (BYDV) strains can be divided into two distinct subgroups (Gill and Chong, 1979a).

e. Bromoviruses. The bromoviruses induce the production of inclusion bodies which can be recognized by light microscopy. The inclusion bodies produced by broadbean mottle virus (BBMV) consist of amorphous material, vesicles, and virions, and their locations may be the sites of viral RNA replication. Viral antigen is also present in the inclusions. The inclusion bodies formed by cowpea chlorotic mottle virus (CCMV) have a fine granular zone, proliferating endoplasmic reticulum, fibril-containing vesicles, and thin flexuous filaments. The CCMV particles are randomly scattered in the cytoplasm and do not aggregate into crystalline structures (Bancroft et al., 1969).

f. Cucumoviruses. Cytological alterations induced by cucumber mosaic virus have been studied in detail. The virus particles, usually found scattered in the cytoplasm, have a tendency to aggregate into crystals in the vacuoles (Fig. 3.12). The presence of viruses in the nuclei of many cells has been frequently observed (Honda and Matsui, 1974). The membrane-bound vesicles associated with the tonoplast may be the site of viral RNA replication (Hatta and Francki, 1981). The ultrastructural changes induced by cucumoviruses include extensive secondary vacuolation resulting in fragmentation of cytoplasm and proliferation of endomembranes. Production of cytoplasmic and intravacuolar crystalline inclusions has been observed more frequently in plants infected by CMV infection than in plants infected by peanut stunt virus (PSV) and tomato aspermy virus (TAV) belonging to the same genus *Cucumovirus* in the family Bromoviridae (Martelli and Russo, 1985; Rybicki, 1995). The strains of PSV have been classified into two

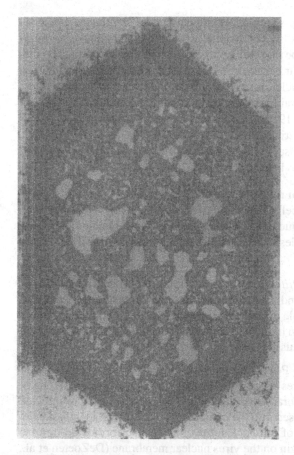

Figure 3.12 An intravacuolar virus crystal consisting of cucumber mosaic virus particles; bar represents 200 nm (Academic Press Inc., USA). (Courtesy of Martelli and Russo, 1984).

major subgroups (I and II) on the basis of serological properties and percent nucleotide sequence homology (Hu et al., 1997, 1998). PSV subgroup II strains induce, in tomato protoplasts or leaf tissues, novel inclusions that appeared as long, thin, densely staining ribbon-like sheet in thin sections. These inclusions were present in the cytoplasm either singly or stacked in small irregular groups. Numerous virus-like particles either singly or in small aggregates were observed adjacent to the surface of the inclusions. The inclusions present in tobacco tissue may be useful to differentiate PSV strain groups, since they are induced only by PSV subgroup II but not by sub group I (Sanger et al., 1998).

g. Sobemoviruses. No group-specific cytological change is induced by sobe-moviruses. The virus particles may occasionally be found to aggregate, forming crystals. Two strains of the type member, southern bean mosaic virus, may be differentiated by their intracellular behavior: Large crystals are produced by cowpea strain in the cytoplasm and nuclei, mainly in infected cells near vascular tissues; on the other hand, the bean strain forms crystals very rarely, and only in phloem cells (Weintraub and Ragetli, 1970). Certain virus-specific changes, such as formation of flexuous tubules in rice yellow mosaic virus-infected plants (Bakker, 1975) and the network of densely stained material in sowbane mosaic-infected plants (Milne, 1967), have also been reported.

h. Dianthoviruses. Carnation ringspot virus (CRSV) induces characteristic cytological changes in infected cells. Large irregular crystalline aggregates are seen in the cytoplasm, nuclei, and nucleolus. The nuclei may also contain tubules, generally without any virus particles, but sometimes tubules may enclose a single row of virions.

i. Tobacco necrosis and satellite viruses. The intracellular behavior of the tobacco necrosis virus (TNV) and that of its satellite show distinct difference. The satellite virus forms large stable crystals in situ, whereas TNV particles are found scattered even when they are in high concentration. Probably the TNV crystals are quite unstable (Martelli and Russo, 1984).

j. Pea enation mosaic virus. Pea enation mosaic virus (PEMV), the only member of the PEMV group, induces specific cytopathological structures consisting of accumulations of single-membraned vesicles in the perinuclear space of infected cells. As the vesicles are released into the cytoplasm, an additional membrane derived from the outer lamella of the nuclear envelope is also formed around the vesicles, which have their origin on the virus nuclear membrane (DeZoeten et al., 1972). Though the vesicular aggregates may resemble those of some luteoviruses, PEMV aggregates are larger than luteovirus aggregates, which may sometimes occur in groups. The PEMV may also occasionally form paracrystalline inclusions in the endoplasmic reticulum (Rassel, 1972).

k. Alfalfa mosaic virus. Alfalfa mosaic virus (AMV) is the sole member of the alfalfa mosaic virus group, and its presence in the ultrathin sections of infected leaves can be recognized by the polymorphic particles, the shape of which varies from spherical to bacilliform. All strains of AMV do not aggregate, but the strains that do may be differentiated on the basis of the manner in which the virus particles aggregate intracellularly. Four types of virus particle aggregates have been recognized: a) short rafts of particles arranged in a hexagonal array; b) long bands of particles aligned side by side, sometimes in a stacked-layer configuration; c) aggregates consisting of whorllike structures; and d) aggregates with four parallel rows of particles packed in either an apparently rhomboid lattice or a hexagonal one (Martelli and Russo, 1984).

l. Reoviruses. Plant reoviruses, consisting of two subgroups with generic status, *Phytoreovirus* and *Fijivirus,* produce three recognizable cytopathic effects: a) proteinaceous material (viroplasm) containing dark spherical bodies (about 50 nm diameter) representing immature inner cores of virus particles accumulates; b) near the viroplasm mature virus particles aggregate into crystals; and c) tubules containing rows of virus particles are also formed. The two subgroups can be distinguished on the basis of the shape and size of viroplasms. *Phytoreovirus* induces small spherical viroplasms, whereas *Fijivirus* produces large elongated viroplasms (Shikata, 1981).

m. Caulimoviruses. The caulimoviruses are found freely scattered in the parenchymatous cells but very often aggregate to form rounded or elongated inclusion bodies in which virus particles are embedded (Fig. 3.13 A and B). The final size

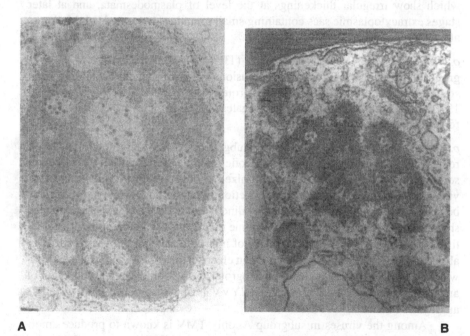

A   B

Figure 3.13 A, Characteristic cytoplasmic inclusion produced by cauliflower mosaic virus; bar represents 200 nm (Academic Press, Inc., USA). (Courtesy of Martelli and Russo, 1984); B, inclusion bodies in peanut leaf cell infected by peanut chlorotic streak virus; bar represents 200 nm. (American Phytopathological Society, Minnesota, USA). (Courtesy of Reddy et al., 1993.)

of the inclusions is determined by the viral genome; hence the individual virus isolates of cauliflower mosaic virus may be differented on the basis of the size of the inclusions (Shalla et al., 1980).

The characteristics of inclusions can be used for diagnosis of caulimoviruses by using a light microscope. Selection of healthy dahlia plants was successfully accomplished by determining the presence or absence of the inclusions in the tissues (Robb, 1963).

n. Hordeiviruses.   The cytological effects of barley stripe mosaic virus (BSMV), the type member of the hordeivirus group, have been studied in detail (McMullen et al., 1978). In the infected cells, small, rounded plastidial vesicles or flask-shaped invaginations of the outer lamella of the boundary membrane containing fine fibrils are formed (McMullen et al., 1978). The virus particles accumulate in either the cytoplasm or the nucleus, forming irregularly shaped or paracrystalline aggregates (Carroll, 1970). Another effect is seen on cell walls, which show irregular thickenings at the level of plasmodesmata, and at later stages extracytoplasmic sacs containing small granules are formed (McMullen et al., 1977).

o. Tobraviruses.   Tobacco rattle virus (TRV), type member of the tobravirus group, induces bulky cytoplasmic inclusions consisting of mitochondria, ribosomes, and electron-dense, possibly proteinaceous material (Harrison et al., 1970). In addition, paracrystalline aggregates of long particles arranged in tiers are formed (Chang et al., 1976).

p. Tobamoviruses.   Two tobamovirus subgroups, tobacco mosaic virus (TMV), representing subgroup A, and beet necrotic yellow vein virus (BNYVV), representing subgroup B, have been recognized. Tobacco mosaic virus and other viruses of subgroup A induce the production of crystalline inclusions which can be seen under a light microscope as hyaline plates with a hexagonal or rounded shape. The characteristics of the crystalline inclusions can be used for virus identification. The inclusions are composed of many stacked layers of virus particles aligned in a parallel array, as seen with an electron microscope (Warmke and Edwardson, 1966). Viruses included in subgroup B do not induce the production of any crystalline inclusions; however, BNYVV is able to produce paracrystalline aggregates (Tamada, 1975).

Among the viruses in subgroup A, only TMV is known to produce amorphous inclusions (X-bodies), which are aggregates of ribosomes, endoplasmic reticulum, and small vacuoles or vesicles; viral protein in granular or tubular form.; and small pockets of virus particles (Esau, 1968). On the other hand, many of the viruses in subgroup B induce amorphous inclusions. In the fully developed inclusions induced by BNYVV, a large number of virions are found scattered throughout the inclusion (Russo et al., 1981). Hibino et al. (1974) reported that nine isolates of soil-borne wheat mosaic virus could be divided into three groups based on the composition and relative abundance of strands of endoplasmic reticulum, tubules, vesicles, membranes, and virions.

Cucumber green mottle virus (CGMV) in subgroup A is distinct from other tobamoviruses in its ability to induce an enlargement and membraneous proliferation of mitochondria which is heavily vesiculated (Sugimura and Ushiyama, 1975).

q. Potexviruses. The characteristic cytopathological effect of potexvirus infection is the formation of irregular, fibrous, or banded aggregates of virus particles. Virions are aligned side by side and aggregated end-to-end, giving spindle shape to the fibrous inclusions, whereas virus particles are aligned in horizontal tiers in banded inclusions which are composed almost entirely of virus particles. This cytoplasmic strands separate successive layers, giving the banded appearance to the inclusions.

Potato virus X (PVX) also induces laminate inclusions consisting of thin sheets of proteinaceous material sometimes studded with beads which are antigenically different from PVX protein (Shalla and Shepherd, 1972). Clover yellow mosaic virus (ClYMV) also produces additional amorphous inclusions in the cytoplasm and vacuoles of cells. These inclusions are composed of a protein antigenically related to the ClYMV protein (Schlegel and De Lisle, 1971).

r. Carlaviruses. The carlaviruses generally do not induce any specific cellular modification that can be considered characteristic of the group. However, many of them produce flexuous bundles of virus particles or bounded aggregates similar to those induced by potexviruses. Red clover vein mosaic virus (RCVMV) appears to be unique and differs from other members in inducing crystalline inclusions composed of polyhedral particles 10 nm in diameter than contain protein and RNA. Though the origin and significance of the constituents of polyhedral particles are not known, they are useful in diagnosing RCVMV infections (Khan et al., 1977).

s. Potyviruses. The potyvirus group is the largest, enclosing 48 definitive members and 67 possible members. Though they differ considerably in particle length and nature of vector, one single unifying ultrastructural feature, the presence of a cylindrical or pinwheel inclusions in the respective host plants infected by different viruses, has been recognized. The pinwheel inclusion has a central core from which rectangular or triangular curved plates radiate. Laminated aggregates may be formed when the plates of adjacent pinwheels may fuse to form a series of stacked laminar structure or scrolls (or tubes) may be formed when the plates roll inward. On the basis of the characteristics of configuration formed as a result of

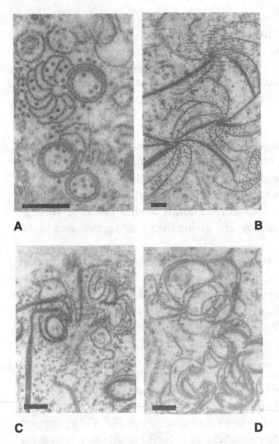

A   B

C   D

Figure 3.14 Cylindrical inclusions of potyvirus; A, pinwheels and scrolls (subdivision I); B, pinwheels and aggregates (subdivision II); C, pinwheels, scrolls, and long straight laminated aggregates (subdivision III); D, pinwheels, scrolls, and short curved laminated aggregates (subdivision IV); bar represents 100 nm (Academic Press Inc., USA). (Courtesy of Martelli and Russo, 1984.)

the association between pinwheels and plates, Edwardson (1981) suggested that potyviruses may be grouped into four subdivisions: a) pinwheels and scrolls; b) pinwheels and long straight laminated aggregates; c) pinwheels, scrolls, and long straight laminated aggregates; and d) pinwheels, scrolls, and short curved laminated aggregates (Fig. 3.14).

Many potyviruses, in addition to proteinaceous pinwheel inclusions, produce complex amorphous cytoplasmic inclusion bodies and secondary vacuolation of the cytoplasm. Monolayers of virus particles in parallel array delimited by two membranes may be seen in the secondary vacuoles.

In the case of some of the potyviruses, the cytological abnormalities are so specific that they may be used for the diagnosis of infection by the respective viruses. Intensely electron-opaque angular or rounded bodies surrounded low mosaic virus infection. Beet mosaic virus induces characteristic satellite bodies consisting of nucleolus-related amorphous accumulation of proteinaceous material (Martelli and Russo, 1969). Accumulation of electron-dense granular material, sometimes consisting of thin rodlike structures resembling virus particles, forms a diagnostic feature of watermelon mosaic virus 1 infection (Martelli and Russo, 1976). The presence of cytoplasmic or intranuclear inclusions consisting of fimbriate bodies is a reliable indicator of infection by zucchini yellow fleck virus (Martelli and Russo, 1984). Tobacco etch virus is known to induce large crystalline inclusions in both the nuclei and the cytoplasm (Kassanis, 1939).

t. Closteroviruses. Among three closterovirus subgroups (A, B, and C) recognized subgroup A does not induce any specific cytopathological effects but the viruses accumulate in high concentrations in companion cells and sieve tubes, filling almost the entire lumen of cells (Bem and Murrant, 1980). On the other hand, plants infected by viruses of subgroups B and C show the presence of virions aggregating into cross-banded structures. The virus particles are closely packed in several tiers (Esau and Hoefert, 1971), in loose wavy paracrystalline aggregates (Esau and Hoefert, 1981), or in irregular fascicles intermingled with membraneous vesicles containing a network of fine fibrils (Martelli and Russo, 1984). The inclusions induced by viruses of the B and C subgroups have diagnostic value as they are produced almost exclusively by these viruses. Tomato infectious chlorosis closterovirus (TICV) produces characteristic inclusion bodies in the phloem of infected *Nicotiana clevelandii,* which can be used as a diagnostic host (Wisler et al., 1996).

u. Tomato spotted wilt virus. Tomato spotted wil virus (TSWV) is a monotypic virus group with only tomato spotted wilt virus, which has a very wide host range. The presence of large spherical particles (70–80 nm diameter) with membrane bound in groups is itself a feature useful for the identification of TSWV infection (Fig. 3.15 A and B). However, viroplasms composed of large cytoplasmic aggregates of dark-staining, probably proteinaceous material (Milne, 1970) or clusters of hollow tubules intermingled with endoplasmic reticulum strands (Francki and Hatta, 1981; Vaira et al., 1993) may also be useful as markers of TSWV infection.

v. Rhabdoviruses. The characteristic particle morphological properties of the rhabdoviruses are useful diagnostic markers. The virus particles appear as spherical, elongated, or bacilliform structures when they are cut transversely or longitudinally (Chen and Shikata, 1971) (Fig. 3.16 A, B, and C). Rhabdoviruses may

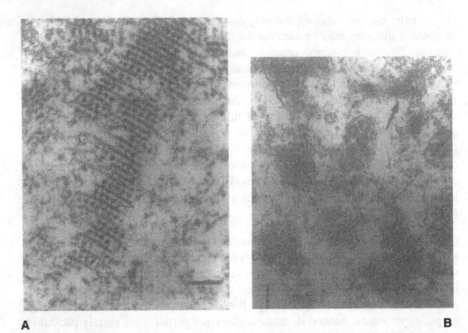

A

B

Figure 3.15 A, Ordered arrays of filaments (type II) in INSV-infected *Nicotiana ben-thamiana* leaf cell; bar represents 100 nm, B, *Nicotiana benthamiana* leaf cells infected by tospovirus with mature virus particles within membranous cisternae; bar represents 200 nm (Blackwell Science Ltd. and British Society for Plant Pathology, UK). (Courtesy of Vaira et al., 1993.)

be divided into three subgroups based on the cytological effects induced by them: In subgroup A, the viruses acquire the membranes from endoplasmic reticulum and accumulate in the cytoplasm as small aggregates within dilated membranous cisternae. Viruses in subgroup B are initiated as buds in the inner membrane of the nuclear envelope and accumulate at the periphery of nuclei as paracrystalline aggregates composed of a high concentration of virus particles. Viruses of subgroup C are associated with the nucleus and are not enveloped. They form characteristic structures known as "spokewheels" at the periphery of nuclei (Martelli and Russo, 1977; Francki et al., 1981).

w. Furovirus. The wheat soil-borne mosaic virus is the type member of this group. Virus aggregates and paracrystals are observed in the cytoplasm of infected cells. Vacuolate inclusions are also produced in the cytoplasm.

x. Tenuivirus. The presence of masses of thread-like inclusions in the cytoplasm is a characteristic feature of infection by tenuiviruses. The structure of the inclusions has a diagnostic value (Christie et al., 1995).

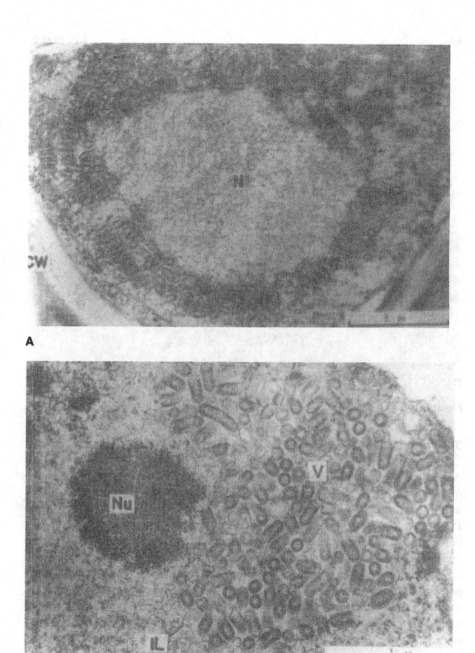

A

B

Figure 3.16 A, Rice transitory yellowing virus particles in large numbers around nuclear membrane of the leaf cell nucleus: N—nucleus, CW—cell wall; B, rice transitory yellowing virus particles in the nucleolus of infected rice leaf cell: Nu—nucleus, IL, inner lamella of nuclear membrane; C, bacilliform particles of rice transitory yellowing virus in a cell of infected rice leaf (Academic Press Inc., USA). (Courtesy of Chen and Shikata, 1971.)

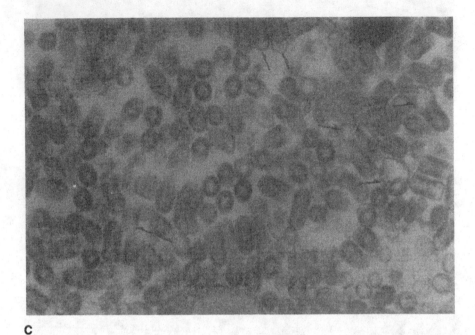

C

Figure 3.16 Continued.

3.3 MOLECULAR GENETICS OF PATHOGENICITY

A great majority of the many thousands of different microbes present in the environment, are strictly saprophytes involved in the degradation of dead organic materials. Less than 10% of them may cause diseases in plants and maybe less than 1% are found to be pathogens of humans and animals (Agrios, 1988). The microbial plant pathogens possess unique genes required for pathogenicity, whereas these are absent in the saprophytes. Host plants do not meekly give way for the establishment of pathogens resulting in different diseases. Even susceptible plants possess various mechanisms to defend themselves, while resistant plants have more efficient surveillance mechanisms for pathogen perception, leading to the rapid activation of cellular defenses that restrict the development of disease. The studies on the molecular genetics of microbial pathogenicity have provided critical information that may be useful for the detection of microbial plant pathogens.

Microbes have to pass through five distinct phases of disease development such as attachment of pathogen to plant surface, germination of spores or pathogenic unit, penetration of host, colonization, and symptom expression for causing diseases. The pathogenicity genes are directly and intrinsically involved in pathogenicity under natural conditions. The pathogenicity gene(s) are required to overcome the structural barriers and the effects of defense-related compounds produced by the plants following infection. The role of pathogenicity genes and their products in pathogenesis is discussed.

3.3.1 Fungal Pathogens

As the spores of fungal pathogens germinate on plant surface, they produce various enzymes that help the adhesion of spores to plant surfaces. The hydrated conidia of *Magnaporthe grisea* produce spore tip mucilage (STM) aiding their adhesion to the hydrophobic rice leaf surface. Among the several fungal genes induced in planta during pathogenesis, *MPG1* gene is highly expressed in infected rice plants and this gene is conserved in many races and host-specific forms of *M. grisea*. In the wild type, the *MPG1* gene is abundantly transcribed during the formation of appressoria, whereas *MPG1⁻* deficient transformants have impaired ability to produce appressoria (Talbot et al., 1993). The action of *MPG1* is required for efficient appressorial formation, as the gene product acts as a component of the surface perception process essential for cellular differentiation. The *MPG1* gene encodes a fungal hydrophobin that is considered to be secreted on the rice leaf surface. The loss of *MPG1* function does not result in loss of pathogenicity entirely, but in dramatic reduction in pathogenicity indicating that expression of *MPG1* is a virulence factor for *M. grisea* (Talbot, 1998). The mutational analysis of pathogenicity in *M. grisea* indicated that seven pathogenicity (*PTH*) genes may have a role in pathogenicity on rice, barley, and weeping love grass (Sweigard et al., 1998).

The formation of appressorium from the germ tube of the conidia is a critical step in fungal pathogenicity. The involvement of a cyclic-AMP (cAMP) signaling mechanism for the differentiation of appressoria was suggested by Kronstad (1997). *CPKA* gene encoding a catalytic subunit of AMP-dependent protein kinase A appears to control appressorium formation by regulating cAMP signaling (Xu et al., 1997). During germination of conidia enhanced levels of a cAMP were observed and appressorial differentiation may be due to such changes (Choi et al., 1998). Another gene, *apf1* has also been demonstrated to be involved in the appressorium differentiation. The *apf1* mutants could not form appressoria even in the presence of cAMP indicating that this gene acts downstream or independently of the cAMP signaling pathway required for differentiation of appressoria in *M. grisea* (Silué et al., 1998).

The requirement of a mitogen-activated protein kinase (MAP kinase)-encoding gene (*PMK1*) for appressorium elaboration was indicated by Xu and Hamer (1996). A null mutation led to loss of appressorium formation and growth of fungus in plant tissue. Liu and Dean (1997) further showed that three G protein α-subunit genes, *magA, magB,* and *magC,* may control growth, development, and pathogenicity of *M. grisea*. Deletion of *magC* resulted in reduced conidiation, but did not affect vegetative growth or appressorium formation, whereas disruption of *magB* significantly reduced vegetative growth, conidiation, and appressorium formation. The results suggest the involvement of G protein α-subunit genes in signal transduction pathways in *M. grisea* that control vegetative growth, conidiation, conidium attachment, appressorium formation, mating, and pathogenicity. The requirement of a G α-protein-encoding gene designated *ctg-1* for the germination of conidia of *Colletotrichum trifolii* was reported by Truesdell et al. (2000). The conidia of the fungal transformants in which *ctg-1* gene was inactivated failed to germinate.

Some fungal pathogens that can penetrate into the host plants are known to produce cutinase, for which a crucial role in pathogenesis has been suggested. Genetic approaches have established the relationship between cutinase activity and pathogenicity in some pathosystems. A mutant of *Nectria haematococca* had reduced (80–90%) cutinase activity and consequently its virulence to pea was also reduced. When the cutinase gene of *N. haematococca* was inserted into *Mycosphaerella* sp., which is a wound pathogen of papaya fruits, the transformants could infect unwounded papaya indicating that cutinase has a role in infection by this pathogen (Dickmen et al., 1989). But later investigation by Stahl and Schäfer (1992) with cutinase-deficient mutants of a highly virulent strain of *N. haematococca* constructed by transformation-mediated gene disruption showed that loss of cutinase activity did not result in loss of pathogenicity, since the pathogenicity of mutants was not altered by the absence of a functional cutinase gene. Hence in this pathosystem cutinase is not essential for pathogenicity. By employing the targeted gene disruption approach, the cutinase gene of *M. grisea* was destroyed. The transformants were found to be pathogenic on three hosts, as the parent type indicating the noninvolvement of the cutinase in pathogenesis. However, a residual cutinase activity observed in the transformants might have assisted the penetration of host plants (Sweigard et al., 1992). The cutinase A (*cutA*) of *Botrytis cinerea* was expressed during early stages of infection of gerbera flowers and tomato fruits. However, the cutinase A–deficient mutants constructed by gene disruption, lacking a functional *cutA* gene, were able to penetrate and cause symptoms on gerbera flowers and tomato fruits like the wild-type strain (van Kan et al., 1997). In the case of *Fusarium solani* f. sp. *curcurbita* race 2, disruption of the *cutA* locus did not result in loss of either virulence or pathogenicity on intact fruits of cucurbits (Crowhurst et al., 1997). It is possible that many enzymes may act synergistically during degradation of cuticle and hence the loss of a single activity may not result in significant change in the infection process.

Fungal pathogens produce cell-wall-degrading enzymes during pathogenesis. It is not clear whether disruption of the activity of one enzyme can cause a change of pathogenicity. Scott-Craig et al. (1990) showed that disruption of a pectin-degrading polygalacturonase (PG) did not affect the pathogenicity of *Cochliobolus carbonum* on maize. On the other hand, *Colletotrichum lindemuthianum* has two genes, *CLPG1* and *CLPG2*, encoding an endopolygalacturonase secreted in the culture filtrate. The antibodies raised against the gene product of *CLPG1* detected the presence of the protein *in planta* and it was associated with extensive degradation of host cell wall. Reverse transcription polymerase chain reaction (RT-PCR) assay showed that *CLPG1*, but not *CLPG2*, was expressed at the beginning of the necrotrophic stage of infection and that *CLPG1* encoded the major secreted endoPG both during saprophytic growth and during pathogenesis (Centis et al., 1997).

The *Bcpg1* gene encodes the endo PG produced by *Botrytis cinerea* and it is expressed during infection on tomato leaves. The *Bcpg1* gene was eliminated by partial gene replacement and the mutants were still pathogenic, causing similar primary infection as the wild type. However, there was a significant decrease in the secondary infection, suggesting that the *Bcpg1* gene is required for full virulence of *B. cinerea* (Have et al., 1998). The pathogenicity in *Botryotinia fuckeliana (B. cinerea)* is controlled by a single gene, *Pat-1*, with a major effect and it is not linked to *Mbc1* (benzimidazole resistance), *Daf1* (dicarboximide resistance), *nit1* (nitrate nonutilising), or *Sel1* (sodium selenate resistance) (Weeds et al., 1999). The *pg1* gene of *Fusarium oxysporum* f. sp. *lycopersici* encoding the major extracellular endo PG in vitro has been cloned and sequenced. The expression of *pg1* in roots and lower stems of infected tomato was detected by RT-PCR. The isolates of *F. oxysporum* f. sp. *melonis* deficient in PG1 were transformed with the cloned gene. The transformants were as virulent as the wild type (Pietro et al., 1998). These studies show that the genes encoding polygalacturonases are required for full virulence of the fungal pathogens studied. The inheritance of virulence of *Pyrenophora teres* f. *teres,* causing barley net blotch disease, was studied by crossing an isolate (0–1) with high virulence with another isolate (15A) with low virulence. The virulence analysis of 82 progenies showed that a single major gene controlled virulence in this pathogen (Weiland et al., 1999).

Fungal pathogens produce different kinds of toxins that may assist in penetration and colonization of host tissues. Some toxins elaborated by the pathogens are host-specific as in the case of *Cochliobolus heterostrophus* race T. The production of T-toxin active against maize with Texas male sterile (tms) cytoplasm is governed by a single locus (Tox1), which is considered as a virulence factor (Yoder and Gracen, 1975). Similar genetic control of toxin production by *C. carbonum* race1 causing leaf spot and ear mold of maize has also been reported by Scheffer et al. (1967). A single locus (Tox2) controls the production of HC-toxin, which is specifically active against maize homozygous for *hm1* locus. The strains that do not produce HC-toxin are not pathogenic and the plants that can inactivate the HC-toxin show resistance (Shäfer, 1994). The gene *TOXC*, which is present only in HC-toxin-producing (TOX 2^+) fungal strain, has been isolated. All three functional copies were on the same chromosome as the gene encoding HC-toxin synthetase. When all copies were mutated by targeted gene disruption, the mutants did not produce HC-toxin and also were not pathogeneic indicating that *TOXC* is required for pathogenicity and virulence (Ahn and Walton, 1997). Another new gene, *TOXE*, in the TOX2 complex of *C. carbonum* was identified by analyzing DNA regions present in the $TOX2^+$ isolated, but not in $TOX\ 2^-$ isolated. Mutation of *TOXE* by targeted gene disruption abolished HC-toxin production and pathogenicity, but there was no effect on growth and sporulation in mutants. *TOXE* was demonstrated to be required for the expression of three genes that have a role in HC-toxin production and this gene was not needed for the expression of *HTS1*, encoding the large multifunctional peptide synthetase, involved as a primary enzyme in HC-toxin biosynthesis (Ahn and Walton, 1998).

Fungal pathogens, after gaining entry into plants, are confronted by preformed plant defense molecules as well as by antimicrobial molecules produced by plants after infection. Cyanogenic plants like sorghum may release cyanide in the tissues damaged by fungal infection. Cyanide hydratase is an inducible fungal enzyme that converts cyanide to nontoxic formamide. The enzyme from *Gloeocercospora sorghi* functions as an aggregated protein consisting of 45-kDa polypeptides (Wang et al., 1992). The mutant constructed by gene disruption caused similar symptoms as the wild type, although it was highly sensitive to cyanide, indicating that cyanide hydratase is not essential for pathogenicity (Wang and Vanettan, 1992).

The ability to detoxify avenacin is a critical factor determining the pathogenicity of *Gaeumannomyces graminis*. Oats plants contain avenacin, which inhibits the growth of fungi in vitro. *G. graminis* var. *tritici* is unable to infect oats, whereas *G. graminis* var. *avenae* can infect oats, because of its ability to enzymatically metabolize avenacin. The gene encoding avenacin was cloned and expressed in *Neurospora crassa,* which became resistant to avenacin. The transformants of *G. graminis* var. *avenae* with disrupted avenacinase gene were more sensitive to avenacin compared with wild type and also lost their pathogenicity to oats drastically. This clearly shows that avencinase gene is essentially required for the pathogenicity of *G. graminis* var *avenae* on oats (Osbourn et al., 1991; Schäfer, 1994).

Tomato and some *Solanum* spp. produce an antifungal, steroidal glycoalkaloid saponin α-tomatine, which is toxic to many fungi, presumably due to its binding to three β-hydroxysterols in fungal membranes. Many tomato pathogens are able to degrade α-tomatine. Sandrock and Van Etten (1998) reported that there was a strong correlation between tolerance to α-tomatine, the ability to degrade this glycoalkaloid, and pathogenicity. However, *Phytophthora infestans* and *Pythium aphanidermatum* capable of infecting tomato were unable to degrade α-tomatine, suggesting the operation of multiple tolerance mechanisms in tomato pathogens to α-tomatine. Another tomato pathogen, *Cladosporium fulvum,* is sensitive to α-tomatine. The effects of heterologous expression of the cDNA encoding tomatinase (capable of detoxifying α-tomatine) from another tomato pathogen, *Septoria lycopersici,* in two races of *C. fulvum* were studied. The transformants producing tomatinase showed enhanced sporulation in susceptible tomato lines and greater severity of symptoms on resistant tomato lines (Melton et al., 1998). The inability of tomato lines to recognize two proteins, ECP1 and ECP2, required for full virulence of *C. fulvum* is considered to be responsible for their susceptibility, whereas the resistant tomato lines recognize the proteins secreted by the pathogen. The capacity to recognize ECP2 leads to induction of hypersensitive response (HR) conferring resistance to *C. fulvum* (Laugé et al., 1998).

Production of phytoalexins following infection is known to occur in many pathosystems and a role for these antimicrobial compounds in the development of resistance has been suggested. The ability to detoxify phytoalexin is reported to contribute to the virulence of *N. haematococca* infecting chickpea. The *MAK1* gene present in *N. haematococca* degrades the phytoalexin maackiain to a less toxic compound. The disruption of *MAK1* gene in a highly virulent MAK$^+$ isolate decreased the virulence to a moderate level, whereas incorporation of multiple copies of *MAK1* to a weakly virulent MAK$^-$ isolate increased virulence to a moderate or high level, demonstrating that detoxification of maackiain is a determinant of virulence of *N. haematococca* (Enkerli et al., 1998).

Fungal pathogens form haustoria in the susceptible host cells. *In planta,* fungal genes are expressed in different structures. A total of 31 different *in planta*–induced genes (PIGs) have been identified. Some of the PIGs are highly expressed in the haustoria of the rust pathogen *Uromyces fabae* infecting broad bean leaves. The PIGs are found to be single- or low-copy-number genes in the rust genome (Hahn and Mendgen, 1997). Screening of a cDNA library constructed from haustoria of *U. fabae* led to the isolation of PIG2, which encodes a protein with high homologies to fungal amino acid exporters. Expression of PIG2 mRNA appeared to be restricted to haustoria. The putative amino acid exporter protein was localized to the plasma membrane of the haustorial bodies as indicated by immunofluorescence microscopy, providing molecular evidence for the role of haustoria in the nutrient uptake from infected cell (Hahn et al., 1997). Molecular biology has provided many powerful tools to determine the role of different molecules of fungi in pathogenesis. Further research on differential cDNA-cloning, promoter-probe libraries and differential display may lead to cloning of fungal genes preferentially expressed during pathogenesis.

3.3.2 Bacterial Pathogens

The bacterial pathogens are known to produce several extracellular compounds, some of which contribute to virulence. The virulence factors include plant cell-wall-degrading enzymes, toxins, hormones, siderophores, and signaling molecules. In addition, the phytopathogenic bacteria have cell-surface-anchored structures such as pili, flagella, lipopolysaccharide, exopolysaccharide slime layers and outer membrane proteins that may assist in bacterial survival within, or ingression of, the host plant. The contribution of molecular genetics in the understanding of pathogenesis has been significant through the studies on *Agrobacterium tumefaciens* and species of *Clavibacter, Erwinia, Pseudomonas,* and *Xanthomonas.* The role of bacterial genes involved in pathogenesis of some of the economically important bacterial pathogens is discussed.

The gram-negative bacterial pathogens secrete macromolecules considered as virulence factors, through four major secretion pathways. Exoenzymes and toxins are secreted through type I pathway which is *sec*-independent. The soft rot pathogen *Erwinia chrysanthemi* produces proteases that are transported from the bacterial cytoplasm directly into the extracellular environment. This one-step secretion is facilitated by Type I secretory apparatus constituted by three accessory proteins (Prt D, E, and F) (Pugsley, 1993). These proteins are membrane-associated and the major signal is located at the C-terminal end of enzymes. The secretion of pectinases and cellulases occurs through a *sec*-dependent, two-step, type II pathway, which is also known as general secretary pathway (GSP). Some of these proteins have been shown to be putative virulence factors. *E. chrysanthemi* and *E. carotovora* subsp. *carotovora* have been reported to secrete virulence factors through GSP (Lindberg and Collmer, 1992; Reeves et al., 1993). Multiple forms of pectate lyase, polygalacturonase, and cellulase are secreted through GSP. When pectinase and cellulase genes of *Erwinia* spp. are expressed in *Escherichia coli,* the enzymes are exported to the periplasm, but they fail to cross the outer membrane. In the second step the movement of the periplasmic form of the enzymes across the outer membrane and into the extracellular environment occurs. The *out* gene cluster is required for the secretion of major pectate lyases, polygalacturanases, and cellulases. Pectinases and cellulases accumulate in the periplasm of *Erwinia out* mutants. The number of *out* genes identified varies with *Erwinia* spp.

Pectate lyases (Pels) can cleave glycosidic bonds (α-1, 4-glycosidic linkages) by β-elimination, producing unsaturated products, whereas polygalacturonase (Peh) cleaves them by hydrolysis, producing saturated products. Pure forms of Pel A, Pel B, Pel C, and Pel E of *E. chrysanthemi* have been used to study modes of depolymerization of polygalacturonate (Preston et al., 1992). *E. chrysanthemi* secretes pectate lyase proteins, which are important virulence factors required for degrading plant cell walls. The bacterial regulator responsible for induction by plant extracts was identified in the promoter region of *Pel E* gene encoding a major pectate lyase. Another bacterial gene, *pir,* has been shown to regulate hyperinduction of this enzyme (Nomura et al., 1998). Degradation of galacturonic acid (Gal UA), a major component of pectin and polygalacturonic acid (PG), is an important early step, since the uptake of molecules derived from them is essential for bacterial development. The GalUA uptake gene (*exuT*) has been cloned and sequenced. Mutants of *exuT* had reduced GalUA utilization and maceration of potato tuber tissue was also delayed (Haseloff et al., 1998). On the other hand, the cell-wall-degrading enzymes from the main virulence determinants of

E. carotovora subsp. *carotovora* and their production are coordinately controlled by a complex regulating network. The ExpS and ExpA proteins encoded by *expS* and *expA* genes are members of two component sensor kinase and response regulator families, respectively, and they might interact in controlling virulence gene expression in this pathogen (Eriksson et al., 1998).

A regulatory locus named *pehSR* controlling polygalacturonase (PG) production and other virulence functions in *Ralstonia solanacearum* was identified by Allen et al. (1997). The *pehSR* mutants could produce only negligible levels of endo-PG activity and exo-PG activity of the mutants was reduced by 50%. Although the mutants grew normally in culture and in planta, their virulence was reduced dramatically. The reporter gene studies indicated that *pehSR* expression increased as the bacteria multiplied in planta and the global virulence gene regulator PhcA was involved in the negative regulation of *pehSR* locus. Tans-Kersten et al. (1998) reported that a pectin methyl esterase (*pme*) gene mutant did not have any detectable Pme activity in vitro and could not grow on 93% methylated pectin as carbon source, but it was as virulent as the wild-type strain on tomato, eggplant, and tobacco, suggesting that Pme is required for the growth of the pathogen, but does not contribute for virulence.

The ability to multiply in the apoplastic fluid of plants, which has low available iron, is a critical condition affecting further steps in pathogenesis. *E. chrysanthemi* is unable to metabolize ferric citrate, the major iron carrier in plant vessels. *E. chrysanthemi* produces a catechol-type siderophore, chrysobactin, which aids in the sequestration of free iron. *E. chrysanthemi* also produces two receptors capable of interacting with bacterial iron-carrier molecules. Pel activity is significantly enhanced under low iron conditions. A regulatory locus *cbrAB* is reported to positively regulate the expression of *pelB*, *pelC*, and *pelE* (Sauvage and Expert, 1994). The role of iron in the pathogenicity of *E. amylovora* was determined by analyzing virulence of mutants obtained by insertional mutagenesis. In this pathogen, the iron transport pathway is mediated by desferrioxamine (DFO). Production of DFO during infection is required for iron utilization by the pathogen and it may also play a role in the oxidative burst elicited by the bacteria (Dellagi et al., 1998). The possible involvement of the ferric uptake regulation (Fur) protein was studied. The signal for coordinated expression of the genes encoding pectate lyases PelB, PelC, PelD, and Pel E and chrysobactin iron transport function is triggered by low iron availability. The expression of *pelA* was not influenced by a *fur* mutation and iron availability, whereas *pelD* and *pelE* transcription levels were higher in the *fur* mutant than in the parental strain in the presence of iron. In *E. chrysanthemi* strain EC 3937, *fur* seems to negatively control iron transport and genes encoding PelD and PelE (Franza et al., 1999).

The type III secretion apparatus is characterized by the functions of *hrp* [for hypersensitive response (HR) and pathogenicity] genes of several gram-negative bacterial pathogens in the genera *Erwinia*, *Xanthomonas*, and *Pseudomonas*. The

molecular mechanism(s) governing bacterial pathogenicity and bacterial elicitation of resistance to plant diseases may involve the same bacterial genes. The *hrp*
genes of *P. syringae* pv. *syringae, P. syringae* pv. *phaseolicola, P. solanacearum,
X. campestris* pv. *vesicatoria,* and *E. amylovora* have been well characterized (van
Gijsegem et al., 1995a). The results of DNA sequence studies indicated that many
hrp genes cloned from different bacterial pathogens causing necrosis are homologous (van Gijsegem et al., 1995b). Many *hrp* genes show striking similarities
with genes of known function. At least 25 *hrp* genes in *P. syringae* pv. *syringae*
have been identified. The *hrp* genes appear to have three biochemical functions:
gene regulation, protein secretion, and production of HR (Gopalan and He, 1996).
Many *hrp* genes exhibit remarkable similarities to those involved in the secretion
of proteinaceous virulence factors in bacteria infecting plants and animals
(*Yersinia* sp., *Shigella* sp., and *Salmononella* sp). Some of the *hrp* gene products
are components of the type III secretion machinery. Mutations affecting type III
protein secretion frequently eliminate bacterial virulence completely and type III
secretion is generally activated in vivo when the bacterial pathogens come in contact with host cells. The type III secretion system is considered to be used by bacterial pathogens to secrete virulence proteins involved in the leakage of plant nutrients in the extracellular space (apoplast) of infected plant tissues (Lindgren,
1997; He, 1998). Appendages known as pili are present on the bacterial cell surface and they are considered to be associated with DNA and/or protein transfer between cells. The Hrp type III protein secretion system of *P. syringae* pv. *tomato*
(strain DC 3000) is responsible for the assembly of Hrp pilus. Wei et al. (2000)
demonstrated that *hrpA* gene encoding the major subunit of the Hrp pilus is required for the secretion of putative virulence proteins such as HrpW and Avr Pto.
Furthermore, the *hrpA* controls the full expression of genes that encode regulatory, secretion, and effector proteins of the type III secretion system. The *hrpA*-
mediated gene regulation appears to be through its effect on the mRNA level of
two regulatory genes, *hrpR* and *hrpS*.

The type IV secretary pathway operates in *Agrobacterium tumefaciens*
causing crown gall diseases in many dicotyledons. The strains of *A. tumefaciens*
are classified based on the kinds of opines such as octopine, nopaline, succi-
namopine, and leucinopine produced in tumors. The crown gall induction is due
to the action of genes present in the Ti plasmid. One half of the Ti plasmid contains the genes for virulence (Vir region) and tumor formation, whereas the
genes for replication, opine catabolism, and conjugation are located in the other
half of Ti plasmid. There are eight operons (*virA* to *virH*) in the Vir region of
the octopine Ti plasmid. Mutations in the *virA, virB, virD,* and *virG* operons result in elimination of tumor formation, whereas restriction of host range occurs
due to mutations in *virC, virE,* and *virH*. The nopaline Ti plasmid has an accessory gene (*tzs*), but not *virF* and *virH* present in octopine Ti plasmid (Hooykaas
and Beijersbergen, 1994). The Vir regions of octopine Ti plasmid (pTi 15955)

and nopaline Ti plasmid (pTiC58) have been fully sequenced (Rogowsky et al., 1990; Beijersbergen, 1993).

The *vir* genes present in the Ti plasmid form a regulon consisting of a set of operons corregulated by the same regulatory proteins (Duban et al., 1993). The VirA and VirG proteins constitute a two-component regulatory system mediating the expression of the *vir* genes. The chemotactic attraction toward wounded plant cells and attachment of the bacteria to the plant are the first steps in pathogenesis. The phenolic compound acetosyringone secreted by wounded tobacco cells can act as an efficient inducer of the *vir* genes (Stachel et al., 1985). The VirA protein acts as a sensor protein acting directly or indirectly as a receptor of acetosyringone (a chemical signal) resulting in the binding of these two compounds. This leads to the phosphorylation of another protein encoded by *virG*, which is inside the bacterium, followed by activation of other genes in the Vir region (Hooykaas and Beijersbergen, 1994; Sheng and Citovsky, 1996).

The T-DNA encodes plant growth regulators that are responsible for tumor production (Zambryski, 1992). The linear fragment of the T-DNA (T-strand) is liberated by the enzyme VirD2 encoded by the *virD2* gene (Stachel et al., 1986). This enzyme is firmly bound to the T-strand, which is guided out of the bacterium and then into the plant cell nucleus. The transfer of the T-complex is dependent on the gene products of the complex *virB* locus of the Ti-plasmid. The *VirB* locus has an operon encoding 11 proteins (Vir B 1–11). The molecular recognition events that lead to T-DNA translocation appear to be exclusively between the *VirB* proteins and the proteins linked to the T-DNA molecules in the secreted complex (Salmond, 1994). *virE* and *virF* genes may play a role in protecting the T-strand, while it is translocated. The functions of Vir E1 and Vir E2 proteins have been elucidated by yeast two-hybrid system. Vir E2 is a single-stranded DNA binding protein and its interaction with Vir E1 was stronger than the Vir E2 self-interaction, which was inhibited by Vir E1. Vir E2 appeared to be preferentially bound to Vir E1. Vir E2 could not be detected in the soluble fraction of *Agrobacterium*, unless it was coexpressed with Vir E1. Analysis of deduced amino acid sequence of Vir E1 protein showed that the Vir E1 protein shared many properties with molecular chaperones that play a role in the transport of specific proteins into animal and plant cells using type IV secretion system. Vir E1 may function as a specific molecular chaperone for Vir E2 protein (Deng et al., 1999). Following the insertion of the T-strand, it is transcribed resulting in the synthesis of two functional groups of proteins involved in the synthesis of opines required for bacterial growth and growth regulators that stimulate cell growth resulting in tumor production (Burr et al., 1998). The structural and organizational differences either in T-DNA or in *vir* loci and differences in chromosomal loci may be reflected in bacterial host range and virulence determinants. The *vir G* region of the Ti plasmid pTi Bo 542 was shown to be responsible for the supervirulence of *A. tumefaciens* strain A281. The importance of using *virG* from pTiBo 542 as an enhancer for plant genetic transformation has been well demonstrated (Cervera et al., 1998).

In addition to *vir* genes present in the Ti plasmid, the involvement of chromosomal genes in the virulence of *A. tumefaciens* has been demonstrated. The attachment of bacteria to the plant cell surface is an early step in the induction of tumor. The genes required for this step, called *att* genes (*att A* to *attH*), are located in the bacterial chromosome. The *att* genes seem to be essential for signaling between the bacteria and host plant (Mathysee, 1994). The attachment of bacteria to plant surface is firmed up by the synthesis of cellulose fibrils, which is regulated by *cel* genes located on the bacterial chromosome near, but not contiguous with, *att* genes. The cellulose-minus mutants exhibit reduced virulence indicating the requirement of *cel* genes for full virulence of this pathogen. The requirement of *cel, attB, attD,* and *attR* genes for the colonization of roots and subsequent infection was confirmed by studies using tomato and *Arabidopsis* (Matthysee and McMahan, 1998). Another new chromosomal virulence gene, *acvB,* was isolated from *A. tumefaciens* strain A208 harboring a nopaline-type Ti plasmid using the transposon (Tn5) tagging (Wirawan et al., 1993). A gene, *virj,* homologous to *acvB* was detected on an octopine-type Ti plasmid (Pan et al., 1995). A Tn5 insertion in the *acvB* gene resulted in an avirulent phenotype, indicating that *acvB* gene is a virulence gene and this gene is involved in the transfer of T-DNA to nuclei of tobacco cells (Fujiwara et al., 1998).

The bacterial pathogens produce diverse symptoms on susceptible plants, as a result of the action of many, sometimes unique, virulence factors elaborated by a bacterial pathogen in addition to *hrp*-controlled pathogenicity factors. Extracellular polysaccharides (EPSs) are involved in the formation of the water-soaking symptom, while tissue disintegration and soft rot symptoms are due to the action of cell-wall-degrading enzymes. Many bacterial pathogens produce large amounts of EPS both in culture and in planta and the ability to produce EPS has been correlated with the virulence of the bacteria concerned. The genes required for EPS synthesis are designated as *eps* genes.

Burkholderia (Pseudomonas) solanacearum, causing wilt diseases in a wide range of crop plants, produce abundant slime containing a high-molecular-weight acidic EPS$^-$. The biosynthesis and transport of this acidic EPS$^-$ are encoded by at least nine structural genes present in the 18-kb *eps* operon (Huang and Schell, 1995). The EPS$^-$ mutants obtained by transposon inactivation of the *eps* operon were able to produce little acidic EPS in planta and had greatly reduced virulence, indicating that the acidic EPS is a necessary wilt-inducing factor of *P. solanacearum* (Denny and Schell, 1994). Two distinct types of EPS are produced by *Clavibacter michiganensis:* a high-molecular-weight acid polymer (type A) and low-molecular-weight neutral EPS (type B). The evidence available to suggest that EPS is responsible for inducing wilt is only circumstantial. Both type A and B EPSs produced in culture by *C. michiganensis* subsp. *michi-*

ganensis could cause wilting in tomato cuttings (Denny, 1995) However, Bermpohl et al. (1996) provided evidence that wilting symptom was not due to the EPS produced by this pathogen, since the EPS⁻ mutants were not altered in virulence and another isolate capable of producing EPS in identitical quantities as the wild type was not able to cause the disease. Additionally, the region of plasmid pCM2 encoding the pathogenicity locus *pat-1* has been demonstrated to be essential for virulence. Endophytic plasmid-free isolates of *C. michiganensis* subsp. *michiganensis* could be converted into virulent forms by introducing the *pat-1* region (Dreier et al., 1997).

    Erwinia amylovora produces both levan and amylovoran, which may be found as either capsule or slime. The gene clusters (*cms* and *cps*) required for their biosynthesis have been sequenced. Amylovoran-negative mutants have been reported to be avirulent commonly, as they did not cause any disease symptoms on inoculated pear seedlings and neither bacterial ooze nor necrosis on immature fruits could be observed (Bernhard et al., 1993). Amylovoran may have an important role in the survival of *E. amylovora* during the early stages of pathogenesis (Tharaud et al., 1994). In *E. chrysanthemi* E.C. 3937 the *eps* genes are clustered on the chromosome and repressed by PecT, a regulator of pectate lyase synthesis. An *eps* mutant was found to be less efficient than wild-type strain in causing maceration, suggesting that full expression of virulence in *E. chrysanthemi* requires the production of EPS (Condemine et al., 1999).

    Xanthomonas campestris pv. *campestris* produces xanthan in culture. Spontaneous and chemically induced mutants produced less or altered xanthan and were found to be less virulent compared with wild type. Likewise, a transposon mutant (Xanthan-minus) was much less virulent and grew very little in mesophyll tissue (Newman et al., 1994). The *pigB* of *X. campestris* pv. *campestris* is required for the production of EPS, xanthomonadin pigments, and the diffusible signal molecule DF (diffusible factor). The production of EPS and xanthomonadin by *pigB* mutant strains can be restored by the extracellular application of DF. The *pigB* mutants could infect cauliflower via wounds but not via hydathodes. A functional *pigB* may be required for epiphytic survival and natural host infection as indicated by Poplawsky et al. (1998) and Poplawsky and Chun (1998).

    A cluster of genes known as *rpf* (regulation of pathogenicity factors) coordinately regulates the synthesis of extracellular enzymes and EPS in *X. campestris* pv. *campestris*. These genes are located within a 21.9-kb region of the chromosome isolated as the cosmid clone pIJ3020. Eight genes or open reading frames (ORFs), based on sequence analysis, were identified: *rpf-D, Orf-1, Orf-2, Orf3, Orf4, recJ, rpfE, greA*. Although the transposon insertions in these genes resulted in changes in the levels of extracellular enzymes and EPS to a relatively moderate extent (two- and threefold), the pathogenicity of *X. campestris* pv. *campestris* on turnip was not affected (Dow et al., 2000). The genes controlling the ferric iron uptake by *X. campestris* pv. *campestris* have been designated *tonB*, *exbB*, and *exb D1*. However, the *exb D2* gene located in the same gene cluster is not essential for ferric iron uptake. Mutational analysis has further shown that *tonB*, *exbB*, and *exb D1* genes are required for the production of black rot symptom on cauliflower and induction of hypersensitive response (HR) on the nonhost pepper (*Capsicum annuum*). In contrast, *exbD2* gene had a role in induction of HR on pepper, but not in symptom production in cauliflower (Wiggerich and Pühler, 2000).

The *lemA* gene has been shown to be conserved among strains and pathovars of *P. syringae* (Barta et al., 1992). Rich et al. (1992) reported that in *P. syringae* pv. *syringae* B728a, the *lemA* gene is required for lesion formation on leaves and pods of beans, but this gene is not required for the pathogenicity of *P. syringae* pv. *phaseolicola*. A *lemA* mutant of *P. syringae* pv. *syringae* lost the ability to induce lesions on bean. The *lemA* gene is also involved in the production of extracellular protease and syringomycin, but is not required for the production of pyoverdin or bacteriocin or for motility of the bacteria. Protease and syringomycin do not seem to have a major role in the process of lesion formation by this pathogen (Hrabak and Willis, 1993). Recently *Pseudomonas glumae* causing rice seedling and grain rot disease has been reported to produce the toxin, toxoflavin containing two acid proteins (TRP-1 and TRP-2). The transposon mutants (TOX) are nontoxigenic, avirulent, and do not produce yellow pigment as the wild-type strain. This clearly indicates that production of toxoflavin is essential for the pathogenicity of *P. glumae* (Suzuki et al., 1998; Yoneyama et al., 1998). It will be useful to determine the gene sequences controlling toxin production and resistance to toxin, since such sequences can be used as probes for disease diagnosis and differentiating the pathovars.

3.3.3 Viruses

The term "virulence," in the case of plant viruses, has often been used to indicate the degree of symptom severity. For example virulent strains of TMV induce severe symptoms, whereas mild symptoms are caused by mild (less virulent) strains. However, the ability to overcome the resistance of host plant is considered as a more appropriate parameter for defining virulence (Fraser, 1990). The studies on tobacco mosaic virus (TMV), potato virus X (PVX), and tomato mosaic virus (ToMV), which exist in the form of many strains with varying virulence, have provided a basic knowledge of the molecular interactions between the virus and host plant.

Among the plant viruses, the molecular genetics of TMV and viral gene functions have been more intensively studied. TMV RNA encodes four proteins, of which two (126 kDa and 183 kDa) are required for viral replication, one (30 kDa) for cell-to-cell movement, and the fourth (17.5 kDa) for the synthesis of viral coat protein. The hypersensitive response of *Nicotiana sylvestris,* when inoculated with many strains of TMV, is due to a single dominant N' gene. The hybrid virus constructed by substituting the coat protein gene of the strain that induced N' gene HR with the coat protein gene of strain that did not induce N' gene HR was unable to induce N' gene HR indicating the involvement of coat protein sequence in the induction of this type of HR (Saito et al., 1989). Later, Culver et al. (1991) demonstrated that induction of N' gene HR is due to the coat protein, but not due to the RNA. On the other hand, the coat protein does not seem to be responsible for induction of HR in plants with the N gene (Saito et al., 1989). A role for the 126-kDa replicase protein of TMV in the induction of HR in NN tobacco has been suggested (Baker et al., 1994). These studies indicate that different viral genes may be involved in inducing HR in *Nicotiana* spp. with different HR genes resulting in necrotic local lesions (incompatible) or systemic infection (compatible).

ToMV is a mechanically transmitted positive-sense RNA virus and it is closely related to TMV. Five virus pathotypes (strains), 0, 1, 2, 1.2, 2^2, have been differentiated by their reactions on four tomato genotypes [(+/+), Tm1/Tm1, Tm2/Tm2, and $Tm2^2/Tm2^2$)]. The study to determine the nucleotide sequences of two ToMV mutants that could overcome Tm2 resistance in tomato and sequence analysis showed that the 30-kDa movement proteins of the mutants had two different substitution of two amino acids when compared with wild-type strain (Meshi et al., 1989). This indicated that ToMV strains with both sets of substitution may overcome Tm2 resistance, which is being widely used. Further studies by Weber and Pfitzner (1998) demonstrated that ToMV-2^2 strain required two amino acids exchanges in the carboxy terminal region of viral 30-kDa movement protein at positions 238 and 244 to overcome Tm-2^2 resistance. They suggested that the carboxy domain of the movement protein of ToMV could serve as a recognition target in respect of the Tm-2^2 resistance gene.

In the case of PVX, all isolates of PVX, except a South American isolate (PVX$_{HB}$), can induce HR in potato cultivars carrying Rx gene. In this interaction, the coat protein has been shown to be the determinant of virulence of the virus isolates. The virulence of the PVX isolates is determined by the amino acids at the 121 and 127 position of the viral coat protein. The natural or mutant isolates with lysine and arginine at positions 121 and 127 could overcome the resistance of the Rx gene, whereas the isolates with threonine and arginine were found to be avirulent (Goulden et al., 1993). The role of viral coat protein as determinant of virulence of PVX was further established by Santa Cruz and Baulcombe (1993). The strain PVX$_{DX}$ induces an HR on potato cultivars carrying Nx resistance gene, whereas the strain PVX$_{DX4}$ is able to overcome Nx-mediated resistance. A single nucleotide difference in the coat protein gene determines the nature of interaction between the PVX strains and potato cultivars with Nx resistance gene. The coat protein (CP) gene of PVX appears to have a direct or indirect effect on virus particle morphology, viral pathogenicity, and type of symptoms induced. Accumulation and rapid spread of PVX even in inoculated leaf depend on production of coat protein and encapsidation of the viral RNA (Spence, 1997). The virulence of cucumber mosaic virus (CMV) is determined by the CP gene on the RNA3 of the virus. Changes in the CP gene of M strain of CMV (M-CMV) at both positions 129 (leucine to proline) and 162 (threonine to alanine) are required to overcome the resistance of maize to this strain. The results of Ryu et al. (1998) suggest that resistance to M-CMV in maize is associated with the inability of the coat protein to accelerate cell-to-cell movement, rather than to adverse effects on virus replication. Evidences indicate that the coat proteins have a major role as a determinant of virulence in the case of TMV, ToMV, PVX, and CMV.

The involvement of viral proteins in the biological activities such as long distance movement, hypersensitive response–like reaction in resistant plant species, and symptom expression has been indicated by recent studies. A role in the long distance movement of CMV has been suggested for the CMV-encoded 2b protein. This protein has been shown to function as a transcriptional activator in yeast. A clone called 2 bip (2b-interacting protein), isolated from tobacco, forms a translation product that may interact with 2b protein (Ham et al., 1999). Evidence indicating that the CMV 2b protein may function as a viral counterdefense factor that interferes with the establishment of virus-induced gene silencing in plants was obtained by Mayers et al. (2000). An antibody to the 2b protein encoded by the Fny strain of CMV (Fny-CMV) was able to recognize the Fny-CMV 2b protein in a 10,000-g pellet fraction of infected tobacco leaf extract. This report appears to be the first demonstration of 2b protein expression by a subgroup I strain of CMV. The 2b protein may be associated with either the nucleus or cytoskeleton of the host cell. Chu et al. (2000) reported that the p19 protein of tomato bushy stunt virus (TBSV) may have a role in symptom expression and virus spread. Substitution of amino acids in the N-terminal portion of the p19 protein affected the production of systemic and local symptoms in *Nicotiana* spp., but did not affect virus spread in spinach. A central region of the p19 protein was required for symptom expression and virus spread.

The virulence of pea seed–borne mosaic potyvirus (PSbMV) seems to be determined by a viral genome-linked protein (VPg). The homozygous recessive *sbm-1* peas are resistant to the pathotype P-1 of PSbMV, but susceptible to the pathotype P-4 of the same virus. The resistance of *sbm-1* genotype to P-1 appears to occur at the cellular level and the inhibition of cell-to-cell movement of virus is not the operating form of resistance, since the presence of viral coat protein or RNA could not be detected by enzyme-linked immunosorbent assay (ELISA) or RT-PCR assay. The specific coding region the 21-kDa, genome-linked protein (VPg) present in the P-4 pathotype is responsible for its virulence to peas with *sbm-1* resistance. The VPg is considered to be involved in the replication of this potyvirus and determinant of infectivity in the pea genotype tested (Keller et al., 1998). This view is supported by the findings of Skaf et al. (2000) on pea enation mosaic disease caused by an obligatory association between pea enation mosaic enamovirus (PEMV-1) and pea enation mosaic umbravirus (PEMV-2). Encapsidated RNA1 and RNA2 are covalently linked to a 3139-Da VPg encoded by the RNA of PEMV-1. The effect of mutations in key amino acids in the VPg was evaluated in pea protoplasts and plants. By using quantitative real-time reverse transcription–polymerase chain reaction (RT-PCR) assay it was shown that the inability of some mutants to infect plants was due to their replicative incompetence rather than their inability to spread. Accumulation of RNA1 of the mutants that produced delayed or less severe symptoms was much less (10–100-fold) compared with RNA1 of wild type (WT-RNA1). The severity of symptoms produced by WT-PEMV was proportional to the amount of RNA1 accumulating in pea plants and appeared to be independent of the amount of RNA2.

    The resistance of the tobacco cv. Virgin A Mutant (VAM) to most potato Y potyvirus (PVY) isolates is considered to be due to a recessive resistance gene, *va*. The resistance of the cultivar seems primarily to be due to reduction in the level of cell-to-cell movement and impairment of replication of PVY. The comparison of amino acid sequences of the mutants and their original isolates indicated that a single amino acid substitution in the VPg domain may result in the formation of VAM-resistance-breaking strains (Masuta et al., 1999). Two mutant strains of beet necrotic yellow vein virus (BNYVV) with deleted regions in RNA3, encoding either 94 or 121 amino acids toward the C-terminal part of the 25-kDa protein (p25), were unable to induce the symptoms of rhizomania disease in beet cultivars. The results show that P25 of RNA3 is required for induction of rhizomania symptoms and it may also facilitate the virus movement from rootlets to taproots in the partially resistant cultivar (Tamada et al., 1999).

    The tomato spotted wilt tospovirus (TSWV) N gene-derived resistance (TNDR) has been used as a model to study how plant viruses defeat resistance genes. The isolate TSWV-D from dahlia and TSWV-10 from peanut were suppressed by TNDR in tobacco cv. Burley 21. When these isolates were subjected to TNDR selection by serial passage in an N-gene transgenic plant, a specific reassortment $L_{10}/M_{10}/S_D$ was formed. The genotype analysis showed that the individual L_{10}/M_{10}, and S_D RNA segments were each selected independently in response to TNDR selection rather than to a mutation of recombination event. It is hypothesized that the genome reassortment is the mode utilized by TSWV to adopt to new host genotypes rapidly and the host resistance is defeated by combination of elements from two or more segments of viral genome (Qiu and Moyer, 1999). The resistance to potato virus Y (PVY) in potato cv. Pito was shown to be due to the expression of the PVY-P1 gene in the sense or antisense orientation and it was found to be specific against PVY° strain. The mechanism of this strain-specific resistance was based on posttranscriptional gene silencing (PTGS) operating in the transgenic potato plants. These transgenic plants were, however, susceptible to PVY^N strain isolates. No distinguishing differences in the P1 gene sequences of PVY° and PVY^N strains were detected. Hence the ability of the PVY^N

strain to overcome the resistance could not be explained based on the P1 gene sequences, suggesting the involvement of factors other than sequence homology (Mäki-Valkama et al., 2000a,b).

SUMMARY

The external symptoms of diagnostic value induced by fungal, bacterial, viral, and mycoplasmal pathogens are described. Histological changes caused by certain viruses have diagnostic value. Cytopathological changes, including formation of intracellular and intranuclear inclusions, are characteristic of virus infections. Some of the virus groups and individual viruses can be reliably diagnosed by using differential stains or by observing ultrathin sections under an electron microscope. The viruses may be readily isolated from the inclusion bodies by new methods that have been developed. The study of molecular genetics of pathogenicity has elucidated some mechanisms for the intrinsic ability of microbes to infect plants and cause diseases in plants. As pathogenicity differentiates the pathogens from saprophytes, the genes determining pathogenicity have great importance for the detection and differentiation of microbial pathogens.

APPENDIX: DETECTION OF INCLUSION BODIES INDUCED BY VIRUSES (CHRISTIE ET AL., 1995).

A. **Preparation of Orange-Green (O-G) protein stain**
    i.   Dissolve Calcomine Orange (1g) and Luxol Brilliant Green BL separately in 2-methoxy-methanol (100 ml); stir well and filter; store in tightly capped brown bottles separately.
    ii. Mix one part of Calcomine Orange, eight parts of Brilliant Green BL, and one part of distilled water (1:8:1, v/v/v). This mixture, designated Orange-Green (O-G), stain is used for staining inclusion bodies induced by plant viruses.

B. **Preparation of Azure A nucleic acid and protein stain**
    i.   Dissolve Azure A (100 mg) in 2-methoxyethanol (100 ml) to have a 0.1% solution.
    ii. Add 0.2 M dibasic sodium phosphate ($Na_2HPO4. 7 H_2O$) at the rate of one part to nine parts of Azure A solution. Always use freshly prepared solution.

C. **Preparation of rinsing solution**
    i.   Add 2-methoxyethylacetate (MeA) (70 ml) to 95% EtOH (30 ml); ethanol (95%) can also be used in place of MeA.

D. **Staining methods**
  a. **Orange-Green or Azure A stain**
    i. Place the epidermal strips in the stain; microwave for 10–15 s; alternatively for 10 min at room temperature.
    ii. Rinse the stained tissues 2 or 3 times with 2 MeA-Et OH solution for 1–2 min at room temperature.
    iii. Place the tissues in MeA (100%) for 1–2 min at room temperature, if necessary.
    iv. Mount the tissues stained with O-G in Euparal (green) and tissues stained with Azure A in Euparal (straw color) and examine under the microscope for inclusion bodies.
  b. **Triton X-100 (Plastid solubilizing agent)**
    i. Place the tissues in Triton X-100 (2%) solution; microwave for 10–15 s or treat for 10 min at room temperature.
    ii. Replace Triton X-100 with O-G stain; follow steps (i) to (iii) as in "a" above. Mount the tissues in Euparal (green) and examine under the microscope for inclusion bodies.

4
Dissemination of Plant Pathogens

Plant pathogens may be disseminated/transmitted in different ways. Fungal and bacterial pathogens may be disseminated largely by wind or water and in some cases with the help of insects. Viruses and phytoplasmas on the other hand, predominantly depend on vectors such as insects, mites, nematodes, and fungi for their natural spread under in vivo conditions. The infected seeds or vegetatively propagated seed materials form the most important primary sources of infection, irrespective of the nature of the pathogen. The infected seeds/seed materials may exhibit some symptoms or show the presence of the fungal or bacterial pathogens when incubated. But infection of seeds by viruses or viroids or phytoplasmas may not be discernible, unless special methods of detection are employed. Transmission of some of the viruses and viroids through pollen, leading to seed infection, is also recognized.

4.1 TRANSMISSION OF PLANT VIRUSES

The plant viruses can be transmitted in several ways under natural and experimental conditions (Matthews, 1991; Narayanasamy and Doraiswamy, 1996). All viruses, viroids, and phytoplasmas may be transmitted to appropriate host plant species by different grafting/budding methods. A leaf-grafting method for the transmission of grapevine viruses has been developed. Pieces of lamina (4–5 cm^2) from mature leaves attached to leaf petioles (4–6 cm) were used for graft inoculation into petioles without lamina of indicator plants. Success rates of 73.3% and 93.3% were obtained with one and two grafts per plant during early spring season. This method is a simple and efficient method of detecting grapevine viruses inducing symptoms in leaves and canes (Kuniyuki et al., 1998). Some of the viruses and phytoplasmas have been experimentally transmitted by using *Cuscuta* spp. This method may be useful when grafting is difficult or not possible, as in monocots. Indexing mother plants and propagative plant materials for the presence of viruses is a general practice adopted for certification and quarantine purposes especially in the case of horticultural crops. Indexing by graft inoculation of indicator plants had been an effective procedure for detection and identification of viruses and their strains as in the case of plum pox viruses (Fuchs et al., 1995), and apple stem pitting, stem grooving, and vein yellows viruses (Ramel et al., 1998). A clone of *Malus micromalus* (GMAL 273a), when graft-inoculated with apple viruses, produced diagnostic foliage symptoms within 2–4 weeks, whereas the recommended cultivar Virginia Crab *(M. domestica)* required 6–8 months to display reliable symptoms (Howell et al., 1996). *Prunus tomentosa* (hybrid selection IR 473 × IR 474) develops diagnostic symptoms when graft-inoculated with plum pox virus (PPV) and several *Prunus* viruses. *P. tomentosa* can also be used as a stock culture plant for maintaining virus cultures in the glasshouses. The strains M and D of PPV could be differentiated based on the symptoms produced on *P. tomentosa* (Damsteegt et al., 1997, 1998).

Many viruses and all viroids may be transmitted by mechanical transmission, whereas attempts to transmit the phytoplasmas by mechanical inoculation from plant to plant have been unsuccessful. However, the phytoplasmas may be transmitted by mechanical inoculation from leafhopper to leafhopper (Maramorosch, 1952). All vectors of phytoplasmas belong to the suborder Auchenorrhyncha and no other types of vector are known. Some plant viruses, such as wound tumor virus (Black and Brakke, 1952), rice dwarf virus (Fukushi and Kimura, 1959), and northern cereal mosaic virus (Yamada and Shikata, 1969), have also been serially transmitted from insect to insect by injecting the juice of viruliferous insects.

The methods of transmission or spread of pathogens per se many not form a sound basis for distinguishing the pathogens. However, the specificity of relationship of the pathogens with their vectors, largely in the case of viruses and phytoplasmas and of some fungi and bacteria, may indicate the identity of the pathogen. The transmission characteristics of the virus may be helpful in the identification of the virus, when considered along with other properties.

4.1.1 Mechanical Transmission

The ability of a virus to be transmitted by mechanical inoculation is considered one of the intrinsic properties of the virus concerned. The properties of the viruses in expressed sap, such as stability in vitro, thermal inactivation point (TIP), and dilution end point (DEP), have been determined for the sap-transmissible viruses. Though the usefulness of these properties for the identification of viruses is somewhat limited, they may be useful in differentiating viruses with widely different physical properties (Francki, 1980) (Table 4.1).

Table 4.1 Physical Properties of Plant Viruses

Name of Virus	Dilution end point (DEP)	Thermal inactivation point (TIP) °C	Longevity in vitro (LIV)	Reference
Agropyron mosaic virus	1:10,000–1:20,000	50	12 days	Bremer, 1964
Alfalfa mosaic virus	1:1,000–1:5,000	50–60	4 hr to 4 days	Smith, 1972
Apple chlorotic leafspot virus	1:1,000–1:5,000	55	7 hr at 24°C	Saksena and Mink, 1969
Apple stem grooving virus	1:1,000	67	2 days at 20°C	Smith, 1972
Arabis mosaic virus	1:1,000–1:100,000	55–60	15–21 days at 18–21°C	Smith, 1972
Barley mosaic virus	1:100–1:500	53–55	6 hr at 19°–20°C	Dhanraj and Raychaudhuri 1969
Barley stripe mosaic virus	1:10,000	63	18 days	Ohmann-Kreutzberg, 1962
Bean (common) mosaic virus	1:1,000–1:10,000	56–58	24–32 hr	Smith, 1972
Bean pod mottle virus	1:10,000	70–75	62 days	Bancroft, 1962
Bean southern mosaic virus	1:500,000	90–95	32 weeks at 18°C	Smith, 1972
Bean yellow mosaic virus	1:800–1:1,000	56–60	24–32 hr	Weintraub and Ragetli, 1966
Beet mosaic virus	1:4,000	55–60	24–48 hr at 21°C	Smith, 1972
Broad bean mottle virus	1:1,000	95	3 weeks	Smith, 1972
Broadbean stain virus	1:1,000–1:10,000	60–65	10–17 days at 18°C	Gibbs et al., 1968
Brome mosaic virus	1:100,000–1:300,000	78–79	14 months at 17°C	Smith, 1972
Cacao mottle leaf virus	1:100	55–60	96 hr	Kenten and Legg, 1967
Cacao yellow mosaic virus	1:10,000	60–65	16–32 days	Brunt et al., 1965
Carnation latent virus	1:1,000	60	2 days	Hollings and Stone, 1965b

Table 4.1 Continued

Name of Virus	Dilution end point (DEP)	Thermal inactivation point (TIP) °C	Longevity in vitro (LIV)	Reference
Carnation mottle virus	1:200,000	85–90	81 days at 18°C	Hollings and Stone, 1964
Carnation ringspot virus	1:100,000	85–90	16 days at 20°C	Kassanis, 1955
Carnation vein mottle virus	1:1,000	50	10 days	Hollings and Stone, 1967
Carrot mottle virus	1:1,000	70	9–24 hr	Murrant et al., 1969
Cauliflower mosaic virus	1:2,000	70–75	14 days	Smith, 1972
Celery mosaic virus	1:100–1:1,000	55–60	6 days	Smith, 1972
Celery (western) mosaic virus	1:4,000	55–60	4–6 days	Purcifull and Shepard, 1967
Cherry chlorotic-necrotic ringspot virus	1:40–1:100	48–50	1–2 days	Kralikova and Kegler, 1967
Cherry leaf roll virus	1:100	52–55	5–10 days at 20°C	Cropley, 1961
Clover (red) mottle virus	1:1,000,000	70–75	2 weeks at 20°C	Valenta and Marcinka, 1971
Clover (red) vein mosaic virus	—	58–60	2–3 days	Osborn, 1937
Clover (white) mosaic virus	1:100,000–1:1,000,000	60	10–99 days	Bercks, 1971
Clover yellow mosaic virus	1:10,000	58–60	6–12 months	Pratt, 1961
Clover yellow vein virus	1:10,000	55	8 days at 18°C	Hollings and Nariani, 1965
Cocksfoot mottle virus	—	65	2 weeks at 20°C	Serjeant, 1967
Cowpea aphid-borne virus	1:4,000	60–62	5 days at 21°C	Lovisolo and Conti, 1966
Cowpea chlorotic mottle virus	1:10,000 1:100,000	65–70	1–2 days	Kuhn, 1964 Bancroft, 1971

Table 4.1 Continued

Name of Virus	Dilution end point (DEP)	Thermal inactivation point (TIP) °C	Longevity in vitro (LIV)	Reference
Cowpea mosaic virus	1:10,000 1:100,000	65–70	1–20 days	Dale, 1949 Vankammen, 1971
Cucumber green mottle mosaic virus	—	80–90	1 year	Bawden and Pirie, 1937
Cucumber mosaic virus	1:10,000	60–70	3–4 days	Smith, 1972
Cymbidium mosaic virus		65–70	7 days	Smith, 1972
Dahlia mosaic virus	1:3,000	85–90	10–14 days	Brierly and Smith, 1950
Eggplant (brinjal) mosaic virus	—	78	Few weeks	Smith, 1972
Elm mottle virus	1:1,000	58–62	7 days	Schmelzer, 1969
Grapevine fan leaf virus	1:5,000–1:10,000	60–62	14–21 days	Cadman et al., 1960
Groundnut (peanut) mottle virus	1:1,000	60	24 hr at 25°C	Kuhn, 1965
Henbane mosaic virus	1:10,000 1:100,000	50–60	About 4 days	Bawden, 1951
Iris mosaic virus	1:100 1:1,000	65–70	3–4 days at 20°C	Brunt, 1968
Lettuce mosaic virus	1:100	55–60	48 hr	Couch and Gold, 1954
Lettuce necrotic yellows virus	1:100	52–54	1–8 hr	Stubbs and Grogan, 1963
Lynchnis ringspot virus	1:2,000	64–68	2–7 days	Bennet, 1959
Muskmelon vein necrosis virus	1:1,000–1:10,000	50–55	2–7 days	Freitag and Milne, 1970
Narcissus mosaic virus	1:100,000	70	12 weeks at 18°C	Brunt, 1966a
Narcissus yellow stripe virus	1:100–1:1,000	70–75	72 hr	Brunt, 1971
Onion yellow dwarf virus	1:10,000	75–80	100 hr at 29°C	Dhingra and Nariani, 1963

Table 4.1 Continued

Name of Virus	Dilution end point (DEP)	Thermal inactivation point (TIP) °C	Longevity in vitro (LIV)	Reference
Papaya mosaic virus	1:10,000	73–76	6 months	Purcifull and Hiebert, 1971
Parsnip yellow fleck virus	1:1,000–1:10,000	57–60	4–7 days	Murant and Goold, 1967
Pea early browning virus	1:10,000–1:100,000	65–70	147 days	Bos and Vander Want, 1962
Pea enation mosaic virus	1:3,000	56–58	3 days	Pierce, 1935
Pea mosaic virus	1:5,000	60–64	3–4 days	Murphy and Pierce, 1937
Pea streak virus	1:1,000,000	58–60	16–32 days	Hagedorn and Walker, 1949
Pelargonium leaf curl virus	1:100	85–90	21 days at 18°C	Hollings, 1962
Plum line pattern virus	1:6,400	65	—	Paulsen and Fulton, 1968
Plum pox virus	1:10–10,000	51–54	1–2 days at 20°C	Smith, 1972
Potato aucuba mosaic virus	1:200–1:500	65	3–4 days at 15°C	Kasssanis, 1961
Potato moptop virus	1:10,000	80	14 weeks at 20°C	Jones and Harrison, 1969
Potato virus A	1:50–1:100	44–52	12–24 hr	MacLachlan et al., 1953
Potato virus S	1:100–1:1,000	50–60	3–4 days	Wetter, 1971
Potato virus X	1:100,000–1:1,000,000	70	Up to a year	Smith, 1933
Potato virus Y	1:100–1:1,000	52–55	24–48 hr	Ross, 1948
Potato yellow dwarf virus	1:1,000–1:10,000	50	2–12 hr	Smith, 1972
Prunus necrotic ringspot virus	1:50–1:100	50–55	—	Smith, 1972
Radish mosaic virus	1:14,000	65–68	14 days at 22°C	Tompkins, 1939
Raspberry bushy dwarf virus	1:10,000	65	4 days at 22°C	Barnett and Murrant, 1970
Raspberry ringspot virus	1:100	60–65	Rapidly inactivated	Roland, 1962
Sowbane mosaic virus	11:100,000– 1:100,000,000	84–86	2 months	Bennett and Costa, 1961

Table 4.1 Continued

Name of Virus	Dilution end point (DEP)	Thermal inactivation point (TIP) °C	Longevity in vitro (LIV)	Reference
Soybean mosaic virus	—	64–66	4–5 days	Gardner and Kendrick, 1921
Squash mosaic virus	1:1,000,000	75	6 weeks	Freitag, 1956
Sugarcane mosaic virus	1:1,000	53–55	2–24 hr	Rafay, 1935 Adsuar, 1950
Tobacco etch virus	1:1,000–1:5,000	54–58	5–8 days	Smith, 1972
Tobacco mosaic virus	1:1,000,000	93	1 year	Smith, 1972
Tobacco necrosis virus	1:10,000–1:1,000,000	70–90	2–3 months	Kassanis, 1970
Tobacco rattle virus	1:100,000	80–85	6 weeks	Cadman and Harrison, 1959
Tobacco ringspot virus	1:1,000–1:10,000	60–65	6–10 days	Henderson, 1931
Tobacco streak virus	1:20–1:100	53	2–3 days at 22°C	Diachun an Valleau, 1950
Tobacco wilt virus	1:1,000	60–65	4 days	Badami and Kassanis, 1959
Tomato aspermy virus	1:100–1:1,000	50–55	24–28 hr	Holdings, 1955
Tomato black ring virus	1:100 1:1,000	58–62	7 days	Gibbs and Harrison, 1964
Tomato bushy stunt virus	—	80	25 days	Smith, 1935
Tomato ringspot virus	1:10–1:1,000	56–58	21–27 hr	Price, 1936
Tomato spotted wilt virus	1:10,000–1:100,000	42	About 5 hours	Best, 1968
Tulip breaking virus	1:100,000	65–70	—	Cayley, 1932
Turnip crinkle virus	1:100,000	80–85	6–7 weeks	Smith, 1972
Turnip mosaic virus	1:1,000	55–60	48–72 hr	Tompkins, 1938
Turnip yellow mosaic virus	1:100,000–1:1,000,000	70–75		Markham and Smith, 1949
Watermelon mosaic virus	1:10,000–1:30,000	55–60	9-10 days	Anderson, 1954
Wheat (soil-borne) mosaic virus	1:100–1:1,000	60–65		Smith, 1972

By selecting an appropriate method of transmission, the viruses may be transmitted to one or a range of plant species which exhibit characteristic symptoms due to local or systemic infection. The symptoms produced on certain host plant species are quite characteristic of the causal virus and are useful in differentiating such viruses. Vaira et al. (1993) reported that the two tospovirus species, tomato spotted wilt virus (TSWV) and impatiens necrotic spot virus (INSV), can be differentiated by the type of symptoms produced on diagnostic plants (Table 4.2). The presence of virus in habitats, other than their plant hosts may be detected by inoculating diagnostic hosts. Petunia asteroid mosaic tombusvirus was detected in water samples taken from the north eastern beach of the Isle of Helgoland, Germany, by inoculating *Chenopodium quinoa,* which produces necrotic local lesions (Fuchs et al., 1996). A new nepovirus causing black currant reversion was identified by slash inoculation of in vitro–propagated young plants (Lemmetty and Lehto, 1999). The local lesion hosts are used for both detection and assay of plant viruses. The plant species which can be used as assay or diagnostic hosts for different viruses are presented in Table 4.3.

4.1.2   Transmission of Viruses by Vectors

Various groups of vectors are involved in the transmission of plant viruses and the viruses transmitted by a particular group of vectors exhibit differences in a number of other properties. The nature of vector involved in transmission and the relationship between the virus and the vector have been studied for the majority of plant viruses and phytoplasmas (Carter, 1973; Harris, 1979; Harris and Maramorosch, 1977).

Table 4.2   Differentiation of Two Tospovirus Species by Reaction of Test Plant Species

| Test plant species | TSWV | | INSV | |
	Local infection	Systemic infection	Local infection	Systemic infection
Nicotiana benthamiana	a*	b	a	a
N. clevelandii	a	c	a	b
N. tabacum cv. White Burley	a	c	a	NI
Lycopersicon esculentum cv. Marmande	a	b	(a)	NI
Datura stramonium	a	b	(d)	NI

* a, Necrotic local lesion; b, mosaic; c, necrosis; d, chlorotic local lesion; (), inconsistent appearance; NI, no infection.
Source: Vaira et al. (1993).

Table 4.3 Assay/Diagnostic Hosts of Plant Viruses

Name of virus	Diagnostic hosts	Infection[a] type (local/systemic)	Reference
Agropyron mosaic virus	*Chenopodium quinoa*	L	Bremer, 1964
Alfalfa mosaic virus	*Chenopodium amaranticolor*	L	Smith, 1972
	C. quinoa	L	
	Gomphrena globosa	L	Hollings, 1959
	Phaseolus vulgaris	L	
	Ocimum basilicum	L	Lovisolo, 1960
Apple chlorotic virus	*C. amaranticolor*	S	Cropley, 1964
	C. quinoa	S	Lister et al., 1964
Apple mosaic	Apple varieties Lord Lambourne, Jonathan, and Golden Delicious	S	Smith, 1972
Arabis mosaic	*Cannabis sativa*	S	Smith, 1972
	C. amaranticolor	S	
	C. murale	S	
	C. quinoa	S	
Barley stripe mosaic virus	*C. amaranticolor*	L	Hollings, 1959
	G. globosa	L	
	Spinacia sp.	L	
Bean common mosaic virus	*Phaseolus vulgaris*	S	Pierce, 1934
Bean southern mosaic virus	*Phaseolus vulgaris*	L	Shepherd, 1971
	P. lunatus	L	
	P. aureus	L	
	Cowpea	L	
Bean yellow mosaic yellow	*Tetragonia expansa*	L	Hollings, 1966
Beet curly top virus	*Datura stramonium*	S	Severin, 1929
	Nicotiana tabacum var. White Burley	S	
Beet mosaic virus	*Gomphrena globosa*	L	Hollings, 1959
	Chenopodium amaranticolor	L	
	Beta vulgaris	S	Russell, 1971
	Spinacea oleracea	L, S	
Broadbean mottle virus	*Coronilla varia*		
	Lourea vespertilionis		Smith, 1972
Brome mosaic	*Zea mays*	L. S	Smith, 1972
Cardamom (greater) mosaic streak virus	*Achorus calamus*	S	Smith, 1972
	Curcuma longa		

Table 4.3 Continued

Name of virus	Diagnostic hosts	Infection[a] type (local/systemic)	Reference
Carnation latent virus	*C. amaranticolor*	L, S	Wetter, 1971
	C. quinoa	L, S	
Carnation mottle virus	*C. amaranticolor*	L	Hollings, 1956
Carnation ringspot virus	*G. globosa*	L	Paludan, 1965
Carnation vein mottle virus	*C. amaranticolor*	L	Smith, 1972
	C. quinoa	L	
Carrot mosaic virus	*Vigna sinensis*	L	Smith, 1972
Carrot mottle virus	*C. amaranticolor*	L	Smith, 1972
	C. quinoa	L	
Cauliflower mosaic virus	*Verbesina encelioides*	L, S	Sheperd, 1970
	Dahlia pinnata	S	
	Ageratum conyzoides	L, S	
	Zinnia elegans	S	
	Amaranthus caudatus	L, S	
	Chenopodium capitatum	S	
Celery mosaic virus	*Apium graveolens* var. *dulce* (celery)	S	Shepard and Grogan, 1971
	Conicum maculatatum	S	
	Pastinaca sativa	S	
	Dacus carota var. *sativa*	S	
Celery (western) mosaic virus	*C. amaranticolor*	L	Wolf, 1969, Smith, 1972
	C. quinoa	L	
	C. amaranticolor	L, S	
Cherry leaf roll virus	*C. murale*	S	Smith, 1972
	N. tabacum cv. *White Burley*	L, S	Smith, 1972
	Vicia faba	S	
	Prunus persica	S	Kegler, et al., 1962
Chili (pepper) mosaic virus	*Beta vulgaris*	L	Nariana and Sastri, 1958
	Nicotiana glutinosa	S	
Citrus infectious variegation virus	*Cucumis sativus*	L	Desjardinis and Wallace, 1962
	C. quinoa	S	
	Crotalaria spectabilis	L	
	Petunia	S	Garnsey, 1968
Citrus tristeza	West Indian (Mexican) lime	S	Wallace, 1951

Table 4.3 Continued

Name of virus	Diagnostic hosts	Infection[a] type (local/systemic)	Reference
Citrus vein enation virus	Rough lemon or sour lemon	S	Wallace and Drake, 1961
Clover (white) mosaic virus	*Vigna sinensis*—cowpea var. Blackeye	L, S	Smith, 1972
	Phaseolus vulgaris		
	Pisum sativum		
Clover (red) mottle virus	*Phaseolus vulgaris*	L	Sinha, 1960
	G. globosa	S	
	C. amaranticolor	L	Smith, 1972
	C. quinoa	L	
Clover yellow mosaic virus	*Antirrhinum majus*, var. Majestic	L, S	Pratt, 1961
Clover yellow vein virus	*Nicotiana clevelandii*	S	Hollings and Narrani, 1965
Cotton leaf curl virus	*Hibiscus cannabinus*	S	Tarr, 1951
Cowpea, aphid-borne mosaic virus	*C. amaranticolor*	L	Lovisolo and Conti, 1966
	O. basilicum	L	
Cowpea chlorotic mottle virus	*Soja max*—soybean	L	Kuhn, 1964
Cowpea mosaic virus	*Canavalia ensiformis*	L	Dale, 1949
Cucumber mosaic virus	*C. amaranticolor*	L	Hollings, 1959
	G. globosa	L	
	Datura stramonium	S	
Dahlia mosaic virus	*Verbesina encelioides*	S	Brierley, 1951
Eggplant (brinjal) mosaic virus	*C. amaranticolor*	L	Smith, 1972
	N. clevelandii		
Grapevine fan leaf virus	*C. amaranticolor*	L, S	Harrison and Nixon, 1960
	G. globosa	L, S	Dias, 1963
Groundnut bud necrosis virus	Cowpea cv. 152	L, S	Ghanekar, et al., 1979
	Dolichos lablab	L	
Groundnut (peanut) mottle virus	Bean (*P. vulgaris*) var. Topcrop	L	Kuhn, 1964
	Cassia occidentalis	S	
Groundnut rosette virus	*C. amaranticolor*	L	Hull and Adams, 1968
	C. hybridum	L	
	C. quinoa	L	
Hop mosaic virus	*N. clevelandii*	L	Bock, 1967
Lettuce mosaic virus	*G. globosa*	L	Rohloff, 1968
	C. quinoa	S	
	Spinacia oleracea	L	Hollings, 1959

Table 4.3 Continued

Name of virus	Diagnostic hosts	Infection[a] type (local/systemic)	Reference
Lettuce necrotic yellows virus	*Nicotiana glutinosa* *Petunia*	L, S S	Stubbs and Grogan, 1963
Lychnis ringspot virus	*Sugar beet* *Callistephus chinensis* *C. amaranticolor* *C. album* *C. capitatum*	L, S L L L L	Bennett, 1959
Maize rough dwarf virus	*Hordeum vulgare*	S	Vidano et al., 1996
Maize streak disease virus	*Digitaria horizontalis*	S	Storey, 1932, 1933
Narcissus mosaic virus	*G. globosa* *Trifolium incarnatum*	L S	Brunt, 1966a
Narcissus yellow stripe virus	*Tetragonia expansa* *Narcissus jonquilla*	L S	Brunt, 1971
Nasturtium ringspot virus	*N. glutinosa* *N. tabacum* var. White Burley *Blackstonia perfoliata*	L, S L, S	Smith, 1950 Juretic et al., 1970
Onion yellow dwarf virus	*Narcissus jonquilla*	S	Henderson, 1935
Papaya mosaic virus	*G. globosa* *C. amaranticolor* *Cassia occidentalis*	L L L	Purcifull and Hiebert, 1971
Parsnip yellow fleck virus	*Chenopodium quinoa*	L	Murant and Gold, 1967
Pea early browning virus	*G. globosa* Cowpea var. Monarch's Black Eye	L, S L, S	Bos and Van der Want, 1962
Pea enation mosaic virus	*Chenopodium amaranticolor*	L	Pierce, 1935
Pelargonium leaf curl virus	*C. amaranticolor* *Phaseolus vulgaris* *N. clevelandii* *Antirrhinum majus*	L L L, S L	Hollings, 1962
Plum line pattern virus	*Nicotiana megalosiphon* *Vigna cylindrica*	S S	Paulsen and Fulton, 1968
Plum pox virus	*Prunus persica* *Chenopodium foetidum*	S L	Smith, 1972

Table 4.3 Continued

Name of virus	Diagnostic hosts	Infection[a] type (local/systemic)	Reference
Potato aucuba mosaic virus	Potato var. Irish Chieftain	S	Clinch et al., 1936
	Tomato var. Kondure Red	S	
	Capsicum annuum	S	Maris and Rozendaal, 1956
Potato leaf roll virus	*Datura tatula*	S	Maramorosch, 1955
	Physalis angulata		
	Potato var. Earlaine	S	
Potato moptop virus	*C. amaranticolor*	L	Harrison and Jones, 1970
Potato spindle tuber viroid	*Solanum rostratum*	S	Singh and Bagnall, 1968
Potato virus A	*N. glutinosa*	S	Sommereyns, 1959
	Solanum demissum	L	Webb and Buck, 1955
	N. tabacum cv. *Samsun*	S	Smith, 1972
	N. tabacum cv. White Burley		
Potato virus S	*C. album*	L	Wetter, 1971
	C. amaranticolor		
	C. quinoa		
	Solanum rostratum		
	Cyamopsis psoraloides		
Rose mosaic virus	*Cyamopsis psoraloides*	L	Fulton, 1952
Sowbane mosaic virus	*C. murale*	L, S	Bennett and Costa, 1961
	C. quinoa	L, S	
Squash mosaic virus	*Cucurbita pepo,*	S	Smith, 1972
	Citrullus lanatus,		
	Cucumis sativum		
Strawberry crinkle virus	*Fragaria vesca*	S	Frazier, 1968
Strawberry mild yellow-edge virus	*Fragaria vesca*	S	Prentice, 1948
Strawberry mottle virus	*Fragaria vesca*	S	Prentice and Harris, 1946
Sugarcane mosaic virus	*Holcus sorghum*	S	Costa and Penteado, 1951

Table 4.3 Continued

Name of virus	Diagnostic hosts	Infection[a] type (local/systemic)	Reference
Tobacco broad ringspot virus	*Helianthus annuus*	L, S	Johnson and Fulton, 1942
Tobacco etch virus	*C. amaranticolor*	L	Holmes, 1946
	Physalis peruviana	L	
	N. tabacum	L, S	Smith, 1972
	C. annuum	S	
	D. stramonium		
	L. esculentum		
Tobacco mosaic virus	*Nicotiana glutinosa*	L	Holmes, 1931
	N. tabacum var. Xanthi	L	
	Phaseolus vulgaris var. Pinto bean	L	Piacitelli and Santilli, 1961
Tobacco necrosis virus A	*Phaseolus vulgaris*	L	Lovisolo, 1966
	Ocimum basilicum	L	Hollings and Stone, 1965c
	Torenia fournieri	L	
Tobacco rattle virus	*C. amaranticolor*	L	Cadman and Harrison, 1959
	Phaseolus vulgaris	L	
Tobacco ringspot virus	*C. amaranticolor*	L	Smith, 1972
	C. quinoa	L	
	Cowpea	L, S	
	Cucumber	L, S	
Tobacco streak virus	*Cyamopsis psoraloides*	L	Fulton, 1948,
	Nicotiana tabacum	L	Smith, 1972
	Vigna cylindrica	L	
Tomato aspermy virus	*Tetragoniaexpansa—* New Zealand spinach	L	Smith, 1972
	C. amaranticolor	L	Hollings and
	N. glutinosa	L, S	Stone, 1971
Tomato black ring virus	*Cucumis sativus*	S	Roland, 1969
	Chenopodium foliosum	S	Roland, 1969
	G. globosa	S	Roland, 1969
	Vicia faba	S	Roland, 1969
	Phlox drummondii	S	Schmelzer, 1970
Tomato bunchy top virus	*N. glutinosa*	S	McClean, 1935a
	Petunia sp.	S	
Tomato bushy stunt virus	Cowpea	L	Schmelzer, 1958
	Datura stramonium	L, S	
Tomato ringspot virus	*C. amaranticolor*	L	Samson and
	C. quinoa	L	Iamle, 1942
	Cowpea		

Table 4.3 Continued

Name of virus	Diagnostic hosts	Infection[a] type (local/systemic)	Reference
Tomato spotted wilt virus	*Petunia* sp.	L	Best, 1968
	N. glutinosa	L	Best, 1968
Turnip crinkle virus	*C. amaranticolor*	L	Hollings and Stone, 1963
Turnip mosaic virus	*C. amaranticolor*	L	Smith, 1972
	N. tabacum	L	
	N. glutinosa	S	
Turnip yellow mosaic virus	*Brassica pekinesis*	L, S	Diener and Jenifer, 1964
	Cleome spinosa	L	
Watermelon mosaic virus (western)	*Lavatera trimestris*	L	Schmelzer, 1965
	C. album	L	Demski, 1968
	C. strictum	L	Demski, 1968
Wheat (soil-borne) mosaic virus	*Triticum spelta*—Red Winter	S	McKinney, 1953
	C. amaranticolor	L	Smith, 1972
	C. quinoa	L	

[a] L, local; S, systemic.

The species of vector and transmission characteristics are useful in differentiating some viruses and their strains. The vector specificity, in certain viruses, may be helpful to establish the identity of viruses with reasonable certainty. The viruses transmitted by nematodes or fungi may be identified through the vector specificity to some extent (Taylor, 1980; Teakle, 1980). A knowledge of the variations in these properties may be useful in differentiating viruses.

4.1.2.1 Viruses Transmitted by Aphids

The viruses transmitted by aphids show differences in the following properties: a) persistence in aphids, b) presence of latent period, c) passage through molt, d) multiplication of virus in aphid, e) mechanical inoculation, f) virus particle morphological characteristics, and g) nature of viral genome. On the basis of these properties the aphid-borne viruses may be grouped as follows:

a. Nonpersistent viruses. The vectors of nonpersistent viruses remain viruliferous for short periods, have no demonstrable latent period, and lose infectivity after molting. Fasting prior to acquisition feeding remarkably increases the period of virus retention and efficiency of transmission by the aphids. They can be mechanically transmitted from plant to plant. All viruses have single-stranded ribonucleic acid (ss-RNA) genome, except caulimoviruses, which have double-stranded deoxyribonucleic acid (ds-DNA) as genome. These nonpersistent viruses are included in the alfalfa mosaic virus, carlavirus, caulimovirus, cucumovirus, fabavirus, and potyvirus groups; they may have mono-, bi-, or tripartite genomes. These viruses fall into four groups based on particle morphological characteristics (Watson, 1972):

i) Rigid rods: carnation latent, potato viruses S and M, and red clover vein mosaic viruses

ii) Long flexible rods: beet mosaic, henbane mosaic, potato virus Y, and turnip mosaic viruses

iii) Isometric: cucumber mosaic and cauliflower mosaic viruses

iv) Isometric-bacilliform: alfalfa mosaic virus

b. Semipersistent viruses.   The semipersistent viruses differ from nonpersistent viruses in having longer periods of retention by aphids, longer periods of acquisition feeding, and the absence of any influence of fasting prior to acquisition feeding. They are not readily transmitted by mechanical inoculation as the nonpersistent viruses are. Beet yellows and citrus tristeza viruses have long flexible rod-shaped particles showing differences in the length of virus particles. Cauliflower mosaic virus is transmitted both nonpersistently and semipersistently by the same aphid species (Brevicoryne brassicae); such transmission is known as bimodal transmission (Chalfant and Chapman, 1962). Pea seed-borne mosaic virus also is transmitted bimodally, by Macrosiphum euphorbiae (Lim and Hagedorn, 1977).

c. Circulative (nonpropagative) viruses.   The circulative viruses have a definite latent period in the aphid vector and pass through molting. Some are transmitted by mechanical inoculation with difficulty. They do not multiply in the vector. Pea enation mosaic virus (monotypic virus group) and luteovirus are transmitted in a nonpropagative circulative manner (Sylvester and Richardson, 1966; Miyamoto and Miyamoto, 1966; Watson and Okusanya, 1967). The barley yellow dwarf, carrot mottle, groundnut rosette, parsnip mottle, pea enation mosaic, and potato leafroll viruses are spherical. The molecular basis of interaction between potato leafroll luteovirus (PLRV) and Myzus persicae has been elucidated by Hoghenhout et al. (1996). Endosymbiotic bacteria synthesize and release a major protein termed symbionin into the haemolymph of the aphids. PLRV has a strong affinity to this protein (Mr = 60,000), which shows high homology with Escherichia coli heat stock protein groEL. Symbionin is considered to determine the persistence of PLRV in M. persicae, since the absence of symbionin antibodies resulted in rapid degradation of major virus capsid protein and concomitant loss of infectivity of the aphids.

d. Propagative viruses.   Many viruses in Rhabdoviridae have been reported to have longer latent periods and to multiply in the aphids. They are not transmitted mechanically from plant to plant, but some viruses may be successfully inoculated into aphids. The Lettuce necrotic yellows virus (LNYV) and sowthistle yellow vein virus (SYVV), transmitted by *Hyperomyzus lactucae* (O'Laughlin and Chambers, 1967; Sylvester, 1973), have bacilliform particles. Eggs of the vector carry SYVV (Sylvester, 1969), which is able to multiply in the primary cultures of aphid cells (Peters and Black, 1970). Successful serial passage of strawberry crinkle virus by needle inoculation to the aphid *Chaetosiphon jacobi* was reported by Sylvester et al. (1974), indicating the possible multiplication of the virus in the vector.

4.1.2.2 Viruses Transmitted by Leafhoppers

According to the transmission characteristics, the leafhopper-borne viruses may be divided into three groups. The properties of the viruses may be useful to differentiate those with similar transmission characteristics. None of the leaf-hopper-borne viruses is transmitted in a nonpersistent manner. Some of the propagative viruses have been successfully inoculated into the leafhoppers, and a few are known to pass through eggs to successive generations (transovarial or congenital transmission).

a. Semipersistent viruses. Rice tungro-associated viruses are transmitted by *Nephotettix virescens* semipersistently (Ling, 1972). Studies by Hibino and Cabunagan (1986) showed that rice tungro spherical virus (RTSV), one component of the rice tungro complex, is independently transmitted by the leafhopper, whereas rice tungro bacilliform virus (RTBV), another component of the complex, can be transmitted only if the leafhopper has already acquired RTSV. These two viruses differ in other properties, such as the virus particle morphological features and the nature of the virus genome. The RTSV has spherical particles with ss-RNA as genome, whereas RTBV particles are bacilliform in shape, containing ss-DNA as genome (Hibino et al., 1991). Maize chlorotic dwarf virus (MCDV) is transmitted in a semipersistent manner by *Graminella nigrifrons* (Choudhury and Rosenkranz, 1983). The virus particles are isometric with ss-RNA as genome (Nault and Ammar, 1989).

b. Circulative (nonpropagative) viruses. The circulative viruses are ingested into the body of the leafhoppers, which transmit them after varying periods of latency. There is no convincing evidence for the multiplication of the virus in the insect body. Beet curly top, *Chloris* striate mosaic, and maize streak viruses belong to the geminivirus group and have paired, small isometric particles with a diameter of 18–22 nm (Mumford, 1974; Goodman, 1981). They are transmitted by different leafhopper species (Bennett, 1963; Goodman, 1981; Harrison, 1985; Nault, 1991).

c. Propagative viruses. Viruses belonging to the Reoviridae, and Rhabdoviridae families, and Marafivirus, and Tenuivirus groups have a definite latent period and are able to multiply in the body of their vectors. Some of them have been reported to be transovarially transmitted to several suscessive generations.

1. Reoviridae. The Reoviridae are spherical in shape but larger than other isometric viruses and have ds-RNA as the genome. The clover wound tumor virus (WTV) particles have ds-RNA enclosed in a capsid with a diameter of 60 nm composed of 32 capsomeres (Bils and Hall, 1962; Black, 1965). The virus induces characteristic cytopathic changes in the nerve cells (Hirumi et al., 1967). By injecting the leafhopper extracts serially into virus-free *Agallia constricta,* Black and Brakke (1952) demonstrated that WTV multiplied in the body of the vector insects. Multiplication of WTV in plant and leafhopper, by causing the formation of viroplasmic loci, has been reported by Shikata and Maramorosch (1967).

Transovarial transmission of rice dwarf virus (RDV) through eggs up to several successive generations was reported by Fukushi (1933, 1940). The RDV particles are spherical with a diameter of 70 nm and have ds-RNA as the genome (Shikata, 1962; Whitcomb, 1972). The presence of virus particles in mycetome and the fat body of leafhoppers was observed by Nasu (1963). Maize rough dwarf vius (MRDV) particles have similar size to RDV. The presence of MRDV in different tissues and its multiplication in the delphacid planthopper were demonstrated by Vidano (1970). Sugarcane Fiji disease virus (FDV) transmitted by *Perkinsiella saccharicada* and rice ragged stunt virus transmitted by *Nilaparvata lugens* are also able to multiply in the vector insects (Conti, 1984).

2. Rhabdoviridae. The Rhabdoviridae have bacilliform nucleoprotein cores enclosed in three layered membranes on which short projections are seen. Potato yellow dwarf virus (PYDV) transmitted by different species of agallian leafhoppers has bacilliform particles with dimensions of 380 × 75 nm (MacLeold et al., 1966). Two strains of the virus present in eastern North America are transmitted by *Agallia constricta* and *Aceratagallia sanguinolenta* (Black, 1941). Transovarial transmission of PYDV is known (Black, 1953). Multiplication of the virus in cells of *A. constricta* was reported by Chiu et al. (1970). By sap inoculation PYDV can be transmitted to *Nicotiana rustica,* which reacts with local lesions initially and with systematic symptoms later (Black, 1953).

Maize mosaic virus, transmitted by delphacids, has bacilliform particles measuring 242 ± (10 × 48) ± 10 nm, and the presence of the virus particles in both plant and vector tissues has been observed (Herold and Munz, 1965). Wheat striate mosaic virus transmitted by deltocephaline leafhoppers also has bacilliform particles measuring 260 × 80 nm (Lee, 1968).

3. Marafiviruses. Maize rayado fino virus is transmitted by the leaf-hopper *Dalbulus maidis.* The presence of the virus in the body of 80% of the vector, after acquisition feed, has been detected by ELISA tests, but only about 10%–34% of the insects are able to transmit the virus. No cytopathological effects due to the virus are known. The virus has isometric particles with 30 nm diameter and ss-RNA as genome (Gamez and Leon, 1988).

4. Tenuiviruses. Maize stripe virus is transmitted by *Peregrinus maidis;* its multiplication in the vector has been indicated by ELISA. The presence of the virus has been detected in different organs and salivary glands of the vector (Nault and Gordon, 1988). Transovarial transmission of rice stripe virus in the vector leafhopper *Laodelphax striatellus* is known (Gingery, 1988). The virus has thin coiled filamentous particles measuring $3 \times 950{-}1350$ nm and ss-RNA as genome (Toriyama, 1983a, 1983b).

4.1.2.3 Viruses Transmitted by Whiteflies

Whiteflies transmit different groups of plant viruses which cause many economically important diseases of crops in tropical and subtropical countries. None of the viruses transmitted by whiteflies has a nonpersistent relationship with its vetor; most seem to be of circulative (nonpropagative) type, and no evidence suggesting the multiplication of the virus in the vector has been obtained so far. Only three species of whiteflies have been reported as vectors of plant viruses; *Bemisia tabaci* is the most frequently reported vector species. Many of the geminiviruses and some viruses belonging to the carlavirus, closterovirus, luteovirus, nepovirus, and potyvirus groups and a DNA-containing rod-shaped virus are transmitted by whiteflies (Duffus, 1985). The virus particle morphological features and nature of the viral genome will be useful to differentiate the viruses transmitted by whiteflies. The differences in the viral genome have been shown to reflect the differences in the transmissibility of geminiviruses. The Kenyan isolate of African cassava mosaic bigeminivirus (ACMV-K) is not transmitted by the whitefly *B. tabaci,* whereas the Nigerian isolate (ACMV-NOg) is rapidly transmissible by *B. tabaci.* The exchange of gene fragments between ACMV-K and ACMV-NOg demonstrated that differences responsible for the loss of transmissibility resided on the viral coat protein and DNA-B C1 gene of ACMV-K (Liu et al., 1999).

4.1.2.4 Viruses Transmitted by Beetles

Beetles have biting mouthparts and differ from other groups of vector insects which suck the sap while feeding on plants. As the beetles do not have salivary glands, the virus along with plant materials is returned to the mouth by regurgitation. During this process the mouthparts are contaminated with virus particles released from plant tissues. The viruses transmitted by beetles do not appear to have a definite latent period or to be able to multiply in the vector. There is a marked degree of specificity between viruses and beetles (Fulton et al., 1987); they can be easily transmitted and remain stable under in vitro conditions. The beetles transmit viruses belonging to the bromovirus, comovirus, sobemovirus, and tymovirus groups. These viruses have small isometric particles with 25–30 nm diameter and ss-RNA as genome.

4.1.2.5 Viruses Transmitted by Thrips

The relationship between the viruses and vector thrips is somewhat unique in the sense that only the larvae can acquire the virus from infected plants, not the adults. Both the thrips species and the tomato spotted wilt virus (TSWV) have wide host range. *Thrips tabaci* and three species of *Frankliniella* transmit TSWV more efficiently (Sakimura, 1962). There is no evidence for the multiplication of TSWV in the vector, but the virus is retained by thrips till their death. The virus is not passed through eggs; TSWV can be transmitted by sap inoculation by taking suitable precautions to prevent inactivation, as the virus has only a short period of longevity in vitro. Transmission by thrips may be lost if the virus is repeatedly transmitted by sap inoculation. Peanut bud necrosis virus, earlier considered as TSWV, has been shown to be a distinct virus belonging to tospovirus group (Reddy et al., 1992).

Other viruses transmitted by thrips include tobacco ringspot virus (nepovirus group) and tobacco streak virus (ilarvirus group) (Bergeson et al., 1964; Keiser et al., 1982). Tobacco ringspot virus is transmitted by a nematode vector. Tobacco ringspot virus and tobacco streak virus have isometric particles without an envelope and have bipartite and tripartite genomes, respectively.

4.1.2.6 Viruses Transmitted by Eriophyid Mites

The relationships between the viruses and their mite vectors have been studied only to a limited extent because of the small size of the mites. Nine viruses have been reported to be transmitted by seven different species of mites (Slykhuis, 1972). Of these, transmission of wheat streak mosaic potyvirus (WSMV) by *Aceria tulipae* has been studied well. *Aceria tulipae* also transmits uncharacterized wheat spot mosaic virus. The WSMV has flexuous rod-shaped particles and is transmitted even after molting by the mite. Only the nymphs can acquire the virus, and there is no convincing evidence indicating the replication of the virus in the mites. The WSMV can be detected in the midgut, body cavity, and salivary glands (Paliwal, 1980). Wheat spot mosaic virus also has a similar relationship with the vector mites. Wheat streak mosaic virus, agropyron mosaic virus, and ryegrass mosaic virus (the latter two viruses are transmitted by *Abacarus hystrix*) may also be transmitted by mechanical inoculation (Slykhuis, 1972).

4.1.2.7 Viruses Transmitted by Nematodes

Nematodes that transmit plant viruses are all free-living ectoparasites and belong to three genera, viz. *Xiphinema, Longidorus,* and *Trichodorus.* The viruses can be acquired and transmitted by all stages of the nematodes. A longer acquisition and inoculation access period results in a greater percentage of transmission. The viruses are retained by the nematode vector for several weeks or months, if the nematodes are kept in fallow soil (Taylor, 1972). Nematode-transmitted viruses are divided into two groups based on particle morphological characteristics. Nepoviruses, with isometric particles, are transmitted by different species of *Xiphinema* and *Longidorus,* whereas tobraviruses, with rod-shaped particles, are transmitted by species of *Trichodorus.* Serologically related viruses, such as grape fan leaf virus and arabis mosaic virus, require *X. index X. coxi,* respectively, indicating specificity of transmission by nematodes (Dias and Harrison, 1963). It is likely that the coat protein of the virus has a role in determining the specificity of transmission. The virus particle morphological features and transmission characteristics may help to differentiate nematode-transmitted viruses.

Many viruses are readily transmitted by mechanical inoculation, and some are seed- and pollen-borne in many host plant species, especially weed plants, which have epidemiological importance.

4.1.2.8   Viruses Transmitted by Fungi

Studying the characteristics of transmission of plant viruses by fungal vectors has been found to be difficult because of the limitations associated with virus-fungus interactions. In some cases adequate experimental evidence of involvement of the fungus in virus transmission is not available. The fungi reported as vectors of viruses belong to three genera, *Olpidium, Polymyxa,* and *Spongospora.* The viruses transmitted by them form a heterogenous group. *Olpidium* spp. transmit isometric viruses, including tobacco necrosis, satellite, cucumber necrosis, and tobacco stunt viruses, whereas rod-shaped viruses, such as wheat mosaic, potato moptop, peanut clump, and potato virus X, are transmitted by the plasmodiophorous genera *Polymyxa* and *Spongospora* (Brunt and Shikata, 1986); some of the viruses may be transmitted by other means also. Pea false leaf roll is also transmitted by aphids and through seeds (Thottapilly and Schumutterer, 1968), whereas potato virus X mainly spreads through leaf contact under field conditions (Teakle, 1972). *Olpidium boronovanus* is the vector of the melon necrotic spot carmovirus (MNSV) and it plays a vital role in acquiring and transmitting MNSV to the roots of germinating seedlings to initiate primary infection. The transmission of the virus present in the seed surfaces does not seem to occur unless the fungal vector is present. Vector-assisted seed transmission (VAST) was reported by Campbell et al. (1996).

4.2   TRANSMISSION OF VIROIDS

Viroids may be transmitted by the following methods under natural or experimental conditions.

Table 4.4   Transmission of Viroids by Dodder

Viroid	Dodder	Reference
Citrus exocortis	*Cuscuta subinclusa*	Weather, 1965
Chrysanthemum stunt	*C. gronovii*	Keller, 1953
Cucumber pale fruit	*C. subinclusa*	Peters Runia, 1974

4.2.1   Grafting

Potato spindle tuber viroid (PSTVd) can be transmitted by side grafting from in-fected *Datura stramonium* to tomato plants (O'Brien and Raymer, 1964). How-ever, it is readily transmitted by mechanical inoculation from potato to other sus-ceptible plant species. Other viroids that are transmitted by grafting are the citrus exocortis and chrysanthemum chlorotic mottle viroids. Mechanical transmission is the preferred method of experimental transmission in the case of most viroids.

4.2.2   Dodder Transmission

Some of the viroids have been transmitted by using parasitic dodders (Table 4.4).

4.2.3   Arthropod Transmission

Though early reports suggested the involvement of arthropods in the spread of vi-roid disease, none of the viroids has been conclusively shown to have any arthro-pod vector. However, the transmission of PSTVd by *Myzus persicae* from source plants doubly infected with PLRV and PSTVd was reported by Salazar et al. (1995). The aphids did not transmit PSTVd from plants infected only with PSTVd. Further studies are necessary to establish whether PLRV helps PSTVd transmission by aphid or the viroid is accidentally transmitted along with PLRV for which the aphid is a natural vector.

4.2.4   Seed Transmission

Potato spindle tuber viroid and tomato bunchy stunt viroid have been found to be transmitted through seeds of infected host plants (Table 4.5). The transmission of grapevine yellow speckle viroid 1 and hop stunt viroid through seeds of eight grapevine cultivars was observed. The presence of the viroids in the seeds was de-tected using a combination of dot-blot hybridization, Northern hybridization, and reverse transcription–polymerase chain reaction (RT-PCR) assays (Wah and Symons, 1999). The modern molecular detection methods help to have confirma-tory results very rapidly.

Table 4.5 Seed and Pollen Transmission of Viroids

Viroid	Host plant	Transmission through seed	pollen	Reference
Tomato bunchy	*Solanum incanum*	+	−	McClean, 1931
top viroid	*Physalis peruviana*	+	−	
Potato spindle	Potato	+	+	Benson and Singh, 1964
tuber viroid	Tomato	+	−	Fernow et al., 1970

4.2.5 Mechanical Inoculation

Most of the plant viroids are readily transmitted by sap inoculation, although some require special methods or appropriate host plant species for successful transmission. Citrus exocortis viroid is inoculated by a contaminated knife or razor blade (Garnsey and Whidden, 1973). Coconut cadang-cadang viroid requires a high-pressure injector for transmission to seedlings (Randles et al., 1977).

4.2.6 Host Range of Viroids

The viroids may have narrow or wide host range indicating the variation in the infecting potential and extent of distribution. The PSTVd has a wide host range distributed in 12 families and 157 species, whereas the burdock stunt (BSVd), avocado sunblotch (ASBVd), coconut cadang-cadang (CCCvd), and chrysanthemum chlorotic mottle (CCMVd) viroids have a narrow host range limited to members of only one family (Table 4.6). Among the 19 families tested, only 3, the Compositae, Cucurbitaceae, and Solanaceae, are susceptible to more than one viroid, and the members of 16 families are susceptible to only one viroid. This specificity of host range may be useful to differentiate the viroids, when considered with other properties. The Solanaceae include members susceptible to a maximum number of viroids (i.e., 8 of 12 viroids).

For the identification of viroids some methods followed for plant viruses cannot be used, because of the absence of the protein coat in viroids. Electron microscopy and serological techniques employed for detection, identification, and assay of plant viruses are not useful for the viroid diseases. Among the methods available diagnostic hosts are commonly used for viroid identification, as in the case of viruses.

4.2.6.1 Diagnostic Hosts

Most of the viroids cause, in their natural hosts, symptoms which are likely to be similar to those induced by other pathogens, and some cultivars do not exhibit any

Table 4.6   Host Range of Plant Viroids

Families susceptible	Viroids											
	ASB	BS	CCC	CCM	CE	CPF	CS	CV	HS	PST	TBT	TPM
Amaranthaceae	—	—	—	—	—	—	—	—	—	1	—	—
Boraginaceae	—	—	—	—	—	—	—	—	—	1	—	—
Campanulaceae	—	—	—	—	—	—	—	—	—	1	—	—
Caryophyllaceae	—	—	—	—	—	—	—	—	—	1	—	—
Compositae	—	2	—	1	2	2	43	—	—	1	1	1
Convolvulaceae	—	—	—	—	—	—	—	—	—	1	—	—
Cucurbitaceae	—	—	—	—	4	30	—	—	10	2	—	—
Dipsaceae	—	—	—	—	—	—	—	—	—	1	—	—
Gesneriaceae	—	—	—	—	—	—	—	1	—	—	—	—
Lauraceae	2	—	—	—	—	—	—	—	—	—	—	—
Leguminosae	—	—	—	—	2	—	—	—	—	—	—	—
Moraceae	—	—	—	—	—	—	—	—	2	—	—	—
Palmae	—	—	4	—	—	—	—	—	—	—	—	—
Rutaceae	—	—	—	—	14	—	—	—	—	—	—	—
Sapindaceae	—	—	—	—	—	—	—	—	—	1	—	—
Scrophulariaceae	—	—	—	—	—	—	—	—	—	6	—	—
Solanaceae	—	—	—	—	21	11	9	2	1	140	24	3
Umbeiliferaceae	—	—	—	—	2	—	—	—	—	—	—	—
Valerianaceae	—	—	—	—	—	—	—	—	—	1	—	—

Source: Peters and Runia (1985).

recognizable symptoms. In such cases use of indicator plants becomes necessary for early detection of infection and identification of the viroid involved, especially for building up disease-free seed stocks, as in the ease of potatoes free of spindle tuber viroid and citrus free of exocortis viroid. Pear blister canker viroid (PBCVd) requires an incubation period of 2–3 years for symptom expression in perry pear *(Pyrus communis)* cv. A 20. But cultivars Fieud 37 and Fieud 110 show symptoms 3–5 months after inoculation with PBCVd. Hence, these two cultivars can be used as indicators of PBCVd infection in perry pear cultivars (Desvignes et al., 1999). Diagnostic hosts for different viroids are listed in Table 4.7.

Hop stunt disease may be more reliably detected by cucumber bioassay than by symptom diagnosis or α-acid analysis. For the cucumber bioassay technique, a temperature-controlled greenhouse or plastic house at 30°C is required and the plants are observed for a period of 1 month. The HSVd can be detected by cucumber bioassay in a sample containing a mixture of one HSVd-infected leaf disc and 200 HSVd-free leaf discs (Sasaki and Shikata, 1980). This technique has been used for HSVd detection in samples from fields (Sasaki et al., 1981).

Table 4.7 Diagnostic Hosts for Viroid Detection and Identification

Name of viroid	Diagnostic host	Metod of testing	Reference
Potato spindle tuber viroid (PSTVd)	*Lycopersicon esculentum*	Sap inoculation	Raymer and O'Brien, 1962
	cv. Rutgers/Allerfruheste		Singh, 1970
	Freil and *Scopolia sinensis*	Sap inoculation	Singh, 1971
Tomato bunchy top viroid (TBTVd)	*L. esculentum* cv. Rutgers	Sap inoculation	Bensen, et al., 1965
Citrus exocortis viroid (CEVd)	*Citrus medica* cv. Etrog	Grafting	Calavan et al., 1964
	Gynura aurantica	Sap inoculation	Weathers and Greer, 1972
Chrysanthemum stunt viroid (CSVd)	*Chrysanthemum morifolium* cv. Mistlefoe	Sap inoculation	Kaller, 1953
Chrysanthemum chlorotic mottle viroid (ChCMVd)	*C. morifolium* cv. Deep Ridge	Sap inoculation	Dimock et al., 1971
Cucumber pale fruit viroid (CPFVd)	*Cucumis sativus* cv. Sporu	Sap inoculation	VanDorst and Peters, 1974
Hop stunt viroid (HSVd)	*C. sativus*	Sap inoculation	Sasaki and Shikata, 1977a

Source: Diener (1979).

4.3 TRANSMISSION OF BACTERIAL PATHOGENS

Bacterial plant pathogens may be transmitted through infected seeds or propagative materials, rain/irrigation water, infected plant debris, and in some cases insects. Infection of seeds by bacterial pathogens may not be detected by visual examination in most cases. It may be necessary to employ special technique(s) for the detection of bacterial infection in seeds. Xie and Mew (1998) developed the leaf inoculation method for detecting *Xanthomonas oryzae* pv. *oryzicola* on rice seeds. Seed washings are inoculated on rice leaf segments placed on 1% agar amended with benzimidazole (75–100 ppm) to maintain freshness of inoculated leaf segments. The agar plates are then incubated in a moist chamber. Typical leaf streak lesions followed by bacterial ooze may be observed on the inoculated rice leaf segments indicating the seed infection by the bacterial pathogen.

Another mechanical inoculation method for transmission of *Xylella fastidiosa*, causing oleander leaf scorch disease was developed by Purcell et al. (1999). A Paraffilm sleeve just below a freshly cut stem apex is formed by wrapping the stretched Parafilm M (American Can Corp., Greenwich, CT). One end of a rubber tube is pushed into the Parafilm sleeve and sealed airtight, while the other end of the rubber tube is connected to a vacuum pump. A suspension of *X. fastidiosa* in phosphate-buffered saline (PBS) is prepared from culture plates. A small drop of the bacterial suspension is pipetted onto a point 5–8 cm below the severed stem apex. Using a No. 2 insect pin, the bacterial suspension is probed repeatedly, while the vaccum pump is kept running, so that the bacterial cells are infiltrated into the stem. The bacterial pathogen could be transmitted from oleander to oleander by this mechanical inoculation procedure. He et al. (2000) demonstrated that *X. fastidiosa* could be transmitted from the infected budsticks of sweet orange, when topgrafted to citrus root stocks that exhibited characteristic symptoms of citrus variegated chlorosis (CVC) disease. The presence of *X. fastidiosa* in the rootlets and main roots of CVC-symptomatic Pera sweet orange root stocks was confirmed by light microscopy, enzyme-linked immunosorbent assay (ELISA), and polymerase chain reaction (PCR) assays. In addition, the transmission of *X. fastidiosa* from infected sweet orange plants to healthy plants through root grafts under pot culture condition was also observed. This appears to be first report of transmission of CVC disease through natural root grafts.

Transmission by insects may be either generalized or specialized. Such association with insects may be useful for differentiating some of the bacterial pathogens. The natural avenues of entry for the bacteria into plant hosts are stomata, water pores, lenticels, and flower nectaries; the feeding and oviposition wounds also form important entry points. The insects may be mechanical carriers of bacteria on their body, and in some cases there may be a mutualistic relationship between the insect and the bacteria. The bacteria may remain in the hibernating insects in the absence of susceptible hosts or during adverse environmental conditions (Carter, 1973). Modern molecular techniques are increasingly employed to demonstrate the presence of bacterial pathogens in/on a wide range of insects present in the agroecosystem. For example, Hildebrand et al. (2000) detected *Erwinia amylovora* causing fire blight disease on insects having different living habits and belonging to at least eight insect families by using semiselective agar plate and polymerase chain reaction (PCR) assays. *E. amylovora* could survive for 5 days on *Chrysoperla carnea* (lace wing) and for 12 days on *Aphis pomi* collected from orchards. The bacterial pathogens which have some kind of association with insects for dissemination are listed in Table 4.8.

4.3.1   Use of Diagnostic Hosts

A simple, fast, and reliable bioassay technique for the detection and isolation of *Agrobacterium tumefaciens* causing tumors in a wide range of plant species was

Table 4.8 Insect Transmission of Plant Bacterial Pathogens

Disease	Bacterial pathogen	Insects associated	Nature of association	Reference
Apple fireblight	Erwinia amylovora	Drosophila melanogaster Musca domestica Lucilia sericata	General	Ark and Thomas, 1936
Walnut blight	Xanthomonas juglandis	Eriophyes tristriatus var. erinea	General	Rudolph, 1943
Beans haloblight	Pseudomonas medicaginis var. phaseolicola	Heliothrips femoralis	General	Buchanan, 1942
Potato	Corynebacterium sepedonicum	Melanoplus differentialis Epicauta pennsylvanica Leptinotarsa decemlineata	General	List and Kreutzer, 1942
Apple bacterial rot	Pseudomonas melophthora	Rhagoletis pomonella	General	Allen and Riker, 1932
Citrus canker	Xanthomonas citri	Phyllocnistic citrella	General	Sohi and Sandhu, 1968
Cucurbit bacterial wilt	Erwinia tracheiphila	Diabrotica vittata D. doudecimpunctata	Specialized: required for inoculation, dissemination, and overwintering	Rand and Cash, 1920; Gould, 1994
Corn bacterial wilt and Stewart's leaf blight	Xanthomonas stewartii	Chaetocneme pulicara C. denticulata	Specialized: required for dissemination and overwintering	Elliott and Poos, 1934
Potato blackleg	Erwinia carotovora	Hylemyia cilicrura	Specialized: mutualistic symbiosis	Leach, 1933
Oliver knot	Pseudomonas savastonoi	Dacus oleae	Specialized: mutualistic symbiosis	Petri, 1910

Table 4.8 Continued

Disease	Bacterial pathogen	Insects associated	Nature of association	Reference
Papaya bunchy top	N.D.	*Empoasia papayae* and *E. stevensi*	N.D.[a]	Davis et al., 1996
Oleander leaf scorch	*Xylella fastidiosa*	*Graphocephala atropunctata, Homalodisca coagulata, and H. lacerata*	N.D.	Purcell et al., 1999

[a] N.D., not determined.

developed by Romeriro et al. (1999). Detached leaves of the highly susceptible *Kalanchoe tubiflora* can be used as biological baits for trapping tumorigenic cells of *A. tumefaciens* from soil and tumor tissues of infected plants. Although a large and heterogeneous microflora may be associated with the soils and infected plant tissues, the bacterial pathogen alone is retrieved and then it may be isolated in pure culture.

4.4 TRANSMISSION OF FUNGAL PATHOGENS

The role played by insects in the transmission of fungal plant pathogens is relatively less important except in a few cases. Most are disseminated by wind, water, humans, and animals. Of course the infected seeds and seed materials transported by humans to different locations form the most important sources of infection.

The soil infested by fungal pathogens is the important source of inoculum for several destructive soil-borne diseases. Baiting technique has been used for the detection of fungal pathogens such as *Rhizoctonia* spp., *Pythium* spp., and *Phytophthora* spp. Angle-spinach seeds placed in soil infected by *R. solani* are colonized by the pathogen. The population of *R. solani* AG-4 in soil could be rapidly assessed by this technique. The detection technique was found to be sensitive and accurate, since there was significant correlation between colonization of angle-spinach seeds by *R. solani* and incidence of damping-off disease in Kale seedlings (Liu et al., 1995). Mung bean *(Vigna radiata)* has been used as a bait plant to detect *R. solani* in soil samples by Castilla et al. (1997). Using sterilized oat seeds, a simple, rapid, reliable, and quantitative technique was developed for assessing the population of *Pythium aphanidermatum* causing leak disease and watery wound rot in potato tubers. This technique is capable of detecting ca. 1 oospore per gram of dry soil (Priou and French, 1997). Apple and lupin seedlings and rhododendron leaves have been reported to be effective baits for the detection of *Phytophthora cactorum, P. cinnamomi, P. citricola, P. cryptogea,* and *P. drechsleri* isolates in water. The bait tests were as efficient in detecting the pathogens as enzyme-linked immunosorbent assay (ELISA) (Themann and Werres, 1997).

The fungi which are adapted to dissemination by insects produce spores in sticky masses which attract the insects. The spores are often deposited in the wounds formed during feeding or oviposition. The fungal pathogens may also initiate infection a) incident to pollination, b) through traumatic injury, c) as a result of internal and external contamination of the insect caused by feeding on fungal masses, d) through feeding punctures, and e) as the result of a symbiotic association between the fungus and insect (Carter, 1973). A biological relationship between a fungal pathogen and an insect vector was reported by Stanghellini et al. (1999). The chlamydospores of *Thielaviopsis basicola,* causing root and stem rot of corn-salad plants *(Valerianella locusta),* were consistently detected in frass excreted by the adults and larvae of shore flies *(Scatella stagnalis).* Internal infestation of the adults and larvae by the pathogen was to an extent of approximately 95% and 85%, respectively. Pathogen-free adult shore flies were able to acquire *T. basicola* by ingestion, after feeding on naturally infected plants. The presence of the chlamydospores was observed in the intestinal tract of the larva under the microscope. These infective insects were able to transmit viable propagules to healthy seedlings, which were subsequently infected. The fungi-insects associations leading to transmission of and infection by the fungal pathogens are presented in Table 4.9.

Table 4.9   Insect Transmission of Fungal Pathogens

Disease	Fungal pathogen	Insects associated	Nature of association	Reference
1. Incident to pollination				
Fig endosepsis	*Fusarium moniliforme* var. *fici*	*Blastophaga psenes*	Incidental to pollination	Caldis, 1927
2. Traumatic injury				
Azalea flowerspot	*Ovulinia azaleae*	*Heterothrips azaleae*	Through wounds produced by insects	Smith and Weiss, 1942

Table 4.9   Insect Transmission of Plant Bacterial Pathogens

Disease	Fungal pathogen	Insects associated	Name of association	Reference
3. Feeding on fungal masses				
Cereal ergot	*Claviceps purpurea*	*Sciara thomae*	Through communication	Mercier, 1911
Corn-salad root-and stem rot	*Thielaviopsis basicola*	*Scatella stagnalis*	Specialized; biological	Stanghellini et al., 1999
4. Feeding and oviposition wounds				
Cucumber anthracnose	*Colletotrichum lagenarium*	*Diabrotica undecimpunct-ata howardii*	Through feeding wounds	Liby and Ellis, 1954
Cotton boll rot	*Fusarium moniliforme*	*Anthonomus grandis* *Heliothis zeae*	Through feeding and oviposition wounds	Bagga and Laster, 1968
	Alternaria tenuis	*Lygus lineolaris*		
Sugarcane red rot	*Colletotrichum falcatum*	*Diatraea saccharalis* *Anacentrinus subnudus*	Through feeding wounds	Abbott, 1955
Apple brown rot	*Sclerotium fructigena*	*Forficula auricularia*	Through feeding injuries	Croxall et al., 1951
Cocoa dieback	*Calonectria rigidiscula*	*Sahlbergella singularis* *Distantiella theobroma*	Through feeding wounds	Crowdy, 1947 Owen, 1956
5. Feeding punctures				
Corn and sorghum downy mildew	*Sclerospora sorghi*	*Rhopalosiphon maidis* *Schizaphis graminum*	Through feeding punctures	Naqvi and Futrell, 1970
Pine burn blight	*Diplodia pinea*	*Aphrophora parallela*	Through feeding punctures	Haddow and Newman, 1942
Cotton internal boll disease	*Nematospora gossypii* *Nematospora* (spp.)	*Dysdercus* (spp.)	Through feeding punctures	Frazer, 1944 Mendes, 1956

Table 4.9 Continued

Disease	Fungal pathogen	Insects associated	Name of association	Reference
Coffee bean rot	*Nematospora* spp.	*Antestia lineaticollis*	Through feeding punctures	LePelley, 1942
Soybean yeast spot	*Nematospora corylii*	*Acrosternum hilare*	Through feeding punctures	Clarke and Wilde, 1971
Rice grain discoloration	*Nematospora corylii*	*Oebalus pugnax*	Through feeding punctures	Daugherty, and Foster, 1966
Cotton lint rot	*Nigrospora oryzae*	*Stieroptes reniformis*	Through feeding punctures	Laemonlen, 1969
6. Symbiotic association				
Apple fruit rot and canker	*Gloeosporium perennans*	*Eriosoma lanigerum*	Symbiotic	Zeller and Childs, 1925
Oak wilt	*Ceratocystis fagacearum*	Many species belonging to Nitidulidae, Scolytidae, and Droso-philadae	Required for spermatization and transmission	Dorsey and Leach, 1956
Dutch elm	*Ceratostomella ulmi*	*Hylurgopinus rufipes* *Scolytus multistriatus*	Required for transmission	Hoffman and Moses, 1940

SUMMARY

The primary method of dissemination/transmission may vary, depending on the nature of the pathogens. However, all of them may be transmitted through infected seeds and seed materials. Fungal and bacterial pathogens are largely disseminated by wind, water, and soil, whereas viruses depend on arthropod vectors and soil-borne nematodes or fungi for their spread, in addition to their transmission through mechanical inoculation, grafting/budding, and parasitic dodders. The methods of transmission or dissemination of pathogens, as such, may not be useful for differentiating them. However, the specificity of the relationship of the pathogens with their vectors may indicate the identity of pathogens in certain cases. Studies on the molecular basis of interaction between plant viruses and their vectors provide convincing evidence for the phenomenon of vector specificity.

Various modes of transmission of plant viruses, physical properties in expressed plant sap, diagnostic hosts, and transmission characteristics associated with different virus-vector interactions as the basis of virus identification are described. The usefulness of special techniques for the detection of bacterial and fungal pathogens in seeds and soil is highlighted. The mechanical/biological relationships of pathogens with their vectors are discussed.

5
Cross-Protection

When a plant already infected with one strain of a virus is inoculated with another strain of the same virus, symptoms due to the challenging strain usually fail to appear. Tobaccos inoculated with a strain of tobacco mosaic virus (McKinney, 1929) or potato virus X (Salaman, 1933) did not develop any additional symptoms when inoculated later with another strain of the respective virus. This phenomenon, known as cross-protection, is one of the biological properties of the viruses used to establish relationships among viruses and their strains.

The degree of cross-protection, as a measure of relatedness of strains, may be determined quantitatively by inoculating a systemically invading strain and local lesion-forming strain on an appropriate host plant species. Thus, relationships between tobacco mosaic virus strains can be assessed by using *Nicotiana sylvestris,* which is systemically infected by type strain, and local lesions are induced by aucuba mosaic strain. The half-leaf method of testing may be followed (Kunkel, 1934). The type strain is inoculated in one-half of the leaf and 2–3 days later the whole leaf is challenge-inoculated with the local lesion-forming strain. The half-leaf inoculated earlier with the systemic strain will develop very few lesions as compared to the unprotected opposite half leaf.

When the systemic spread of the virus is slow, an indicator local-lesion-forming strain may be employed, as in the case of cucumber mosaic virus (CMV) for the identification of its strains. Very young plants of *Zinnia elegans* are inoculated with systemic strains of CMV and 12–14 days later the leaves are challenged with local-lesion-forming strains. Relationships between two systemic strains may be established, if the macroscopic symptoms caused by them are quite distinct. Cross-protection between strains of potato virus X (PVX) may be demonstrated by inoculating the mild systemic strain on *Datura stramonium* plants followed by severe strains causing necrotic mottle mosaic symptoms (Loebenstein, 1972).

Cross-protection may be observed in persistent viruses transmitted by vectors, but not by mechanical inoculation. Potato plants inoculated with a mild strain of potato leaf roll virus (PLRV) are protected against severe strains of PLRV introduced by *Myzus persicae* (Harrison, 1958).

When the external symptoms induced by strains of a virus are not markedly different, differences in cytopathic effects caused by strains of the virus may form the basis to distinguish them. Bean yellow mosaic virus forms intranuclear inclusions in the epidermal cells of leaves of *Vicia faba,* and the inclusions can be readily seen by staining with trypan blue. The occurrence and size of the inclusions differ, depending on the strain of bean yellow mosaic virus. If plants are already infected by one strain of the virus, formation of inclusion by another strain is inhibited, indicating the cross-protection offered by the challenged strain against the challenging strain. On the other hand, such protection is not afforded to other unrelated viruses affecting bean (Mueller and Koenig, 1965).

Although cross-protection against strains of many plant viruses has been reported, some viruses do not exhibit such protection against their strains. Strains of sugar beet curly top virus do not afford protection in either sugar beet or water pimpernel *(Samolus parviflorus)* (Bennett, 1955). Tobacco (potato) veinal necrosis virus infection does not offer protection to tobacco or potato plants against serologically related strains (Klinowski and Schmelzer, 1960). Only some strains of PVY, not all strains, can protect plants from other severe strains of the virus (Loebenstein, 1972). None of the strains of tobacco streak virus is able to protect tobacco against other strains (Fulton, 1978).

Interference between strains of a viroid and also between unrelated viroids has been observed (Niblett et al., 1978). Tomato plants inoculated with a very mild strain of potato spindle tuber viroid (PSTVd) were protected against its severe strain when challenged 2 weeks after inoculation with mild strain of PSTVd (Fernow, 1967). When both mild and severe strains were inoculated simultaneously, the severe strain suppressed the development of mild strain, though the inoculum contained very high concentrations (100-fold) of the mild strain (Branch et al., 1988). Interference between a mild exocortis agent (CVd-IIa) and the cachexia agent (Cvd-IIb) has been observed (Semancik et al., 1992). The mild isolate CVd-IIa could interfere with the replication and/or accumulation of the severe cachexia agent CVd-IIb in citron *(Citrus medica)*, indicating the possibility of employing CVd-IIa for the control of cachexia in commercial plantings.

The degree of cross-protection provided by the challenged strain to susceptible plants depends on the relatedness of the strains involved, the protection being greater with more closely related strains. Thus, cross-protection as a practical disease management strategy has been successfully employed in the case of some virus diseases affecting perennial crops. In the case of citrus tristeza disease, commercial exploitation of cross-protection has been achieved in several countries (Costa and Muller, 1980; Rocha-Pena et al., 1995; Ieki et al., 1997). A combination of a mild strain and benign satellite RNA, when used for preinoculation of susceptible plants, provided effective protection to pepper and melons against two severe strains of cucumber mosaic cucumovirus in both glasshouse and field experiments (Montasser et al., 1998).

In sugar beet plants a high degree of protection offered by beet soil-borne mosaic virus (BSBMV) was observed. Reciprocal cross-protection between these two closely related viruses was also seen. The interval between the time of infection by the protecting virus and the time of infection by challenging virus appears to be a critical factor determining the degree of cross-protection. The longer the interval, the greater will be the effectiveness of protection. With longer interval, the protecting virus can reach greater concentration resulting in a higher degree of protection. The cross-protection affected the accumulation of the coat protein of the challenging virus resulting in the possible inactivation of the RNA that could be detected by reverse transcription–polymerase chain reaction (RT-PCR) assay indicating that the accumulation of coat protein of protecting strain or virus is essential for the development of cross-protection (Mahmood and Rush, 1999).

In addition, the protecting strain also has to be present in sufficient concentration in the plants to offer effective cross-protection. This is revealed by the study of Wang and Gonsalves (1999) on zuccini yellow msoaic virus (ZYMV) infecting zucchini squash *(Cucurbita pepo)*. By using the monoclonal antibody against ZYMV-TW strain, the development of cross-protection in zucchini squash inoculated with the mild strain was studied. In cross-protected plants that did not exhibit severe symptoms, only a low concentration of ZYMV-TW (challenging) strain was detected by enzyme-linked immunosorbent assay (ELISA). On the other hand, when cross-protection was not effective, ZYMV-TW strain rapidly increased in concentration as the severe symptoms developed, while the concentration of the protecting strain significantly decreased.

The avirulent and virulent strains of potato X potexirus (PVX) were used to characterize the viral elicitor involved in the hypersensitive resistance in potato conferred by *Nb* gene. The *Nb* avirulence determinant was mapped in the PVX 25K-gene coding for the 25-kDa movement protein (MP). The isoleucine residue at position 6 of this protein was shown to be required for activation of the *Nb* response in potato (Malcult et al., 1999)

The results of cross-protection tests have to be interpreted cautiously, because of interference between unrelated viruses (Bos, 1970b) and absence of cross-protection between related strains of some viruses. It is desirable to use reciprocal inoculation procedures by introducing the viruses or strains sequentially into one series of plants and reversing the order of inoculation in the second series of plants. High or complete protection can be considered good evidence for a close relation between the strains or viruses tested. Absence or a low degree of protection may not give a conclusive picture, and the relationship or lack of relationship between the viruses has to be established by other tests.

In the case of bacterial and fungal pathogens, a major limitation seems to be the lack of suitable techniques to differentiate and characterize strains of these pathogens within a morphological species. Avirulent or nonpathogenic isolates of a pathogen, have been tested for their ability to protect plants against virulent or pathogenic strains. Naturally occurring avirulent strains of *Erwinia amylovora*, causing apple fire blight disease, offered protection against virulent strains. Insertion mutants of *E. amylovora* also protected apple seedlings against virulent strains (Tharaud et al., 1993). An avirulent strain isolated from *Strelitzia reginae* protected tomato plants against virulent strains of *Ralstonia solanacearum* causing bacterial wilt at 18°–25°C, but the protection was abolished by higher temperatures (30–37°C) (Arwiyanto et al., 1996).

The avirulent strains of some fungal pathogens have been reported to protect the susceptible plants against virulent strains. The cross-protection provided by an avirulent isolate of *Fusarium oxysporum* f.sp. *cucumerinum* against virulent strains causing cucumber wilt disease was demonstrated under field conditions by Yang and Kim (1996). Reduction in wilt disease incidence from 56% to 18% in 1993, from 11% to 1% in 1994, and from 35% to 8% in 1995 was observed following prior inoculation with the avirulent strain. Simultaneous inoculation of tomato cultivars possessing resistance gene 1, with races 1 and 2 of *F. oxysporum* f.sp. *lycopersici* resulted in a high level of resistance to race 2. This type of resistance did not develop in tomato cultivars lacking the resistance gene 1. Similar protection was provided to muskmelon plants inoculated with avirulent strains of *F. oxysporum* f.sp. *melonis* against virulent races of this pathogen. The results suggest that the cross-protection provided against a virulent race may be mediated by recognition of a specific elicitor from the avirulent race by the host plant resistance gene product and by subsequent induction to initiate plant defense reactions (Huertas-Gonzalez et al., 1999). The possibility of using a nonpathogenic strain of *F. oxysporum* Fo47 to protect tomato plants against *F. oxysporum* f.sp. *lycopersici* under field conditions was reported by Fuchs et al. (1999). The avirulent isolates of *Pyricularia oryzae* and a nonpathogen *Bipolaris sorokiniana* have been shown to reduce rice blast disease significantly (Manandhar et al., 1998).

SUMMARY

Interference occurs generally between related viruses or strains, resulting in protection of plants against the challenging virus or strains. This phenomenon of cross-protection is used as a basis to establish relationships between viruses or strains. The extent of reduction in disease intensity or failure of formation of intracellular inclusions induced by challenging strains will indicate the degree of cross-protection, which may be used as a measure of relatedness between strains. The potential of using avirulent isolates/strains of bacterial and fungal pathogens to protect plants against severe strains has been demonstrated in some pathosystems. The possibility of using cross-protection for the management of crop diseases is indicated.

6
Chemodiagnostic Methods

6.1 DETECTION OF PLANT VIRUSES

Different kinds of chemicals have been used to distinguish plants infected by viruses and phytoplasmas. Lindner (1961) has reviewed the various chemical tests that have been used for the diagnosis of virus infection in plants. These tests depend on the reaction between the chemicals used for detection and the compounds present in infected plants, which results in the development of a recognizable color either in the extracts or in tissues. These tests can be grouped according to the nature of the compounds involved.

6.1.1 Protein Test

Virus-infected plants have protein profiles different from those of healthy plants. Accumulation of free amino acids and amides in infected plants has been reported by many workers (Diener, 1963; John, 1963; Narayanasamy and Ramakrishnan, 1966). Potato tubers infected by leaf roll virus had two- to three-fold higher concentrations of glutamine than comparable healthy tubers (Allison, 1953). On the basis of the higher glutamine content of diseased tubers, Cornuet (1953) developed a color test using Nessler's reagent to distinguish diseased tubers. Chiu et al. (1958) reported that apple trees infected by rough skin disease could be identified by a test based on higher arginine content. Such a test is, however, not used because of the inconsistent results obtained.

6.1.2 Carbohydrate Test

The starch iodine lesion test was developed by Holmes (1931) to detect the diffused lesions formed by potato virus X and tomato spotted wilt virus. This test is based on the reduced photosynthetic activity and poor translocation of starch from infected tissues. The infected leaves are decolorized by hot alcohol and then dipped into iodine in potassium iodine solution. This test was modified and used for the detection of rice tungro virus infection. The rice leaves are excised before sunrise, and the cut ends of the leaves are dipped into iodine in potassium iodide solution. Dark blue streaks appear in the infected leaves (IRRI, 1983). The iodine test, though simple, cannot be reliably applied for the detection of rice tungro virus infection under varied conditions (Narayanasamy, 1989).

The accumulation of sugars in different tissues of virus-infected plants has been the basis of detection of virus infection in certain host plants. Starch accumulates after citrus tristeza virus infection in the parenchyma above the bud union of sweet orange grafted on sour orange root stock. Applying 3% alcohol solution of iodine after scraping the bark slightly at the bud union helps to identify infected plants (Bitancourt, 1944). Mature potato leaves infected by leaf roll virus contain higher concentrations of glucose, which reacts with aniline phthalate, producing a distinctive blackish brown color (Martin, 1954).

The presence of pentoses in cells of tissues of virus-infected plants is revealed by the color formed by reaction with phloroglucinol and concentrated HCl. Free hand sections of stem or petiole are immersed in a 1% alcoholic solution of phloroglucinol for 1 min followed by treatment with the concentrated HCl until the xylem stains red. In healthy tissue xylem cells alone are stained, whereas the phloem cells, in addition to xylem, turn pinkish in infected plants. These tests have been found to be useful in detecting potato leaf roll infection in potato, exocortis infection in *Poncirus trifoliata,* and bud necrosis virus infection in groundnut (Childs et al., 1958; Lindner, 1961; Narayanasamy and Natarajan, 1974). The accumulation of the intermediary compound 4-methyl-*d*-glucuronic acid in leaves of grapevine infected by leaf roll virus could be detected by paper chromatography in 625 infected plants but not in 700 healthy plants or in plants infected by other diseases, indicating the specific nature of the test (Ochs, 1960). The presence of two unidentified chemical compounds in citrus leaves infected by decline disease could be detected by paper chromatography. The abnormal compounds were detected prior to the appearance of virus symptoms, making this technique useful for disease diagnosis (Newhall, 1975).

6.1.3 Enzyme Test

Stimulation of oxidases in virus-infected plants is frequently observed (Diener, 1963). A solution of 2,6-dichlorophenol indophenol is decolorized with sodium bisulfite and the extracts from diseased and healthy leaves are added. The blue color reappears much earlier in tubes containing the extracts of diseased plants. Symptomless infection of dahlia mosaic virus infection in dahlia could be detected with a high degree of accuracy by this technique. Tests with potato shoots infected by potato virus Y had 95% agreement with serological tests on 1000 tubers (Lindner, 1961).

Rapid reduction of the dye 2,3,5-triphenyl tetrazolium chloride by the virus-infected tissues forms the basis for identification of plants infected by southern bean mosaic virus, common bean mosaic virus, bean pod mottle virus, sugar beet yellows and curly top viruses, and tobacco mosaic virus (Beal et al., 1955) and banana bunchy top and infectious chlorosis viruses (Summanwar and Mazama, 1982). Association of greater specific activity of acid phosphatase with clover yellow mosaic virus infection was reported by Tu (1976). The infected leaves and root nodules exhibited a 100% more acid phosphatase activity than healthy controls.

6.1.4 Polyphenol Test

Virus infection is known to increase the polyphenol contents in infected plants. The test developed by Lindner et al. (1950) consists of heating the leaf discs in acidulated alcohol to give a red color. Whole leaves are tested by heating in alcohol containing formaldehyde and then in sodium hydroxide. The polyphenols remain fixed in leaves, giving a blue color. The color that develops may not, however, be specific for any particular substance.

6.1.5 Nucleic Acid Test

The ultraviolet (UV) absorption spectra of leaf extracts form the basis of differentiating virus-infected plants. Maximum absorption occurs at 260 nm with leaf extracts of plants infected by mosaic viruses, whereas extracts of tissues infected by yellows type disease have the peak at 280 nm (Lindner, 1961). After removal of chlorophylls by heating in alcohol, leaf discs are extracted with acidulated alcohol. The absorption curves are prepared for the extracts. This technique was employed for rapid assay of tobacco mosaic virus (TMV) in tomato and quantitative assay of stone fruit ringspot virus in cucumber and for detection of TMV and potato virus X in tobacco. Kimura and Black (1972) reported that UV absorbance of extracted RNA was correlated with the counts of purified clover wound tumor virus particles determined by electron microscopy.

6.1.6 Fluorescence Test

Fluorescence associated with the extracts of certain plants infected by viruses seems to have diagnostic value, and this property has been used to detect carnation mosaic virus infection. A portion of the vegetative shoot in distilled water is autoclaved for 45 min at 1.06 kg/cm^2 pressure. The extract is then mixed with n-butanol and a drop of ammonium hydroxide is added to the mixture. A light pink color appears between the water and *n*-butanol phases when viewed under UV light. Extracts from healthy shoots do not show any fluorescence. When the results of the fluorescence and infectivity tests were compared, it was observed that the fluorescence test detected infection in 35,670 plants, whereas 35,678 plants were found infected out of a total of 35,763 plants tested over a period of 3 years (Thomas et al., 1951).

The possibility of using a remote sensing infrared thermometer for previsual diagnosis of citrus young tree decline was tested by Edwards and Ducharme (1974). It was observed that the temperature of trees in early decline was virtually identical to that of healthy trees. They concluded that tree remote temperature measured within the spectral range of 6.5–20 μm was not a reliable indicator for making these diagnoses.

6.1.7 Hastening Symptom Expression by Chemicals

A method for rapid in vitro indexing of grapevine leafroll virus was developed by Tanne et al. (1996). Diseased explants placed in Murashige-Skoog medium poorly expressed the symptoms of the disease. On the other hand, when the explants were placed under mild stress condition, using sorbitol (4%), the symptoms appeared within 4–8 weeks as against 18–30 months required for conventional indexing procedure. Under sorbital stress, symptom severity was accentuated, enabling visual detection of infection easier.

6.2 DETECTION OF PHYTOPLASMAS

6.2.1 Histochemical Methods

Use of Dienes's stain for the detection of phytoplasmal infection was first reported by Deeley et al. (1979). The presence of phytoplasmas in different plant species was detected by using Dienes's stain: rice yellow dwarf (Srinivasan, 1982; Reddy, 1986; Rao, 1988), brinjal little leaf and sandal spike (Srinivasan, 1982), and coconut root (wilt) (Solomon and Govindan Kutty, 1991). Dark blue color is retained in the phloem cells of phytoplasma-infected plants, indicating the positive reaction.

The presence of *Spiroplasma citri,* phytoplasmas causing aster yellows (AY), pear decline (PD), and tomato big bud (TB); or clover club leaf (CCL) agent could be observed in the sieve elements of infected plants by treating the midribs with macerating enzymes (cellulase and macerozyme) and then separating the sieve elements. The helical motile spiroplasma, pleomorphic phytoplasmas and slender rod-shaped CCL agent could be visualized under dark field microscope (Lee and Davis, 1983) (Figs. 6.1A and B and 6.2). This simple technique is useful to study morphological and other properties of viable cells of these pathogens present in the sieve elements.

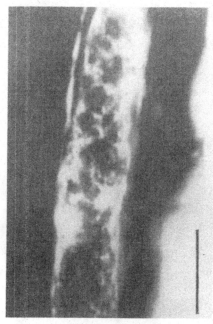

A   B

Figure 6.1 A, Typical cells of *Spiroplasma citri* in infected sieve element of *Catharanthus roseus;* bar represents 10 μm; B, single and clumped phytoplasma released from osmotically shocked sieve tube elements of pear decline infected plant; bar represents 10 μm (American Phytopathological Society, Minnesota, USA). (Courtesy of Lee and Davis, 1983.)

Hiruki and da Rocha (1986) developed a histochemical method for diagnosing phytoplasmal infections in *Catharanthus roseus* by using a fluorescent DNA binding stain, 4,6-diamidino-2-phenyl indole-2 HCl (DAPI). The rapid compression technique developed by Dale (1988) can be employed for the rapid detection of phytoplasmas Excised leaf midrib sections are firmly compressed and crushed by a small spatula. The vascular system is then removed with fine forceps and fixed in Karnovsky's fixative for 20 min. The vascular tissue, after rinsing in phosphate buffer for 5 min is placed into DAPI solution for 3–5 min. After staining, the tissue is mounted in a drop of DAPI solution on a glass slide and examined under a fluorescence microscope. By this technique AY phytoplasma in aster, blue berry stunt phytoplasma in periwinkle, and *Spiroplasma citri* in periwinkle can be detected.

Using DAPI, infection of phytoplasmas in several plant species has been detected. Sinclair et al. (1989) found the DAPI fluorescence test superior to Dienes's stain for histological detection of aster yellows phytoplasma in lilac witches'-broom. In different plant species phytoplasmas also could be diagnosed by using

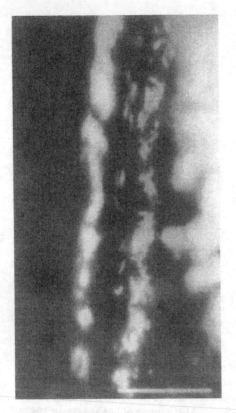

Figure 6.2   Sieve elements from clover club leaf-infected *Catharanthus roseus* plants containing rod-shaped bacterial cells. bar represents 10 µm (American Phytopathological Society, Minnesota, USA). (Courtesy of Lee and Davis, 1983.)

the DAPI fluorescence test (Hibben et al., 1991). Sinclair et al. (1992) reported that the DAPI fluorescence test detected ash yellows phytoplasma in white ash trees as consistently as the DNA probes in the dot hybidization technique. Griffiths et al. (1994) used the DAPI fluorescence test to detect phytoplasmas in six species of ash and lilac in 13 locations in the United States. The tissue-culture-generated mulberry plants grown under greenhouse conditions for 3 years were tested by DAPI fluorescence test for the presence of the mulberry dwarf phytoplasma. The regenerated mulberry plants (70%–90%) were found to be apparently free of the phytoplasma during the period of testing and similar results were obtained by polymerase chain reaction (PCR) (Dai et al., 1997). The histochemical methods are rapid and relatively sensitive but are nonspecific in certain cases. They can be, however, used for preliminary diagnosis of phytoplasmal infection, when good-quality antiserum is not available.

6.3 DETECTION OF FUNGAL PATHOGENS

Fungal pathogens present in plant tissues are examined directly under microscopes either by making sections or after isolating them in appropriate media. In most cases, conventional staining procedures or selective media for isolation may be sufficient and no special histochemical methods are necessary. Moreover, it is possible for sufficiently trained plant pathologists to make a rapid presumptive identification of the pathogens from characteristic disease symptoms. However, some fungal pathogens may not induce distinctive symptoms, and it may be difficult to isolate and identify them in culture. Some fungi may grow very slowly in culture, taking weeks or even months to produce spore-bearing structures required for identification. Under such conditions alternative reliable methods are required for detection and identification of fungal pathogens.

6.3.1 Use of Selective Media

The growth and development of a fungal pathogen may be specifically encouraged by providing a selective medium. Cultural differences and enhanced micromorphological features were observed when Czapek solution agar containing 20% saccharose was used to grow *Fusarium moniliforme, F. proliferatum,* and *F. subglutinans. Gibberella fujikuroi* could be readily distinguished from the other two species on the basis of the differences in colony color and texture. Lowering of the pH of the medium from 7.7 to 4.4 intensified these differences without affecting micromorphological characteristics adversely (Clear and Patrick, 1992). Duffy and Weller (1994) developed a semiselective medium (R-PDA) consisting of dilute potato-dextrose agar amended with 100 μm/ml of rifampicin and 10 μg/ml of tolclofosmethyl, useful for isolation and diagnosis of the presence of *Gaeumannomyces graminis* var. *tritici,* which was able to alter the color of rifampicin in R-PDA from orange to purple in about 24 hr. It was found that R-PDA was more effective in isolating *G. graminis* var. *tritici* than the semiselective medium for *Gaeumannomyces graminis tritici* (SM-GGT3), another selective medium used earlier.

A selective medium containing malt extract agar (33.6 g/L), tannic acid (3000 μg/ml), benodanil (50 μg/ml), and tridemorph (0.5 μg/ml) was more effective than the simple malt extract agar medium for isolation of *Lasiodiplodia theobromae* from *Pinus elliottii* seeds, woody tissue, and infested soil. This selective medium suppressed all other fungi present along with the pathogen (Cilliers et al., 1994). Detection and quantification of ascospores of *Sclerotinia sclerotiorum* causing collar rot disease of tobacco seedlings were possible by placing petriplates containing a semiselective medium inside and outside commercial greenhouses. The level of pathogen inoculum could be determined by assessing the amount of ascospores trapped in the selective medium and this parameter may be used to develop a forecasting system for this disease (Gutierrez and Shew, 1998).

The use of tannic acid, as a marker, for easy identification of *Rhizoctonia solani* was suggested by Hsieh et al. (1996). A simple differential medium consisting of water agar containing tannic acid (300 ppm) can be used for isolation and quantification of mycelium of *R. solani*. The color of the agar plates or Czapeck's liquid medium containing tannic acid changes from light yellow to dark brown after growth of the pathogen. There is a good positive correlation between the amount of mycelial growth and absorption values at 363 nm.

6.3.2 Biochemical Test

The presence and viability of spores of fungal pathogens on seeds can be determined by fluorescein diacetate assay (FDA). Lipase activity is detected consistently in the extracts of viable teliospores of *Tilletia controversa,* but not in extracts of autoclaved spores. Detection of lipase activity is consistent when 4-methyl-umbelliferyl-palmitate is used as a substrate. Fluorescein diacetate assay can be used to get the results rapidly, and the time required to assess the viability of a teliospore population may be reduced to 1 hr from the 2 months needed for other methods of assay (Chastain and King, 1990).

Oliver et al. (1993) produced strains of *Cladosporium fulvum,* which infects tomatoes, and *Leptosphaeria maculans,* pathogen of brassica crops, that constitutively expressed β-glucuronidase, the activity of which was used to detect histochemically the presence of fungal hyphae in host plant tissues. In addition, the β-glucuronidase activity of *C. fulvum* could be used to quantify fungal biomass in the cotyledons of infected seedlings. The enzyme β-glucuronidase is stable, and its activity can be quantified by fluorimetric assays by using the substrate 4-methyl-umbelliferyl β-D-glucuronide (MUG). It can also be detected histochemically by using 5-bromo-4-chloro-3-indolyl-β-D-glucuronide as the substrate. Yates et al. (1999) developed a method employing visual markers that can be recognized by histochemical staining for the detection of endophytes such as *Fusarium moniliforme*. Three strains of *F. moniliforme* were transformed with a plasmid PHPG, containing the *gusA* reporter gene encoding β-glucuronidase (GUS) and the *hph* gene for hygromycin resistance as a selectable marker. As this pathogen is present in plants and grains, its detection in plant tissue is of great importance to animal and human health. This approach is useful, because this enzyme is essentially absent from all fungi and plants tested.

Wheat loose smut pathogen *(Ustilago nuda)* can be detected in seed embryos by extracting them in 5% NaOH and concentrating them. The embryos are stained with trypan blue to reveal the presence of fungal mycelia (Khanzada and Mathur, 1988). This technique was modified for the detection of *U. tritici* and *U.*

nuda in wheat and barley seeds. The seeds are incubated in NaOH (10%) containing trypan blue (added at the rate of 1 g/L) for 12–16 hr. The embryos are then separated from endosperm by passing the seeds through a series of sieves with different mesh size and boiled in alkali solution for 15 min. The embryos are washed and then boiled in acetic acid or lactic acid (45–50%) for 1 min. The embryos are examined under microscope for the presence of stained mycelium (Feodorova, 1997). The presence of intercellular mycelium and haustoria in smut-infected sugarcane tissues could be visualized by staining with trypan blue. This test is useful for rapid diagnosis of smut infection under field conditions (Sinha et al., 1982; Padmanabhan et al., 1995). Another staining technique developed by Hood and Shaw (1996) has been reported to be useful for the study of interactions between plants and fungi belonging to Deuteromycotina, Ascomycotina, Basidiomycotina, and Mastigomycotina. Fresh specimens (plant tissues) are autoclaved for 15 min at 121°C in 50 ml of 1 M KOH and then rinsed in deionized water. The cleared tissues are then stained with aniline blue dye (0.05%) in 0.67 M K_2HPO_4 at pH 9.0 and examined under microscope with UV light source. A high degree of resolution and contrast between fungal hyphae and plant tissues is the clear advantage provided by this procedure.

Latent fungal infections in other plant parts such as leaves and fruits may be detected by treatment with chemicals. Detached winter wheat leaves showing no visible symptoms are washed, surface sterilized, and treated with paraquat, 0.03–0.32% active ingredient (ai). The treated leaves are then incubated in nutrient agar. Microscopic examination shows the presence of many fungal pathogens infecting wheat leaves. Treatment with paraquat also induces the development of latent infection of *Botrytis cinerea* in grapevine berries (Gindrat and Pezet, 1994). Treatment of plums with paraquat has been reported to reveal latent infection by *Monilinia fructicola,* providing the opportunity to take up corrective measures to reduce brown rot disease in mature plums (Northover and Cerkauskas, 1994). Likewise, treatment of banana fruits with paraquat (500 ppm ai) revealed latent infection by *Colletotrichum musae* much earlier compared with untreated controls (Fig. 6.3) (Rajeswari et al., 1997). Another method applicable to immature banana fruits for early detection and quantification of inoculum level of *C. musae* was developed by Bellaire et al. (2000). The banana fruits are treated with ethylene (1200 µl/L) for 24 hr at 25°C and kept at 32°C for 5 days. The lesions on treated fruits whose age is 5–6 weeks after the emergence of inflorescence become visible earlier. The concentration of CO_2 has to be maintained at optimal level, since high concentrations of CO_2 inhibit the development of lesions.

Doster and Michailides (1998) reported that bright greenish yellow fluorescence (BGYF) of dried figs *(Ficus carica)* under longwave was a diagnostic characteristic for the colonization of dried fig, by *Aspergillus* spp. BGYF observed in naturally infected fig was associated with decay caused by *A. flavus* and *A. parasiticus* capable of producing aflatoxin and also by *A. tamarii* and *A. albiacens,*

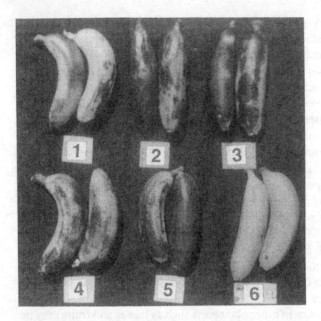

Figure 6.3   Detection of latent infection of *Colletotrichum musae* in banana fruits using different concentrations of paraquat: 1, 100 ppm; 2, 500 ppm; 3, 1000 ppm; 4, 1500 ppm; 5, 2000 ppm; 6, Untreated control. (Courtesy of Rajeswari and Palaniswami.)

which do not produce aflatoxin. Assessment of BGYF may help to remove afla-toxin-contaminated figs under certain specific conditions existing in California.

The effectiveness of an electronic nose in recognizing the odor of wheat grains infected by common bunt pathogen *(Tilletia caries)* was assessed in comparison with the ability of a panel of grain assessors. The infected samples were separated from uninfected samples with equal efficiency by both the panel and electronic nose, when the samples were artificially inoculated. The electronic nose was much more efficient than the panel, when the samples from farmers were tested. The electronic nose appears to sense a different characteristic not related to common bunt odor (Börjesson and Johnson, 1998).

Recognition of infection of plants by soil-borne pathogens, before symptom expression, has been found to be very difficult. The efficacy of visual and infrared assessment of root colonization of apple trees by *Phymatotrichopsis omnivora* was compared. Watson et al. (2000) reported that the differences between the in-frared readings of canopy temperature and air temperature could be used as the basis for predicting infection of asymptomatic, infected apple trees, since these differences were significant ($p < 0.01$). The asymptomatic infected trees showed extensive taproot decay and infection of lateral roots that could not be recognized by visual examination. Infrared technique seems to have the potential for early identification of the infected trees leading to formulation of suitable management strategies.

6.3.3 Isozyme Analysis

The usefulness of isozyme electrophoresis for the detection and identification of fungal pathogens such as *Peronosclerospora* spp. (Bonde et al., 1984), *Phytophthora cinnamomi* (Old et al., 1984, 1988), *Fusarium oxysporum* (Bosland and Williams, 1987), and *Trichoderma* spp. (Stasz et al., 1989) has been assessed. The enzymes are separated by electrophoresis in a horizontal starch gel. Isozyme patterns are recorded according to their relative mobility, and each band is considered as an allele of a specific locus. The bands are then labeled alphabetically from the slowest to the fastest.

Bonde et al. (1989) reported that isozyme patterns could be used to distinguish between teliospores produced by *Tilletia indica* and *T. barclayana* present in stored grains, storage facilities, and transportation vehicles. The isozyme patterns were determined by horizontal starch-gel electrophoresis and staining. Oudemans and Coffey (1991a, 1991b) suggested the use of isozyme analyses for classifying 12 papillate *Phytophthora* species. Oudemans and Coffey (1991a) reported that in terms of isozyme analysis *P. cambivora, P. cinnamomi*, and *P. cactorum* were clearly separated and each species could be further subdivided into electrophoretic types (ETs). Three enzymes, viz. phosphoglucose isomerase, malate dehydrogenase, and lactate dehydrogenase, when fractionated by cellulose acetate electrophoresis, were found to have diagnostic potential, permitting clear differentiation of these three species. Further studies by Oudemans and Coffey (1991b) to examine intraspecific diversity and interspecific relatedness of different papilate species of *Phytophthora* indicated interspecific ralationships which cannot be predicted on the basis of morphological comparisons alone. *Phytophthora medii* and *P. botryosa* clustered together, indicating a very close genetic relatedness; *P. katsurae* and *P. heveae* also formed a single cluster; *P. capsici* and *P. citrophthora* formed another distinct cluster.

Similar approach was made by Sippell and Hall (1995) to differentiate highly virulent and weakly virulent strains of *Leptosphaeria maculans* causing black leg or stem canker disease in canola *(Brassica napus)*. Of the 92 isolates of *L. maculans* collected from six countries, 68 isolates contained a single isozyme of glucose phosphate isomerase (GPI) that moved 70 mm in starch gel after 11 hr of electrophoresis. The isozyme of GPI present in other isolates could move 65 mm. Thus the isolates of *L. maculans* were divided into two electrophoresis types (ET) 1 or 2 reflecting the fast or slow movement of isozymes of GPI. The highly virulent strains had fast isozyme and belonged to ET1, whereas the weakly virulent strains had slow isozyme and they were placed in ET2. Isozyme analysis may be useful to study the intraspecific population diversity in fungal pathogens as in the case of *Colletotrichum gloeosporioides,* which infects a variety of plant species. The isozymes of nicotinamide adenine dinucleotide dehydrogenase (NAOH) and diaphorase (DIA) yielded the greatest number of electrophoretic phenotypes that clustered on the basis of host origin. Three major groups (I, II, and III) and four subgroups (IA, IB, IIIA, and IIIB) were delineated with *C. gloeosporioides* (Kaufmann and Weidemann, 1996).

Using the cellulose-acetate electrophoresis (CAE), excellent resolution of allozyme genotypes of *Phytophthora infestans* causing potato late blight disease, at two loci glucose-6-phosphate isomerase (GPI) and peptidase (Pep) could be achieved. Moreover, the cellulose-acetate system is more rapid, requiring only 15–20 min as against 16–18 hr required for starch gels. The CAE system offers a rapid and accurate method of predicting mating types and metalaxyl sensitivities of *P. infestans* isolates present within fields (Goodwin et al., 1995). Based on a comparison of allozyme banding patterns at two loci GPI and Pep with markers for mating type, metalaxyl sensitivities, and cultural morphology, 726 isolates of *P. infestans* collected in Canada were grouped into eight genotypes. Seven of these eight genotypes could be distinguished by differences in the banding patterns for the allozymes of the GPI locus alone. There were similarities between five of these genotypes and the genotypes [US-1, US-6, US-7, US-8, and g11 (or US-11)] recognized in the United States. Correlations between allozyme banding patterns and mating type, metalaxyl sensitivity, and cultural characteristics were also observed (Peteers et al., 1999). Eighty-five isolates of *P. infestans* collected from tomato and potato fields in North Carolina, were grouped into four allozyme genotypes at GPI and Pep loci. These isolates were predominantly found to belong to US-7 genotype (55 isolates) and US-8 genotype (24 isolates) (Fraser et al., 1999).

The GPI electrophoresis on starch gels performed directly on leaf lesion extracts has been shown to be a useful and reliable method to identify *Leptosphaeria maculans* infecting oil seed rape *(Brassica napus)* and to differentiate from *Pseudocercosporella capsellae,* Four different electrophoretic patterns, ET1, ET2, ET3, and ET4, of allozymes were recognized. Highly virulent isolates (group A) of *L. maculans,* had ET1 pattern, whereas group B isolates (weakly virulent) showed ET2 pattern. ET3 allozyme also was recovered from a few typical and atypical leaf lesions caused by *L. maculans.* But the fastest ET4 allozyme could be specifically detected in lesions induced by *P. capsellae* only (Braun et al., 1997). When 27 isolates of *Fusarium oxysporum* were subjected to isozyme analysis, polymorphisms in five enzymes were detected leading to the identification of 26 different electrophoretic groups (Paavanen-Huhtala et al., 1999).

6.3.4   Cell Wall Glycoprotein Analysis

Snow rot disease of winter cereals is caused by *Pythium graminicola, P. iwayamai, P. okanoganense, P. paddicum, P. vanterpoolii,* and *P. volutum.* The cell wall glycoproteins of these six species (about 25–40 kDa) form the major component among cell wall proteins of each species. The electrophoretic patterns of the glycoproteins, detected with Coomassie brilliant blue, lectin, and antibody, exhibited sufficient interspecific polymorphism and intraspecific stability to allow identification and classification of the six *Pythium* spp. that cause snow rot disease (Takenaka and Kawasaki, 1994).

6.3.5 Analysis of Enzyme Activities

The enzyme activities show significant variations in plants infected by some fungal pathogens and such changes have been used for the diagnosis of diseases caused by them. Among the activities of peroxidase (PO), polyphenoloxidase (PPO), phenylalanine-ammonia lyase (PAL), β-1,3-glucanase (Glc), superoxide dimutase (SOD), and amylase (Amy) tested, the activities of Glc and Amy were significantly higher in the leaves of egg plants infected by *Verticillium dahliae* (Kawaradani et al., 1994). Later studies by Kawaradani et al. (1998) showed that *Verticillium* wilt of eggplants (aubergines) may be diagnosed by measuring Glc activity, which is increased up to 5 times that in healthy plants, using the PNPG4 degradation method. This method, using *p*-nitrophenyl - βD-laminariantetraoside as a substrate, is easier and more precise.

Although some earlier reports indicated that *Fusarium subglutinans* could be involved with mango malformation disease, unequivocal evidence for its infection potential was demonstrated by using *F. subglutinans* transformed with the β-glucuronidase (GUS) reporter gene. The GUS activity was monitored microscopically in inoculated infected and noninoculated mango floral and vegetative buds that were removed 6–8 weeks after inoculation. The GUS-stained mycelium present in the infected mango tissue was observed after infiltration of a mixture containing X-Gluc (50 µg/ml), 0.1 M NaPO$_4$ (pH 7.0), 10 mM EDTA, and 0.5 mM each of Kferri and ferrocyanide in 0.05% Triton-X (v/v) into the tissues for 5 min. After the tissues were cleared of chlorophyll by washing twice over a period of 48 hr in chloral hydrate (120 g/100 ml), the tissues were examined under a stereomicroscope for the presence of GUS transformants. The presence of GUS-stained mycelium of *F. subglutians* in the infected tissues provided evidence for its ability to infect mango (Freeman et al., 1999).

6.4 DETECTION OF BACTERIAL PATHOGENS

Bacterial pathogens induce variable symptoms, and some diseases have long incubation periods before symptoms are expressed. It is not unusual to see close similarities between symptoms of bacterial infection and other biotic and abiotic factors. Identification of bacteria requires, as a first step, the isolation of bacteria, followed by a series of physiological or biochemical tests and a pathogenicity test. To distinguish pathovars, type of symptoms induced by the bacterial isolate in a particular host plant species/cultivar and host range of the bacterial isolate have to be studied. The identification of bacterial pathogens by these studies is very time-consuming and laborious and sometimes inconclusive.

6.4.1 Use of Selective Media

It is possible to isolate different bacterial pathogens by using specific media (Schaad, 1988). Suitable selective medium has to be identified for each bacterial species. Of the 13 media tested, Nauman et al. (1988) found that nutrient agar containing 5% saccharose (NASA), NASA + 2 ppm crystal violet, Kings agar B, and Kings agar Bf were most suitable for isolating and detecting *Pseudomonas syringae* pv. *phaseolicola* in *Phaseolus vulgaris* seeds. The pathogen could be detected at a concentration of $<10^2$ cells/ml in the water used to rinse the seeds for 6 hr at 22°C, whereas immunofluorescence and enzyme-linked immunosorbent assay (ELISA) tests could detect it only at a minimum concentration of 10^5 cells/ml.

Jansing and Rudolph (1990) developed a sensitive and quick test to determine the bean seed infection by *Pseudomonas syringae* pv. *phaseolicola*. Seeds are soaked in physiological saline solution for 20 hr at 4°–6°C. Saline solution, concentrated by centrifugation, is streaked on a semisolid selective medium which supports the growth of all strains of the pathogen. The pathovars are distinguished by determining phaseotoxin production by *Escherichia coli* bioassay. The pathogen may be detected even when 1 out of 50,000 seeds is infected by the bacterial pathogen.

Pseudomonas viridiflava causing bacterial streak and bulbrot of onion was isolated using a semiselective diagnostic agar medium (T-5) [Chapter 2, Appendix 2(iv)], in conjunction with low-temperature incubation. *P. viridiflava* was selectively recovered from field soil with reduction of 99.99% of nontarget bacteria. Incubation at 5°C still further reduced the recovery of contaminating microflora by 1000–10,000-fold (Gaitaitis et al., 1997). Another semiselective medium (PCCG) [Chapter 2, Appendix 2(iv)] has been found to be useful in detecting viable cells of *Ralstonia solanacearum* in infested soils (Ito et al., 1998). A new semiselective medium has been developed for the isolation of *Xanthomonas axonopodis (campestris) manihotis,* causing cassava bacterial blight disease, which grows slowly on general plating media and is easily overgrown by saprophytic bacteria, making the isolation of the pathogen in pure culture difficult. This new semiselective medium, Cefazolini trehalose agar (CTA) medium [Chapter 2, Appendix 2(iv)], was more efficient in the isolation of *X. axonopodis* pv. *manihotis* from infected cassava leaves and infested soil compared with starch-based semiselective medium (SXM) (Fessehaie et al., 1999).

6.4.2 Biochemical Tests

In addition to the morphological features, several other biochemical/physiological properties have to be studied for the proper identification of bacteria. The following are the important characteristics useful for identification and classification.

6.4.2.1 Properties of Bacteria in Culture

a. Size of bacterial cells.   The size of the bacteria may vary, depending on cultural conditions. However, the size measurements will be useful, if the age of culture, constituents and pH of the nutrient medium, incubation temperature, and staining method are similar, when comparisons are made between different bacterial isolates/species.

b. Location of flagella.   The presence or absence of flagella and their location/distribution in the bacterial cells are characteristic of the bacterial species. The procedure of Zettonow as modified by Stapp (1966) can be employed for staining the flagella. The flagella may be located as a single polar flagellum at one end of the bacterial cell or a group of flagella either at one end (lophotrichous) or at both ends (amphitrichous) of the bacterial cells. In some cases the flagella may be distributed all over the bacterial cell (peritrichous).

c. Gram staining.   The retention of Gram's stain by the bacterial cells is one of the important properties used for the identification of the bacteria. The Gram's complex, consisting of magnesium-ribonucleoprotein, which retains the stain is present in the gram-positive bacteria, whereas magnesium is present in the ionic form in gram-negative bacteria. Among the bacterial plant pathogens bacteria belonging to the genus *Corynebacterium* are gram-positive, whereas those in other genera are gram-negative.

d. Presence of food reserve materials in bacteria.   Volutin, fat, glycogen, and iogen are the reserve materials present in bacterial cells. Of these, volutin and fat alone have been observed in plant pathogenic bacteria. Volutin is colorless, viscous, and less refractive than the fat globules seen in bacteria. It is not stained by the fat stains and is different from glycogen in its reaction with Lugol. Presence of volutin in *Agrobacterium tumefaciens* can be observed. Fat in the bacterial cells may be seen as highly refractive globules varying in size. When stained with dimethyl paraphenyl diamine, globules are stained blue.

e. Reduction of nitrates.   Nitrates are reduced to nitrates, ammonia, and finally free nitrogen by bacterial activities. The ability to reduce nitrates accompanied by visible gas formation is another property used to classify the bacteria.

f. Hydrogen sulfide production.   Production of hydrogen sulfide in the liquid cultures of bacteria is tested by using a strip of lead acetate paper which turns brown to black after incubation.

g. Production of indole.   Production of indole in the liquid medium containing tryptophan by the activities of bacteria is determined by adding Ehrlich reagent after incubation of the culture for 1–7 days. Appearance of cherry red color indicates the production of indole.

h. Utilization of carbon and nitrogen compounds.   The ability of different species of bacteria to utilize different carbon compounds as sources of energy varies, and such differences may be useful in the identification of pathogenic bacteria. Production of acid or gas or both or neither from a certain compound by the bacteria is determined. Quantitative determinations of gas production are made by using ferment tubes with a liquid substrate. Acid production is detected by indicators such as litmus bromothymol blue and bromocresol purple in 0.4% alcoholic solutions.

    The nitrogen sources that support the multiplication of bacteria are also studied by providing different organic and inorganic sources. Different amino acids are also incorporated to find out the suitable amino acids required for the growth of the bacteria. The differences in the requirements of carbon and nitrogen sources for optimal growth of bacteria may help in the identification of bacterial species.

i. Starch hydrolysis.   The extent of diastase activity is assessed by using Lugol's iodine after incubation of the bacterial culture in medium containing starch. Phytopathogenic bacteria such as *Xanthomonas begoniae* have strong diastatic activity.

j. Lipolytic activity.   The bacterial culture is inoculated on agar medium in a petri dish, and olive oil, cotton seed oil, or castor oil is sprayed with an atomizer to produce fine droplets. The transparent oil droplets turn granular and opaque if the bacterial species has lipolytic activity. This property has been used for differentiating the bacterial species.

k. Action on litmus milk.   The litmus solution is prepared by dissolving litmus granules (80 mg) in 40% alcohol (300 ml). A mixture of skimmed milk with low butter content and 2% litmus solution is taken in sterilized test tubes, inoculated with the test bacterial culture, and incubated for 10 days at room temperature. Change of color to red indicates a positive reaction.

l. Gelatin liquefaction.   Gelatin medium consisting of beef extract (3.0 g), peptone (5.0 g), gelatin (120.0 g), and distilled water (1000 ml) is transferred to test tubes at 10 ml/tube after sterilization. The test bacterium is stab-inoculated and incubated for 5 days at 20°C. The medium is liquefied, indicating a positive reaction.

m. Ammonia production.   The test bacterium is inoculated on peptone nitrate broth, placed into test tubes, and incubated for 72 hr at room temperature. Formation of reddish brown precipitate when Nessler's reagent is added indicates the production of ammonia by the bacterial pathogen.

n. Kovocs's oxidase test (Kovocs, 1956). Place Whatman No. 1 filter paper in a sterilized petri dish and add 3–4 drops of freshly prepared 1% aqueous solution of tetramethyl paraphenylenediamine dihydrochloride at the center of the filter paper. Transfer a loopful of concentrated bacterial growth from a 24–48-hr-old culture and rub as a strip of 1 cm in length across the impregnated filter paper. The appearance of purple color indicates oxidase activity of the test bacteria.

o. Thornley's arginine dehydrolase activity. Thornley's arginine medium, containing peptone (1.0 g), NaCl (5.0 g), K_2HPO_4 (0.49 g), L-arginine HCl (10.0 g), phenyl red (0.01 g), phenol red (0.019 g), agar (3.0 g), and distilled water (1000 ml), pH 7.2, is stab-inoculated with a loopful of bacterial culture, covered with sterile molten petroleum jelly (Vaseline), and incubated for 72 hr at room temperature. A positive reaction is inferred from the appearance of red color.

p. Tyrosinase activity. Dye's medium (Appendix 2(iv)B) in slants is inoculated with the test bacterium. Tyrosine in the medium is converted into brown colored melanin by the activity of tyrosinase if the bacterial pathogen has the enzyme.

The important characteristics of the genera of plant pathogenic bacteria are presented in Table 6.1.

6.4.2.2 Properties of Bacterial Cells

a. Direct colony thin layer chromatography (Matsuyama et al., 1993a). One loopful of bacterial culture is applied directly to the origin line on silica gel thin layer chromatography (TLC) plate and dried completely. The plate is developed at 25°C for 10 min, using chloroform methanol (2:1, v/v) till the solvent moves up to 6 cm from the origin line and dries. After scraping out the bacterial cells the TLC is run in the same direction, using chloroform methanol water (60:25:4, v/v/v), for about $1\frac{1}{2}$ hr. After drying the plate, ninhydrin is sprayed and dried at 100°C for 10 min for the development of spots. The lipid profiles of various bacterial species tested show distinct differences (Fig. 6.4). *Erwinia chrysanthemi* and *E. carotovora* subsp. *carotovora* show marked differences in their lipid profiles. On the basis of lipid profile differences three major types of pseudomonads can be distinguished: The chromatograms of *P. gladioli* pv. *gladioli*, *P. glumae*, *P. plantarii*, *P. caryophylli*, and *P. cepacia* resemble each other, and they are designated Cepacia type. The *P. solanacearum* is categorized as Solanacearum type, whereas other pseudomonads are included in the Syringae type (Matsuyama and Furuya, 1993b). Striking differences are seen between the chromatograms of gram-positive (*Clavibacter* spp.) and gram-negative bacteria. The chromatograms of *Erwinia* spp. are quite distinct from those of *Xan-*

Table 6.1 Important Characteristics of Important Genera of Plant Pathogenic Bacteria

Characteristics	*Agrobacterium*	*Corynebacterium*
1. Morphology	Small rods with 1–4 peritrichous flagella	Straight or slightly curved rods with club-shaped swellings frequently
2. Gram staining	Negative	Positive
3. Motility	Motile	Nonmotile
4. Oxygen requirement	Aerobic	Generally aerobic; some may be microaerophilic or even anaerobic
5. Gas production	No visible gas production	
6. Acid production	Not detectable by litmus	Slight
7. CO_2 production	Produced in synthetic media detectable with bromothymol blue or bromocresol purple	Oxidizes glucose, producing $CO_2 + H_2O$ without visible gas production
8. Liquefaction	Liquefies gelatin very slowly or not at all	May or may not liquefy gelatin
9. Nitrogen utilization	No fixation of free nitrogen; nitrates or ammonium salts are utilized	May or may not produce nitrites from nitrates
10. Optimum temperature	25°–30°C	
11. Habitat	Soil, plant roots, and stems	Mostly pathogens of humans and domestic animals; some are plant pathogen
12. Chief symptoms	Hypertrophy, galls	Rotting of tissues
13. Important species	*A. tumafaciens* *A. rhizogenes* *A. radiobacter*	*C. sepedonicum* *C. michiganense*

thomonas spp. and *Agrobacterium tumefaciens* (Matsuyama et al., 1993a). (Fig. 6.5). The usefulness of the direct colony TLC method for rapid identification of different *Pseudomonas* spp. in ribosomal RNA (rRNA) homology group II was reported by Matsuyama (1995). By employing the high-performance liquid chromatography (HPLC) technique, two phytopathogenic bacteria, *Clavibacter* and *Erwinia,* could be rapidly differentiated. Distinct differences were also observed at species level in *Erwinia* (Matsuyama, 1995). By applying HPLC technique in conjunction with direct colony TLC method, *Burkholderia gladioli* isolates, which showed a distinct peak at Rt 6.2 min, could be differentiated from the isolates of *B. glumae* and *B. plantarii,* which did not exhibit such a peak (Matsuyama et al., 1998).

Table 6.1 Continued

Characteristics	Erwinia	Pseudomonas	Xanthomonas
1. Morphology	Rod-shaped	Rods with monotrichous, lophotrichous flagella; presence of diffusible fluorescent pigments of greenish, bluish pink, yellowish colors or nondiffusible bright red or yellow pigments	Rods usually monotrichous with yellow water-insoluble pigments
2. Gram's staining	Negative	Negative	Negative
3. Motility	Motile	Motile	Motile
4. Oxygen requirements		Aerobic	Aerobic
5. Gas production	Visible gas may or may not be produced		Gas produced by some species
6. Acid production	Acid produced	Acid produced	Acid produced from sugars
7. Liquefaction	May or may not liquefy gelatin		Some species may liquefy pectin
8. Utilization of nitrogen	May or may not produced nitrites from nitrates	Nitrates reduced to nitrites, ammonia, or free nitrogen	
9. Habitat	Plants	Soil, water, pathogens of plants and animals	Many plant pathogens
10. Chief symptoms	Dry necrosis galls, wilts, and soft rots	Wilts	Necroses
12. Important plant pathogens	Erwinia amylovora E. cartovora carotovora	P. solanacearum P. phaseolicola P. pisi P. tabaci	X. campestris pv. oryzae X. c. malvacearum X. juglandis

Source: Stapp (1966).

The direct colony TLC method was modified to reduce the time for drying and to avoid development of chromatograms in two different solvent systems. The lipid extract obtained using a mixture of chloroform and methanol (2:1, v/v) is spotted on the origin of silica gel TLC plate and dried well. The plate is developed with chloroform-methanol-0.2% calcium chloride (55:35 : 8, v/v/v) for 1 hr at 25°C, followed by spraying of ninhydrin and successive heating at 100°C for 10

F -

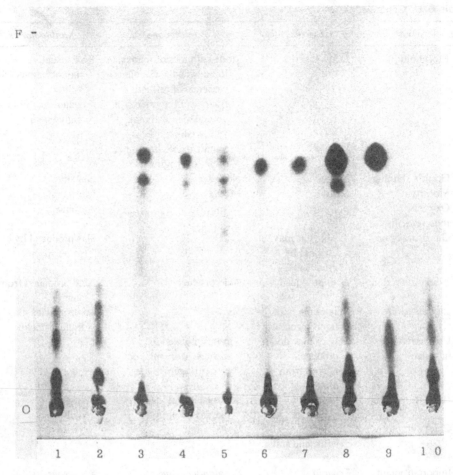

O

1   2   3   4   5   6   7   8   9   1 0

Figure 6.4   Thin layer chromatogram of lipids from phytopathogenic bacteria: 1, *Clavibacter michiganensis* subsp. *sepedonicus* 1; 2, *Clavibacter* sp. 5215; 3, *Agrobacterium tumefaciens* 1sk; 4, *A. tumefaciens* Uh; 5, *A. tumefaciens* Ku7411; 6, *Xanthomonas campestris* pv. *citri* Ku 7512-2, N 6829-1-3; 7, *X. campestris* pv. *citri* N 6829-1-3; 8, *Erwinia carotovora* subsp. *carotovora* 489-4; 9, *E. chrysanthemi* pv. *chrysanthemi* Ku8601-L1; 10, *Clavibacter michiganensis* subsp. *michigannesis* N 6206. (Courtesy of Matsuyama et al., 1993b.)

min. The pseudomonads belonging to rRNA homology group II could be differentiated based on the Rf values (0.42–0.83). *B. caryophylli*, *B. cepacia*, *B. gladioli*, *B. glumae*, *B. plantarii*, and *B. vandii* could be differentiated at species level and they exhibited distinct differences from *Ralstonia solanacearum*, *B. andropogonis*, and *Herbaspirillum rubrisubalbicens*. The lipid profile of *H. rubrisubalbicens* was quite unique (Khan and Matsuyama, 1998).

The chloroform-methanol (CM) extracts of bacterial lipids exhibited characteristic differences in the states of bacterial cells of different genera. Whereas suspensions of some bacteria were homogeneous, cells of *B. gladioli* agglutinated and remained in a fibrous state. Whitish-blue, purple, yellow, red, or pink fluorescence was observed upon UV irradiation depending on bacterial species. The bluish CM solution containing $CoCl_2$ (0.6%) in which *B. gladioli* cells were suspended changed to pinkish color (Matsuyama, 1998). This diagnostic test appears to be simple, but its reliability has to be assessed by testing a wide range of bacterial species.

The fatty acid composition of phytopathogenic bacteria has been demonstrated as a basis for their identification. The dendrogram based on the fatty acid composition showed that all pathovars of *Pseudomonas syringae, P. viridiflava,* and the pear and radish strains were closely related and lauric acid and palmitoleic acid were their major fatty acids. Physiological and biochemical tests also provided similar results indicating that strains of *P. syringae* infected pear blossoms, while strains of *P. viridiflava* caused rotting of radish leaves (Khan et al., 1999).

b. Polyacrylamide gel electrophoresis. Two major phenotypic groups could be identified by using polyacrylamide gel electrophoresis (PAGE) and silver staining of sodium dedecyl sulfate of lysed cells of *Xanthomonas campestris* pv. *vesicatoria.* Broad dark gray bands with molecular weight (MW) 32–35 kDa and 25.57 kDa designated α and β are present in different strains. The α band is present in 192 or 197 tomato race 1 strains and the β-band is seen in all 55 strains of tomato race 2. Moreover race 1 strains expressing the α band cannot hydrolyze starch (Amy$^-$) or degrade pectate (Pec$^-$), whereas most race 2 strains are Amy$^+$ and Pec$^+$. Silver staining of protein profiles and testing for amylolytic activity may be useful to differentiate strains of *X. campestris* pv. *vesicatoria* (Bouzar et al., 1994). It is difficult to differentiate the pathovars of *Pseudomonas syringae* merely by biochemical tests and O-antigen serological reactions. The envelope protein profiles of bacterial cells have provided a reliable basis of differentiation. Three protein bands, 60, 65, and 150 kDa, present in *P. syringae* pv. *pisi* strains differentiated them from 29 strains of *P. syringae* pv. *syringae* in which they were absent (Malandrin et al., 1996).

Pulse-field gel electrophoresis (PFGE) technique was used for the genomic analysis of *Erwinia amylovora* strains from the Mediterranean region and European countries. The PFGE patterns were determined by assaying the *Xba*1 digests of bacterial genomic DNA. The strains from Austria and Czechia were placed along with the central European type (Pt1), whereas the strains from the eastern Europe and Mediterranean region belonged to another group Pt2. The strains from Italy showed patterns of all the three types (Zhang et al., 1998).

c. Conductimetric assay. The measurements of conductance of the media in which phytopathogenic bacteria are grown may be useful for the detection of metabolically active soft rot pathogens infecting potato such as *Erwinia chrysan-themi* (Ech) and *E. carotovora* subsp. *atroseptica* (Eca). The pectate medium con-taining pectate as the sole carbon source was found to be useful for the specific au-tomated conductimetric detection of *Erwinia* spp. in potato peel extracts. The threshold of detection for Eca by this method varied between 10^2 and 10^3 cfu/ml of peel extracts, respectively, at 20°C and 26°C, whereas Ech was detected at con-centrations of 10^4–10^5 or 10^3–10^4 cfu/ml of peel extracts respectively at 20° and 26°C. Confirmation of results of this assay by serological or nucleic acid assay may, however, be required (Fraije et al., 1996).

d. Isozyme analysis.   Isozyme profiles of bacterial pathogens may provide a reli-able basis for their identification. The patterns of enzymes of esterase (EST) and superoxide dimutase (SOD) of *Pseudomonas syringae* pv. *pisi* strains were stud-ied for their usefulness for the purposes of identification and diagnosis of diseases caused by this pathogen. Two EST zymotypes specific for this pathogen could be used for their identification. Moreover, these two EST patterns were correlated to race structure of this pathovar. Zymotype 1 corresponded to races 2, 3, 4, and 6 whereas zymotype 2 was correlated to races 1, 5, and 7 (Malandrin and Samson, 1998). The specificity of isozyme analysis and its usefulness for the identification of pathovars of other bacterial pathogens have to be assessed.

e. Bacteriophages.   Bacteriophages infect bacteria, causing lysis of the susceptible-ble bacterial cells that leads to the formation of plaques in bacterial cultures. Bac-teriophages that infect plant pathogenic bacteria have been isolated from infected leaves, irrigation water, and soil. Most of the phages that infect *Xanthomonas campestris* pv. *oryzae* are tadpole-shaped with a polyhedral head and tail. A fila-mentous phage Xf was reported by Kuo et al. (1967).

The OP_1 phage that infects *X. campestris* pv. *oryzae,* which causes rice bac-terial blight disease, can be employed for detection of the presence and quantita-tive estimation of bacterial population (Wakimoto, 1957). On the basis of sensi-tivity to phages, *X. campestris* pv. *oryzae* could be differentiated into 15 lysotypes. However, the sensitivity of the bacterial strains to phages and their serological reactions are neither related to one another nor to the virulence of the strain (Ou, 1985). On the other hand, Freigoun et al. (1994) reported that phage sensitivity of *Xanthomonas campestris* pv. *malvacearum,* causative agent of cot-ton bacterial blight, was related to the pathogenic potential of two races prevalent in Sudan. These races were found to be quite distinct in their phage sensitivity. Race 1 was lysed by three or rarely four of the six phages used for typing, whereas race 2 was sensitive to all six phages. The use of bacteriophages CP115 and CP122 isolated from canker lesions on grapefruit and sweet orange, respectively, for the detection and identification of *X. axonopodis* pv. *citri* has been reported by the Wu et al. (1993). By using a membrane filter (pore size 0.22 μm) after cen-trifugation of the solution/sample to be tested, to exclude the contaminating bacteria, the effectiveness of phage technique for the detection of *X. axonopodis* pv. *citri* was improved (Ebisugi et al., 1998). The phages may be utilized for both ecological and epidemiological studies to detect and determine the population of pathogenic bacteria.

6.4.2.3 Biochemical Tests with Host Plant Tissues

a. Fluorescent markers. Indexing of plant materials for the presence of citrus greening and exocortis, using fluorescent marker substance, was reported by Schwarz (1968). Alcohol extracts of leaves or stem bark are concentrated, after filtering through a sintered glass funnel. The extracts and the fluorescent marker gentisic acid (GeA) are spotted on silica gel thin layer chromatography plates and run with water-saturated *n*-butanol. Buffered sodium borate (pH 8.7) is used to develop color, and the plates are examined under UV light at 365 nm for the presence of fluorescent violet spots corresponding to the marker. Hooker et al. (1993) reevaluated the reliability of this procedure as a method of diagnosing greening disease. The severity of foliar symptoms was correlated significantly with the amount of GeA in young and old bark tissues. This method is found to be reliable when GeA levels are more than 300 μg/g of tissue and when it is used along with diagnostic criteria under greenhouse conditions.

Papaya bunchy top disease, earlier considered as a phytoplasmal disease, was shown to be caused by a bacterial pathogen. Transverse sections of petiole tissues from infected plants exhibited the presence of fluorescing materials on the periphery of the pholem, between the phloem and xylem and sometimes extending along the phloem rays up to cortex, when examined under epifluorescence microscope (Davis et al., 1996). The phloem transport and unloading of the fluorescent dye carboxy fluorescein (CF) in galls induced by *Agrobacterium tumefaciens* and the symplsastic movement of potato X potexvirus (PVX) expressing a green fluorescent protein-CP fusion (PVX GFP-CP) were compared. A clear symplastic pathway between the pholem of host stem and cells of the tumor and also a considerable capacity for subsequent cell-to-cell transport between tumor cells was revealed. Both CF and PVX could also move through the vascular rays of the host stem toward the stele. This study indicated that host and tumor tissues in *A. tumefaciens* galls are in direct symplastic continuity (Pradel et al., 1999).

6.5 DETECTION OF VIROIDS

Methods that depend on the possibility of locating the viroid itself or the presence of specific products of host-viroid interaction have been suggested for detection of infection by viroids.

6.5.1 Phloroglucinol Test

The ray cells in the bark tissue of citrus infected by citrus exocortis viroid (CEVd) contain compounds that can react with aldehyde decoupling reagents such as phloroglueinol-concentrated HCl. The reaction leads to the development of characteristic color which may be visible under the microscope. Cross-sections of the bark are immersed in alcoholic phloroglucinol solution then treated with concentrated HCl for a few seconds. A positive color reaction may be seen well before the onset of scaling of bark, which is the early visible symptom in CEVd-infected plants. A very high correlation (98.6%) between color reaction and presence of CEVd was reported by Childs et al. (1958). Such viruses as psorosis, xylopsorosis, and tristeza do not seem to affect the test results. However, use of diagnostic hosts for CEVd detection appears to be preferred by researchers (Diener, 1979).

6.5.2 Polyacrylamide Gel Electrophoresis Technique

Potato spindle tuber viroid (PSTVd) is detected in potato and tomato plants by extracting the cellular nucleic acids from both healthy and infected plants. The nucleic acids are then separated by the conventional polyacrylamide gel electrophoresis (PAGE) system, and the gel is stained to reveal the presence of different bands of nucleic acids. The PSTVd, as a nucleic acid, appears as a separate distinct band only in the samples from infected tissue; it is absent in the comparable healthy tissues. This technique can be used to detect the presence of both mild and severe strains of PSTVd. Several elite or basic seed stocks of potato have been freed of PSTVd by eliminating infected tubers by the PAGE system in certification programs (Morris and Wright, 1975). The presence of PSTVd in true seeds could be detected by return electrophoresis, which was found to be comparable with or superior to nucleic acid hybridization (Singh et al., 1988). The return gel electrophoresis (RGE) has been found to be a rapid and relatively simple technique for the detection of apple scar skin viroid (ASSVD) in apple leaf tissue, hopstunt viroid (HSVd) in plum bark, and citrus exocortis viroid (CEVd) in citrus leaves (Asai et al., 1998).

For the detection of coconut cadang-cadang viroid (CCCVd), leaf samples are blended with 0.1 M Na_2SO_3 and precipitated with polyethylene glycol (PEG) 6000 and ammonium sulfate. The PEG-insoluble fraction is then extracted in phenol-SDS-chloroform and the nucleic acids are recovered by ethanol precipitation and further fractionated with 2 M LiCl. The nucleic acids are then subjected to either PAGE or blot hydridaztion. In the PAGE system, toludene blue used for staining the gels detects about 0.1 μg of cadang-cadang RNA (ccRNA) per band. The differences in the electrophoretic mobility of cadang-cadang isolates can be used as the basis for distinguishing them (Randles, 1985).

The viroid molecules differ distinctly from normal host RNAs in their electrophoretic mobility in nondenaturating and partially denaturating gels. Schumacher et al. (1983) used this property for PAGE analysis. The leaf extracts are run first in nondenaturating gels in one direction and are run either in the reverse direction or at 90° to the first direction under denaturating conditions. Viroid bands can be separated easily from host nucleic acid. Using the silver staining procedure, viroids can be detected at 600 pg level, and with purified ccRNA1 as low as 0.4 to 1.6 ng may be detected.

6.5.3 Analytical Agarose Gel Electrophoresis

A simple and reliable method of detecting coconut tinangaja viroid (TiVd) in coconut leaf samples was developed by Hodgson et al. (1998). The coconut leaf extract was subjected to analytical agarose gel electrophoresis. Two-dimensional PAGE analysis revealed the TiVd band containing circular molecules typical for viroids. The identity to TiVd was confirmed by either diagnostic oligonucleotide probe (DOP) hybridization assay or reverse transcription–polymerase chain reaction (RT-PCR) technique (Chapter 9).

SUMMARY

Chemodiagnostic tests depend on the reaction between chemicals and certain compound(s) present in the infected plants, resulting in the development of a visible color reaction. Detection of plant viruses, phytoplasmas, fungi, and bacteria by using different chemodiagnostic tests involving the use of dyes, different kinds of chemicals, selective media, and fluorescent markers has been reported by many workers. The chemodiagnostic tests, in general, are less specific and sensitive than tests that depend on serological properties and variations in genomic nucleic acid sequences of test pathogens. However, isozyme analysis, polyacrylamide gel electrophoresis, and direct colony thin layer chromatography have been found to yield reliable results for the detection and differentiation of certain pathogens.

APPENDIX 6(i): STAINS FOR BACTERIAL PATHOGENS

    A. Gram's stain (Hucker's modified method)

Solution A:	Crystal violet (90% dye content)	2.0 g
	Ethyl alcohol (95%)	20.0 ml
Solution B:	Ammonium oxalate	0.8 g
	Distilled water	80.0 ml

        Mix solutions A and B.

    B. Lugol's solution (mordant)

Iodine	1.0 g

Potassium iodide	2.0 g
Distilled water	300.0 ml
C. Counterstain	
Safranin (2.5% solution in 95% alcohol)	10.0 ml
Distilled water	100.0 ml
D. Flagella stain	
20% Aqueous tannic acid	100.0 ml
Ferrous sulfate	20.0 g
10% Basic fuchsin in alcohol	10.0 g
Distilled water	40.0 ml

Ferrous sulfate is dissolved in water by warming, followed by addition of other ingredients.

E. Loeffler's flagella stain	
1% Basic fuchsin in alcohol	20.0 ml
3% Aniline water	80.0 ml
F. Spore stain	
Malachite green	5.0 g
Distilled water	100.0 ml
Safranin	
Safranin 0 (2.5% solution in 95% ethyl alcohol)	10.0 ml
Distilled water	100.0 ml
G. Negative stain	
Nigrosin (water-soluble)	10.0 g
Distilled water	0.5 ml
Formalin	0.5 ml

APPENDIX 6(ii): STAINING METHODS FOR BACTERIAL PATHOGENS

A. Gram's staining
    i)   Prepare thin smears of the bacteria on grease-free clean slides; air
         dry and heat the smears.
    ii) Cover the smears with crystal violet solution for 30 sec; pour off
         the stain and wash the slides in distilled water for a few seconds.
    iii) Cover the slide with iodine solution (mordant) for 30 sec; wash
         the slides with 95% ethyl alcohol by adding in drops till no more
         color appears in the fluid from the slide; wash the slides in dis-
         tilled water and drain the water.
    iv) Cover the slides with safranin (counterstain) for 30 sec; wash

with distilled water, dry the slides by blotting with filter paper, and air dry.

    v) Examine the bacteria under the oil immersion lens; gram-positive bacteria retain crystal violet stain (dark purple to blue color), and gram-negative bacteria are stained by safranin and appear pink in color.

B. Flagella staining

    i) Use new grease-free slides; place a loopful of freshly prepared bacterial suspension on one end of the slide, tilt the slide slowly, and spread the bacterial suspension as a thin film and air dry.

    ii) Cover the slide with mordant solution for 10 min and wash the slide gently with distilled water.

    iii) Flood the slide with carbolfuchsin for 5 min; wash gently with distilled water and air dry.

    iv) Examine the slide under the oil immersion lens.

C. Bacterial spore staining

    i) Prepare smears of the bacteria on grease-free glass slides; air dry the slide and fix smears by heating.

    ii) Cover the slides with malachite green solution; heat the slides to steaming for 5 min, adding more stain as it is evaporated; wash the slides gently in running tap water for a few minutes.

    iii) Counterstain with safranin solution for 30 sec; wash the smear with distilled water and dry the slide with filter paper by blotting.

    iv) Examine the slide under an oil immersion lens; endospores may be green; vegetative cells appear red.

D. Negative staining

    i) Transfer a drop of nigrosin solution to one end of a grease-free clean glass slide; place a loopful of bacterial suspension on the stain drop and mix well.

    ii) Using another slide, prepare a smear of the bacteria-nigrosin mixture on the slide and allow the smear to air dry.

    iii) Examine the smear under an oil immersion lens; bacterial cells appear colorless on a dark blue background.

APPENDIX 6(iii): DIRECT COLONY THIN LAYER CHROMATOGRAPHY FOR IDENTIFICATION OF PHYTOPATHOGENIC BACTERIA (MATSUYAMA ET AL., 1993)

    i) Take a loopful of bacterial colony from the slant; paste the culture directly on the origin line in the silica gel thin layer plate of 20 $\times$

   20 or 10 × 20 cm and dry completely; place samples at a spacing
   of 1.5 cm.

ii) Incubate the plates on moistened glass vessels and immerse the
     sample side of the plate in the chloroform-methanol mixture (2:1
     v/v) to a depth of less than 1 cm; run the chromatogram at 25°C in
     the incubator for 10 min until the solvent front reaches the 6 cm line
     from the origin spot.

iii) Scrape the bacterial cells in the origin spot completely; run the
     chromatogram in the same direction again; using a chloroform-
     methanol-water (60:25:4, v/v/v) solvent system for about 90 min at
     25°C in the incubator.

iv) Dry the plates well, spray ninhydrin, and develop the color by heat-
     ing in chromatographic chamber at 100°C for 10 min.

v)   Determine the ratio between solvent and solute fronts (Rf values)
     for each spot in the chromatogram.

7

Electron Microscopy

The electron microscope (Fig. 7.1) is analogous to the light microscope (Chapter 2) in principle. The solid ground glass lenses present in the light microscope are replaced by magnetic field lenses in the electron microscope: the light source of the light microscope is replaced by an electron beam in an electron microscope (Fig. 7.2). The wavelength of visible light (white light) is 540 nm and that of ultraviolet is 260 nm, whereas the wavelength of the beam of electron is usually less than 0.1 nm, depending on the accelerating voltage. The wavelength of electrons in an electron microscope using 60 kV will be about 0.005 nm. The resolving power of the microscope is inversely proportional to the wavelength of light being used. This is the reason for the high resolving power of the electron microscope, which can magnify objects about 200,000 times, whereas even the best light microscope may magnify only up to 1500 times. Electron microscopes that are now available can be used to distinguish objects 1 nm apart with sufficient density to hold back electrons. The image of the object is focused on a fluorescent screen, photographed, and then enlarged for detailed examination. The image of the object cannot be viewed directly as in the case of light microscope. Because of the high resolving power of the electron microscope, it has been possible to study the ultrastructure of biological materials such as individual cells and molecules such as proteins and nucleic acids whose structure had been known only through biochemical analyses.

Electron microscopy has been found to be useful to study the histopathological characteristics and ultrastructural changes induced by pathogens. Virus particle morphological features and characteristics of intracellular inclusions have been studied to differentiate viruses. In certain cases, the cytopathic effects of viruses are so characteristic that the causative virus can be reasonably identified. Geminiviruses cause characteristic cytopathic effects such as formation of macrotubules in the nuclei of infected cells, as in *Euphorbia* mosaic virus-infected *Datura stramonium* plants (Kim and Lee, 1992). Electron microscopy can be used

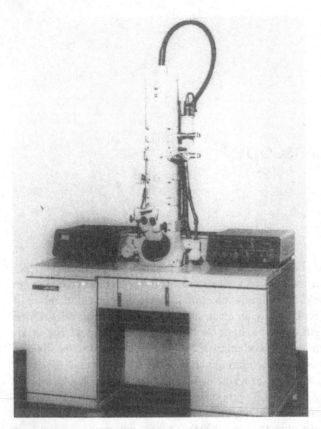

Figure 7.1 Transmission electron microscope (TEM) model Jeol—JEM 100 SK.

to reveal the presence of virus particles in extracts of individual local lesions caused by the transcripts of a full-length complimentary DNA (cDNA) copy of clover yellow mosaic potex virus RNA (Holy and Abou Haider, 1993).

The presence of virus particles in extracts from plants to be indexed can be detected by electron microscopy. Peach mosaic virus (PMV) particles were observed in the extracts from symptomatic leaves of *Chenopodium amaranti-color* mechanically inoculated with PMV. The results were confirmed by using polyclonal antibodies in Western blot analysis (Gispert et al., 1998). Of course, electron microscopy is useful as a tool to track the virus during the process of purification. It has also been possible to recognize unusual type of virus-like particles by examining crude extracts from plant organs showing no visible symptoms. James et al. (1999) observed the presence of coiled virus-like particles (spirions) in the crude extracts of flowers, leaves, and/or roots of graft-in-

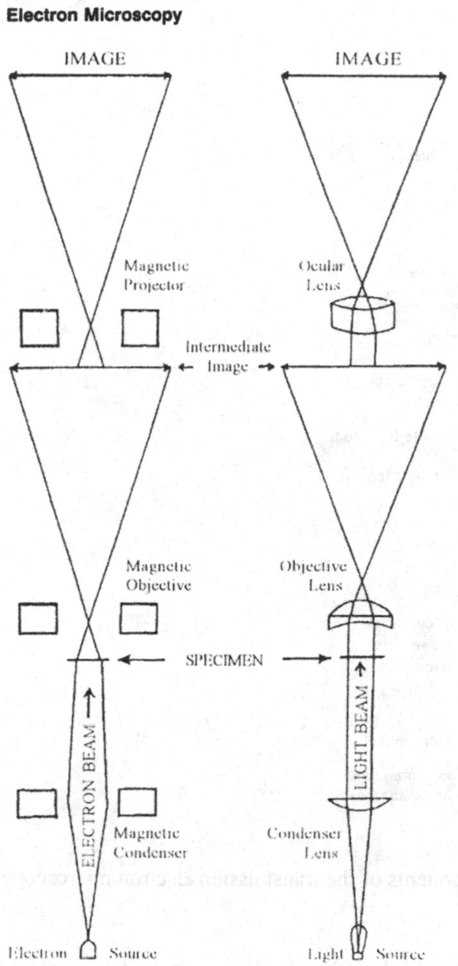

Electron Microscopy

Figure 7.2 Comparison of image formation in light and electron microscopes.

oculated *Prunus* spp. The individual spirions were observed in lower epidermis, palisade, and spongy mesophyll cells of leaves of *P. avium* and *P. mume*. They measured 132×134 nm and appeared to be coiled forms of a filamentous virus.

The essential features of major components of the electron microscope (Fig. 7.3) are described in the following discussion.

The electron gun is the illumination source of the electron microscope. The electron source should be small and have high brightness and stability. The elec-

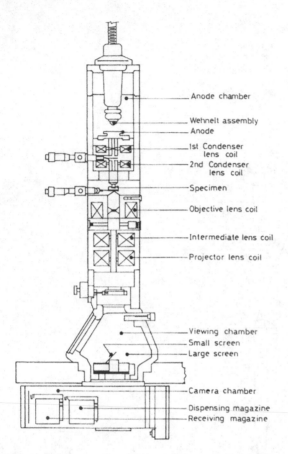

Figure 7.3 Schematic diagram of components of the transmission electron microscope.

trons are accelerated by the electric potential difference between the filament and anode. The electrons are concentrated at a "cross-over point," from which they are later emitted. A condenser lens is used to cause the electron beams to converge from the electron gun and to illuminate the specimen to a desired level. Generally the field of view in high magnification is limited to a very small area and the illumination area also has to be small. A double-condenser lens is provided to achieve this requirement.

The specimen chamber contains a stage that can be used easily for rapid exchange of specimens. This stage holds the specimen holder in a stable condition and moves smoothly for easy selection of the field of view. External vibration, which adversely affects the resolving power of the high-performance electron microscope, should be eliminated. The specimen chamber is so constructed as to accommodate many attachments for wider applications.

The image-forming lens system consists of three lenses, namely, the objective lens, intermediate lens, and projector lens. The objective lens, located immediately under the specimen, forms a first-stage image which is further enlarged by the intermediate lens. The final image is formed on a fluorescent screen or photographic film by the high-magnification projector lens. In certain models (e.g., JEM) a four-stage image-forming lens system with a two-stage intermediate lens is provided for high performance. Focusing is done by altering the exciting current for the objective lens. Variations in magnifications are due to the intermediate lengs. But in the case of low magnification, it can be altered by deactivating one of the system lenses or by changing the projector lens. The optical axes of both lenses must align with each other, and misalignment of the axes must be rectified immediately.

The viewing chamber and camera chamber are required to observe the images and to record them in film. The images of the specimen cannot be directly observed by the naked eye. They are converted into light images and focused on a fluorescent screen for observation or on films with high resolving power. A camera equipped with an airlock mechanism and capacity for rapid film exchange should be used to prevent breaking of the vacuum column.

The electrons are easily scattered by air. Hence it is essential to maintain a high degree of vacuum, which is achieved by operating oil rotary pumps or oil diffusion pumps. A poor vacuum may cause high-tension electrical discharge, specimen contamination, contrast reduction, or damage to the specimen. The electron microscope has an electric circuit consisting of four main parts: a) a high-voltage power supply for electron beam acceleration, b) a lens power supply for electron excitation, c) a deflecting coil power supply for electron beam deflection, and d) other circuits, including a vacuum control circuit, vacuum pump power supply, camera drive circuit, automatic exposure circuit, and electron gun filament heating circuit. For further details, see Siegel (1964), Kay (1965), and Sjostrands (1967).

7.1 PREPARATION OF SUPPORT FILMS

Electroplated copper or nickel grids (400 mesh) are covered with carbon alone or with a plastic film (e.g., Formvar, polyvinyl formaldehyde) alone or are reinforced with carbon. Parlodion, pyroxylin, and necoloidine films supported with carbon are equally good. It is better to use the support films soon after they are coated with carbon, since during storage they become hydrophobic, resulting in unsatisfactory specimen adhesion and negative staining.

7.2 CALIBRATION OF THE ELECTRON MICROSCOPE

The electron microscope has to be calibrated for determination of size measurements and magnification. For this, different standards have been used. A diffraction grating replica with 2160 lines/mm can be used as an external standard to calibrate magnifications up to 40,000×. Beef liver catalase is available as a crystal lattice which may be negatively stained and used as an external or internal standard. Tobacco mosaic virus particles whose modal length has been already determined can be used as an internal standard for calibration of the electron microscope. There are other standards (Milne, 1972), which are not commonly used.

7.3 TECHNIQUES FOR IDENTIFICATION OF PLANT VIRUSES

7.3.1 Preparation of Ultrathin Sections

The steps in the process of fixation, embedding, sectioning, and staining are time-consuming, and these procedures have limited use for rapid detection of plant viruses and phytoplasmas.

7.3.1.1 Fixation

Several fixatives have been used for fixing plant materials infected by viruses. Buffered 1%–2% solution of osmium tetroxide or 2%–5% solution of glutaraldehyde has frequently been used. Phosphate buffer (0.2 M) with pH 5.8–7.4 and cacodylate buffer with pH 7.4 are used to prepare the fixatives. Fixation time depends on the tissue and fixative, but short durations are preferable. Thin leaves are fixed with glutaraldehyde for 1 hr after infiltration, and 1 mm strips of leaves are further fixed in osmium tetroxide for 3 hr in the cold or at room temperature. Excess fixative is washed out of the tissue with buffer.

7.3.1.2 Dehydration

The fixed tissue has to be dehydrated by using a graded series of 50%, 70%, 90%, and 100% acetone or ethanol. The tissue is soaked in each concentration of acetone or ethanol for 10 min and finally in three changes of water-free solvent.

7.3.1.3 Embedding

Different kinds of embedding resins, such as methacrylates, epon, or araldite, have been used. Epon or araldite is used more frequently. In the case of epon, two stock solutions, viz., A, which contains 63 ml Epon 812 (Epicote 812) plus 100 ml dodecenyl succinic anhydride (DDSA), and B, which contains 100 ml Epon 812 plus

89 ml methyl nadic anhydride (MNA), are prepared. They are mixed in different proportions to produce hard or soft blocks; solution B increases hardness. The optimal proportion is 4 parts of A and 6 parts of B. An accelerator, such as 1.5% tridimethyl aminomethyl phenol (DMP 30) or 3% benzyldimethylamine (BDMA), is added to hasten curing. The blocks are hardened at 40°C–60°C for 2–3 days in closed capsules.

Different formulations of araldite have been used, depending on the plant materials. The embedding formulation contains araldite CY 212 (10 ml), DDSA (10 ml), dibutyl phthalate (1.0 ml), and DMP 30 (2.4% by volume). The resin is cured for about 24 hr at 60°C in closed capsules.

7.3.1.4 Sectioning

Diamond or glass knives are used for cutting ultrathin sections in an ultramicrotome. The knife may be set up permanently with its water bath. The gelatin capsule is removed by soaking in water. The sections, as they are cut, float onto a water bath, which is in contact with the knife edge. Generally the water bath is made of a strip of adhesive tape or metal foil. The water bath is filled with a mixture of acetone and water or water only. The thickness of the sections is roughly determined by interference color. Gray-, silver-, and gold-colored sections may have a thickness of 50 nm, 50–80 nm, and 150 nm, respectively. Thinner sections are required for better resolution. The sections are then transferred to grids coated with Formvar or parlodion and stained.

7.3.2 Negative Staining

Though negative staining is simple in principle, several factors, including individual skill, may appreciably affect the results. Many negative stains are available.

7.3.2.1 Phosphotungstate

Sodium or potassium phosphotungstate (PTA) aqueous solution at 2% concentration adjusted to pH 7.0 with NaOH or KOH has been used to examine many plant viruses. However, it is not useful for PTA-labile viruses such as alfalfa mosaic virus, tomato spotted wilt virus, cucumoviruses, rhabdoviruses, fijiviruses, and some members of the geminivirus, ilarvirus, and closterovirus groups (Francki et al., 1984). Floating the grids with adsorbed virus particles on a drop of 0.1% glutaraldehyde for 5 min before staining with PTA may reduce the damage due to PTA (Milne, 1984).

7.3.2.2 Uranyl acetate

Aqueous solution of uranyl acetate (UA) usually at 2% concentration (sometimes as low as 0.1%) without adjusting the pH (which may be around 4.2) is used as a

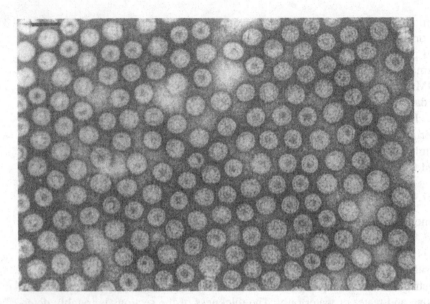

Figure 7.4 Purified peanut chlorotic streak virus particles stained with uranyl acetate; bar represents 100 nm (American Phytopathological Society, Minnesota, USA). (Courtesy of Reddy et al., 1993.)

negative stain for most of the plant viruses (Fig. 7.4). But rhabdoviruses are sometimes sensitive to UA and may be stripped of their envelopes, leaving the helical nucleocapsid. The stain should be stored in a dark bottle, as UA is unstable in strong light. Uranyl acetate gives better contrast and higher resolution than PTA. Prestain rinsing of the grids with 0.01 M phosphate buffer, pH 7.0, or with 0.1 M $CaCl_2$ is recommended.

7.3.2.3 Uranyl formate

The stain uranyl formate (UF) has properties similar to those of UA and the stain solution has to be prepared afresh every day. Uranyl formate is preferred for helical viruses and alfalfa mosaic virus, since it is found to be more effective than UA.

7.3.2.4 Ammonium molybdate

Ammonium molybdate (AM) is used at 2% concentration with varying pH levels (4–9) adjusted with HCl or ammonia and is stable at room temperature. It gives relatively poor contrast generally. However, AM has been used extensively because it may be mixed directly with virus suspensions and used to stain PTA-labile viruses. In the case of barley yellow striate mosaic rhabdovirus, AM was found to be better than PTA and UA (Milne, 1984).

7.3.2.5 Methylamine tungstate

The stain methylamine tungstate (MT) is used at 2% concentration and pH 6.5 (Oliver, 1973); it can be mixed directly with a virus sample. In the case of such viruses as cucumber mosaic virus, cauliflower mosaic virus, and bean common mosaic virus, more virus particles (10–20 times) may be required when MT is used than when UA is used, after a water rinse. However, the contrast is poor and there is frequent stain precipitation.

7.3.2.6 Sodium silicotungstate

A 2% aqueous solution of sodium silicontungstate (SST) adjusted to pH 7.0 is used. It is found to be preferable for some high-resolution studies of antibodies and antigen-antibody binding (Harris and Horne, 1986). It may produce images that give contrast to upper side of virus particles.

The virus particles are negatively stained with any of the stains described as follows: The grid is held with forceps and the filmed face is rinsed with a few drops of glass-distilled water, followed by 5 drops of negative stain. The grid is then dried by holding a piece of absorbent paper (filter paper) to the edge. Excess stain may be washed with drops of distilled water as before. Alternatively the grid may be briefly floated on a series of small water drops (Milne, 1993a).

7.3.3 Metal Shadowing or Shadow Casting

The metal shadowing or shadow casting method is more laborious, severely limits resolution, and obscures inner details. The grid with the virus preparation is placed face up and horizontal in a vacuum. The specimen is coated with a thin layer of electron-dense material (heavy metals such as gold, platinum, or palladium or alloys) by evaporation of thin filaments placed above and to one side. The evaporated atoms travel in approximately straight lines parallel to each other and coat the exposed surfaces of virus particles. Because metals are opaque to the electron beam, the beam casts a shadow, giving a three-dimensional effect to the material to be photographed under the electron microscope (Fig. 7.5). With the development of negative staining methods, metal shadowing has limited use as a diagnostic or taxonomic aid for determination of shape and fine structures of virus particles. The metal shadowing technique is particularly useful for coating nucleic acid molecules, which cannot be contrasted easily with negative or positive staining. Rotary shadowing is done to coat the specimen at different angles (Milne, 1972).

7.3.4 Dip Method (Brandes, 1957; Hitchborn and Hills, 1965)

The dip method is a simple technique that can be used to check the presence of virus in plants quickly. A freshly cut leaf or an epidermal strip is passed through

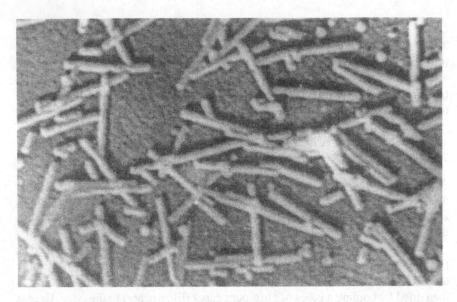

Figure 7.5 Purified tobacco mosaic virus particles shadowed with chromium. (Courtesy of Eishiro Shikata.)

the surface of a drop of water placed on a glass slide. Some of the heavier cellular contents flow into the drop; particulate materials such as virus particles form a film on the surface of the water drop. A filmed grid is then touched to the surface and negatively stained. In a variant of this method, the cut leaf may be passed directly through a drop of negative stain placed on the grid. The grid is then drained and dried. Dip preparation may help to identify viruses quickly, as in watermelon infected by a strain of tomato spotted wilt virus (Honda et al., 1989).

7.3.5. Leaf-Dip Serology

Leaf-dip serology, developed by Ball and Brakke (1968), is based on the principle of the leaf-dip method of Brandes (1954). The antiserum is diluted with 0.001 M ammonium acetate solution to produce a 1:1000 dilution. A drop of this serum is placed on a carbon-backed collodion film on a specimen grid. Pieces of leaf tissue (1.5–2.0 × 0.5 cm) are cut from diseased leaves, and the narrowest end is drawn through the antiserum drop for 1 or 2 sec. The drops are then air-dried and a drop of a solution containing 1 part of 1% vanadatomolybdate, pH 3, plus 3 parts of 2% potassium phopshotungstate is used to stain the virus-antiserum mixture and then dried. The infection of tobacco mosaic virus (TMV) and barley stripe mosaic virus can be detected by observing the virus particles (Ball, 1971). This method can also be used for determining the dilution end point of the antiserum and the relationship between strains of elongated viruses (Langenberg, 1974). Though this method is simple, the immunosorbent electron microscopy, decoration, and gold-labeling techniques have many advantages and are being widely used.

7.3.6 Immunosorbent Electron Microscopy

Derrick (1972, 1973) developed the method of serologically specific electron microscopy (SSEM), later redesignated immunosorbent electron microscopy (ISEM) by Roberts and Harrison (1979). The grids are coated with (Formvar [polyvinyl formaldehyde] or Parlodion) strengthened with a layer of evaporated carbon. Crude antiserum raised against the virus to be detected or assayed or fractionated α-globulins may be used to sensitize the coated grids. The optimal dilution of crude antiserum may be around 1:1000 to 1:5000. The antisera and virus preparation may be diluted using 0.05 M Tris-HCl buffer, pH 7.2 (Brlanksy and Derrick, 1979). Other buffers, viz., 0.1 M phosphate, pH 7.0 (Milne and Luisoni, 1977); 0.06 M phosphate, pH 6.5 (Roberts and Harrison, 1979); and sodium carbonate buffer, pH 9.6 (used in enzyme-linked immunosorbent assay [ELISA] for coating) (Thomas, 1980), have also been used. The grids are then coated with the antibodies by an appropriate dilution of antiserum for 5 min at room temperature (Milne and Luisoni, 1977). Incubation of grids at 37°C results in firm attachment of virus particles to the grids and better spreading of negative stain (Roberts, 1981). Sugarcane bacilliform virus was detected in suspected sugarcane clones by trapping virus particles on sensitized grids, followed by staining with 2% uranyl acetate. The grids were carefully rinsed with either buffer or water (Viswanathan et al., 1996) (Fig. 7.6). Further studies by Viswanathan and Premachandran (1998) revealed the extensive infection of sugarcane germplasm by sugarcane bacilliform virus as detected by ISEM. Thottappilly et al. (1998) reported that a Nigerian isolate of banana streak badnavirus (BSV) was detected in greater number of asymptomatic BSV-infected banana plants by ISEM than by triple antibody sandwich (TAS) enzyme-linked immunosorbent assay (ELISA).

The virus preparations are commonly diluted in 0.05 M Tris-HCl buffer, pH 7.2, or 0.1 M phosphate buffer, pH 7.0. The pH of the buffer is an important factor influencing the number of particles trapped on the grids, and optimal pH is determined for different host-virus combinations (Cohen et al., 1982). Addition of 0.1 M EDTA to 0.1 M phosphate buffer, pH 7.0, is necessary for trapping turnip mosaic virus isolates from chinese cabbage (Lesemann and Vetten, 1984) and papaya ringspot virus (Gonsalves and Ishii, 1980). Some plants may contain substances that may prevent trapping of virus particles on the grids. Addition of 2.5% nicotine to the sap of plum or grapevine helped to detect plum pox virus (Noel et al., 1978) and grape fan leaf virus (Russo et al., 1982). Likewise, detection of viruses in saps of plants in Rosaceae, grapevine, and poplar required the incorporation of 2% polyvinyl pyrrolidone (Milne, 1981).

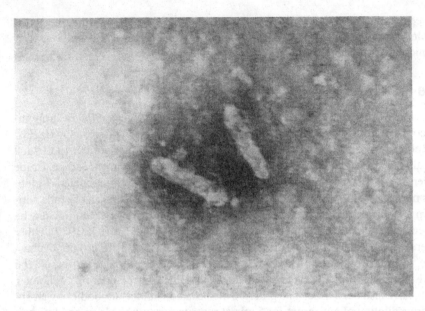

Figure 7.6 Sugarcane bacilliform virus (SCBV) particles trapped with antiserum to SCBV; bar represents 400 nm. (Courtesy of Viswanathan.)

The periods for which the grids are incubated with virus preparation may vary from 15 min to several hours (or overnight); for rapid diagnosis incubation for 15 min may be sufficient (Milne and Luisoni, 1977). The sensitivity of the test, however, increases with increase in the incubation period at room temperature or preferably at 37°C (Roberts and Harrison, 1979). Incorporation of sodium azide (1:5000) is recommended to prevent bacterial growth during long incubation periods (Milne and Lesemann, 1984). The grids are rinsed after incubation with antiserum.

Metal shadowing (Derrick, 1973) and positive staining (Derrick and Brlansky, 1976) were the methods formerly used to provide good contrast. Milne and Lesemann (1984) have recommended the use of negative stains for high resolution and good particle preservation. Uranyl acetate (2%) and sodium phosphotungstate (2%), neutral or pH 6.5, are effective in the case of many viruses. For viruses such as alfalfa mosaic virus, tomato spotted wilt virus, ilarviruses, cucumoviruses, and some of the rhabdoviruses and geminiviruses which are labile in PTA, 2% ammonium molybdate at pH 6.0–7.0 can be used for staining the virus particles (Roberts, 1981).

The comparative efficacy of detection of arabis mosaic virus, *Prunus* necrotic ringspot virus, and strawberry latent ringspot virus by ISEM, ELISA, and infectivity assay was assessed by Thomas (1980). Thomas found that ISEM was 10 to 20 times more sensitive than ELISA in detecting the virus in purified preparations and extracts from infected plants (Table 7.1).

Table 7.1 Comparative Efficacy of Detection of Plant Viruses by ISEM, ELISA, and Infectivity Assay

Virus	Mininum concentration of detectable virus (ng/mL)		Maximum reacting dilution of sap (reciprocal)		
	ISEM	ELISA	ISEM	ELISA	Infectivity
Arabis mosaic virus	0.5	5.0	320,000	40,000	100,000
Prunus necrotic ringspot virus	4.0	4.0	80,000	10,000	400
Strawberry latent ringspot virus	50.0	10000.0	128,000	5,500	64,000

Source: Thomas (1980).

Two modifications of the regular ISEM procedure have been reported to increase the sensitivity and usefulness of the technique. The grids are precoated with protein A obtained from *Staphylococcus aureus,* which has the specific affinity of binding to the Fc portion of the immunoglobulin G (IgG) molecule. A significant increase in the number of virus particles trapped by protein A–coated grids is seen (Gough and Shukla, 1980). This procedure is useful when the antiserum is of low titer. In another modification, known as the decoration step, the virus particles already adsorbed on the grid are coated with IgG molecules. The grids are incubated with the same antiserum at a dilution between 1:10 and 1:100 for 15 min at room temperature. The virus particles are decorated with a halo of IgG molecules (Milne and Luisoni 1975). The virus particles may be more easily identified by this additional step introduced into the regular ISEM procedure.

Immunosorbent electron microscopy has helped to detect and differentiate viruses rapidly. Detection of cymbidium mosaic virus in crude sap of infected orchid leaves by ISEM was about twice as sensitive as detection by ELISA (Hsu et al., 1992). Virus particles resembling those of geminiviruses were observed in extracts of plants infected by bhendi yellow vein mosaic, croton yellow vein mosaic, dolichos yellow mosaic, horsegram yellow mosaic, Indian cassava mosaic, and tomato leaf curl viruses. These viruses positively reacted with the MABs generated against African cassava mosaic geminivirus (Harrison et al., 1991; Muniyappa et al., 1991). The chickpea chlorotic dwarf virus was identified as a new leafhopper-borne geminivirus, since it did not react with the antiserum against sugar beet curly top virus (Horn et al., 1993). Characteristic cross-banded filamentous particles (about 805 nm in length) of sweet potato sunken vein virus (SPSVV-Ke), a Kenyan isolate, were detected by ISEM using the antiserum to virions of an Israeli isolate (SPSVV-Is) (Hoyer et al., 1996). ISEM has been employed for the detection of cocksfoot mottle virus in oat plants by Truve et al. (1997).

Detection of virus infection and assay of viruses by ISEM have certain advantages: a) Direct visualization of virus particles is possible; b) sensitivity of detection is comparable to and better than that of ELISA, which may sometimes give false-positive results in certain cases; c) crude antiserum without fractionation or with low titer can also be used; d) very small volumes of antiserum (5–10 μl) or antigen samples (1 μl) are sufficient for the tests; e) the results may be obtained within 2 hr; f) the presence of antibodies against host material will not interfere with the results; g) using the decoration step, it is possible to detect the contaminating virus, if any, present in the plant sample. Though there are several advantages of using ISEM for detection of virus infection, it is not being used widely, because it is labor-intensive and requires costly equipment and expertise to handle the equipment. So it may not be suitable for the routine testing of large number of samples.

7.3.7 Decoration Technique

Milne and Luisoni (1975) developed the method of "decoration" of plant viruses for their rapid identification. A combination of immunosorbent electron microscopy (ISEM) and decoration resulted in greater resolution. This procedure, in principle, consists of trapping the virus particles on sensitized grids, followed by staining of the viruses with a negative stain, and incubation with the same or a different antibody to give a visible antibody halo (decoration). The decoration may be further refined with gold particles conjugated either to a secondary antibody or to protein A (PA) (Milne, 1992).

The decoration technique may be used for detection and quantitative estimation of viruses. The virus particles extracted or suspended in an appropriate buffer are adsorbed to sensitized filmed grids. If necessary, the virus particles on the grid may be fixed with 0.1% buffered glutaraldehyde (GA) for 15 min at room temperature. The grid is rinsed with 0.1 M phosphate buffer or 0.05 M borate buffer, pH 8.1, then incubated with a suitable dilution of antibody at room temperature for 15 min, followed by rinsing with water and staining with 1% aqueous uranyl acetate. The excess stain is drained at the edge with filter paper and allowed to dry.

The decoration technique was found to be useful for the detection of several potyviruses and carlaviruses. This technique may be employed for a) serological identification of viruses, b) detection of virus mixtures, c) detection of partial degradation of coat proteins, d) quantitative estimation of degrees of relationship between viruses, e) measurement of antiserum titer, f) localization of particular antigens (viral gene products on the viral surface), and g) enhancement of the virus particle's conspicuousness by increasing its size and electron density as an aid for rapid diagnosis. Xu and Li (1992) reported that ISEM combined with decoration was a suitable method for distinguishing strains of ribgrass mosaic virus and establishing the extent of the serological relationship between related viruses. Employing the combined immunosorbent and double-decoration method, TMV and PVY were detected at 0.184 ng/ml and 0.128 ng/ml, respectively, in purified preparations within 1 hr and using as low a magnification as 2000 times (Chen et al., 1990).

ISEM with some modification has been reported to be useful for the detection of onion yellow dwarf potyvirus, leek yellow stripe potyvirus, and carnation latent carlavirus infecting garlic. The presence of these viruses in tissue-culture-derived garlic plants could be detected demonstrating the potential usefulness of this technique for the production of virus-free garlic stocks (Maeso et al., 1997). Using the antiserum raised against viral coat protein, garlic mite–borne filamentous virus (Gar V-A) was detected in Argentine garlic cultivars (Helguera et al., 1997). The grapevine A trichovirus (GAV) and another virus with filamentous, flexuous particles that can be mechanically transmitted to *Nicotiana benthamiana* and *N. occidentalis* were detected by decoration with the antibodies raised against GAV and grapevine B trichovirus (GBV), respectively. This test helped to identify these viruses infecting grapevine rapidly (Credi, 1997). The serological relationship of shallot yellow stripe (SYSV) and Welsh onion yellow stripe (WoYSV) viruses with other potyvirus was established by decoration technique using the monoclonal antibodies raised against SYSV as well as the antisera to onion yellow dwarf (OYDV) and leek yellow stripe (LYSV) viruses. SYSV was identified as a distinct potyvirus and WoYSV as a strain of SYSV (van der Vlugt et al., 1999).

7.3.8 Gold Labeling

Though the decoration technique is sufficient for the detection of many plant viruses, labeling the decorating antibody with gold offers additional advantages in the case of larger viruses such as rhabdoviruses and tospoviruses. Pares and Whitecross (1982) introduced the gold label antibody decoration (GLAD) procedure for gold labeling of viruses in suspensions. This technique was further refined by Louro and Lesemann (1984) and van Lent and Verduin (1985). Gold labels with varying diameters ranging from 5 to 20 nm have been used. Gold may be conjugated directly to the primary or coating antibody, but more usually, it is attached to protein A (PA) or a secondary antibody (goat antirabbit IgG).

The virus particles are adsorbed on filmed nickel grids and rinsed with 0.1 M phosphate-buffered saline solution containing 0.05% Tween 20 (PBST). The grids are incubated for 15 min with 0.1%–1.0% aqueous bovine serum albumin (BSA) to block nonspecific protein adsorption sites and rinsed again with PBST. The grids are then incubated for 15 min with drops of antiserum (raised against the particular virus) diluted in 0.1 M phosphate buffer, pH 7 (PB), and rinsed with PBST. The grids are incubated with protein A gold (PAG) in PBS for 60 min, followed by rinsing successively in PBST and water. The preparation is then stained with uranyl acetate (1%) to enhance the contrast (Milne, 1992, 1993).

The presence of alfalfa mosaic virus antigen could be detected in the cytoplasm and vacuoles of ovule integuments, microspores, mature pollen grains, and anther tapetum cells of infected alfalfa plants. Raftlike aggregates of virus particles and large crystalline bodies were observed in the cytoplasm of the pollen grains and anther tapetum cells, whereas nonaggregated virus particles were detected in the vacuoles and cytoplasm of ovule integument cells (Pesic et al., 1988). The virus associated with pear vein yellow disease was considered to belong to the closterovirus group, because the antiserum raised against the apple stem pitting virus produced positive gold labeling of the virus aggregates (Giunchedi and Pollini, 1992). The presence of virus in infected tissues can be detected by labeling capsid structural proteins or entire coat protein. By treating with polyclonal antibodies (PABs) to three major structural proteins G, N, and M of *Festuca* leaf streak rhabdovirus, followed by protein A gold labeling, all the structural proteins could be detected in virions present at the periphery of viroplasms (Lunsgard, 1992). Gold-labeled antibodies to coat protein (Cp) of tobacco vein mottling potyvirus (TVMV) were bound to cylindrical inclusions and individual virions produced in protoplasts inoculated with TVMV-RNA. In inoculated leaves also, antibodies to TVMV-Cp were bound to cylindrical inclusions and aggregates of virions in the cytoplasm (Ammar et al., 1994).

Using polyclonal antibodies to the nucleoprotein of tomato spotted wilt virus (TSWV) expressed as a recombinant fusion protein in *Escherichia coli* and labeled with protein A-gold, TSWV was detected in both greenhouse and field plant materials. The labeled antibodies showed good detectability and specificity (Vaira et al., 1996). Cucumber mosaic cucumovirus (CMV) is known to be transmitted in the seeds of spinach. Immunogold labeling revealed the presence of CMV in the cytoplasm of ovary wall cells, ovule integuments and nucellus, anther and seed coat cells, and also in fine-fibril-containing vesicles and electron-dense inclusions of amorphous aggregates in the central vacuoles of these cells (Yang et al., 1997). The immunogold localization studies using the antiserum raised against the N-terminal P1 proteinase protein of PVY expressed in *Escherichia coli* showed that P1 protein was associated with the cytoplasmic inclusions that have diagnostic value for potyvirus infection. No other plant cell organelles or cell wall or plasmodesmata exhibited any significant P1 antibody binding (Arbatova et al., 1998).

The clues for location of the virus particles in vector tissues and evidence for multiplication of plant viruses in the vector may be observed by using immunochemical methods. Ullman et al. (1992) reported that although adults of *Frankliniella occidentalis* ingested TSWV, the virus was not retained, whereas TSWV could be detected serologically in the midgut epithelium and hemocoel of larvae fed on TSWV-infected plants, indicating that the midgut of the adult thrips may be a barrier preventing the virus from reaching the hemocoel. Using a specific antibody to nonstructural proteins (NSs) of TSWV demonstrated the presence of NSs in thrips cells, and the immunochemical evidence observed by Ullman et al. (1993) indicates that TSWV replicates in vector cells. In situ immunolabeling of salivary glands and other tissues of adult *F. occidentalis* showed the accumulation of large amounts of nucleocapsid and nonstructural proteins and the presence of several vesicles with virus particles in the salivary glands, indicating that the salivary glands may be a major site of virus replication (Wijkamp et al., 1993). The presence of potato leaf roll virus particles in the intestinal epithelium could be visualized by the immunogold labeling and immunofluorescence techniques, suggesting that intestinal cells might be the pathway for PLRV transport from gut lumen into hemocoel (Garret et al., 1993). Immunogold labeling was employed to locate barley mild mosaic virus particles in the zoospores of the fungal vector *Polymyxa graminis*. Labeled bundles of presumed virus particles were observed in about 1% of the zoospores released from the plant roots, and in zoospores inside zoosporangia (Jianping et al., 1991). Localization of beet necrotic yellow vein virus (BNYVV) in the fungal vector *Polymyxa betae* was studied using immunogold labeling technique. The presence of labeled virions inside the young zoosporangia, mature zoosporangia, immature zoospores, and outside resting spores was observed. Though some labeling could be noted in the internal wall and vacuoles of resting spores, no virion was seen directly (Peng et al., 1998).

Aphid transmission of potyviruses requires the presence of helper component (HC) protein. Immunogold labeling technique provides convincing evidence for the role of HC in aphid transmission of potyviruses. In the sections of aphids (*M. persicae*) that fed on a mixture of tobacco etch virus (TEV) or tobacco vein mottling virus (TVMV) and HC, the presence of filamentous, virus-like particles (VLP) embedded in a matrix material associated with the epicuticle, predominantly in the maxillary food canal and also in the precibarium and cibarium of the foregut, was observed. On the other hand, no VLP could be seen at any of the sites in the aphids fed on purified virions only. The VLP was found to be either TEV or TVMV depending on the virus-Hc mixture on which the aphids fed. Helper component appears to be directly involved in binding or attachment of virions to the cuticle of the maxillary food canal and foregut of aphid vectors (Ammar et al., 1994b).

7.3.9 Gel Double-Diffusion Precipitin Bands

The precipitin bands developed when the antigen and antibody interact in agar gel can be examined after negative staining under the electron microscope. Two procedures may be followed.

7.3.9.1 Excavation Method (Watson et al., 1966)

A small amount of precipitin band formed in an agar plate (0.5 × 2 × 1 mm) is scooped out by a scalpel or razor blade and homogenized in 5 μl buffer. The homogenate is adsorbed to a filmed grid, which is gently rinsed with water and negatively stained. Clumps of virus particles linked by antibody bridges are visible under the electron microscope.

7.3.9.2   Direct Printing (Milne, 1993b)

A filmed grid is gently placed over the preciptin band formed in an agar plate and left for about 15 min. The grid is then taken, rinsed gently with water, and examined.

7.4   TECHNIQUES FOR DETECTION OF PHYTOPLASMAS

The plant pathogenic phytoplasmas are generally larger than viruses with varying shapes and sizes ranging from 50 nm to 1000 nm in diameter. When extracted from infected plant tissues, the phytoplasma bodies may become flattened or fragmented, resembling degraded host debris derived from chloroplasts, mitochondria, or endoplasmic reticulum, thus posing a formidable problem for establishing the identity of the phytoplasma reliably. Moreover, none of the plant phytoplasmas barring the spiroplasma, has been isolated in pure culture, making it difficult to study the distinct morphological features of different phytoplasmas, which are morphologically indistinguishable in the characteristics determined by electron microscopy (Hiruki, 1988; Whitcomb and Tully, 1989; Clark, 1992). Hence detection and identification of phytoplasmas require the use of immunosorbent electron microscopy and immunogold labeling, and in recent years, antisera have been raised against several phytoplasmas, facilitating the application of these techniques.

The possible association of another kind of pathogenic agent rickettsia-like organism (RLO) with carrot proliferation disease was reported by Fránová et al. (2000) based on observations under transmission and scanning electron microscopes. The presence of the RLO bodies in sieve tube elements of symptomatic carrot plants was detected and the RLO bodies had a diameter of 0.2–0.3 μm and a length of 2.2 μm.

7.4.1   Immunosorbent Electron Microscopy

Derrick and Brlansky (1976) showed that corn stunt spiroplasma could be trapped by ISEM; this finding led to the successful demonstration that phytoplasmas could be trapped and identified by ISEM (Milne, 1992). Sinha and Benhamou (1983) could trap aster yellows (AY) phytoplasma bodies, using the homologous antiserum, from partially purified extracts of AY-infected aster plants. The AY phytoplasma and peach X phytoplasma were detected by ISEM in individual vector leafhoppers, *Macrosteles fascifrons* and *Paraphelpsius irroratus,* respectively. However, some of the hoppers that had access to infected plants did not transmit the phytoplasma and were ISEM-negative (Sinha, 1988). The presence of the phytoplasma that causes flavescence dorée (FD) disease of grapevine was detected by ISEM in extracts of both infected grapevine and vector leafhoppers *Euscelidius variegatus* (Caudwell et al., 1983). The antiserum raised against FD-enriched material from infective leafhoppers was used to detect the phytoplasma in plants, whereas the antiserum to material from infected *Vicia faba* plants was used to detect the phytoplasma in leafhoppers to prevent the effects of host-directed antibodies (Lherminier et al., 1990).

7.4.2 Immunogold Labeling Technique/Gold Label Antibody Decoration

The availability of both polyclonal and monoclonal antibodies specific for several phytoplasmas has helped to improve and refine serological techniques that can be employed for reliable detection and identification of phytoplasmas. The protein A gold labeling procedure has been found to yield reliable and convincing results. Lherminier et al. (1990) reported that flavescence dorée phytoplasma infecting grapevine could be detected by gold labeling of trapped phytoplasma bodies. Vera and Milne (1994) developed the protocol for immunotrapping, gold labeling, and electon microscopy of phytoplasmas from crude preparations of infected plants or vector insects. The European aster yellows (EAY) phytoplasma in periwinkle plants (*Catharanthus roseus*) and in vector leafhopper. *Macrosteles quadripunctulatus* and Australian tomato big bud (TBB) in infected tomato plants were detected and differentiated. The phytoplasmas of EAY and TBB are morphologically indistinguishable, but by using the immunogold labeling technique these two phytoplasmas can be distinguished (Figs. 7.7 and 7.8) (Appendix 7 [i]). Milne et al. (1995) reported that by immunogold labeling of thin sections (postembedding labeling) and labeling of bulk tissues before embedding and sectioning (preembedding labeling), primula yellows (PY), tomato big bud (TBB), and bermudagrass white leaf (BGWL) phytoplasmas could be reliably differentiated (Fig. 7.9) (Appendix 7 [ii]).

7.5 TECHNIQUES FOR DETECTION OF FUNGAL PATHOGENS

Immunosorbent electron microscopy and protein A gold labeling have been used to detect surface antigens of fungal pathogens. Colloidal gold was linked to chitinase or lectins to detect the presence of chitin on fungal cell walls in root cells infected by *Fusarium oxysporum* f.sp. *radicis-lycopersici* (Chamberland et al., 1985). When two antisera, one against the surface components of *Botrytis cinerea* and the other raised against the fimbriae of the smut fungus *Ustilago violacea,*

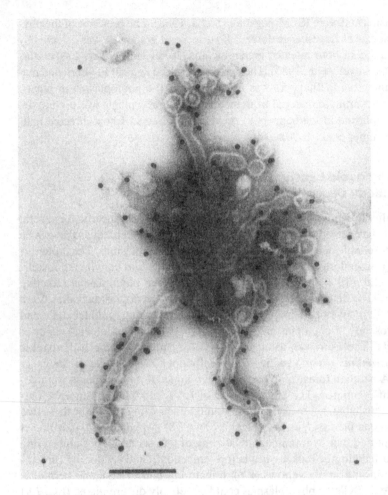

Figure 7.7 European aster yellows (EAY) labeled with gold and negatively stained with 0.5% ammonium molybydate; bar represents 200 nm (Blackwell Science Ltd. and British Society for Plant Pathology, UK). (Courtesy of Vera and Milne, 1994.)

labeled with protein A-gold complex were used, heavy gold labeling of host cells in infected tissue, but not cells in healthy tissue, was observed (Appendix 7 [iii]). The labeled host cells, in many cases, were not penetrated by the pathogen and the nearest hypha was some distance away from labeled cells. This may be the result of movement of the surface antigen of *B. cinerea* into host tissue in advance of the pathogen (Svircev et al., 1986). Ultrathin sections prepared from maize tissues infected by *Fusarium moniliforme* were examined by immunoelectron mis-

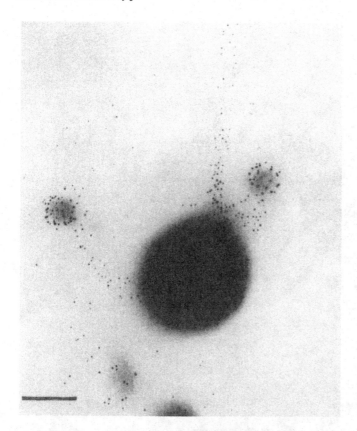

Figure 7.8 EAY phytoplasma in sap from *Catharanthus roseus* labeled with gold without phytoplasma staining; bar represents 200 nm (Blackwell Science Ltd. and British Society for Plant Pathology, UK). (Courtesy of Vera and Milne, 1994.)

croscopy. Using the goat anti-rabbit IgG conjugated to colloidal gold, cellular and subcellular immunolocalization of pathogenesis-related maize seed (PRms) protein was monitored. PRms accumulated at very high levels in those cell types, representing the first barrier of fungal penetration such as aleurone layer of germinating seeds and the subcellular epithelial cells of isolated germinating embryos. The presence of large number of fungal cells with abnormal shape with PRms-specific labeling was also observed. Accumulation of PRms in clusters over the fungal cell wall could also be noted. The possibility of PRms having a role in plant defense response is suggested because of the accumulation of PRms in cell types that first establish contact with *F. moniliforme* as well as in papillae and fungal cell walls (Murillo et al., 1999).

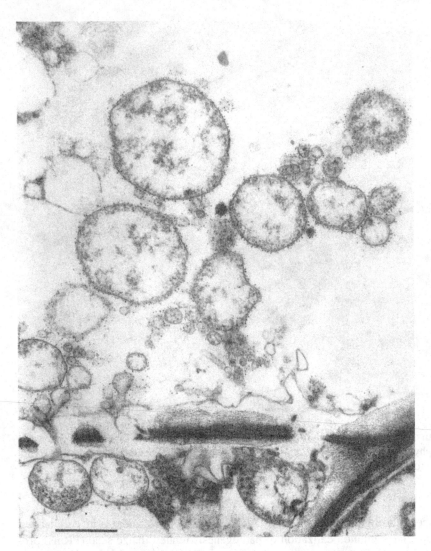

Figure 7.9 Primula yellows phytoplasma external to sieve plate are labeled and cells be-low sieve plate that are not exposed to antibody remain unlabeled; bar represents 500 nm (Kluwer Academic Publishers Netherlands). (Courtesy of Milne et al., 1995. Reprinted by permission of Kluwer Academic Publishers.)

7.6 TECHNIQUES FOR DETECTION OF BACTERIAL PATHOGENS

The immunogold labeling procedure for the detection and identification of bacterial pathogen has been used only to a limited extent. Fuerst and Perry (1988) developed the immunogold labeling procedure to demonstrate the presence of lipopolysaccharide antigens on both the sheathed flagellum and cell surface of *Vibrio cholerae*. Li et al. (1993) employed the immunogold labeling technique to determine the location of protein epitopes in *Pseudomonas andropogonis,* using monoclonal antibodies conjugated with protein A-gold complex. Gold particles were observed around lysed cells and within the cytoplasm of thin sections of bacterial cells (Appendix 7 [iv]).

7.7 SCANNING ELECTRON MICROSCOPY

The transmission electron microscope (TEM) (Fig. 7.1) is analogous to a compound light microscope, whereas the scanning electron microscope (SEM) is quite similar to a binocular stereoscopic microscope. In principle, the electron beam focused in an SEM is scanned across the specimen. When the electron beam strikes the specimen, secondary electrons are emitted and they are attracted toward a positively charged grill and detection system, generating electronic signals with variable intensity. These signals are then amplified and processed to produce the final image on a cathode-ray tube screen. The images on the screen can be photographed. SEM is generally used to study the surface features of a specimen up to 2 cm wide. Fungal and bacterial pathogens have been examined to study the morphological characteristics and ultrastructure of the spores and reproductive structures. Penetration of host tissues, intracellular development, and formation of haustoria in host cells (for nutrient absorption) have been clearly demonstrated by freeze fracturing frozen host tissue to expose internal hyphae and other fungal fructifications.

7.7.1 Bacterial pathogens

Colonization of pear blossoms by *Pseudomonas syringae* pv. *syringae* causing blossom blast diseases was studied by SEM. Numerous bacteria were observed between papillae, in the underlying tissue of stigmas, and on the nectariferous tissue at 1 day after inoculation. Bacterial masses were seen in receptacle a day later (Mansvelt and Hattingh, 1987). The evidence suggesting that *P. syringae* pv. *savastanoi* causing olive knot enter through wounds and leaf abscission scars was obtained by Surico (1993). *Agrobacterium tumefaciens* strain B6S3 causing tumors in potatoes was seen in the nondifferentiated cells of tumors, in xylem cells at the site of inoculation (of 5 weeks) as well as in xylem cells of the adjacent stem.

The bacteria were found attached by fibrillar aggregates to tumor cell walls (Stark-bauerová and Srobarová, 1993).

7.7.2 Fungal pathogens

The development and morphology of conidiophores and conidia of *Stemphylium vesicarium*, causing blight disease, on the leaf surface were studied. The emergence of conidiophores singly or in groups through the epidermis was observed. Smooth, round, bud-like conidial initials were formed at the apical end of verucose conidiophores and the matured conidia were oblong to ovoid and densely verucose (Aveling and Rong, 1994). The process of infection of *Botrytis cinerea* on nectarine and plum fruit was monitored using SEM, light and fluorescence microscopy. Production of a variety of infection structures on nectarine and plum fruit was observed. Protoappressoria and multicellular lobate appressoria were the predominant infection structures formed (Fourie and Holz, 1995). In the case of *Colletotrichum capsici*, causing anthracnose diseases, infections were initiated after production of appressoria. Infection hyphae penetrated cuticles and began to grow beneath cuticles in the anti- and periclinical walls of epidermal cells causing extensive wall degradation (Pring et al., 1995).

The early stages of pathogenesis for the powdery mildew pathogen (*Oidium* sp.) infecting poinsettia (*Euphorbia pulcherrima*) were examined by SEM. Primary germ tubes were initiated within 2 hr of inoculation. Appressoria arising from primary germ tubes were lobed and they were observed on 90% of the germinated conidia at 6 hr after inoculation. Haustoria were seen as globose structures with the haustorial sacs clearly visible (Calico and Hausbeck, 1998). [Appendix 7(v)]. The process of infection by *Colletotrichum actuatum* on guava fruits was visualized by SEM. The infection hyphae, growing directly from the conidia, either entered the fruit wall through stomata or directly penetrated the cuticle. Further development of the fungal pathogen by producing intercellular hyphae led to rapid disintegration of cell wall membrane and subsequent cell collapse, facilitating the acervuli to pierce the fruit cuticle (Das and Bora, 1998). Observations using SEM revealed the apparent morphological differences in the sporodochia produced by *Microdochium nivale* var. *nivale* on detached leaves of wheat, barley, and oat inoculated with the pathogen (Diamond et al., 1998).

Ultrastructural changes during early stages of pathogenesis have been observed using SEM. The germination of uredospores of *Hemileia vastatrix* (race III), causing coffee rust, appressoria differentiation on leaf stomata, and colonization of the mesophyll tissues inter- and intracellularly were tracked. The intercellular hyphae including haustorial mother cells and haustoria contained β-1,3-glucans and chitin in their walls. The interface of *H. vastatrix* with host cell was characterized by the presence of adhesive pectin material, which was detected by using anti-galacturonic acid monoclonal antibodies (Silva et al., 1999).

Direct penetration by *Fusarium moniliforme* hyphae through epidermal cells of the maize seedling and colonization of the host tissue by forming inter- and intracellular hyphae could be visualized using scanning microscope. Induction of defense-related ultrastructural changes as reflected by the formation of appositions on the outer host cell wall surface, the occlusion of intercellular spaces, and the formation of papillae is the noteworthy phenomenon observed following the pathogen ingress into the maize tissues (Murillo et al., 1999). *Polymyxa graminis* infects the roots of cereals and the zoospores of the pathogen act as vectors of wheat soil-borne mosaic virus. The cystosori produced by *P. graminis* were distributed only in the epidermal cells of barley roots. The cystosori contained hundreds of resting spores, which, when mature, released primary zoospores (Cheng et al., 1999). These investigations show that scanning electron microscopy is an effective tool to study the early stages of host-pathogen interaction leading to susceptible or resistance reactions.

SUMMARY

The basic principle involved in electron microscopy, methods of preparing support film, calibration of the electron microscope, and negative staining and different stains employed are described. The standard electron microscopic techniques, such as metal shadowing, dip method, leaf-dip serology, immuno-sorbent electron microscopy, decoration, and gold label antibody decoration, have been employed for rapid detection of viruses and phytoplasmas in plant materials. The usefulness of scanning electron microscopy to study the morphological characteristics and ultrastructure of the spores and reproductive structures that may be utilized for the differentiation of fungal and bacterial pathogens is discussed.

APPENDIX 7(i): DETECTION OF PHYTOPLASMAS IN CRUDE EXTRACTS OF PLANTS AND VECTOR INSECTS BY GOLD LABEL ANTIBODY DECORATION (GLAD) (VERA AND MILNE, 1994)

   i)   Coat carbon-fronted Formvar-filmed 400 mesh nickel grids with F(ab')2 of MLO-specific IgG at 6 μg/ml in 0.1 M maleate buffer, pH 6.8 (MB), for 15 min at room temperature.
   ii) Extract the sap of infected/healthy plant stem in 0.3 M glycine buffer, pH 8, containing 20 mM MgCl (GMgB), at 200 μl/100 mg of tissue, using a roller press; in case of infective insects, anesthetize with CO_2 and homogenize individual insects in 50 μl of GMgB.
   iii) Rinse coated grids in MB and float them on phytoplasma extracts for 3 hr at room temperature and rinse the grids with MB.

iv) Float the grids on 1% glutaraldehyde in MB for 15 min and rinse the grids with MB.

v) Float the grids on intact anti-phytoplasma or control IgG at 20 μg/ml in MB for 15 min and rinse with MB.

vi) Float the grids on goat antirabbit IgG bound to 5 nm or 15 nm diameter gold particles at 1:50 dilution (v/v) in MB for 15 min and rinse first with MB and then with distilled water.

vii) Negatively stain with 0.5% (w/v) ammonium molybdate, pH 6.5.

APPENDIX 7(ii): IDENTIFICATION OF PHYTOPLASMAS BY IMMUNOGOLD LABELING OF THIN SECTIONS OR BULK TISSUES FROM INFECTED PLANTS (MILNE ET AL., 1995).

A. Post embedding procedures for thin section labeling

i) Fix pieces of leaf in 4% formaldehyde (FA) (w/v), 0.1% glutaraldehyde (GA) (w/v), and 0.5 mM calcium chloride in 0.1 M potassium phosphate buffer, pH 7 (PB), at room temperature by vacuum infiltration for 1 hr; remove the edge tissue and treat the small veins with fresh fixative on ice in a shaker for another 1 hr.

ii) Wash four times, 10 min each, in PB plus calcium chloride, containing 3.5% sucrose, and shake for 1 hr in the same solution plus 50 mM ammonium chloride.

iii) Wash the tissues four times, 15 min each, in cold 0.1 M maleate buffer, pH 6.8 (MB), containing sucrose; change the maleate buffer and leave the tissues overnight in the buffer.

iv) Soak the tissues for 2 hr in MB-sucrose containing 2% uranyl acetate (UA), pH 6.4.

v) Transfer the tissues to 50% acetone (v/v) at 0°C and then to a rotor in a freezer at −20°C.

vi) Perform the following steps at −20°C: transfer tissues to 90% acetone and infiltrate with LR gold resin (LRG); then transfer the tissues to LRG plus 0.5% Lowicryl initiator (w/v); embed in closed BEEM capsules and polymerize under indirect UV light.

vii) Perform further processing at room temperature; block nonspecific sites for 15 min in NGS-TBS-BSA (normal goat serum (NGS) at 1/30 (v/v) dilution in 0.05 M Tris-HCl, 0.15 M NaCl (TBS) containing 0.2% bovine serum albumin (BSA) (w/v), pH 7.6).

viii) Incubate with primary antibody (50–200 μg/ml) diluted in NGS-TBS-BSA overnight at 4°C and rinse five times, 3 min each, on NGS-TBS-BSA.

ix) Incubate on GAR-G5 or GAM-G5 at 1/50 dilution in TBS, pH 8.2, containing 1% BSA; rinse on TBS, pH 8.2 five times, 3 min each, and rinse again on deionized water five times, 1 min each.

x) Stain with 5% aqueous uranyl acetate (w/v) for 60 min; rinse on dcionized water five times, 3 min each, and transfer the sections on filmed (Formvar) copper grids.

B. Preembedding procedures for labeling bulk tissues

i) Chill young leaves infected with phytoplasma on ice; isolate the veins and slice transversely into disks of 0.2–0.4 mm thickness.

ii) Immerse a group of disks in 50–200 μl of primary antibody diluted in PB placed in a well of a microtiter plate; agitate at 4°C for 60 min and rinse the disks in cold PB four times, 3 min each.

iii) Incubate the disks in 200 μl of GAR-G5 or GAM-G5 diluted in PB for 60 min and rinse them in cold PB four times, 3 min each.

iv) Fix the tissues in the cold on a shaker in 2.5% GA (w/v) in PB for 60 min; then fix in 0.1% osmium tetroxide (w/v) in PB for 60 min and embed in Epon, after proper orientation of the tissues for longitudinal sectioning.

v) Cut longitudinal sections through the entire disk of tissue, exposing both ends to antibody and strain in lead citrate, and observe under the electron microscope.

vi) Maintain controls in heterologous combinations.

APPENDIX 7(iii): DETECTION OF FUNGAL ANTIGENS BY PROTEIN A-GOLD LABELING (SVIRCEV ET AL., 1986)

A. Electron microscopy

i) Cut leaf disks (1 mm diameter) from inoculated and control leaves at predetermined intervals after inoculation.

ii) Fix in 4% glutaraldehyde in cacodylate buffer, pH 6.8, for 2 hr and rinse twice in buffer.

iii) Postfix with 2% osmium tetroxide for 1 hr and rinse twice in water.

iv) Stain in 5% uranyl magnesium acetate for 20 min at room temperature; dehydrate in a graded acetone series; infiltrate with Epon-Araldite; cut sections by using a diamond knife on an ultramicrotome; and mount the sections on coated grids.

B. Protein A-gold labeling

i) Add 4.0 ml of 1% aqueous sodium citrate to 100 ml of a boiling solution of 0.01% chloroauric acid; cool for 5 min until a wine red color develops and store the colloidal gold suspension in the dark.

ii) Adjust the pH of colloidal gold suspension (10 ml) to 6.9 with potassium carbonate; add the suspension to 0.3 mg of protein A in 0.2 ml of distilled water and centrifuge at 48,000 g at 4°C.

iii) Resuspend the dark red protein A-gold complex pellet in 10 ml of 0.01 M phosphate-buffered saline solution, pH 7.4, and store at 4°C.

iv) Place thin sections of test plant tissues on a saturated solution of sodium periodate for 2–3 min; wash the sections three times in distilled water and treat with 1% ovalbumin for 5 min to prevent nonspecific binding.

v) Float the sections on specific antiserum for 30 min at room temperature and wash the grids through a series of water droplets.

vi) Treat the sections with protein A-gold solution for 30 min; wash as before; stain with 3% uranyl acetate for 20 min and observe under the electron microscope.

APPENDIX 7(iv): IMMUNOGOLD LABELING OF BACTERIAL PATHOGENS (FUERST AND PERRY, 1988; LI ET AL., 1993)

A. Immunolabeling of whole cells

i) Cover a copper grid with nitrocellulose and stabilize with carbon; place the grid on a bacterial cell suspension, carbon-side down, for 5 min in a plastic petri dish or parafilm sheet.

ii) Place the grid successively on 1% glutaraldehyde in phosphate-buffered saline solution (PBS), pH 7.2, for 5 min; on PBS-20 mM glycine for 10 min; on PBS-1% bovine serum albumin (PBS-BSA) for 5 min; on MAB-culture supernatant fluid for 30 min; on 4 drops of PBS-BSA for 1 min each; on goat antimouse IgM and IgG gold at 1/20 dilution in PBS-BSA for 30 min; and finally on 3 drops of sterile distilled water (SDW) for 5 min each.

iii) Blot excess water with filter paper, air-dry, and observe under the electron microscope without staining or stain first with 4% uranyl acetate for 1 min and then with citrate for 30 sec before examining the grids under the electron microscope.

B. Immunolabeling of thin sections

i) Fix the pellet of bacterial cells, after centrifugation with 5% glutaraldehyde in 0.05 M N-2-hydroxyethylpiperazine-N'-2-ethanesulfonic acid (HEPES) buffer, pH 8.0, for 1 hr at room temperature.

ii) Wash the cells with buffer, suspend in 3% agarose, and fix in 1% osmium tetroxide in HEPES buffer for 1 hr at room temperature.

iii) Treat with 2% aqueous uranyl acetate for 30 min; dehydrate through two changes of 70% ethanol for 30 min each; infiltrate with LR white resin overnight at 4°C and embed in LR white resin by polymerization at 50°C for 24 hr.

iv) Cut thin sections by using an ultramicrotome and collect them on carbon-stabilized nitrocellullose-filmed copper grids.

v) Place the grids successively on PBS-BSA for 5 min on MAB diluted in PBS-BSA for 30 min, on 4 drops of PBS-BSA for 1 min each, of protein A-gold at 1/20 dilution in PBS-BSA for 30 min, and on 3 drops of SDW for 5 min each; remove excess fluid with filter paper and air-dry.

vi) Stain the sections with uranyl acetate for 5 min and then with Reynolds lead citrate for 2 min and observe under the electron microscope.

APPENDIX 7(v): SCANNING ELECTRON MICROSCOPY OF FUNGAL PATHOGENESIS (CELIO AND HAUSBECK, 1998)

i) Excise leaf disks with a cork borer (1 cm diameter): surface-sterilize with sodium hypochlorite solution (0.525%) for 1 min; rinse leaf disks in sterile distilled water and dry them under a laminarflow hood.

ii) Inoculate the leaf disks by using a plastic bristle paint brush to dislodge the conidia from infected leaf tissues kept 4 cm above the leaf disks to be inoculated.

iii) Place the inoculated leaf disks' abaxial surface upon on a piece of water agar (7 cm diameter, 20 g/L) suspended by a plastic mesh grid (1.7 squares/cm) in a glass humidity chamber containing saturated KCl salt solution (100 ml) for maintaining 82–85% relative humidity; incubate for required periods (2–48 hr).

iv) Trim leaf disks, after incubation; mount on copper specimen stubs using Tissue-Tek; process for observation using a sputter cryo system (Emscope SP 200).

v) Plunge the mounted specimens into liquid nitrogen slush for 30 sec; transfer under vacuum to the specimen chamber of a scanning electron microscope (JSM-35CF; JEOL, Peabody, MA) fitted with a cold stage attachment; Thermally etch the samples at −65°C to remove ice crystals.

vi) Transfer the samples under vacuum to the work chamber and sputter-coat with gold and observe the images at an accelerating voltage of 10 KV at a specimen temperature of −145°C.

8
Serodiagnostic Methods

Early detection of pathogenic infection is essential to eliminate all infected seeds and planting materials which form primary sources of infection and to rogue out infected plants present in the field whenever possible. The detection and characterization of plant pathogens usually depend on isolation of the pathogen and recording of the symptoms it induces on susceptible host plants. This procedure is a time-consuming process and often cannot be used to distinguish reliably between closely related species and strains of the pathogen, necessitating the use of faster and more discriminating detection methods. The methods of detecting plant pathogens may be divided into two groups: a) specific methods which may be used to detect a particular species or group of pathogens after preliminary diagnosis suggesting the presence of the particular pathogen, b) nonspecific methods which may be employed in detection of unknown pathogens or when the presence of a number of pathogens is to be detected, as in plant quarantines (Chu et al., 1989). However, the use of the term *specific* or *nonspecific* to indicate the extent of the reliability of a method to be employed for the detection of pathogens will be more appropriate. Thus serological and nucleic acid hybridization methods can be considered as more specific and reliable than most of chemical methods, which are often found to be nonspecific and less reliable, though these methods are simpler.

8.1 DETECTION OF PLANT VIRUSES

Production of antiserum, either polyclonal or monoclonal, is the first requirement for performing any of the several serological assay techniques. Different antibody preparations are available for the detection of various plant viruses. Polyclonal antibodies can detect several isolates or strains of a virus, whereas monoclonal antibodies can be used to differentiate between isolates or strains. The plant viruses for which monoclonal antibodies (MABs) have been produced are listed in Table

8.1 (Fig. 8.1). The choice of serological technique depends on specificity, sensitivity, and accuracy of results; ability to detect even at low concentrations of antigens; adaptability to field conditions; relatively inexpensive nature; and amenability to testing of a large number of samples within short periods, so that it can be employed in certification programs. Methods of production of antisera and various serological techniques have been described in detail by Matthews (1957), Bercks et al. (1972), Van Regenmortel (1982), Van Regenmortel and Dubs (1993), and Narayanasamy and Doraiswamy (1996).

In recent years significant advancements have been made in the production of antibodies specific for plant viruses. A novel method of producing antisera has been developed when purified viruses or viral proteins are not readily available. The coat protein (CP) gene of the virus is cloned and expressed in *Escherichia coli* cells and the polypeptide(s) produced in the bacterial cells are used as immunogen for the production of polyclonal antibodies in rabbits. Polyclonal antiserum against the nucleoprotein of tomato spotted wilt virus (TSWV) expressed as a recombinant fusion protein in *E. coli* was raised (Vaira et al., 1996). The protein p25 encoded by RNA3 of beet necrotic yellow vein furovirus (BNYVV) was cloned into a bacterial expression vector. The antiserum raised against the expressed p25 fusion protein was used to detect the subcellular location of p25 in mechanically inoculated sugar beet leaves (Li et al., 1996). The coat protein (CP) gene of a strain of garlic virus A (GarV-A) was cloned and expressed in *E. coli* cells and the polypeptide was employed to produce the polyclonal antibodies in rabbits (Tsuneyoshi and Sumi, 1996; Helguera et al., 1997) [Appendix 8(i)]. The movement protein (MP) and CP genes of grapevine virus A (GAV) were also cloned and expressed in *E. coli*. As the MP protein appeared earlier in infected plants, detection of the nonstructural MP may be a more effective means of detecting GAV infection in grapevines (Rubinson et al., 1997). The N-terminal P1 proteinase of PVY, associated with inclusion bodies induced by PVY, was expressed in *E. coli*. The antiserum raised against the expressed protein detected the P1 protein associated with the cytoplasmic inclusion bodies, characteristic of PVY infection (Arbatova et al., 1998). The polyclonal antibodies thus produced are used in different serological techniques whose sensitivity, specificity, and reliability are enhanced. Moreover, these antibodies combine specificity to the target proteins and versatility, in addition to the absence of undesirable reactions resulting from immunization using complex virus purified from infected host plant materials, as in the case of TSWV.

The development of phage-displayed recombinant antibodies specific for plant viruses is yet another remarkable achievement. This technique was first described by McCafferty et al. (1990). M13 phage particles are flexuous with dimensions of 6×870 nm and have ss DNA as the genome. It infects strains of *E. coli* that carry the f' episome (plasmid). The infected bacterial cells act like a factory producing M13 phage components continuously without undergoing lysis.

Table 8.1 Plant Viruses for Which MABs Have Been Produced

Virus group	Virus	References
1. Alfalfa mosaic virus	Alfalfa mosaic virus	Halk et al., 1984 Hajimorad et a., 1990
2. Carlavirus	Carnation latent virus Potato virus M	Jordan, 1989 Saarma et al., 1989
3. Carmovirus	Carnation mottle virus	Saarma et al., 1989
4. Caulimovirus	Carnation etched ring virus	Hsu and Lawson, 1985a, 1985b
5. Closterovirus	Apple chlorotic leaf spot virus	Poul and Dunez, 1989
	Carnation necrotic fleck virus	Saarma et al., 1989
	Citrus tristeza virus	Vela et al., 1986
	Pineapple wilt–associated closterovirus	Hu et al. 1996
6. Comovirus	Bean pod mottle virus	Joison and Van Regenmortel 1991
	Cowpea mosaic virus	Kalmar and Eastwell, 1989
	Cowpea severe mosaic virus	Kalmar and Eastwell, 1989
7. Cucumovirus	Cucumber mosaic virus	Haase et al., 1989 Maeda et al., 1988 Porta et al., 1989
8. Dianthovirus	Sweet clover necrotic mosaic virus	Hiruki et al., 1984
9. Furovirus	Beet necrotic yellow vein virus	Sukhacheva et al., 1996
	Peanut clump virus	Huguenot et al., 1989
	Soil-brone wheat mosaic virus	Bahrani et al., 1988
10. Geminivirus	African cassava mosaic virus	Thomas et al., 1986.
	Bean golden mosaic virus	Cancino et al., 1995
	Croton yellow vein mosaic virus	Swanson et al. 1998
	Maize streak virus	Dekker et al., 1988
	Tobacco leafcurl virus	Swanson et al., 1998
11. Hordeivirus	Barley stripe mosaic virus	Sukhacheval et al., 1996

Table 8.1 Continued

Virus group	Virus	References
12. Ilarvirus	Apple mosaic virus	Halk et al., 1984
	Citrus variegation virus	Hsu et al., 1983
	Prune dwarf virus	Jordan, 1984
	Prunus necrotic ringspot virus	Halk et al., 1984
	Tobacco streak virus	Halk et al., 1984
13. Luteovirus	Banana bunchy top virus	Wu and Su, 1990
	Barley yellow dwarf virus	Diaco et al., 1983
		D'Arcy et al., 1990
	Beet western yellows virus	D'Arcy et al., 1989
		Ohshima and Shikata, 1990
	Potato leaf roll virus	Martin and Stace-Smith, 1984; Ohshima et al., 1988; Ohshima and Shikata, 1990
	Soybean dwarf virus	Damsteegt et al., 1999
	Subterranean clover red leaf virus	Hewish et al., 1983
	Tobacco necrotic dwarf virus	Ohshima et al., 1989
14. Nepovirus	Arabis mosaic virus	Dietzgen, 1983
		Tirry et al., 1988
	Grapevine fanleaf virus	Huss et al., 1987
	Satsuma dwarf virus	Nozu et al., 1983, 1986
	Tomato ringspot virus	Powell and Marquez, 1983; Powell, 1990
15. Potexvirus	Cymbidium mosaic virus	Vejaratpimol et al., 1998
	Potato virus X	Torrance et al., 1984; Lizarraga and Fernandez-Northote, 1989
16. Poty virus	Bean common mosaic virus	Wang et al., 1985
	Bean yellow mosaic virus	Scoft et al., 1989
	Clover yellow vein virus	Scoft et al., 1989
	Grapevine leaf roll virus type III	Hsu et al., 1989; Zimmerman et al., 1990
	Lettuce mosaic virus	Hill et al., 1984
	Maize dwarf mosaic virus	Hill et al., 1984
	Papaya ringspot virus	Baker and Purcifull, 1984
	Pea mosaic virus	Scoft et al., 1989
	Peanut mottle virus	Sherwood et al., 1985; Sherwood et al., 1987

Table 8.1   Continued

Virus group	Virus	References
	Peanut stripe virus	Culver and Sherwood, 1988
	Plum pox virus	Himmler et al., 1988; Myrta et al., 1998
	Potato virus A	Gugerli, 1983; Bonnekamp et al., 1990
	Potato virus Y	Gugerli and Fries, 1983 Rose and Hubbard, 1986
	Shallot yellow stripe virus	Van der Vlugt et al., 1999
	Soybean mosaic virus	Hill et al., 1984; Nutter et al., 1998
	Sugarcane mosaic virus	Cheng et al., 1993
	Tobacco etch virus	Dougherty et al., 1985
	Tropaeolum mosaic virus	Soria et al., 1998
	Tulip breaking virus	Hsu et al., 1984
	Watermelon mosaic virus II	Somowiyarjo et al., 1988
	Welsh onion yellow stripe virus	Van der Vlugt et al., 1999
	Zucchini yellow mosaic virus	Somowiyarjo et al., 1988; Wisler et al., 1989
17. Reovirus	Rice dwarf virus	Harjosudarmo et al., 1990
	Rice ragged stunt virus	Ohshima et al., 1990
18. Sobemovirus	Southern bean mosaic virus	Tremaine and Ronald, 1983
19. Tenuivirus	Rice stripe virus	Omura et al., 1986
20. Tobamovirus	Beet necrotic yellow vein virus	Merten et al., 1985; Torrance et al., 1988
	Cucumber green mottle mosaic virus	Takahashi et al., 1989
	Odontoglossum ring spot virus	Dore et al., 1987
	Tobacco mosaic virus	Dietzgen and Sander, 1982; Briand et al., 1982
	Tomato mosaic virus	Dekker et al., 1987 Takahashi et al., 1989
21. Tospovirus	Tomato spotted wilt virus	Sherwood et al. 1989 Huguenot et al., 1990 Hsu et al., 1990 Adam et al., 1991
22. Trichovirus	Apple chlorotic leafspot virus	Malinoswki et al. 1997
23. Unclassified viruses	Grapevine fleckvirus	Schieber et al., 1997

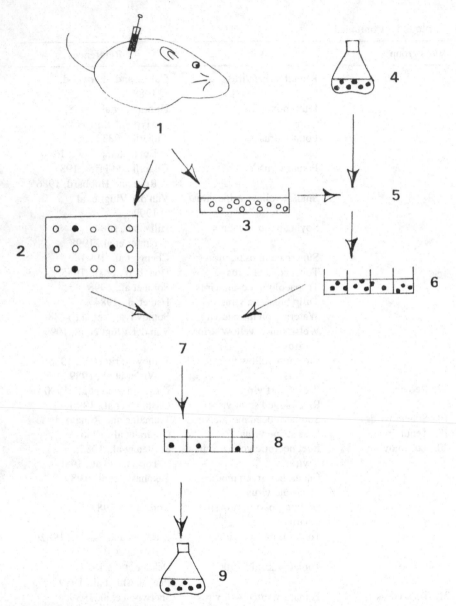

Figure 8.1 Steps in the production of monoclonal antibodies: 1, Immunization of mouse with antigen; 2, testing of screening methods with test bleed antisera; 3, collection of β-lymphocytes from spleen; 4, preparation of myeloma cell suspension; 5, fusion of myeloma cells and β-lymphocytes; 6, selection of hybridoma in multiwell plates; 7, determination of antibody contents secreted by selected hybridomas; 8, dilution of hybridomas to yield 1 cell/well; 9, cultivation of selected hybridoma cells and storing by freezing (CAB International, U.K.). (Adapted from Leach and White, 1991.)

The phage components include phage DNA, gene 8 coat proteins, gene 3 attachment proteins, and other proteins that may be linked or fused to these proteins. The gene 3 proteins (3–5 copies/phage particle) located on the tip of the phage facilitate the attachment of phage particles to receptors on the *E. coli* cells. The receptor sites are present in the hair-like projections (pili) expressed on the surface of *E. coli* strains carrying the f' episome.

For the production of M 13 phage-displayed recombinant antibodies, the DNA from antibody producing β-lymphocytes or hybridomas is genetically linked to the phage gene 3 DNA. The proteins encoded by the antibody DNA and gene 3 DNA are coexpressed or fused to one another resulting in the production of an antibody–gene 3 fusion protein. When the bacteriophage carrying the gene fusion infects *E. coli*, expression of an antibody molecule occurs on its tip. Any antibody DNA linked to phage DNA and any antibody proteins fused to phage proteins will be assembled and secreted in a manner similar to that of phage proteins.

The approach of using phage-displayed recombinant antibody technology is found to be useful for the detection of several plant viruses. A single-chain Fv antibody fragment (scFv) was prepared from a synthetic phage-antibody library after four rounds of selection against a purified preparation of potato leafroll luteovirus (PLRV). The DNA encoding the scFv was subcloned into pDAP2 in such a way that a scFv-alkaline phosphatase fusion protein was produced by transformed *E. coli* following induction by isopropyl-β-D-thiogalactopyranoside (IPTG). This fusion protein preparation could be directly used in enzyme-linked immunosorbent assay (ELISA) to detect PLRV in leaf extracts from infected plants (Haiper et al., 1997). The suitability of 11 different single-chain variable fragment antibodies (scFv) for diagnosis of potato leaf roll infection was assessed. These antibodies can be produced in large quantities cheaply in bacterial fermenters and they can be used in different ELISA formats such as standard triple antibody sandwich (TAS) ELISA (Toth et al., 1999). Likewise, scFv antibody fragments highly specific for potato Y potyvirus (PVY) were selected and used for the detection of PVY by ELISA technique. The strains PVY^O (common strain) could be differentiated from other strains PVY^N and PVY^{NTN} (Boonham and Barker, 1998). The detection and identification of two geminiviruses, African cassava mosaic virus (ACMV) and tomato yellow leafcurl virus (TYLCV), was achieved by employing phage-displayed peptides selected from the Cys 1 random phage display peptide library that bound strongly to the monoclonal antibody (SCR20) raised against ACMV. The production of peptides displayed on phage can be done quickly and it is less expensive (Ziegler et al., 1998). The phage-displayed peptides that specifically bound to coat protein (CP) of plant viruses can be selected for their rapid identification and differentiation [Appendix 8(ii)].

The PABs prepared against glutaraldehyde-fixed cucumber mosaic virus (CMV) virions were used to screen a random peptide library of heptamers displayed on the surface of phage. Among the 36 sequenced phage clones, eight had inserts similar to a putative virion surface domain of the CMV-CP. This region had sequences corresponding to amino acids 194–199 in the Fny-CMV CP, whereas another six clones contained sequences corresponding to C-terminal sequences of the Fny-CMV CP. The third group of four clones carried sequences that matched a portion of the sequences corresponding to amino acids 89–96 in the Fny-CMV CP (He et al., 1998). The phage-displayed antibodies can be effectively used to identify epitopes on the coat protein that determine the serological relationship between plant viruses. The phage-displayed peptides capable of specifically binding to the CP of CMV were used to detect CMV both in purified preparations and in crude leaf extracts from infected plants (Gough et al., 1999). The feasibility of employing phage display system for the detection of beet necrotic yellow vein furovirus (BNYVV) has also been reported by Griep et al. (1999). BNYVV-specific scFvs were produced in *E. coli* as a scFv fusion protein with alkaline phosphatase. The fusion protein can be used in ELISA as specific ready-to-use antibody-enzyme conjugates, providing an additional advantage for preferring this system for large-scale application. The detection of BNYVV in stored sugarbeets was difficult because of the loss of immunodominant C-terminal epitope of the viral coat protein due to proteolysis occurring during storage. This problem could be overcome by selecting clones that produce scFv specific for protease-stable BNYVV epitopes. By using the fusion proteins of the scFv with a human IgG kappa chain or with alkaline phosphatase, BNYVV was detected in stored sugarbeets in ELISA. The sensitivity of the tests was often higher than when polyclonal antibodies were employed for BNYVV detection (Uhde et al., 2000). These findings clearly demonstrate the usefulness of phage-displayed recombinant antibodies for the detection and differentiation of plant viruses and their strains.

The possibility of developing a genetic immunization system by engineering scFv fragments capable of recognizing epitopes on plant virus capsids has been indicated by Franconi et al. (1999). A scFv fragment was engineered to recognize an epitope of the glycoprotein G1 conserved among a large number of tospoviruses. An epichromosomal expression vector derived from potato X potexvirus (PVX) was used to assess the functional expression in the plant. The scFv gene was cloned to produce a cytosolic or secretory protein in this vector. The gene encoding the secretary scFv was used to transform *Nicotiana benthamiana* plants. These engineered scFvs may become inexpensive tools for virus disease diagnosis in addition to the possibility of development of a "plantibody"-mediated resistance to the tospoviruses such as tomato spotted wilt virus causing significant losses in several crops.

8.1.1 Labeled Antibody Techniques

By attaching a label to either the antigen or antibody, the sensitivity of detection of antigen-antibody reactions can be substantially increased. Three types of markers, viz. enzymes, fluorescent dye, and radioactive materials, have been most commonly used for labeling the antibodies specific for the antigen to be detected.

8.1.1.1 Enzyme-Linked Immunosorbent Assay

Introduction of enzyme-linked immunosorbent assay (ELISA) (Engvall and Perl-mann, 1971; Clark and Adams, 1977) has been an important landmark in serological detection and assay of plant viruses. It has become a preferred method, because of its sensitivity, economical use of antiserum, availability of quantifiable data, and the capacity for handling large numbers of samples rapidly. There are many variations in ELISA; the most common form is the double-antibody sandwich (DAS) method (Fig. 8.2). Other widely used variations are direct antigen coating (DAC)-ELISA, protein A (extracted from the cells of *Staphylococcus aureus*) coating (PAC)-ELISA, and indirect ELISA using virus antibody, in addition to labeled antiglobulin conjugate. The methods are described in Appendix 8 (iii) A, B, C, and D. The DAS-ELISA method is highly strain-specific and requires the preparation of different antibody conjugate for each virus to be tested. On the other hand, the indirect ELISA procedure has an important advantage in that a single conjugate can be used for different viruses and the antirabbit globulin conjugate or antigoat globulin conjugate commercially available can be employed. All serotypes of a virus may be detected with an antiserum to one strain (Van Regenmortel and Burckard, 1980). However, Su and Wu (1989), using monoclonal antibody specific to banana bunchy top virus (BBTV), reported that direct ELISA was 16 times more sensitive than indirect ELISA in detecting BBTV. When the PAC-ELISA procedure is followed, care should be taken to use the optimal concentration of protein A, as higher concentrations may cause nonspecific reactions, and lower concentrations may result in negative results. Under a short incubation period (1–2 h at 35°C) DAC-ELISA and PAC-ELISA were as sensitive as DAS-ELISA in detecting groundnut mottle and clump viruses (Hobbs et al., 1987). Lin et al. (1991) reported that avidin-biotin-peroxidase complex indirect sandwich ELISA was more sensitive than DAS-ELISA and protein A indirect sandwich ELISA in detecting cucumber mosaic virus. The use of streptavidin-polymeric horseradish peroxidase (HRP) conjugates enhanced the detection limit by about 12–25 times over the monoclonal DAS-ELISA system. With the streptavidin-polymeric HRP, it was possible to detect barley stripe mosaic hordeivirus (BSMV) even when a single infected seed was present among more than 104 healthy barley seeds. Likewise, the beet necrotic yellow vein furovirus (BNYVV) could be detected in leaf extracts diluted up to 1:12,000 (Sukhacheva et al., 1996).

The sensitivity of the different ELISA formats may vary depending on several factors such as the host-virus combinations, tissue tested, and level of susceptibility of the cultivar to the virus concerned. DAS-ELISA was found to be more sensitive in detecting potato X potexvirus than the double assay consisting

A   B

1    **1**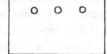

2 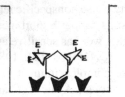   **2**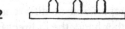

                             3

3 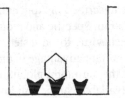   **4**

4
 5 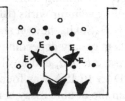   **5**

of DAS-ELISA and amplified ELISA performed in sequence, using the same samples in the same microtitre plate (Darda, 1998). A modified DAS-ELISA developed by Wu et al. (1998) is reported to detect both apple chlorotic leaf spot virus (ACLSV) and apple stem grooving virus (ASGV), whereas conventional DAS-ELISA could detect only ASGV and not ACLSV. However, both formats helped to shorten the detection time of the viruses.

Many signal enhancement techniques have been used in conjunction with the basic ELISA procedure.

a. Biotin-avidin. The protein avidin has very high affinity for biotin, which is chemically coupled to the IgG. Avidin is coupled to the enzyme used for detection. The procedure increases the number of enzyme molecules trapped on antigen, increasing the sensitivity (Zrein et al., 1986).

b. Fluorogenic substrate. A fluorogenic substrate, 4-methyl-umbelliferyl phosphate, for alkaline phosphate is used instead of the standard chromogenic substrate (Torrance and Jones, 1982).

c. Radioactive labels. In the radioimmune ELISA ^{125}I-labeled IgG is substituted for enzyme-linked IgG and the bound and labeled IgG is, after dissociation, assayed (Ghabrial and Shepherd, 1967).

d. Polystyrene beads. Polystyrene beads (6.5 mm diameter) are used as solid-phase instead of ELISA plates. Chen et al. (1982) found this procedure to be more sensitive in detecting the differences among isolates of soybean mosaic virus.

Van Regenmortel and Dubs (1993) have reviewed the merits of various other variations in ELISA suggested by different researchers. Modifications to improve sensitivity and to reduce assay time have been suggested by Goodwin and Banttari (1984) and Stobbs and Barker (1985).

Detection of virus infection in asexually propagated planting materials such as cuttings, tubers, and bulbs is important to have virus-free propagative materials. In certain cases, special requirements have to be satisfied to obtain reliable detection limits, even when sensitive assay methods such as ELISA are adopted. Wounding by removing a small piece of the bulbs just after lifting, followed by

Figure 8.2 Schematic representation of different steps in enzyme-linked immunosorbent assay (ELISA) and immunoblot assay (DIBA): ELISA: (A) 1, Adsorption of specific antibody to plate; 2, trapping of antigen by antibody; 3, addition of enzyme-labeled specific antibody; 4, addition of enzyme substrate; 5, color development indicating positive reaction. DIBA: (B) 1, Spotting of antigen sample onto paper; 2, addition of specific antibody; 3, addition of protein A-enzyme conjugate; 4, addition of enzyme substrate; 5, color development indicating positive reaction (CAB International, U.K.). (Adapted from Leach and White, 1991.)

storage at 17°C or 20°C for 3 weeks, markedly increased the detectability of iris severe mosaic virus by ELISA (van der Vlugt et al., 1993). The lily mottle potyvirus (LMoV) can be detected by ELISA only when the scales are stored at 20°C for 2–3 weeks under white fluorescent lights (for 12–16 hr/day; 13–16 W/m^2) prior to the testing, although lily symptomless carlavirus (LSV), cucumber mosaic virus, and lily X potexvirus are easily detected by ELISA without any pretreatment (Derks et al., 1997). Likewise, storage of potato tubers at 20°C for 3 weeks prior to testing resulted in doubling of the number of sample sites on tubers at which potato moptop virus (PMTV) could be detected by ELISA. Pretreatment of tubers at higher temperature may improve the accuracy of tuber tests, especially when tubers may carry viuses that are both symptomless and irregularly distributed in tuber tissues (Sokmen et al., 1998). In the case of tomato spotted wilt virus (TSWV) infection in dahlia, leaf tests did not give reliable results, since the virus was not uniformly distributed in the foliage. Hence the procedure involving storage of tubers for a minimum period of 4 weeks and taking mixed samples of three roots per tuber resulted in ELISA detection of TSWV with sufficient reliability to facilitate the removal of infected tubers. The combination of inspection, detection, and selection of virus-free tubers has been primarily responsible for significant reduction in TSWV infection in Dutch dahlia crops (Schadewijk, 1996).

The serological activity of banana bunchy top virus could be preserved for 1 year by drying the leaf samples in air oven at 50°C for 14–16 hr followed by storage at −70°C. Thus samples collected in one place can be sent for testing in another country where facilities for diagnosis are available (Wu, 1998). Sufficient reduction in the reaction volume and work time can be achieved by adopting the fluorogenic ultramicro enzyme immunoassay (DAS-UM-ELISA) developed by Peralta et al. (1997). UM-ELISA requires as little as 10 μl of reactants and the assay is completed in 5 hr. There was no significant difference in the sensitivity of detection of citrus tristeza between ELISA and UM-ELISA.

Attempts have been made to find less expensive chemicals to reduce the cost of testing without losing the sensitivity of the alkaline phosphatase (ALP) system. An ELISA test conducted with horseradish peroxidase (HRP) was as sensitive and specific as the ALP system for detection of turnip yellow mosaic virus, potato virus Y, and potato leaf roll virus in infected plants (Polak and Kristek, 1988). Neustroevx et al. (1989) reported that β-galactosidase isolated from *Escherichia coli* was more effective in labeling antibodies to potato virus X (PVX) than horseradish peroxidase, as it decreases background reaction because of the lack of this enzyme in plant sap. Penicillinase-(PNC)-based ELISA had similar sensitivity to ALP- and H RP-based ELISA for detecting maize mosaic virus, peanut mottle virus, and tomato spotted wilt virus. Penicillinase and penicillin are readily available and substantially less expensive (Sudharsana and Reddy, 1989).

It was found that DAS-ELISA with urease conjugate was superior to the ALP system, as the color change from yellow to violet could be more easily visually recognized than the change from colorless to yellow in the ALP system. The visual detection threshold was 0.5 ng/ml of TMV with urease and 500 ng/ml with phosphatase (Gerber and Sarker, 1988). Mizenina et al. (1991) developed the protocol for using the inorganic pyrophosphatase (PPase) from *Escherichia coli* conjugated with antibodies and tetrazolium pyrophosphate as the substrate in ELISA. The advantages of this method are high sensitivity, negligible level of background reactions in controls, bright blue-greenish color allowing visual detection of reaction, and high stability of PPase-conjugated antibodies. Twelve viruses, such as potato viruses X, Y, M, S, and leaf roll; carnation mottle virus; barley stripe mosaic virus; tobacco necrosis virus; soybean mosaic virus; and cucumber mosaic virus, could be detected by this method. The use of inorganic pyrophosphatase provides greater sensitivity for the detection of raspberry ringspot, strawberry latent ringspot, tomato black ring, and arabis mosaic nepoviruses, when compared to DAS-ELISA using horseradish peroxidase (HRP). These viruses could be more reliably detected in crude extracts of infected raspberry leaves (Surguchova et al., 1998).

e. Applications of ELISA.   Enzyme-linked immunosorbent assay tests have been performed to detect the viruses in different plant parts, seeds, and vectors which transmit the plant viruses even when the viruses are present in very low concentrations and in very early stages of disease development. Arabis mosaic virus could be detected at concentrations as low as 80 ng/ml and in extracts of infected leaves at dilutions up to 10^{-16}. Plum pox virus, an elongated virus, was detectable at 10 ng/ml in purified virus preparations and in leaf extracts at 10^{-4} dilution (Voller et al., 1976). Enzyme-linked immunosorbent assay has been used to detect several viruses, including citrus tristeza virus (Bar-Joseph et al., 1979); groundnut viruses (Nolt et al., 1983; Konate and Barro, 1993); squash mosaic virus in extracts from seed coats, papery layers, and distal halves of embryos from individual curcurbit seeds (Nolan and Campbell, 1984); rice tungro-associated viruses in rice plants (Bajet et al., 1985); strawberry latent ringspot virus in many plant species (Kristek and Polak, 1990); rice grassy stunt virus in infected rice plants and individual viruliferous planthopper (*Nilaparvata lugens*) (Iwasaki et al., 1985); and maize streak virus in infected maize plants and inoculative *Peregrinus maidis* (Falk et al., 1987) and in tomato leaf curl virus-infected tomato plants and viruliferous whiteflies (Ragupathi, 1995). Using DAS-ELISA, MacKenzie and Ellis (1992) observed that there was no systemic accumulation of TSWV in transgenic tobacco plants expressing viral nucleocapsid protein, since these plants were resistant to infection after mechanical inoculation with virus. Dahal et al. (1992) employed ELISA to determine the amount of viral coat protein of rice tungro-associated viruses in agroinfected rice plants and found that it was highly correlated with viral nucleic acid contents determined by DNA hybridization.

Using the polyclonal antiserum raised against African cassava mosaic virus (ACMV), eight geminiviruses were detected by ISEM in leaf extracts of plants infected by them, indicating their serological relationship with ACMV. In DAS-ELISA tests, two of the eight geminiviruses did not react with ACMV antiserum, whereas the strength of the reaction of other viruses varied widely (Table 8.2). Among the panel of 10 MABs, there was no reaction between 5 MABs and the 8 geminiviruses tested. The reaction strength with the rest of the MABs employed varied from 0 to 4, as determined by the A_{405} nm values after overnight incubation (Table 8.3) (Harrison et al., 1991).

The presence of viruses in vectors has been detected by ELISA tests. Tomato spotted wilt virus (TSWV) could be detected in 210 individuals of 340 *Frankliniella occidentalis* and in 24 of 120 of *F. schultzei* laboratory-grown adult thrips which had access to infected plants as larvae (Cho et al., 1988; Marchoux et al., 1991; Chamberlain et al., 1993). The accumulation of two proteins—the nu-

Table 8.2 Detection of Geminiviruses by ISEM and DAS-ELISA in Leaf Extracts

		Technique	
Virus	Source plant	ISEM[a]	DAS-ELISA[b]
Bhendi yellow vein mosaic virus (BYVMV)	*Abelmoschus esculentus*	+	0
Croton yellow vein mosaic virus (CYVMV)	*Croton bonplandianum*	+++	+++
Dolichos yellow mosaic virus (DYMV)	*Lablab purpureus*	+	E
Horsegram yellow mosaic virus (HYMV)	*Phaseolus lunatus*	+	−
Horsegram yellow mosaic virus (HYMV)	*P. vulgaris*	++	0
Indian cassava mosaic virus (ICMV)	*Manihot esculenta*	+	+++
Malvastrum yellow vein mosaic virus (MYVMV)	*Malvastrum coromandelianum*	0	E
Tomato leaf curl virus (TomLCV)	*Lycopersion esculentum*	+	+
Thailand mungbean yellow mosaic virus (TMYMV)	*P. vulgaris*	+++	+++

[a] Virus particles abundant (+++), common (++), few (+), or not found (0).
[b] Reaction strong (+++), moderate (++), weak (+), equivocal (E), not detected (0), or not tested (−).

Source: Harrison et al. (1991).

Table 8.3 Reactions of Geminiviruses with a Panel of MABs to African Cassava Mosaic Virus

Virus	Monoclonal antibodies (MABs)[a]				
	SCR 15	SCR 17	SCR 18	SCR 20	SCR 23
Bhendi yellow vein mosaic virus (BYVMV)	—	0	4	1	0
Croton yellow vein mosaic virus (CYVMV)	3	4	4	4	3
Dolichos yellow mosaic virus (DYMV)	0	4	3	3	4
Horsegram yellow mosaic virus (HYMV)		0	4	3	3
Indian cassava mosaic virus (ICMV)	1	0	4	4	0
Malvastrum yellow vein mosaic virus (MYYVMV)	—	0	1	1	0
Tomato leaf curl virus (TomLCV)	—	0	2	0	0
Thailand mungbean yellow mosaic virus (TMYMV)	0	4	2	4	—

[a] Reaction strengths classified on A_{405} nm values as 4 (> 1.8), 3 (1.2–1.8), 2 (0.6–1.2), 1 (0.3–0.6), and 0 (< 0.3).
Source: Harrison et al. (1991).

cleo capsid (N) and a nonstructural (NSs) proteins—of TSWV in the larvae and adults of *F. occidentalis* after ingestion of the virus for short periods on infected plants was detected by ELISA. Within 2 days the amounts of both proteins increased above the levels ingested, indicating the possible multiplication of TSWV in the thrips (WijKamp et al., 1993). Bandla et al. (1994) also reported that monoclonal antibodies specific to NSs proteins could be employed to detect TSWV in thrips vectors. The changes in the accumulation of TSWV nucleoprotein (N) in different growth stages of *Thrips setosus* were studied using ELISA. The N-protein concentration registered progressive increase from the first instar to second instar larval stage following acquisition feeding on infected plants and the concentration attained its peak on the fifth day. Virus titer declined dramatically commencing from the second instar larval to pupal stage. Though the N-protein concentration in the adult thrips was low, they could transmit the virus, indicating that the thrips may be efficient transmitters of TSWV (Tsuda et al., 1996). The efficiency of transmission of chrysanthemum stem necrosis tospovirus (CSNV) by three species of thrips, *Frankliniella schultzei, F. occidentalis,* and *Thrips tabaci,* was assessed, using a leaf disc assay combined with DAS-ELISA. *F. schultzei* was found to be more efficient than *F. occidentalis,* while *T. tabaci* did not transmit CSNV. However, low amounts of CSNV could be detected in 75% of the tested population of *T. tabaci* (Nagata and Avila, 2000). Tospovirus subgroups I and II are considered distinct species and designated as species-tomato spotted wilt virus (TSWV) and impatiens necrotic spot virus (INSV) on the basis of results obtained with DAS-ELISA (Vaira et al., 1993). Virus contents of individual aphids carrying sugar beet western yellows, barley yellow dwarf virus, and alfalfa mosaic virus could be determined by ELISA (Kastirr, 1990; Ahoonmanesh et al., 1990). Vector indexing of the aphids by employing ELISA revealed that a greater percentage of *Myzus persicae* trapped in sugarbeet crop carried beet mild yellowing virus (BMYV), when compared with aphids trapped in oilseed rape crop. In contrast, there was no significant difference in the percentage of *M. persicae* carrying beet Western yellows virus trapped in both sugarbeet and oilseed rape crops (Stevens et al., 1995). The enzyme-amplified ELISA technique was found to be more sensitive in detecting BMYV in the aphids than the standard DAS-ELISA procedure. The reliability of BMYV detection was increased when the aphids were indexed 48 hr after acquisition feeding. However, BMYV could not be detected in individual aphids (Polák, 1998). On the other hand, faba bean necrotic yellows virus was detected in individual aphids (*Aphis craccivora* and *Acyrthosiphon pisum*) using triple antibody sandwich (TAS) ELISA procedure (Franz et al., 1998).

Chickpea chlorotic dwarf geminivirus (CCDV) was detected in individual leafhoppers (*Orosius occidentalis*) by DAS-ELISA. The concentration of the virus decreased when the leafhoppers were fed on a nonhost for the virus, indicating the absence of virus multiplication in the leafhopper vector (Horn et al., 1994). The pineapple wilt–associated closterovirus (PCV) was detected by indirect ELISA using MABs produced against PCV. The reactivity of MABs with PCV was observed by the decoration of PCV particles in immunosorbent electron microscopy (ISEM). In the indirect ELISA test, a PAB was employed to trap PCV particles and this was followed by allowing the MABs to react with the virus. The pineapple root tissues were found to be the most suitable plant tissue for the detection of PCV by indirect ELISA. The presence of PCV in the mealy bugs (*Dysmicoccus brevipes*) collected from wilted pineapple plants was detected by this ELISA procedure (Hu et al., 1996).

The availability of virus reservoirs in weed hosts is an important epidemiological factor. Cho et al. (1989) analyzed more than 9000 samples by ELISA and found that 44 plant species in 16 families were infected with TSWV. Among these, 25 plant species were important reservoirs and 24 species were new hosts for TSWV. The presence of sowbane mosaic sobemovirus (SoMV) in the seeds of *Chenopodium album* and *C. quinoa* was detected using DAS-ELISA. Seed transmission of SoMV can result in reduction in weed population and consequent reduction in sources of virus infection (Kazinczi and Horváth, 1998). An indirect ELISA procedure was employed to detect the papaya lethal yellowing virus (PLYV) in the papaya lethal yellowing virus (PLYV) in the papaya seeds, soil, and also water from the rhizosphere of infected papaya plants. PLYV present on the seed surface may become a source of virus in the field (Camarco et al., 1998). Enzyme-linked immunosorbent assay can be used for rapid field tests to screen a large number of samples, and it is of great benefit for epidemiological studies and for indexing of seeds and vegetatively propagated plant materials (Huguenot et al., 1990). Indirect ELISA using monoclonal antibody (MAB) gave accurate results for the detection of bean common mosaic virus in individual bean seeds. Klein et al. (1992), using indirect ELISA to test flour samples of seeds, reported that incidence of bean common mosaic virus-infected seedlings could be predicted in germplasm accessions. Konate and Barro (1993) also detected the peanut clump virus in 16.5% of seeds by ELISA. Indian peanut clump furovirus (IPCV-H) was detected in the seeds of groundnut (peanut) by DAS-ELISA. A positive correlation between the results of ELISA and growing-on test, the conventional seed health test, was observed (Reddy et al., 1997), indicating the reliability of ELISA results to determine the seed transmission of IPCV-H. ELISA also detected IPCV-H in all seeds of infected wheat plants, though natural transmission of the virus through wheat seeds occurred at a frequency of 0.5–1.3%. IPCV-H could not be detected in the seeds of barley, which is another natural host for the virus (Delfosse et al., 1999). Pelargonium line pattern virus could be reliably detected by ELISA, if petioles of fully expanded leaves were tested. Virus distribution in infected pelargonium plants was studied by assaying the virus concentration by ELISA (Bouwen and Maat, 1992).

Using the standard DAS-ELISA procedure and specific MABs, the barley yellow mosaic virus (BaYMV) was detected in leaves and stems of infected plants at dilutions of 1/2560 and 1/160, respectively. The differences in the detection thresholds may be due to variation in the virus titer in different plant tissues (Ma Hong et al., 1997). The MAB specific to Indian tobacco leafcurl virus was found to provide more sensitive and reliable detection than the polyclonal antibodies raised against this virus (Swanson et al., 1998). A sensitive DAS-ELISA method for the detection of cymbidium mosaic virus in orchids was developed by using virus-specific MAB as the coating antibody and egg yolk immunoglobulin G (IgY) as the detecting antibody. The virus was detected in leaf extracts diluted up to 1024-fold (Vejaratpimol et al., 1998). The antiserum raised against the coat protein (CP) of garlic mite–associated filamentous virus (Gar V-A) specifically reacted in ELISA with extracts from plants infected by Gar V-A but not with plants infected by onion yellow dwarf potyvirus, leek yellow stripe potyvirus, or carnation latent carlavirus (Helguera et al., 1997).

Routine indexing of fruit trees and asexually propagated plants in the field has been done for the presence of several viruses, such as apple mosaic virus (Clark et al., 1976), plum pox virus (Adams, 1978), prune necrotic ringspot virus (Barbara et al., 1978), and citrus tristeza virus (Bar-Joseph et al., 1979), and the testing of potato leaves, tubers, and sprouts for the presence of potato virus S (Richter et al., 1977), potato virus Y and potato virus A (Maat and DeBokx, 1978b), potato leaf roll virus (Maat and DeBokx, 1978a; Smith et al., 1993), and PVX (De Bokx et al., 1980). Tomato ringspot virus, naturally transmitted by nematode, was transmitted by dodder to *Cucumis sativa,* and the presence of the virus was detected by ELISA test (Welliver and Halbrendt, 1992). By using a specific MAB, bean golden mosaic, euphorbia mosaic, rhynchosia mosaic, squash leaf curl, soybean yellow mosaic, and tomato mottle geminiviruses were detected by Cancino et al. (1995). Biological indexing for detection of viruses infecting fruit crops requires considerably long periods. Use of molecular methods such as ELISA provides a distinct advantage by reducing the time required, in addition to the possibility of handling large number of samples as demonstrated for germplasm screening programs. In the case of citrus psorosis-associated viruses, all plants that were psorosis positive by indexing on orange cv. Madame Vinous and Dweet Tangor were also found to be positive when tested by ELISA. Additionally, application of ELISA helped to detect infection in four more citrus accessions that indexed negative. A high percentage of plants were ELISA positive when compared to biological indexing, suggesting that ELISA may be used for the detection of psorosis-associated virus reliably (D' Onghia et al., 1998).

Resistance to peanut stripe virus in different *Arachis* spp. was tested by detecting the virus by ELISA test. Certain lines of *A. diogi, A. halodes,* and *Arachis* spp. did not have detectable amounts of the virus (Culver et al., 1987). Screening of plum germplasm entries for the presence of viruses was done using ELISA. Flowers, leaves, and bark of clones were tested and the presence of plum pox potyvirus, apple mosaic virus, and *Prunus* necrotic ringspot virus was confirmed rapidly, facilitating the selection of resistant entries (Karesová and Paprstein, 1998). The incidence of sugarcane bacilliform virus in the sugarcane gemplasm collection has been assessed by ELISA to facilitate the selection of resistant lines (Viswanathan and Premachandran, 1998). The effectiveness of elimination of peanut mottle virus, peanut stripe virus, and TSWV from vegetatively maintained groundnuts was tested by DAS-ELISA (Dunbar et al., 1993). Tiongco et al. (1993) employed the ELISA test or latex test to monitor the incidence of rice tungro disease. Rice tungro spherical virus (RTSV) was detected 1 week earlier than rice tungro bacilliform virus (RTBV). The infectivity of green leafhoppers collected from the field corresponded to disease development. Induction of resistance to rice tungro disease in susceptible cultivars by using the antiviral principles (AVPs) from the seed sprouts of pigeonpea and mungbean was demonstrated by determining the titers of RTBV and RTSV, after the application of AVPs (Muthulakshmi and Narayanasamy, 1996).

Transgenic cultivars or clones expressing viral genes have been developed as an effective management strategy against virus diseases. The level of resistance provided by the transgenic plants has been assessed by determining the virus concentration using different ELISA formats. A significant reduction in the incidence of watermelon mosaic and zucchini yellow mosaic viruses in transgenic lines of squash and cantaloupe expressing the coat protein genes was revealed by ELISA tests (Clough and Hamm, 1995). Transgenic plum trees (*Prunus domestica*) expressing CP gene of plum pox virus (PPV) were inoculated by viruliferous aphids or chip budding and tested by DAS-ELISA to determine the virus infection. Among the five transgenic clones tested, clone C-5 was found to be resistant to PPV (Ravelonandro et al., 1997). The transgenic tomato lines expressing CP gene of cucumber mosaic virus (CMV) showed attenuated symptoms of infection, because of the lower concentration of CMV in transgenic plants as determined by ELISA (Murphy et al., 1998). Potato cv. Pito plants transformed with P1 sequence of potato virus Y (PVY°) in antisense orientation were highly resistant to the PVY° strain. Neither detectable amounts of PVY in the inoculated and the upper uninoculated leaves as determined by ELISA nor the symptoms of infection were found at 21 and 35 days after inoculation. The results indicate that effective strain-specific resistance to PVY can be achieved by expressing the P1 sequence in antisense orientation in potato and that the level of resistance can be determined by ELISA based on the concentration of virus present in transgenic plants (Mäki-Valkama et al., 2000a).

The temporal increase and spatial spread of soybean mosaic virus (SMV) strain G-5 released from a point source were assessed employing strain-specific MABs, so that non-G5 SMV isolates could be discriminated under field conditions. The virus isolates from external sources have the potential to change the temporal and spatial patterns of within-field virus spread. SMV strain G-5 spread at random within the field in some years, but the pattern was mostly aggregated in certain years, indicating the possible interplay of the isolate from external sources. Non-G-5-SMV isolates arising from exogenous sources showed a random spatial pattern over a period of time. This study used strain-specific MABs for detection of virus infection and disease incidence to monitor and model SMV spread in time and space (Nutter et al., 1998).

Though the ELISA test is found to be useful for detection of many plant viruses, Ramsdell et al. (1979) reported that it was less sensitive than infectivity tests on *Chenopodium quinoa* for indexing immature grape tissue for peach rosette mosaic virus. Hughes and Ollennu (1993) also found that ELISA was less sensitive for the detection of cocoa swollen shoot virus than the virobacterial agglutination (VBA) test involving *Staphylococcus aureus* cells for conjugation with antivirus antibodies. The VBA test alone detected the virus in all cocoa trees infected by swollen shoot virus and additionally identified infection in many symptomless trees. Moreover, the need for careful interpretation of the results obtained by ELISA is indicated by Mink et al. (1985), who observed that extracts from rapidly growing shoot tip leaves of apparently healthy apple rootstock and scion trees showed absorbance values A_{405} nm that were similar to those of tomato ringspot virus-infected samples. No virus could be detected in these tissues by bioassay or partial purification. The rapidly growing apple shoots appear to contain nonviral antigens capable of reacting with antibodies in several tomato ringspot virus antisera.

Among the several applications of ELISA, rapid identification of new viruses or strains has been important. Grapevine leaf roll disease is a complex with which five viruses are known to be associated. Gugerli et al. (1997) reported the association of the sixth virus designated grapevine leaf roll–associated virus-6 (GLRa V-6) with leaf roll complex in a Chassela clone. GLRaV-6 was detected using a new MAB in ELISA in the extracts of leaf blades and petioles of infected plants. Another GLRaV-associated virus (GLRaV-2), occurring in low concentration in infected grapevines, was detected by DAS-ELISA using virus-specific antibodies (Köklü, 1999). The serological relationship between GLRaV-1 and GLRaV-3 was established by using MABs that recognize the coat proteins (CPs) of these viruses in DAS-ELISA. One MAB (IG10) reacted specifically with the CPs of GLRaV-1 and GLRaV-3 in ELISA, immunosorbent electron microscopy (ISEM), and Western blotting. GLRaV-1 could be differentiated by employing either the MAB IC4, which detected 25 of the 33 GLRaV-1 isolates, or MAB 1B7, which reacted positively with 32 isolates including the isolates that were not recognized by MAB IC4. These MABs are being routinely used for the detection of GLRaV-1 (Seddas et al., 2000). A new mechanically transmissible virus infecting mashua (*Tropaeolum tuberosum*) was identified and named *Tropaeolum* mosaic potyvirus (TropMV) based on the results of ELISA tests. TropMV was found to be distinct, but closely related to potato virus Y (common strain), tobacco etch virus, potato virus A, watermelon mosaic virus, and bean common mosaic virus and it shared the buried epitopes with maize dwarf mosaic virus (Soria et al., 1997). A new bacilliform virus infecting *Dioscorea alata* was identified and its serological relationship to sugarcane bacilliform virus and banana streak virus was established. The *Dioscorea* bacilliform virus (DaBV) was found to be a distinct badnavirus (Philips et al., 1999).

Another new virus infecting tobacco was identified as a phytoreovirus using indirect ELISA. The PAB raised against the wound tumor virus (WTV), the type species of the genus *Phytoreovirus,* showed strong cross-reaction to the virus infecting tobacco indicating the relationship with WTV (Rey et al., 1999). A new mechanically transmissible virus with icosahedral particles (approximately 28–30 nm diameter) and positive sense single-stranded RNA, infecting maize in Nigeria, was identified by ELISA and immunoblot assays as maize mild mottle virus (Thottappilly et al., 1999). Two serologically distinct tospoviruses, chrysanthemum stem necrosis virus (CSNV) and zucchini lethal chlorosis virus (ZLCV), occurring in Brazil were identified by ELISA and Western immunoblot analyses. The nucleocapsid (N) protein amino acid sequence analysis of CSNV and ZLCV showed similarities with other tospoviruses ranging from 20 to 75%, but these newly identified tospoviruses were more closely related showing 80% similarity between them (Bezerra et al., 1999).

Using ELISA, serological differentiation and identification of plant viruses and strains have been performed. The serological relationships of five isolates of Indian peanut clump virus were established by ELISA (Nolt et al., 1988). Sherwood et al. (1989) reported that the monoclonal antibody (MAB) against TSWV reacted positively with isolated nucleocapsid proteins, but not with envelope-associated proteins in the ELISA test. Twenty isolates of TSWV were grouped into three serotypes based on their reactions to the antisera raised against virus nucleocapsid protein and glycoprotein (de Avila et al., 1990). On the basis of the serological reactivity and infectivity on *Capsella bursa-pastoris,* the common strain and a new strain of beet mild yellowing luteovirus could be distinguished (Table 8.4) (Stevens et al., 1994). The common strain reacts with all three MABs positively and also infects *C. bursa-pastoris,* whereas the new strain reacts positively only with two of the three MABs and does not infect *C. bursa-pastoris.* Hill et al. (1994) proposed a system of antigenic signature analysis for the rapid differentiation of soybean mosaic virus isolates by employing a panel of nine MABs. The antisera may be prepared against disrupted virus particles or other proteins present in virus-infected plants such as helper components, and these antisera can also be used for detection and differentiation of plant viruses. Joisson et al. (1992) demonstrated that by using MABs prepared with proteolysed tobacco etch potyvirus and five other potyviruses, viz., PVY, pepper mottle virus, papaya ringspot virus, watermelon mosaic virus, and bean yellow mosaic virus, could be detected. Canto et al. (1995) reported that MABs and PABs raised against the helper component-pro-

Table 8.4 Differentiation of Beet Mild Yellowing Luteovirus Strains by ELISA Test and Infectivity[a]

Strain of		MABs		
BMYV	Infectivity	MAFF24	BWYVV-BC-510H	BYDV-PAV-IL-1
Common strain	+	+	+	+
New strain	−	+	+	−

[a] +, positive; −, negative.
Source: Stevens et al. (1994).

tease (HC-Pro) purified from plants infected with a nonaphid transmissible strain
of PVY can be used to differentiate strains of PVY.

The use of MABs in different ELISA formats has provided a reliable basis
for the differentiation of and grouping strains of viruses into serogroups. Smith et
al. (1996) produced a monoclonal antibody that differentiated a sugarbeet-infect-
ing isolate of beet mild yellowing luteovirus (BMYV) with differing host range
and serological properties from other commonly occurring strains under field conditions. Isolates of wheat yellow mosaic virus from France and Japan were differentiated by using an MAB in antigen-coated plate (ACP) ELISA and indirect
DAS-ELISA procedures (Hariri et al., 1996). Apple chlorotic spot virus strains
could be identified by using different MABs. MAB5 was suitable for routine
screening by ELISA, while a nontypical strain required the use of MAB1, MAB2,
and MAB9 (Malinowski et al., 1997). Both MABs and PABs prepared against
Chinese wheat mosaic virus (CWMV) and soil-borne wheat mosaic virus Okla-
homa isoalte (SBMV-OKI) were employed to differentiate the wheat and oat
furoviruses [WMV, SBWMV, oat golden stripe virus (OGSV)] and European
wheat mosaic virus (EWMV). By using ELISA and Western blotting techniques,
the dominant epitope(s) of CWMV was shown to share partially with OGSV,
while those of SBWMV were found to share with CWMV, OGSV, and EWMV
in different degrees (Ye et al., 2000).

The phenomenon of cross-protection operating in plants between strains or
closely related viruses has been exploited for protecting plants against several
strains in some crops such as citrus (Chapter 5). The development of cross-protection has been studied at molecular level by employing ELISA and Western
blotting techniques. The soil-borne beet mosaic virus protects beet plants against
the closely related, but serologically distinct beet necrotic yellow vein virus
(BNYVV). The accumulation of the coat protein of the challenging virus,
BNYVV, was affected as cross-protection developed. The coat protein of
BNYVV could not be detected by ELISA in the protected beet plants, although the
RNA of the BNYVV was detected by reverse transcription–polymerase chain reaction (RT-PCR) technique. This study suggests that replication of the challenging virus/strain may be affected if the coat protein synthesis is reduced or pre-
vented by protecting virus (Mahmood and Rush, 1999).

New derivatives of viruses may arise by selection or mutation with different infection potential. Such strains have to be recognized rapidly so that necessary changes in the management practices may be effected. A new mutant strain
of potato virus M, designated potato virus M-ID (PVM-ID), differing in the coat
protein sequences in the aminoterminus region was identified using ELISA by
Cavileer et al. (1998). Virus-like symptoms appeared in wheat and corn. The antiserum raised against a protein purified from the symptomatic corn reacted positively with the extracts from wheat. This virus, named high plains virus (HPV),
was not related to wheat streak mosaic virus and it was transmitted by wheat curl

mite *Aceria tosichella* (Seifers et al., 1997). A new strain of potyvirus infecting *Dioscorea alata* was recognized based on its biological characteristics and serological properties (Odu et al., 1999).

The serological affinities of five cereal viruses transmitted by fungal vectors, viz., barley yellow mosaic (Ba YMV), barley mild mosaic (BaMMV), oat mosaic (OMV), wheat yellow mosaic (WYMV), and oat golden stripe (OGSV) viruses, were studied by using F(ab')2 and protein A ELISA tests. Within the group, BaYMV and WYMV were serologically related. Barley yellow mosaic virus reacted with antiserum against one isolate of bean yellow mosaic potyvirus (BYMV-G) among the antisera against 29 other elongated viruses tested. Oat golden stripe virus showed affinities with BYMV-G, potato virus M, red clover vein mosaic virus, and possibly hordeum mosaic virus. Barley mild mosaic virus, OMV, and WYMV did not exhibit any affinity with different viruses tested (Jianping and Adams, 1991). Reddy et al. (1992) reported that groundnut bud necrosis virus (BNV) did not react with antisera to TSWV obtained from different sources and in reciprocal tests TSWV antigen did not react with antiserum to BNV infecting groundnut in India. On the basis of the serological differences, BNV is considered distinct and this serotype appears to be restricted to Asia. Shalitin et al. (1994) identified two serogroups of Israeli citrus tristeza virus strains that could be differentiated by monoclonal antibodies, and these groups were correlated with groups differentiated by sequencing of their coat protein genes. Ellis and Wieczorek (1992), using selected monoclonal antibodies specific either to beet western yellows (BWYV) or to potato leaf roll virus (PLRV); isolates of beet mild yellowing virus; turnip yellows virus; and the *Rhopalosiphum padi*-transmitted (RPV) strain of barley yellow dwarf found that they were closely related to BWYV; *Solanum* yellows virus and four isolates from potato were identified as PLRV. Abutilon mosaic virus, tobacco leaf curl virus, and tomato yellow leaf curl virus, belonging to the geminivirus group, could be detected by selected MABs specific to African cassava mosaic virus or Indian cassava mosaic virus by indirect ELISA. These viruses could be differentiated by using two other MAB clones (MacIntosh et al., 1992). The geminiviruses African cassava mosaic (ACMV), okra leaf curl (OLCV), tobacco leaf curl (TobLCV), and tomato leaf curl (TYLCV) viruses could be distinguished by determining the epitope profiles by using panels of MABs to ACMV, OLCV, and Indian cassava mosaic virus. African cassava mosaic virus and OLCV had similar, but distinguishable profiles (Konate et al., 1995).

Based on the reactivity of MABs, the plant viruses and their strains have been assigned to different serotypes, which may have similar biological or structural properties in common. Four serotypes of plum pox potyvirus, M, D, C, and El Amar, have been differentiated using DAS-ELISA with a combination of the universal MAB5B and MABs specific for the serotypes (Myrta et al., 1998). Using both PABs and MABs raised against the West African isolates of rice yellow mottle sobemovirus (RYMV), 73 isolates of RYMV were placed into three distinct serogroups, which were correlated to two RYMV pathotypes based on their reaction on a set of differential rice varieties. (Konate et al., 1997). Pathotypes (I–IV) of bean yellow mosaic virus, recognized based on variations in biological properties, showed serological variations when tested with a panel of 16 MABs in TAS-ELISA (Sasaya et al., 1998). By employing different combinations of PABs and MABs for trapping and as intermediate detecting antibodies in indirect DAS-ELISA, the isolates of citrus tristeza virus (CTV) that induced stem pitting in sweet orange were differentiated from other isolates of CTV that did not cause stem pitting. This differentiation is important because the sweet oranges suffer severely due to stem pitting syndrome (Nikolaeva et al., 1998). A panel of 30 MABs were used in plate-trapped antigen (PTA) ELISA for differentiation of 41 isolates of turnip mosaic virus (TuMV). The coat protein (CP) amino acid sequence homology groups of TuMV isolates were correlated with some serotypes of the virus. No correlation could be observed between serotypes and pathotypes and also no evidence was available to indicate that the epitopes recognized by MABs might act as elicitors for resistance genes present in *Brassica napus* lines (Jenner et al., 1999).

Epitopes or antigenic determinants have the capacity to undergo specific binding with antibodies or lymphoid cell receptors. Monoclonal antibodies react with different epitopes present on the viral coat protein surface. Maeda et al. (1997) developed a rapid, sensitive, and simplified ELISA protocol for the detection of CMV, by using two MABs that recognized different epitopes of the virus. Differences in the epitope profiles as determined by the reactivity with MABs may be used for the differentiation of plant viruses. The three MABs prepared against an isolate of soil-borne wheat mosaic furovirus (SBWMV) from Okhlahoma showed different reactivities when tested against isolates of SBWMV from Nebraska, France, and Japan. One MAB SCR 133 reacted positively with oat golden stripe furovirus. It appears that three amino acids of the viral coat protein differ between isolates of SBWMV and these are located near the coat protein surface and determine antigenic reactivity (Chen et al., 1997). A strain of potato virus M (PVM) differing in the coat protein sequences in the amino terminals region was differentiated from other strains of PVM by employing ELISA (Cavileer et al., 1998). Two mutant strains of beet necrotic strains of beet necrotic yellow vein virus (BNYVV) containing deletions in RNA3 were constructed, the deleted regions encoding either 94 or 121 amino acids toward the C-terminal part of the 25-kDa protein (p25). The mutant strain did not induce any symptoms of rhizomania disease in either susceptible or resistant sugarbeet cultivars and the virus contents of roots was much lower compared with that of the wild-type virus, as determined by ELISA test (Tamada et al. 1999). ELISA formats have been employed for serological characterization and immunopurification of some viruses rapidly. Among the panel of 10 MABs prepared against grapevine fleckvirus (GFKV) that reacted with samples from diverse geographical origins, one MAB, 2B5, was found to be more sensitive than other MABs. Using MAB 2B5, this virus was purified by immunoaffinity chromatography (Schieber et al., 1997).

Assessment of the relative efficacy of detection techniques is made to select a suitable technique for the detection of viruses in different host plants. Generally, the nucleic acid–based techniques are more sensitive than serological techniques. However, in the case of banana bunchy top nanavirus (BBTV) DAS-ELISA tests were apparently more sensitive than dot-blot technique using digoxigenin-labeled DNA probe in detecting BBTV in diluted crude extracts from infected plants (Shaarawy et al., 1997). Although the immunocapture-polymerase chain reaction (IC-PCR) assay has been demonstrated to be more sensitive in detecting several viruses, this assay was found to be less sensitive than ELISA in detecting pea seed–borne mosaic virus (Phan et al., 1997). The citrus psorosis-associated virus (CPsAV) was detected in four citrus accessions by ELISA, whereas these accessions indexed negative by graft indexing, demonstrating the greater reliability of ELISA test results (D' Onghia et al., 1998).

Joisson et al. (1992) demonstrated that by using MABs prepared with proteolysed tobacco etch potyvirus, five potyviruses, viz., PVY, pepper mottle virus, papaya ringspot virus, watermelon mosaic virus, and bean yellow mosaic virus, could be detected. Richter et al. (1994) developed a polyclonal antiserum that showed high reactivity with 20 aphid-borne potyviruses as well as with the mite-borne ryegrass mosaic virus and the fungus-borne barley mild mosaic virus with indirect ELISA or ISEM and Western blotting tests. The antiserum raised against tomato leaf curl virus (TLCV) reacted only with TLCV, but not with Indian cassava mosaic (ICMV) and African cassava mosaic (ACMV) viruses, whereas the antiserum against ACMV reacted positively with ICMV, indicating the serological relationship between ICMV and ACMV (Ragupathi and Narayanasamy, 1996) (Fig. 8.3).

8.1.1.2 Dot Immunobinding Assay

Dot immunobinding assay (DIBA), in principle, is similar to ELISA; the polystyrene plates are replaced by nitrocellulose- or nylon-based membranes on which the antigen is immobilized. As these membranes have high affinity for proteins, the free protein-binding sites present in the membranes necessarily have to be blocked by using bovine serum albumin (BSA) or nonfat dry milk powder or gelatin. Of the blocking agents, nonfat dry milk powder is less expensive, readily available, and as effective as other substances. The unconjugated virus-specific antibody is then allowed to react with the immobilized antigen. The trapped antibody is then probed with alkaline phosphatase, horse-radish peroxidase-labeled protein A, anti-Fc, or anti-IgG. Appropriate substrate is provided to allow visual detection of a colored product (Appendix 8 [iv]) (Fig. 8.2).

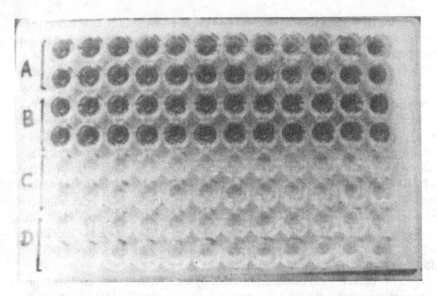

Figure 8.3   Cross-reactivity of antisera raised against tomato leaf curl virus (TLCV) and African cassava mosaic virus (ACMV) with TLCV, ACMV, and Indian cassava mosaic virus (ICMV), as determined by indirect ELISA: A rows, TLCV + TLCV antiserum; B rows, ICMV + ACMV antiserum; C rows, TLCV + ACMV antiserum; D rows, ICMV + TLCV antiserum. Note positive reactions in wells of A and B rows.

The DIBA technique could detect seed-borne infection of barley stripe mosaic virus in barley and bean common mosaic virus in French bean when single seed or flour was tested (Lange and Heide, 1986). The assay method is also useful for the detection of TMV, tobacco ringspot virus, and tomato ringspot virus in different plant species (Powell, 1987). The DIBA is satisfactory in detecting peanut mottle virus only when monoclonal antiserum is used, whereas ELISA is more effective when either monoclonal or polyclonal antiserum is used (Sherwood et al., 1987). Using electroblot immunoassay, Lenardon et al. (1993) showed that maize dwarf potyvirus strains A, D, E, and F could be serologically differentiated. These strains were differentiated earlier on the basis of their biological properties. Dot immunobinding assay has been employed to detect virus coat proteins in transgenic plants (Barker et al., 1992; Hibrand et al., 1992). Cherry mottle leaf trichovirus (CMLV) was more efficiently detected by DIBA using PABs or MABs than TAS-ELISA or Western blot analysis. Young leaves and flowers of cherries were the most suitable tissues for CMLV detection (James and Mukerji, 1996). The DIBA test has been found to be effective in detecting lily symptomless virus (LSV), tulip breaking virus–lily (TBV-L), and cucumber mosaic virus (CMV) in the scale segments of *Lilium* sp. (Nümi et al., 1999).

In asymptomatic leaves or stems that produced negative results in ELISA tests, positive reactions could be observed by a dissecting microscope. This assay was found to be eight times more sensitive than ELISA for TSWV and five times more sensitive for purified potyviruses (Berger et al., 1985). Dot blot immunoassays can be performed by using a chemiluminescent substance (disodium 3-(4-methoxy spirol 1,2-dixoetane-3-2'-tricyclo-(3,3,1,1)decan-(4,4,1) phenylphosphate) or a chromogenic compound (nitroblue tetrazolium or 5-bromo-1-chloro-3-indobutyl phosphate) as substrate to detect TSWV, bean yellow mosaic virus, and lily symptomless virus in infected leaf extracts or purified virus preparations (Mansky et al., 1990; Chahal and Nassuth, 1992; Makkouk et al., 1993).

Dot immunobinding assay has the following advantages over ELISA: a) It detects viruses in extremely small volumes, as in insect or plant extracts; b) the membranes can be easily stored and transported as required during disease surveys; c) membranes can be more easily processed than plates; d) DIBA requires a shorter period than ELISA; e) direct tissue blotting does not require any sample preparation: the sensitivity can be increased appreciably by limiting the area of application and/or by applying a higher concentration of antigen on the membrane; f) quantitative measurements are possible by using a reflectance densitometer (Banttari and Goodwin, 1985). g) DIBA has been further simplified by Lange et al. (1991), who showed that immunoblots could be prepared by using plain paper, which was as effective as nitrocellulose membrane. Five seed-borne viruses, viz. pea seed-borne mosaic virus, pea early browning virus, bean common mosaic virus, barley stripe mosaic virus, and squash mosaic virus, were detected by DIBA on paper.

8.1.1.3 Tissue Blot Immunoassay

The tissue blot immunoassay (TBIA) involves the transfer of viral antigens from freshly cut plant tissue surface to niltrocellulose membranes. By applying slight pressure on the cut tissue surface a tissue imprint is made on the membrane. After blocking the unoccupied protein binding sites on the nitrocellulose matrix, the blots are probed with specific antibodies raised against test viruses labeled with enzymes. In the indirect method, the antigens are allowed to react first with primary antibodies specific to the virus, and then with enzyme–labeled secondary antibodies.

The direct tissue blotting technique was used for the detection of tomato spotted wilt virus (TSWV) in infected leaves of *Nicotiana benthamiana* and leaf and stems of *Eustoma (Lisianthus)* and *Impatiens* plants showing virus-like symptoms (Hsu and Lawson, 1991) [Appendix 8(v)]. Using both direct and indirect methods, apple chlorotic leaf spot trichovirus, prunus necrotic ilarvirus, prune dwarf ilarvirus, plum pox potyvirus, citrus tristeza closterovirus, tomato spotted wilt tospovirus, and lettuce mosaic potyvirus were detected reliably. This technique is simple, sensitive, and rapid. Results can be obtained in 3 hr and imprinted membranes can be stored for long periods (Cambra et al., 1994). Using biotin-goat antibodies labled with alkaline phosphatase, TSWV infection in field samples of tomato, capsicum, chrysanthemum, gladiolus, hydrangea, and oleander was detected, when ELISA tests gave negative or erratic results. This technique was found to be suitable for both small- and large-scale testing especially for critical diagnosis in quarantine and certification programs (Louro, 1995). When the sensitivity of TBIA was compared with ELISA and DIBA for the detection of lily symptomless virus (LSV) in lily bulb scales, TBIA detected LSV in more number of samples. LSV was detected by TBIA in some samples that gave negative reaction in tests with the other two methods indicating the greater sensitivity of TBIA test (Gera et al., 1995) (Table 5).

Direct TBIA was used, along with reverse transcription–polymerase chain reaction (RT-PCR) for the survey of virus disease incidence on garlic and TBIA was found to be suitable for large-scale and routine diagnosis of garlic mosaic potyvirus (GMV), garlic latent caralavirus (GLV), and three unclassified viruses (Gar V-A, Gar V-B, and Gar V-C) (Tsuneyoshi and Sumi, 1996). The TBIA was employed in a survey to test more than 20,000 Hawaiian pineapple samples for the presence of pineapple closterovirus (PCV). The virus was detected in symptomless pineapple plants in the field as well as in the USDA pineapple germplasm collections. The involvement of PCV in mealybug wilt of pineapple (MWP) was indicated by TBIA tests (Hu et al., 1997). Direct TBIA was employed for the detection of carnation mosaic carmovirus (CarMV) in leaves, stems, and incompletely opened flowers of carnation plants. The TBIA was found to be simple, but it was as sensitive as ELISA or dot-ELISA tests (Zheng, 1999).

8.1.1.4 Immunoblot Analysis

The basic principle involved in this system is the transfer of electrophoresed viral proteins from the gel matrix onto a membrane and making them accessible for

Table 8.5 Comparative Efficacy of TBIA, DIBA, and ELISA for the Detection of Lily Symptomless Virus (LSV) in Lily Bulb Scales

| Nature of assay | No. of scales giving[a] | | |
	Positive reaction	Weak positive reaction	Negative reaction
ELISA	8	5	13
DIBA	17	2	7
TBIA	20	0	6

[a] Total number of scales tested: 26.
Source: Gera et al., 1995.

subsequent analysis by specific immunoprobes. Using this technique, detection of viral capsid protein or nonstructural proteins present in infected plants has been achieved. Polyclonal and monoclonal antibodies have been used as immunoprobes. The nonstructural protein p25 associated with infection by beet necrotic yellow vein furovirus (BNYVV) was detected in the subcellular location in infected sugarbeet, *Chenopodium quinoa,* and *Tetragonia expansa* plants mechanically inoculated with BNYVV (Li et al., 1996). The coat protein (CP) and movement protein (MP) of grapevine viruses (GAV) were detected by immunoblot analysis in rugose wood disease–affected grapevines. The MP was detected earlier than the CP; hence detection of MP may provide a basis for early detection of GAV in grapevine (Rubinson et al., 1997). Using Western blot analysis with PABs and MABs, the P1 protein and a protein of approximately 25 kDa (P1-C25) of potato leafroll luteovirus (PLRV) accumulating readily in detectable levels in PLRV-infected plants could be detected. P1-C25 represents the C-terminus of P1 and it is a proteolytic cleavage product produced during P1 processing (Prüfer et al., 1999).

The serological relationship between viruses and strains has been established by using immunoblot analysis. The identity of the protein expressed by CP gene of sweet potato sunken vein closterovirus (SPSVV-Ke), Kenya isolate was established as the viral CP by Western blot analysis. It was also demonstrated that using bacterially expressed CP antiserum, SPSVV-Ke was closely related serologically to similar closterovirus isolates infecting sweet potato in Israel, Nigeria, and the United States (Hayer et al., 1996). The existence of serologically distinct strains of wheat streak mosaic virus (WSMV) was demonstrated, indicating the necessity of precise identification of strains in breeding for resistance to WSMV (Montana et al., 1996). Immunoblot analysis revealed the mixed infections in grapevine by serologically related and unrelated grapevine leaf roll–associated viruses (GLRaVs) and grapevine corky bark–associated virus (GcBaV) resulting in complex symptoms (Monis and Bestwick, 1997). The association of a flexuous filamentous rods with an average length of 888 nm with peach mosaic disease and its serological relationship with cherry mottle leaf trichovirus were established by Western blot analysis by Gispert et al. (1998) and James and Howell (1998). Isolates of soybean dwarf luteovirus dwarfing (SbDV-D) and yellowing (SbDV-Y) have been shown to be serologically related based on the results of immunoblots probed with SbDV-D PAB antiserum. A single 26-kDa CP band was observed, confirming close serological relationship to SbDV (Damsteegt et al., 1999) [Appendix 8(vi)].

Proteins associated with virus infection in plants and vectors have been detected and identified using Western blot analysis. The presence of P1 protein associated with cytoplasmic inclusion bodies, characteristic of PVY infection, and in the cytoplasm of infected plant cells was detected by Arbatova et al., (1998). The 5'-terminal open reading frame (ORF)-1a of beet yellows virus (BYV) encodes a 295-kDa polyprotein with domains of papain-like cystein proteinase, methyl transferase (MT), and helicase (HEL), while ORF 1b encodes an RNA-dependent RNA polymerase. Monoclonal antibodies (MABs) derived from mice injected with the bacterially expressed fragments of the BYV-1a product encompassing the MT and HEL domains were employed to detect MT and HEL in BYV-infected *Tetragonia expansa* plants. Four MABs specific for MT recognized a = 63-kDa protein and two MABs against HEL reacted with a = 100-kDa protein on immunoblots of proteins from BYV-infected plants. Both MT-like and HEL-like proteins were associated with mainly in the fractions of large organelles (PI) and membranes (P30) of the infected plants. These virus-related enzymes appear to be associated with membrane components in the host cells (Erokhina et al., 2000). Immunoblot analysis indicated that a 94-kDa thrips protein exhibited specific binding of tomato spotted wilt tospovirus (TSWV) particles. This 94-kDa protein was present in all developmental stages of both *Frankliniella occidentalis* and *Thrips tabaci* vectoring TSWV (Kikkert et al., 1998). A method for cloning virus-binding polypeptides (receptor candidates) in *F. occidentalis* was developed by Medeiros et al. (2000). By using purified TSWV for screening a lambda-phage cDNA expression library, several virus-binding polypeptides were identified in a far-Western assay. This procedure eliminated the need for (i) a cellular infection or binding system and (ii) the identification, cloning, and expression of a functional virus attachment protein.

8.1.1.5 Filter Paper Seroassay

Filter paper seroassay (FiPSA), developed by Haber and Knapen (1989), is simpler, faster, and cheaper than ELISA. In this test, ELISA plates are replaced by no. 1 filter paper disks (Whatman) as solid phase and it seems to be suitable only for viruses which are present in high concentrations, since the filter paper disks retain lower quantities of proteins than membranes (Goodwin and Nassuth, 1993). Barley stripe mosaic virus was detected in infected barley leaves, barley embryos from infected seeds, and partially purified sucrose gradient fractions of the virus suspension. This technique can detect viral antigens in 1–2 μl samples containing more than 2 ng, and the results may be obtained within 2–3 hr of spotting the samples. Another advantage of the method is that it can be used for the detection of plant pathogenic bacteria and seed-borne fungi with suitable modifications (Appendix 8 [vii]).

8.1.1.6 Rapid Immuno-Filter Paper Assay

The Rapid immuno-filter paper assay (RIPA) was developed by Tsuda et al. (1992, 1993) for the detection of plant viruses using white and colored latex beads coated with antibodies. Quantitative assays can be made by using a chromatoscanner. This technique is simpler, less time-consuming, and less expensive than ELISA, but it is as sensitive as ELISA and can be used as easily as pH test paper (Cabauatan et al., 1994) (Appendix 8 [viii]) (Fig. 8.4).

Using RIPA, purified preparations of TMV and CMV could be detected at 5 ng/ml and 50 ng/ml, respectively, and in extracts of infected tissues at dilutions of 10^{-7} and 10^{-5}, respectively, with the naked eye. In 13 species in six families

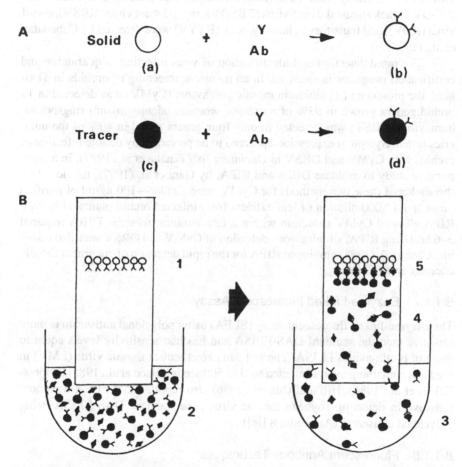

Figure 8.4 Rapid immuno filter paper assay (RIPA): A, White latex particle (a) as solid phase coated with antibody (b): pink latex particle (c) as tracer coated with antibody (d); B, 1, Immobilized solid phase as a strip on a filter paper strip: 2, bottom end of the filter paper strip dipped into a mixture of dyed latex particles (tracer) coated with antibody and sample extract or purified virus; 3, reaction between virus and antibody linked to tracer; 4, movement of virus-antibody-tracer complex by capillary action; 5, virus particles on tracer sandwiches with solid phase, forming a pink line (American Phytophathological Society, Minnesota, USA). (Courtesy of Tsuda et al., 1992.)

CMV was detected and it could be detected, using chromatoscanner, at 10 to 100 times lower concentrations than detected with the naked eye (Tsuda et al., 1992). Simultaneous detection of several viruses, viz. TMV, CMV, PVY, and turnip mosaic virus, using different colors of sensitized beads, was reported by Tsuda et al. (1993). The rice viruses rice tungro bacilliform virus (RTBV), tungro spherical virus (RTSV), grassy stunt virus (RGSV), stripe virus (RStV), gall dwarf virus (RGDV), black streaked dwarf virus (RBSDV), ragged stunt virus (RRSV), dwarf virus (RDV), and transitory yellowing virus (RTYV) were detected by Cabauatan et al., (1994).

The rapid detection and identification of virus infection in quarantine and certification programs is essential. In an intensive screening of orchids in Thailand, the presence of cymbidium mosaic potexvirus (CyMV) was detected in 17 orchid genera grown in 93% of nurseries, whereas odontoglossum ringspot tobamovirus (ORSV) was detected in only four genera raised in 40% of the nurseries tested. Symptom expression appeared to be promoted by double infection of orchids with CyMV and ORSV in *Oncidium* sp. (Tanaka et al., 1997). In a comparative study to evaluate DIBA and RIPA, by Gara et al. (1997), the detection thresholds of these two methods for CyMV were similar—100 ng/ml of purified virus or 1:10,000 dilution of leaf extracts from infected orchid plants. However, RIPA allowed CyMV detection within a few minutes, whereas DIBA required 5–6 hr. Using RIPA, simultaneous detection of CyMV and ORSV was also possible. Choi et al. (1998) employed RIPA for the rapid detection of cucumber mosaic cucumovirus (CMV).

8.1.1.7 Enzyme-Linked Fluorescent Assay

The enzyme-linked fluorescent assay (ELFA) using polyclonal antiserum is more sensitive than the standard DAS-ELISA and has had sensitivity levels equal to those of biotin-avidin-ELISA. The test can detect lettuce mosaic virus (LMV) in seed lots containing even 1 infected seed of 500 seeds (Diaco et al., 1985; Dolores-Talens et al., 1986). Hill and Durand (1986) also reported that ELFA was more sensitive in detecting soybean mosaic virus than standard DAS-ELISA using polyclonal antiserum (Appendix 8 [ix]).

8.1.1.8 Fluorescent Antibody Techniques

Tests involving the use of antibodies labeled with fluorescent dyes, ferritin, and radioactive iodine have been employed to study the intracellular location and distribution of plant viruses in infected plants and in insect vectors and vector cell monolayers. The labeling compounds used as specific tracers form a valuable tool for histochemical and cytochemical studies designed to elicit details of virus replication. The labeled antibodies present in the tissues are detected by fluorescent microscopy, electron microscopy, and radioautography (Appendix 8 (x) A, B, and C).

The distribution and synthesis of TMV in various plant tissues have been studied by Schramm and Rottger (1959), Nagaraj (1965), and Schonbeck and Spengler (1979). Otsuki and Takebe (1969) employed the immunofluorescence test to determine the multiplication of TMV in tobacco protoplasts. Multiplication of wound tumor virus in the leafhopper vector was demonstrated by using fluorescent antibodies (Nagaraj et al., 1961; Sinha, 1965; Chiu and Black, 1969). The development of WTV in vector cell monolayers was studied by Peters and Black (1970), Reddy and Black (1972), and Hsu (1978). By conjugating globulin with ferritin, which is a small protein with high iron content, the specific reaction between antigen and labeled antibody can be detected in the electron microscope as ferritin itself acts as an electron-dense stain. The distribution of TMV protein in tomato leaf cells has been studied, and it is possible to detect even a single virus particle by using this method (Shalla and Amici, 1967). By fluorescent antibody labeling, southern bean mosaic virus (SBMV) and bean pod mottle virus were detected in veins leading from the feeding wound caused by beetles (*Epilachna varivestis*) fed on purified virus preparation and primary infection sites in mesophyll cells at 2–3 days post feeding on plants (Field et al., 1994). The presence of potato leaf roll virus in intestinal epithelium was revealed by immunofluorescence technique (Garret et al., 1993). Southern bean mosaic virus and the cowpea strain of TMV were detected in the gut lumen and epithelial cells of the beetle vector (*Diabrotica undecimpunctata howardii*) by using immunofluorescent- and electron microscopy (Wang et al., 1994).

Localization of plant virus particles in different tissues of other kinds of vectors has been monitored by using immunofluorescence assay methods. The tomato spotted wilt tospovirus (TSWV) nucleoprotein (N) was found localized in the first- and second-instar larvae and pupae by indirect immunofluorescence. Specific fluoescence signals were observed, 2–4 days after acquisition access, in the anterior midgut and then in the whole midgut. With lapse of time, N protein was detected throughout the larval midgut and later in salivary glands. Fluroescence signals within the pupal midgut were less intense compared to that in larvae (Tsuda et al., 1996). Wheat streak mosaic virus (WSMV)–specific fluorescence was detected near the anterior end of viruliferous mite (*Aceria tosichella*). This technique was used to determine the percentage of viruliferous mites present in the field collections (Mahmood et al., 1997). By employing indirect immunofluorescent labeling of tomato mottle virus (ToMoV) and cabbage leafcurl virus (CabCCV)– specific PABs and MABs, the localization of both begomoviruses in the anterior region of the midgut and filter chamber of adult whiteflies was observed. ToMoV was detected also in salivary glands of whiteflies (Hunter et al., 1998) [Appendix 8(x) C]. The PABs labeled with fluorescein isothiocyanate (FITC) were used to detect the tobacco ringspot nepovirus (TRSV) in the lumen of the stylet extension of esophagus of the nematode vector *Xiphinema americanum*. The virus-specific fluorescence increased with increase in the acquisition access period given to the nematodes on the infected plants from 0 to 22 days (Wang and Gergerich, 1998) [Appendix 8(x) D)].

8.1.1.9 Time-Resolved Fluoroimmunoassay

The time-resolved fluoroimmunoassay (TR-FIA) was developed by Hemmila et al. (1984) and Halonen et al. (1986) and modified by Siitari et al. (1986) for the detection of plant viruses. This technique is based on the determination of fluorescence of antibody-lanthanide (europium) conjugates. Using lanthanide molecules such as europium tags, TR-FIA provides higher specific activity than conventional radioactive labels, but the drawbacks of radioactivity and hazards are eliminated and it also entirely eliminates the nonspecific background fluorescence originating from the sample. The lanthanide molecules, when excited, have much larger decay times (100–1000 μs) than background fluorescence (<1 μs) and hence show a larger shift between the excitation and emission wavelengths. By determining the emission, after a delay period and at a higher wavelength, the background due to autofluorescence may be reduced (Appendix 8 [xi]).

Using the TR-FIA procedure potato viruses M, S, X, and Y and potato leaf roll virus have been detected in leaf and tuber extracts (Siitari and Kruppa, 1987; Sinijarv et al., 1988). With purified PVX, TR-FIA was found to be 5–100 times more sensitive than DAS-ELISA (Siitari and Kruppa, 1987). In the case of potato virus M, 0.5 ng/ml of the virus could be detected by TR-FIA; ELISA could detect the virus only at 10 ng/ml (Jarvekulg et al., 1989). By using monoclonal antibodies labeled with europium and samarium, two potato viruses were detected simultaneously (Saarma et al., 1989).

8.1.1.10 Cytofluorimetric Method

A new assay method based on flow cytometry was developed by Iannelli et al. (1996) for the detection and quantification of plant viruses. Flow cytometry has been used to translate biological properties into measurable fluorescence intensity. Determination of DNA contents of animal and plant cells and detection of intracellular antigens have been carried out by using this principle. Since the measurement of fluorescence is possible only on a population of single cells, plant virus particles present in plant extracts are adsorbed onto latex particles and then analyzed by flow cytometry.

The cytofluorimetric method was first employed for the detection of cucumber mosaic cucumovirus (CMV) in purified preparation and extracts from infected plants. Extracts from either healthy or CMV-infected leaves are incubated with latex particles, followed by washing and incubation in succession with rabbit anti-CMV antibodies and anti-rabbit IgG labeled with fluorescein. This treatment allows CMV particles to become visible to the laser of the cytometer. The detection limit of cytofluorimetric method for CMV in purified preparation was 10 pg/ml as against 2.5 ng/ml for ELISA, indicating the greater sensitivity of cytofluorimetric method. The coat protein (CP) of CMV was detected by cytofluorimetric method in the extracts of transgenic plants at a dilution of 1:1000, whereas ELISA detected viral CP only at a dilution of 1:500 of similar plant extracts (Iannelli et al., 1996).

The cytofluorimetric method can be also used for simultaneous detection of two or more viruses in infected plants, by using latex particles with different diameters. Leaf extracts from healthy or CMV- or potato Y potyvirus (PVY)-infected plants are incubated with latex particles each with a diameter of 3 μm. For the detection of tomato mosaic virus (ToMV) latex particles with 6 μm diameter are used. The latex particles are then washed and incubated in succession with primary and secondary antibodies labeled with phycoerythrin (PE) or fluorescein (FITC). The fluorescence emitted by FITC or PE is used for distinguishing CMV and PVY, whereas the size of the latex particles (6 μm) will differentiate ToMV [Appendix 8(xii)]. These three viruses can be distinguished by following another procedure. The latex particles with 3, 6, and 10 μm diameter are separately sensitized with antibodies specific for CMV, PVY, and ToMV. A mixture of sentitized latex particles (in equal volumes) is then incubated with extracts from plants infected by all three viruses followed by incubation with antibodies labeled with FITC. The cytofluorimetric method can also be employed for the purification of these viruses (Iannelli et al., 1997).

8.1.1.11 Virobacterial Agglutination Test

Different types of serological agglutination tests have been employed to detect and identify plant viruses. Although these tests are simple, they require antisera of relatively high titer. This limitation may be overcome by using the bacterium *Staphylococcus aureus* for sensitization with the antiserum raised against the test virus(es). Protein A, which occurs in high concentration and is naturally covalently linked to the bacterial cell wall, forms the binding sites for immunoglobulins, especially the IgG type. The Fc portion of IgG is linked to the binding sites, leaving Fab arms for trapping antigens.

Walkey et al. (1992) reported that by using the VBA test (Appendix 8[xiii]) seven potyviruses—bean common mosaic virus, lettuce mosaic virus, maize dwarf mosaic virus, papaya ringspot virus, potato virus Y, turnip mosaic virus, and zucchini yellow mosaic virus—could be differentiated in both homologous and heterologous reactions. However, the VBA test did not differentiate strains of bean yellow mosaic virus. The test was found to be as sensitive as ISEM and local lesion assay, but less sensitive than the direct ELISA test. On the other hand, Hughes and Ollennu (1993) found the VBA test more useful in detecting cocoa swollen shoot virus (CSSV) than ELISA, which failed to detect latent infection of

CSSV in cocoa. The relationships among seven isolates of CSSV that cause a range of mild to severe symptoms could be established. A field study for comparing the efficacy of the VBA and ELISA (direct and indirect) tests showed that the VBA test alone could detect CSSV in all cocoa trees known to be infected and in many symptomless trees, indicating its usefulness in the rapid detection of CSSV in cocoa. Further study to differentiate the CSSV group using the VBA test showed that among the eight groups differentiated using seven antibody types, the largest group comprising those isolates closely related to CSSV 1A could be further subdivided into four groups (Hughes et al., 1995).

Two strains of *S. aureus* were compared for their effect on the sensitivity of VBA tests. Strain No. 1800 was found to be 2–10-fold more sensitive in detecting tomato mosaic virus (ToMV), cucumber mosaic virus (CMV), and potato virus X (PVX). The detection thresholds of the test were 3.4–13.0 mg/ml for purified virus preparations and leaf extract dilutions of 1:1000–1:10,000, which were almost equal to PAS-ELISA tests (Feng et al., 1998).

8.1.1.12 High-Density Latex Flocculation (HDLF) Test

The antigen or antibody may be attached to the surface of red blood cells or carrier particles of similar size such as latex, bentonite, or barium sulfate. By sensitizing the carrier particles in this manner, it is possible to induce visible clumping, when the sensitized carrier particles react with the other reactant (antigen or antibody) that is present in the solutions. Using antibody-coated latex, several viruses have been detected even at 100–1000-fold lower concentrations than required by microprecipitin or immunodiffusion tests (van Regenmortel, 1982). A simple detection procedure using high-density latex was developed by Kawano and Takahashi (1997). Rice stripe virus was detected in the infected rice plants as well as in the viruliferous leafhoppers. Other viruses detected include carnation mottle virus, cucumber mosaic virus, cymbidium mosaic virus, odontoglossum ringspot virus, rice dwarf virus, and turnip mosaic virus. These viruses were detected in leaf sap of dilutions at 1:8000–1:16,000.

8.1.2 Immunocapillary Zone Electrophoresis

Immunocapillary zone electrophoresis (I-CZE) technique combines the specificity of serological methods with the sensitivity, rapidity, and automation provided by capillary zone electrophoresis (CZE) for the detection of plant viruses in purified preparations and plant extracts. Eun and Wong (1999) applied I-CZE assay for the detection of cymbidium mosaic potexvirus (CyMV) and odontoglossum ringspot tobamovirus (ORSV), which cause serious losses in various orchid species. The labeled antibody techniques require the antibody or antigen to be

Table 8.6   Effect of Periods of Inoculation of Virus-Antibody Mixture on the Detection of CyMV and ORSV in Purified Virus Preparation and Leaf Extracts from Infected Plants by Immunocapillary Zone Electrophoresis (I-CZE) Technique

	Incubation period (min)							
	10		20		30		50	
Nature of samples	CyMV	ORSV	CyMV	ORSV	CyMV	ORSV	CyMV	ORSV
Purified virions	−	+	+	+	+	+	+	+
Nicotiana benthamiana	+	+	+	+	+	+	+	+
Oncidium orchid	−	−	−	−	−	+	+	+

+ = positive; − = negative
Source: Eun and Wong, 1999.

immobilized on a solid support, whereas CZE-based immunoassays can analyze antigen-antibody reaction in free solution without the need for immobilization of antigen or antibody. An incubation period up to 50 min may be required for the viruses to reach equilibrium in binding to their antibodies and to be detected (Table 8.6). I-CZE assays can detect as little as 10 fg each of CyMV and ORSV in purified preparations and in crude sap from infected *Nicotiana benthamiana* and *Oncidium* orchid [Appendix 8 (xiv)].

I-CZE technique has several advantages: i) only a small amount of test material is required; ii) as the antibody-virus peak is present in the same place for the same virus-infected plant sample, reproducible results can be obtained; iii) the use of high-potential fields leads to extremely efficient separation of molecules within minutes; iv) crude plant extracts are incubated with virus-specific antibodies in sample vials for 10–50 min and automatically injected into the capillary; v) about 30 test samples can be analyzed automatically; and vi) results can be recorded in a computer for further analysis.

I-CZE technique can be employed for routine mass indexing of plant materials required by certification, quarantine, germplasm screening, and breeding programs. It has the potential to become an alternative to polymerase chain reaction (PCR) and digoxigenin-labeled cRNA probes currently employed widely for the detection of orchid viruses.

8.2 DETECTION OF PHYTOPLASMAS

The necessity of developing sensitive detection methods is probably greater for plant pathogenic mollicutes than for any other kind of plant pathogens, since only three helical spiroplasmas have been cultured and adequately characterized. All the phytoplasmas resemble true mycoplasmas, but they cannot be classified because of lack of information on other properties. Characterization based on type of symptoms caused, host range, and vector transmission characteristics is found to be unreliable. Serological assays employing polyclonal or monoclonal antibodies have been found to be more reliable and sensitive in detecting and establishing relationships among phytoplasmas (Lin and Chen, 1985; Clark et al., 1989; Lin et al., 1993). Jiang et al. (1988) reported that by using an affinity column consisting of protein A covalently linked to Sepharose matrix and coupled with MABs specific to phytoplasma, the aster yellows phytoplasma could be purified and intact, undamaged cells were observed under the electron microscope.

8.2.1 Growth Inhibition Test

Twofold dilutions of the polyclonal antiserum and normal serum are prepared and a drop of spiroplasma culture is added. The dilutions are incubated at 32°C for 4 days. The lowest dilution that permits acid production representing the growth of spiroplasma is recorded. The acid production is detected by using phenol-red, which turns yellow, as indicator. The correlation between acid production and growth can be established by examination under a microscope (Markham et al., 1974). A growth inhibition test showed that the spiroplasma causing brittle root symptoms in horseradish was a strain of *Sprioplasma citri* (Davis and Fletcher, 1983).

8.2.2 Rapid Slide Test

McIntosh et al. (1974) developed the rapid slide test on the basis of the reduction of spiral forms to oblong and round forms in the presence of specific antibodies. The antiserum prepared against *Spiroplasma citri* (California strain) and the 72-hr-old *S. citri* broth culture (diluted to 1:10 with growth medium) are mixed. A drop of this mixture is placed on a microscope slide and covered with a cover slip and observed under the dark field of a microscope. If the reduction in spiral forms is 50% or more, the reaction is considered positive. The antiserum against California strain reacted positively with Morocco strain. It was also shown that *S. citri* was antigenically related to corn stunt spiroplasma, but not to aster yellows agent.

8.2.3 Enzyme-Linked Immunosorbent Assay

Monoclonal (MABs) and polyclonal (PABs) antibodies have been prepared against several phytoplasmas (Lin and Chen, 1985, 1986). Production of monospecific polyclonal antibodies capable of reacting with phytoplasma associated protein was reported by Errampalli and Fletcher (1993). Using MABs specific for *S. citri* in indirect ELISA, the specific reaction of nine clones of hybridoma cell lines with different strains of *S. citri* was demonstrated. The MABs are able to distinguish *S. citri* from corn stunt spiroplasma, whereas PABs could not distinguish them (Lin and Chen, 1985). The relative sensitivities of PABs and MABs against aster yellows (AY) phytoplasma were tested by indirect ELISA. Monoclonal antibody reacted specifically with AY phytoplasma-infected plants and differentiated AY phytoplasma from other phytoplasma.

The spiroplasma that causes brittle root symptoms in horseradish plants was identified as a strain of *Spiroplasma citri* by using PAB in ELISA tests (Davis and Fletcher, 1983). A polyclonal antiserum prepared against groundnut witches'-broom phytoplasma was used to detect the phytoplasma infection by employing a protein A indirect ELISA procedure. The phytoplasma could be detected in crude extracts of leaves, stems, and pegs of infected groundnut plants. The extracts of tissues infected with eggplant (brinjal) little leaf, *Catharanthus roseus* witches'-broom, and *Datura* witches'-broom diseases of presumed phytoplasma origin did not react with the groundnut witches'-broom phytoplasma antiserum (Hobbs et al., 1987). The phytoplasma associated with faba bean (*Vicia faba*) was detected by indirect ELISA by coating directly either the whole antigen or $F(ab')_2$ fragments of the IgG (Saeed et al., 1993). Sesamum phyllody disease phytoplasma, could be detected in infected sesamum plants and inoculative leafhoppers by indirect ELISA. Vector indexing carried out for 14 months by using ELISA showed that the percentage of leafhoppers remaining inoculative varied from 16% to 60% and that there was a significant positive correlation between percentage of inoculative leafhoppers and phyllody incidence at 45 days after sampling (Srinivasulu and Narayanasamy, 1995 a,b). Using standard DAS-ELISA and DAC-ELISA, the grassy shoot disease-associated phytoplasma was detected in infected sugarcane (Viswanathan and Alexander, 1995). In a later study, Viswanathan (1997a) reported that indirect ELISA was more efficient than other ELISA formats in detecting the grassy shoot disease (GSD) phytoplasma in different plant tissues. The antiserum against the sugarcane white leaf disease (WLD) occurring in Taiwan cross-reacted with the extracts from GSD-infected plants, indicating a possible serological relationship between these two phytoplasmas infecting sugarcane. X-disease phytoplasma was detected using an MAB produced by employing an enriched antigen from an infected chokecherry (*Prunus virginiana*) plant. This MAB reacted in ELISA, with five of seven tested phytoplasmas included in the X-disease phytoplasma cluster and one of four phytoplasmas (pigeonpea wiches' broom phytoplasma) outside that cluster. This ELISA protocol using the MAB has the potential for commercial diagnosis of X-disease phytoplasma infection in stone fruits nurseries and orchards and for screening germplasms for breeding programs (Guo et al., 1998). The MABs prepared against tagetes witches' broom (TWB) disease agent detected the phytoplasma in artificially inoculated *Catharanthus roseus* and *Tagetes* and also in naturally infected *Tagetes* plants when ELISA and DIBA tests were used. Cross-reaction was observed with a phytoplasma isolated from grapevine infected with yellows diseases (Loi et al., 1998). Using the specific PABs in ELISA tests, *Spiroplasma citri* causing citrus stubborn disease was detected in infected citrus plants and leafhopper vector *Circulifer haematoceps* (Najar et al., 1998). The corn stunt disease appears to be a complex caused by mollicutes. The involvement of *Spiroplasma kunkelii* as one of the causative agents was demonstrated by employing specific F $(ab')^2$ protein A-ELISA (Henriquez et al., 1999).

In partially purified preparations from AY phytoplasma-infected plants, proteins associated with phytoplasma were detected by Western blotting. The PABs produced against AY-phytoplasma recognized a specific protein (23 kDa) in infected but not in healthy plants. The antibodies specific for phytoplasma-associated protein were purified by trapping them on AY-phytoplasma protein obtained by electrophoresis of infected plant extracts, and they were then transferred to a nitrocellulose membrane. The monospecific antibodies eluted from nitrocellulose reacted specifically with the AY phytoplasma-associated proteins. The monospecific PABs reacted positively with AY isolates from carrots and lettuce, but not with other phytoplasma and spiroplasma tested (Errampalli and Fletcher, 1993).

8.2.4 Immunofluorescent Technique

The immunofluorescent technique can be used for in situ detection of AY phytoplasma in the midribs of AY-infected lettuce plants. The acetone fixed sections are stained with fluorescein isothiocyanate conjugated antimouse IgG. When MABs are used, they are bound specifically to AY phytoplasma in the sieve tubes of diseased plants, whereas use of PABs results in fluorescence throughout the sections in both healthy and diseased plants, indicating the nonspecific binding of PABs to the cell wall and membrane (Lin and Chen, 1985). Immunofluorescent staining technique was utilized for detecting, identifying, and establishing the genetic relatedness of phytoplasmas associated with geographically diverse grapevine yellows diseases (Chen et al., 1994).

8.2.5 Immunosorbent Electron Microscopy and Gold-Labeled Antibody Decoration

The aster yellows (AY) phytoplasma bodies are trapped from the extracts of infected periwinkle plants or infected leafhoppers (*Macrosteles quadripunctulatus*) by using electron microscope grids coated with the (Fab')2 portion of specific rabbit IgG. The phytoplasma bodies are then decorated with intact IgG-labeled goat antirabbit IgG conjugated with gold particles (5 or 15 nm diameter). The grids are negatively stained with 0.5% ammonium molybdate. The European AY phytoplasma could be distinguished from serologically unrelated tomato big bud phytoplasma, which is morphologically indistinguishable from AY phytoplasma (Vera and Milne, 1994).

8.3 DETECTION OF PLANT PATHOGENIC BACTERIA

As the bacterial antigens are complex and have not been well characterized, much difficulty is experienced in producing antisera against bacterial pathogens. Polyclonal antisera have been raised against some important bacteria (Mushin et al., 1959; Choi et al., 1980; Zeigler et al., 1987; Quimio, 1989; Reddy and Reddy, 1989). Reproducibility of the reaction, loss of reactivity of a particular strain after subculturing, and colony-type variants with different serological properties are some of the problems associated with PABs. Development of monoclonal antibody technology helped to overcome these problems. MABs have been produced for many important genera of bacterial pathogens (Alvarez et al., 1985; De Boer and Mc Naughton, 1987; Magee et al., 1986; Alvarez et al., 1989; Benedict et al., 1989). Using a modified ELISA, the loss of bacterial cells from plates is prevented by drying the bacteria on the walls of microtiter plates (Benedict et al., 1989).

The MABs produced by using whole cells of *Xanthomonas campestris* pv. *campestris* (Xcc) were specific at the genus, pathovar, and strain levels (Alvarez et al., 1985). The MABs specific to certain cellular or extracellular fractions used as immunogens have been employed to differentiate serogroups of *Erwinia carotovora* subsp. *atroseptica* or *E. chrysanthemi* (De Boer and McNaughton, 1987) and *Corynebacterium sependonicum* (DeBoper and Wieczorek, 1984). The MABs generated after immunization with formalinized bacteria are found to be specific to antigens located on the surface of *Xanthomonas orzyae* pv. *oryzae*. A pathovar-specific MAB thus developed was used in an ELISA test to identify the pathogen in an outbreak of bacterial blight in the United States (Benedict et al., 1989; Jones et al., 1989). A highly specific MAB was used to detect latent ring-rot infections in potato by immunofluorescence assay (DeBoer and McNaughton, 1986).

8.3.1 Agglutination Test

The agglutination test was used to distinguish different species or strains of *Xanthomonas* spp. (Fang et al., 1950; Patel et al., 1951). The eight species studied were differentiated into five serogroups based on serological relationships (Patel et al., 1951). *Xanthomonas translucens* that causes bacterial stripe blight in cereals and grasses was differentiated into five special forms (strains) by Fang et al. (1950). Lyons and Taylor (1990) developed a rapid slide agglutination test that uses polyclonal antisera conjugated to *Staphylococcus aureus* cells which have high concentrations of protein A on their surface. This test could be used for the detection of *Pseudomonas syringae* pv. *phaseolicola* and *P.s.* pv. *pisi* in lesions on bean and pea, respectively. *Pseudomonas gladioli* pv. *alllicola* and *Lactobacillus* sp. were detected in rotted onion bulbs. The presence of specific strains of *Rhizobium phaseoli* in bean root nodules could also be detected (Appendix 8[xiii]). Using a monospecific antiserum (MSA) raised against a specific protein associated with the bacterium causing blood disease in banana, an agglutination test and a colony blot test were performed. The MSA reacted specifically with all virulent strains of blood disease bacterium, but not or only weakly with *Pseudomonas solanacearum* or other bacterial species tested (Baharuddin et al., 1994).

8.3.2   Gel Diffusion Test

The Ouchterlony gel diffusion test can be used to distinguish pathovars; *Pseudomonas syringae* pv. *syringae* and *P. syringae* pv. *pisi* may be reliably distinguished by inoculating them on susceptible cultivars. Mazarei and Kerr (1990) developed a more rapid and convenient serological test for distinguishing these pathovars. The antiserum raised by using glutaraldehyde-fixed bacterial cells showed a high level of specificity in the Ouchterlony gel double-diffusion test. The antiserum against *P. syringae* pv. *pisi* may be used to detect the pathogen in pea seeds. Bragard and Verhoyen (1993), using phenol-treated cells, reported that the specificity of PABs could be improved in double-diffusion tests for differentiating pathovars of *X. campestris*. By using MABs, the pathovars *X. c. undulosa, X. c. translucens, X. c. hordei, X. c. cerealis,* and *X. c. secalis* were positively distinguished.

8.3.3   Enzyme-Linked Immunosorbent Assay

The ELISA tests have been widely employed for the detection and differentiation of pathogenic bacteria by using PABs and MABs. Zhu et al. (1988) detected *Xanthomonas campestris* pv. *oryzae* in leaves and seeds of 60 rice accessions. The MAB (Xco-1) specific for *X. c. oryzae* reacted positively with all 178 tested strains of *X. c. oryzae* from diverse geographical locations, but not with *X. c. oryzicola* or other xanthomonads in modified ELISA tests (Benedict et al., 1989). Using a specific MAB generated against a Florida citrus nursery strain of *X. campestris* in ELISA and microfiltration enzyme immunosorbent assay, the existence of at least two serologically distinct populations of the bacteria could be observed (Permar and Gottwald, 1989). The monoclonal antibody specific for *Pseudomonas andropogonis,* which causes leaf spot disease, could be employed to check the presence of the pathogen in carnation cuttings by indirect ELISA and immunofluorescence assay (Li et al., 1993) (Appendix 8[xv]).

The types of antigens recognized by different MABs generated by *E. amylovora* were determined. Six of the MABs reacted with protein antigens, as determined by loss of reactivity in indirect ELISA, after treatment of sonicated bacterial cells with proteinase K, whereas two MABs reacted with purified polysaccharide from *E. amylovora*. This indicates that the MABs may be bound to different epitopes (McLaughlin et al., 1989). Lipp et al. (1992), by employing 6 MABs, identified 12 major serogroups in *X. c.* pv. *diffenbachiae*.

Benedict et al. (1989) showed that 178 strains of *X. o. oryzae* could be differentiated and classified into groups I, II, III, and IV on the basis of their reaction with four monoclonal antibodies (Table 8.7). Two of these MABs reacted positively with *X. c. oxyzicola*, but there was no reciprocal reaction between the MAB specific for *X. c. oryzicola* and any of the four groups of *X. c. oryzae* strains. One of the four MABs (X1) positively reacted with other xanthomonads in addition to *X. c. oryzicola*.

On the basis of the serological reactivity of MABs and PABs, 63 strains of *X. c.* pv. *oryzae* were grouped into 9 reaction types consisting of 4 serovars and 7 subserovars (Huang et al., 1993). Gnanamanickam et al. (1994) showed that pathovar-specific MABs could be used for the identification of *X. o.* pv. *oryzae* in rice seeds which were contaminated by other seed-borne microflora, and the reactivity with MABs was found to be correlated with pathogenicity tests. Alvarez et al. (1994) reported that *X. campestris* pv. *campestris* and *X. c.* pv. *armoraciae* could be rapidly identified in field and seed assays by employing specific panels of MABs.

The sensitivity of the ELISA test can be enhanced by using a semiselective enrichment broth (SSEB) (Appendix 8[xvi]). The lower limit of detection of *Xanthomonas campestris* pv. *undulosa*, which is seed-borne in wheat, was 5×10^3 colony-forming units (Cfu)/ml. By enrichment, *X. c. undulosa* could be detected in samples that originally had less than 5×10^2 Cfu/ml. Percentages of seed infection determined by SSEB-ELISA were highly correlated with potential seed infection (PSI), determined by greenhouse tests (Table 8.8). The SSEB-ELISA method may be a convenient tool for rapid initial screening of wheat seed lots in wheat certification programs. The antiserum raised against somatic antigens of *X. c. undulosa* can also be used to detect other pathovars: *X. c. cerealis*, *X. c. translucens*, and *X. c. phleipratensis* (Frommel and Pazos, 1994) (Fig. 8.5).

Visual criteria had been primarily used to select seed potatoes free from infection by *Erwinia* spp. resulting in nonrecognition of latent infections. To prevent

Table 8.7 Grouping of Strains on *Xanthomonas oryzae* pv. *oryzae* with Monoclonal Antibodies

| MAB | X. o. oryzae group | | | | X. c. oryzicola | Other xantho- monads | Nonxantho- monads |
	I	II	III	IV			
X₁	+	+	+	+	+	+	−
Xco-1	+	+	+	+	−	−	−
Xco-2	+	−	+	−	−	−	−
Xco-5	−	−	+	+	+	−	−
Xccola	−	−	−	−	+	−	−
Total no.	140	9	15	14	8	130	89

Source: Benedict et al. (1989).

Table 8.8 Comparative Efficacy of Determination of Wheat Seed Infection with *X. campestris* pv. *undulosa* by Three Different Laboratory Methods and Seedling Infection Under Optimal Greenhouse Conditions

Seed lot	Percentage of seedling infection	Percentage of infected seeds		
		XTS	ELISA-ST	SSEB-ELISA
1	12.0	5.2	2.72	17.93
2	14.0	6.5	2.73	16.30
3	1.0	0.1	7.07	3.26
4	8.75	8.0	15.78	10.30
5	2.5	6.6	14.72	5.00
6	14.0	5.8	1.10	18.29
7	6.5	3.6	2.10	8.54
8	8.75	4.5	2.45	10.97
9	6.5	8.0	2.48	6.10
10	12.0	3.6	1.22	7.60
11	9.0	4.1	2.28	12.19
12	6.5	2.6	4.88	8.50

Source: Frommel and Pazos (1994).

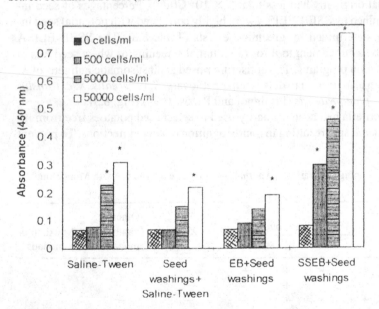

Figure 8.5 Effect of semiselective enrichment (4 hr incubation period) on the detection of *Xanthomonas campestris* pv. *undulosa* strain LP_2 by ELISA (DAS), as affected by seed bacterial saprophytes (Blackwell Science Ltd. and British Society for Plant Pathology, U.K.). (Courtesy of Frommel and Pazos, 1994.)

the escape of infected seed potatoes, an ELISA protocol was employed after anaerobic amplification of soft rot bacterium *Erwinia chrysanthemi* in DPEM medium. This procedure enhanced the sensitivity of the test, and the level of detection improved to from 10^5 bacteria/ml to $< 10^3$ bacteria/ml. When this procedure was used for testing, the latent infection in 10% of 133 seed lots that had been certified by visual examination was detected. This procedure was recommended for large-scale use for the detection of latent infection and to forecast disease outbreaks in Switzerland (Cazelles et al., 1995). The bacterial ringrot pathogen *Calvibacter michiganensis* subsp. *sepedonicus* was detected in field-grown potatoes by ELISA, the sensitivity of detection being equal to that of polymerase chain reaction (PCR) assays. The effectiveness of detection was affected by several factors such as inoculum dose ($>10^9$ colony-forming units, cfu), cultivar, and interval after planting (Slack et al., 1996). The variability of ELISA tests for the detection of *C. michiganensis* subsp. *sepedonicus* in potato tissues was analyzed to assess the extent of repeatability (within analyst variation) and reproducibility (among analyst variation). The standard deviation (SD) for repeatability of ELISA was small, but increased at higher absorbance values, whereas the SD for reproducibility was greater and also increased at higher absorbance values, indicating the need for adjusting the bacterial concentration at optimal levels (De-Boer and Hall, 2000).

ELISA tests using a new MAB specific for *Xanthomonas campestris* pv. *pelargonii* were applied for the detection of the pathogen in geranium. The intensity of response was moderately correlated ($r = 0.56$) with symptom severity caused by 14 strains of *X. campestris* pv. *pelargonii* (Chittaranjan and Boer, 1997). The effectiveness of ELISA for the detection of *Xanthomonas albilineans* causing sugarcane leaf scald disease was demonstrated by Comstock et al. (1997). The bacterial infection in 99% of the symptomatic stalks and in 14% of the asymptomatic stalks was revealed by ELISA tests.

The identity of the bacterial pathogen causing coffee leaf scorch disease was established by ELISA tests. The leaf extracts from infected coffee plants showed positive reaction with the antiserum raised against *Xylella fastidiosa*. The presence of the bacteria in the xylem of artificially inoculated plants was detected using the antiserum against citrus variegated chlorosis (CVC) strain of *X. fastidiosa* indicating the relationship between the two pathogens (Lima et al., 1998). The monoclonal antibodies specific for *Xylella fastidiosa,* causal agent of pear leaf scorch disease, were used for detection of the pathogen. These MABs did not cross-react with 14 other bacterial strains belonging to nine genera (Leu et al., 1998). ELISA test was employed for the detection of *X. fastidiosa,* causing leaf scorch disease in oleander (Purcell et al., 1999).

The infection of sugarcane by bacterial pathogens has been detected by ELISA tests. The sugarcane ratoon stunting disease caused by *Clavibater xyli* subsp. *xyli* could be diagnosed by employing indirect ELISA. Infection by this bacterial pathogen could reliably be detected in asymptomatic sugarcane plants/cultivars that may be carriers of the disease (Viswanathan, 1997b). Indirect ELISA technique was more sensitive in detecting *Xanthomonas albilineans,* causing sugarcane leafscald disease, than immunoblot assay. The sugarcane internode tissues contained greater concentrations of the pathogen than leaf samples (Viswanathan et al., 1998). The ELISA tests were successfully applied for the detection of *Pantoea stewartii* subsp. *stewartii,* causing Stewart's wilt disease in corn, and the vector, corn flea beetles (*Chaetocnema pulicaria*) (Khan et al., 1997).

Latent infection of groundnut (peanut) by *Ralstonia* (*Pseudomonas*) *solanacearum* causing bacterial wilt disease, in the main root, hypocotyl, and stems, was detected by using ELISA technique (Shan et al., 1997). Polyclonal antiserum developed against virulent bacterial cells encapsulated with mucin was highly sensitive in detecting as few as 100 cells/ml. This antiserum specifically reacted with tomato isolate of *R. solanacearum* and it did not react with isolates from chili (pepper) or aubergine (eggplant). The reactivity of the PAB in ELISA correlated well with the degree of infection in tomato seeds and plants. The ELISA protocol was found to be suitable for the detection of all tomato isolates in tomato seeds (Rajeshwari et al., 1998). Recombinant single-chain (scFv) antibodies produced against the bacterial lipopolysaccharide were selected by phage display [Appendix 8(xvii)]. These antibodies were more efficient in ELISA for the diagnosis of potato brown rot caused by *R. solanacearum* (Griep et al., 1998).

8.3.4   Dot Immunobinding Assay

The citrus canker pathogen *Xanthomonas axonopodis* pv. *citri* was detected using PABs in dot immunobinding assay (DIBA). The detection threshold was 1×10^4 cfu/ml. All isolates (26) of *X. axonopodis* pv. *citri* tested reacted positively and it was possible to detect the pathogen, on an average, in 38.4% of the asymptomatic samples tested by DIBA, which reflected the actual conditions (Wang et al., 1997). Coffee leaf scorch disease was demonstrated to be due to a bacterium related to *Xylella fastidiosa* by DIBA test. Antisera raised against the bacteria cultured from infected coffee plants and the strain of *X. fastidosa* causing citrus variegated chlorosis disease cross-reacted with each other in DIBA tests. The threshold of detection was 5×10^5 bacteria/ml (Lima et al., 1998).

8.3.5   Immunofluorescence Tests

Using the PABs and MABs either individually or as a mixture, Franken (1992) reported that *X. c.* pv. *campestris* could be detected in crucifer seeds. However, Franken et al. (1992) found that all six MABs tested cross-reacted with other pathovars of *X. campestris,* such as *X. campestris* pv. *vesicatoria* and *X. campestris* pv. *amoraciae.* Bragard and Verhoyen (1993) successfully applied the indirect immunofluorescence test for the detection of *X. c.* pv. *undulosa* in infected wheat seed lots. Gnanamanickam et al. (1994) employed the immunofluorescence test to identify *X. o.* pv. *oryzae* in highly contaminated rice seeds. Immunofluorescence tests indicated that the MABs Xco-1 and Xco-2 detected surface antigen in *X. c.* pv. *oryzae* and their epitopes were heat-sensitive and heat-resistant, respectively. The Xco-2 epitope was present in the lipopolysaccharide fraction (Benedict et al., 1989). Three serovars were distinguished among 215 strains of *Xanthomonas albilineans,* causal agent of sugarcane leaf scald disease, by immunofluorescence assay. Serovar I is the largest, including strains from Australia, the United States, Guadeloupe, India, Mauritius, South Africa. Serovar II consists of strains from Africa—Burkino Faso, Cameroon, Kenya, and Ivory Coast—whereas serovar III, the smallest group, has strains from Caribbean islands, Oceania (Fiji), and Asia (Sri Lanka) (Rott et al., 1994). Microscopic identification of *Clavibacter michiganensis* subsp. *sepedonicus* could be improved by using the rhodamine-labeled oligonucleotide probe in conjunction with an indirect immunofluorescence protocol based on specific MAB detected with a FITC-labeled conjugate. Bacterial cells labeled simultaneously with the oligonucleotide and antibody probes could be accurately identified by microscopic examination. This procedure is particularly useful when isolation and other methods of establishing the identity of the bacterial pathogens is difficult or not possible (Li et al., 1997). The potato brown rot disease caused by *Ralstonia solanacearum* was efficiently diagnosed using the monoclonal antibodies prepared from phage display library, against the lipopolysaccharide of *R. solanacearum* (biovar 2, race 3) (Griep et al., 1998)

8.3.6 Immunosorbent Electron Microscopy

Immunosorbent electron microscopy, along with the immunofluorescence test, was used for the characterization of MABs generated against *X. c.* pv. *oryzae* (Benedit et al., 1989). The immunogold staining technique was found to be a highly specific and rapid method of detecting the bacterium associated with citrus greening disease (Ariovich and Garnett, 1989).

8.3.7 Immunomagnetic Fishing/Isolation of Bacterial Pathogens

Immunomagnetic "fishing" can be used to improve recovery ratio between the target and nontarget bacteria present together in samples. Many seed samples are contaminated commonly in addition to pathogenic bacterial species with many saprophytic bacteria, which may predominate. The suspension containing the bacterial pathogen *X. campestris* pv. *pelargonii* and nontarget bacteria is incubated with rabbit PAB raised against *X. campestris* pv. *pelargonii* for 1 hr. Paramagnetic iron oxide particles coated with goat antirabbit IgG are mixed with the suspension and incubated. The polished surface of a neodymium super magnet (14 nm diameter) is placed at the air-water interface, so that magnetic particles-bacteria-antibody complex is attracted to the magnet. After all the magnetic particles are fished out, the magnet is dipped in sterile buffer to remove the nontarget bacteria. Target bacteria attached to the magnet are dislodged by rubbing gently over the surface of appropriate nutrient agar. The population of the nontarget bacteria is reduced to 11.4% of the initial population, facilitating the isolation of target pathogenic bacteria (Jones and Vuurde, 1996).

Immunomagnetic separation (IMS) procedure was developed for the selective separation of *Erwinia carotovora* subsp. *atroseptica* from potato peel extracts. A combination of selective medium, crystal violet pectate medium supplemented with 100 μg/ml of streptomycin and streptomycin tolerant strain of *E. carotovora* subsp. *atroseptica* was used to assess the recovery level of target bacteria. Using the advanced magnetics (AM) protein A particles–antirabbit IgG particles, potato soft rot bacteria could be enumerated in potato tuber peel extract consistently, the detection limit being 100 target bacterial cells/ml. On the other hand, polymerase chain reaction (PCR) assay could detect *E. carotovora* subsp. *atroseptica* only when the concentration was at least 10^5 cells/ml if the IMS procedure was not carried out prior to PCR assay (Wolf et al., 1996). Immunomagnetic separation of *Xylella fastidiosa* from the leafhopper tissues prior to nested PCR assay enabled the identification of two leafhoppers *Graphocephala coccinea* and *G. versuta* as the vector of the bacterial leaf scorch disease of American elms (Pooler et al., 1997).

8.4 DETECTION OF FUNGAL PATHOGENS

Fungal pathogens, in general, are relatively easily diagnosed by the symptoms induced in infected plants and characteristics of spores and mycelium. In certain cases alternative methods are required for their detection and identification. Contamination by fast-growing saprophytes poses a difficult problem in detecting the slow-growing seed-borne fungi in many crops, for example, rice. The fungi are complex antigens, and the lack of characterization of antigens is a major problem in the application of serological techniques for the early detection and identification of fungal pathogens. However, many researchers have attempted to make suitable modifications to improve the sensitivity and reliability of the tests.

Tempel (1959) used the gel diffusion test to differentiate *formae speciales* of *Fusarium oxysporum*. Later more sensitive techniques were employed for the detection of fungal pathogens (Holland and Choo, 1970; Nachmias et al., 1979; Fitzell et al., 1980; Savage and Sall, 1981; Nachmias et al., 1982; Gerik et al.,

1987). Polyclonal antisera using culture filtrate, cell fractions, whole cells, cell walls, and extracellular components, as immunogens have been prepared, and production of highly specific monoconal antibodies has been found to be difficult in many fungi. Species-specific and subspecies-specific monoclonal antibodies have been generated for pathogens such as *Phytophthora cinnamomi* (Hardham et al. 1986), and inclusion of glutaraldehyde in the fixative improves the specificity of reaction (Hardham et al., 1991). Specific detection of *Humicola languinosa* and *Penicillium islandicum* causing discoloration of stored rice is possible by using monoclonal antibody probes (Dewey et al., 1989, 1990). *Penicillium islandicum* is known to produce mycotoxins that cause liver lesions, cirrhosis, and primary liver cancer; hence its rapid detection is very important to eliminate the affected grains in storage.

Currently, simple diagnostic kits are being developed to conduct plant "side testing." Turf diseases due to *Pythium* sp., *Rhizoctonia solani*, and *Sclerotina homoecarpa* can be diagnosed by using visible immunodiagnostic assay kits developed commercially by Agri-Diagnostics, Cinnaminson, USA (Rittenberg et al., 1988).

Pathogen propagules present in soil or other complex substrates have not been quantified satisfactorily. Immunoassays have the potential to detect and quantify the soil-borne pathogens more accurately. Seroassays when used in conjunction with baiting assay can give results rapidly (El-Nashaar et al., 1986).

8.4.1 Enzyme-Linked Immunosorbent Assay

Standard ELISA and its variants have been employed to detect and quantify the pathogenic fungi and to assay their metabolic products. Banks and Cox (1992) developed a method to immobilize fungal hyphae onto microplate walls by precoating the walls with poly-L-lysine and glutaraldehyde, then attaching the hyphae to walls by overnight drying. The attached hyphae were uniformly coated and remained reactive. The plates could be stored at $-20°C$ (Fig. 8.6 A, B, and C) (Appendix 8[xviii]).

Using the antiserum produced with β-D-galactosidase-labeled antirabbit IgG as the secondary antibody and cell fragments of the strain of *Fusarium oxysporum* f. sp. *cucumerianum* attached to balls (Amino Dylark) as the solid-phase antigen, a highly specific and sensitive ELISA test was developed for the detection of homologous strains (Kitagawa et al., 1989). The PABs generated against purified mycelial proteins from *Verticillium dahliae* reacted positively with 11 of 12 *V. dahliae* isolates from potato, cotton, and soil, but not with the isolate from tomato. Using DAS-ELISA, *V. dahliae* and *V. albo-atrum* in infected roots and stems of potato could be detected (Sundaram et al., 1991). The sugarcane red rot pathogen (*Colletotrichum falcatum*) was detected using the antisera raised against

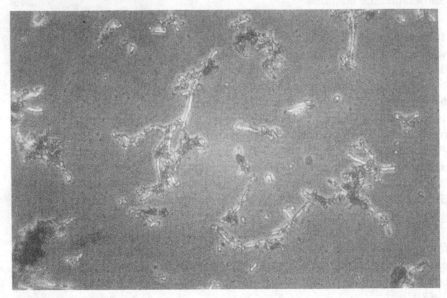

A

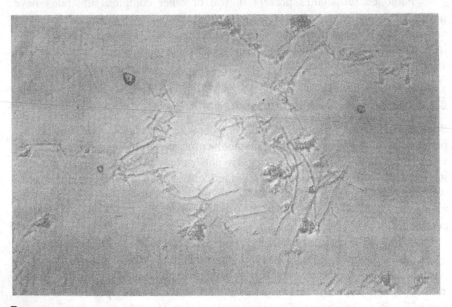

B

Figure 8.6 Detection of *Penicillium aurantiogriseum* var. *melanoconidium* by ELISA: A, *P. aurantiogriseum* var. *melanoconidum* hyphae after drying overnight on poly-L-lysine-bound to glutaraldehyde-pretreated microplate well; B, before addition of enzyme substrate; C, *P. aurantiogriseum* var. *melanoconidium* dried on a well not treated with poly-L-lysine or glutaraldehyde (Kluwer Academic Publishers, Netherlands). (Courtesy of Banks and Cox, 1992. Reprinted by permission of Kluwer Academic Publishers.)

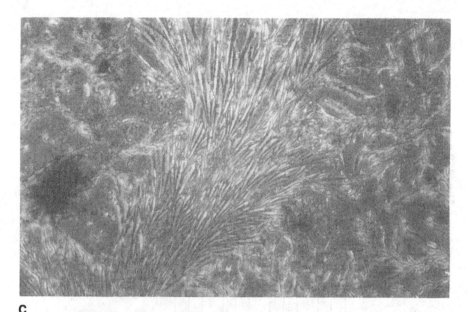

C

Figure 8.6   Continued.

the unfractionated protein and a 101-kDa polypeptide present in all pathotypes in ELISA tests. The presence of the pathogen was detected in root eyes, buds, leaf scar, and pith regions of the stalk (Viswanathan et al., 1998). Infection of young petals of rapeseed by *Sclerotinia sclerotiorum* was detected by the DAS-ELISA test by Jamaux and Spire (1994). In a further study, antisera against the mycelium and ascospores of *S. sclerotiorum* present on the petals were prepared separately. The sensitivity of the antimycelium serum (Smy) was greater than the antiascospore antiserum (Ssp) in detecting the pathogen in mycelial extract. However, both antisera were equally reactive when exposed to the ascospore antigen. These antisera showed cross-reaction with *Botrytis cinerea* (Jamaux and Spire, 1999). The presence of *Pseudocercosporella herpotrichoides* could be detected by using a specific MAB in plants infected by the pathogen based on the absorbance value (Fig. 8.7). *Rhizoctonia solani* and other *Rhizoctonia* spp. in poinsettia cuttings were rapidly detected by ELISA (Benson, 1992). Lyons and White (1992) employed PAB to detect *Pythium violae* in cavities developed in field-grown carrots. Bossi and Dewey (1992) reported that MABs specific for *Botrytis cinerea* could be used to detect the pathogen in strawberries (Fig. 8.8). The MABs recognized mycelial fragments, saline extracts of mycelia, and germinating conidia by both ELISA and immunofluorescence. Thornton et al. (1993) developed four lines that produced species-specific monoclonal antibodies capable of recognizing the antigen from *Rhizoctonia solani* by the ELISA technique. This assay technique can be

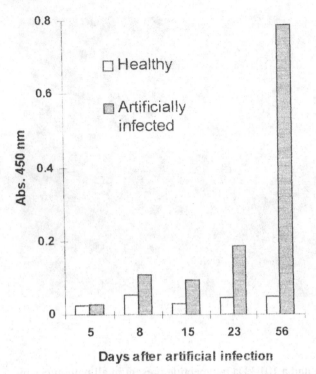

Figure 8.7 Absorbance values from DAS-ELISA tests of extracts from stem bases of healthy wheat plants and wheat plants 5, 8, 15, 23 and 56 days after infection. (British Society for Plant Pathology, U.K.) (Courtesy of Priestley and Dewey, 1993.)

used to detect live propagules of the pathogen present in the soil. In the further study, Thornton et al. (1999) demonstrated the usefulness of the combined-baiting, double monoclonal antibody ELISA. This technique was rapid requiring only 3 days from the receipt of soil samples containing *R. solani*. The format permits recovery of *R. solani* isolates from colonized baits for the determination of their anastomosis group affiliation and pathogenicity. The isolates pathogenic to lettuce were identified as AG4 group. Hardham et al. (1986) showed that MABs raised against components on the surface of glutaraldehyde-fixed zoospores and cysts of an isolate of *Phytophthora cinnamomi* could be used as isolate-specific, species-specific, and genus-specific markers without any ambiguity. These MABs have a valuable spectrum of taxonomic specificities. Stace-Smith et al. (1993) using specific MABs to distinguish highly virulent and less virulent strains of *Leptosphaeria maculans* that cause black leg of canola. Polyclonal antibodies raised against *Phomopsis longicolla* from soybeans using culture filtrate (cf) and mycelial extact (me) were tested by DAS-ELISA and indirect ELISA. The DAS-ELISA test was found to be more specific and 100 times more sensitive in detect-

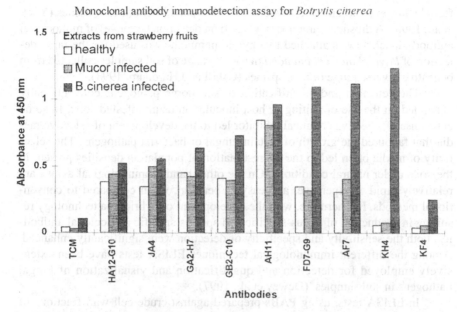

Figure 8.8 Antibodies tested for specificity by ELISA against extracts from healthy strawberries and strawberries infected by *Botyrtis cinerea*. (Association of Applied Biologists, U.K.) (Courtesy of Bossi and Dewey, 1992.)

ing the fungus in the *Diaporthe-Phomopsis* complex, and the variability in specificity was less when DAS-ELISA was employed (Brill et al., 1994).

Using an indirect ELISA with PABs raised against mycelial extracts of *Pestalotiopsis theae* causing tea gray blight disease, the infection by the pathogen could be detected as early as 12 hr after inoculation. The antiserum did not react with other pathogens such as *Glomerella cingulata* and *Corticium invisum*, indicating the specificity of the PABs. The ELISA technique can be used for the early detection of gray blight pathogen even at a very low level of infection (Chakraborty et al., 1996). *Bipolaris sorokiniana* infecting wheat can also be detected using specific PAB by following indirect ELISA procedure (Lopes et al., 1998).

An MAB specific to *Pythium ultimum* was highly reactive to 21 isolates of *P. ultimum* and did not react with any of the 16 species of *Pythium* tested by ELISA. The test was effective in detecting the pathogen in sugar beet seedling roots, with more than two infections/10 cm of root (Yuen et al., 1993). A PAB prepared against the cell walls of *P. ultimum* as the capture antibody and a MAB specific for recognition were used in indirect DAS-ELISA test for the detection and quantification of this fungal pathogen. Strong positive reactions were observed when culture filtrates of seven isolates of *P. ultimum* were tested. The pathogen was detected in the roots of sugarbeet, beans, and cabbage seedlings grown in infested soils, even when there was one infection per 100 cm of root tissues (Yuen et al., 1998). A dipstick immunoassay based on detection (Azodye) of monoclonal antibody-labeled cysts attached to a nylon membrane was used for the rapid detection of *Phytophthora cinnamomi* in a wide range of soil samples collected from beneath a diverse range of host species (Cahill and Hardham, 1994).

The detection and quantification of soil-borne pathogens conventionally depended on the use of baiting or host inoculation using infested soils. In some cases, use of selective chemical inhibitor led to the development of selective media that favoured the growth of certain fungal or bacterial pathogens. The selectivity of media often led to the overestimation of population densities present in the soils under natural conditions. On the other hand, immunological assays are relatively rapid, inexpensive, and easy to perform, when compared to conventional methods. Furthermore, with the development of hybridoma technology resulting in production of genus-specific and species-specific monoclonal antibodies, both the sensitivity and specificity of detection were significantly enhanced. Among the different immunological techniques ELISA tests have been extensively employed for detection and quantification and visualization of fungal pathogens in soil samples. (Dewey et al., 1997).

In ELISA tests, using PABs prepared against crude cell wall fractions of *Pythium aquatile* or *P. coloratum* associated with root rot of tomato, specific reactions with closely related isolates were observed (Rafin et al., 1994). A fungal capture sandwich ELISA was developed by using the polyclonal antiserum raised against soluble protein extracts of chlamydospores and mycelium of *Thielaviopsis basicola* and the IgG was labeled with biotin. The test detected both brown and gray cultural types of *T. basicola* and the antibodies showed negligible cross-reactivity with other soil-borne fungi found in cotton field soil. The minimum detection limit was between 1 and 20 ng of *T. basicola* protein, and the pathogen could be detected in cotton roots at 2 days after inoculation when initial symptoms were not apparent (Holtz et al., 1994). Using the PAB produced by employing a homogenate of spore balls (cystosori) of *Spongospora subterranea,* the presence of the pathogen could be detected in dilute tuber extract containing the equivalent of as little as 0.08 spore balls/ml (Harrison et al., 1993) (Fig. 8.9).

Commercial immunoassay kits have been developed for the detection of fungal pathogens in plant tissues as well as in soil. Miller et al. (1994) compared the efficacy of the Albert *Phytophthora* "flow through" immunoassay and multiwell ELISA kits (Agri-Screen) for the detection of *Phytophthora* capsici and *P. cactorum.* The former was easy to perform and rapidly (within 10 min) detected *P. capsici* in pepper and cucurbit crops; the latter was effective in detecting the pathogen in pepper tissues but had higher absorbance values for healthy samples. There was excellent agreement between results of ELISA tests and isolation of pathogens in semiselective medium. The presence of *Phytophthora* in soils also could be detected by the kits. More specific detection and characterization of *P. capsici* are possible by using MABs. Kimishima and Kobayashi (1994) produced

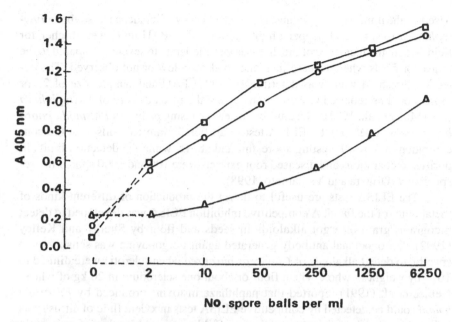

Figure 8.9 Effect of potato tuber sap concentration on detection of spore balls of *Spongospora subterranea* by PTA-ELISA. (British Society for Plant Pathology, U.K.) (Courtesy of Harrison et al., 1993.)

three MABs (PC1A6, PC1B2, and PC1C5) against mycelial suspension of *P. capsici,* which were tested by ELISA and Western blot analysis. The study suggested that among the three MABs, PC1C5 could recognize one epitope common to all isolates tested and PC1A6 recognized a carbohydrate epitope, whereas PC1B2 and PC1C5 were able to recognize protein epitopes. Commercial ELISA kits have been developed to detect and to estimate propagule densities of fungal pathogens such as *Phytophthora citrophthora* in plant roots and soil samples (Timmer et al., 1993). Likewise, *P. fragariae* var. *rubi* was detected in the root tissues of raspberry using a commercial multiwell assay kit from 4 days after inoculation. This ELISA procedure had a detection limit of about 0.25% of simulated infection level (percentage of infected tissue/healthy tissue, w/w) (Olsson and Heiberg, 1997). *Phytophthora infestans,* causing potato late blight disease, was detected by indirect ELISA before the appearance of first symptoms. The pathogen could be detected in potato shoots of 5–9 weeks old plants about 39 days before disease outbreak in the field. However, there was no correlation between ELISA results and symptom development later on single plants in the field, possibly because of infection caused by zoospores transported in soil water following heavy rainfall (Schlenzig et al., 1999). The population densities of *Phytophthora* spp. were assessed by ELISA in field soils. *Phytophthora* antigen units (PAU), calculated based on the immunoassay values, reflect the history of incidence of soybean *Phytophthora* root rot and pepper blight diseases. The PAU units were higher for fields where the incidence of the diseases was moderate to severe, compared to the index for fields where disease incidence had been low or not observed. The detection threshold value was determined as 11.3 PAU and sample size of 20 or more would be required to estimate precisely the mean density of *Phytophthora* spp. (Miller et al., 1997). The clubroot disease causing pathogen *Plasmodiophora brassicae* was detected by ELISA test in artificially infested soils with the concentration of 1×10^4 resting spores/ml and above, while the detection limits in purified preparations and diseased root extracts were 100 and 1000 spores/ml, respectively (Orihara and Yamamoto, 1998).

The ELISA tests are useful to detect the production of different kinds of metabolites of the fungi. A competitive inhibition ELISA was developed to detect picograms/gram of ergot alkaloids in seeds and flour by Shelby and Kelley (1992). A monoclonal antibody generated against ergonovine was sensitive and specific to detect alkaloids of *Claviceps purpurea* when sclerotia were diluted to 10^{-5} by weight in whole wheat flour or about one sclerotium in 20 kg of wheat. Nemec et al. (1991) reported that naphthazarin toxins produced by *Fusarium solani* could be detected by competitive ELISA tests in xylem fluid of citrus roots infected by *F. solani* and symptomless scaffold roots and branches of apparently healthy and diseased citrus trees.

Fungal pathogens secrete several enzymes that may degrade host constituents resulting in tissue disintegration. The reduction in the content or absence of specific host protein following infection by *Pseudocercopsorella herpotrichoides* causing eyespot disease of wheat has been used as an indicator of disease severity. In the stem bases of healthy wheat plants a specific protein (Pc) is abundantly produced, whereas in the tissues affected by the pathogen this specific protein cannot be detected. An ELISA format was developed by Coff et al. (1998) for the quantitative estimation of Pc protein. The content of Pc protein was negatively correlated with the degree of tissue degradation by *P. herpotrichoides* and the ELISA readings could be used as estimates of disease severity. The Pc protein was identified as plastocyanin by amino acid sequence analysis. The antiserum containing PABs against purified exopolygalacturonase (exoPG) produced by *Fusarium oxysporum* f.sp. *radicis lycopersici* (FORL) was used to detect the enzyme produced by FORL in the roots of infected tomato plants. The production of exo PG in planta was confirmed. Moreover, the progressive enhancement of exoPG expression level together with development of root rot symptom during infection process was also observed (Platiño Álvarez et al., 1999).

The fungal pathogen biomass present in plants can be estimated by using MABs in ELISA tests. An MAB prepared against the surface washings of *Cladosporium herbarum* could recognize an epitope present on *C. fulvum* causing tomato leaf mold and other *Cladosporium* spp. This epitope present in two races of *C. fulvum* is constitutively expressed both in culture and in infected plants. By using the specific MAB, in ELISA, *C. fulvum* was detected in infected tomato leaf tissues at levels starting from around 1 mg fresh weight of the pathogen per gram fresh weight of leaf tissue. The assessment of biomass of *C. fulvum* by ELISA was in agreement with measurements of β-glucuronidase (GUS) activity in tissues infected with a transgenic isolate of *C. fulvum* race 4 (Chapter 6.3.2).

Many seedborne fungi are known to produce metabolites toxic to human beings and animals. Rapid detection of fungi such as *Aspergillus, Penicillium,* and *Fusarium* becomes necessary to avoid consumption of contaminated foods and feeds. Banks et al. (1994) prepared two MABs capable of reacting with 12 field and 27 storage fungi. The indirect ELISA using an MAB and extracts of barley inoculated with the antigen (*Penicillium aurantiogrioseum* var. *melanoconidium*) exhibited a positive correlation between absorbance and the amount of fungal growth. The mold fungi *Aspergillus parasiticus, Penicillium citrinum,* and *Fusarium oxysporum* associated with rice and corn could be detected in ELISA tests using PABs raised against them. There was a clear positive relationship between the absorbance values of ELISA and the amount of mold growth. The detection limit of the assay was 1 μg/ml (Chang and Yu, 1997). The presence of nine toxigenic *Aspergillus* spp. in rice and corn was detected by DAS-ELISA, using PABs produced against the respective fungus (Wang and Yu, 1998). *Aspergillus* spp are known to produce aflatoxins in the food and feedstuff posing great danger to the consumers. Among the several immunoassays tested, direct ELISA formats have been found to be more sensitive for their detection and quantification. Horseradish peroxidase (HRP) is used to prepare the antibody-enzyme conjugate employed as the marker. The color intensity formed in the wells is inversely proportional to the aflatoxin concentration (Wilson et al., 1998). Using high-affinity PABs against afflatoxin B, a direct competitive ELISA protocol was developed employing HRP for labeling the antibodies. The detection limit was 0.25–5.0 ng/ml (Chen and Chen, 1998).

Spore traps have been used to monitor the population of spores released by different fungal pathogens into the airstream and the pathogens are identified by microscopic examination. Considerable experience is required to count the spores trapped and to establish the identity of the pathogens. Applications of immunoassays may help to monitor pathogen populations rapidly and reliably, providing data for forecasting the possible disease outbreaks. Schmechel et al. (1997) developed a spore trap, designated strip rotorod trap (SRT), which can provide samples that can be analyzed by indirect ELISA protocol. The airborne conidial populations of *Alternaria brassicae* were assessed by such immunomonitory technique, using monoclonal antibodies prepared against the pathogen.

Among the several applications of diagnostics, the ELISA formats are being used as a forecasting tool, since the tests provide rapid and reliable results for the assessment of disease intensity based on which recommendation for the fungicidal applications are made. The standard multiwell ELISA kit, providing a highly sensitive and accurate system for the identification and quantification of *Septoria tritici* and *S. nodorum* causing *Septoria* diseases in wheat, was used in a nationwide, diagnostic survey called 'Septoria Watch' in the United Kingdom. Assessment of disease by the immunoassays formed the basis for the presymptomatic warning resulting in good yield benefits as well as saving of fungicidal application costs (Smith et al., 1994). The usefulness of immunodiagnosis as an aid to determine the timing of fungicide sprays for the management of *Mycospherella graminicola* (*Septoria tritici*) in winter wheat was demonstrated by Kendall et al. (1998). Presymptomatic detection of *M. graminicola* on leaf 3 was shown to be an accurate indicator of the optimum timing of a second fungicide application for the protection of top two leaves, whose health accounts for over 85% of the grain-filling capacity of wheat plants. This study clearly brings out the need for rapid, reliable, and sensitive diagnostic tests for the crop disease management tactics to be effective and economical.

The ELISA test was used to quantify fungal antigen in resistant and susceptible cultivars after inoculation with the pathogen. The polyclonal antiserum developed by using mycelia and zoospores of *Aphanomyces euteichus* as immunogens reacted positively with *A. euteichus*, but not with *Phytophthora, Fusarium,* and *Pythium.* When the roots of resistant lines were exposed to 100 zoospores/ml, the buildup of *A. euteichus* was slower when compared to that of susceptible lines as determined by ELISA test using the antiserum, indicating inhibition of the growth of the pathogen within the inoculated tissues (Kraft and Boge, 1994). In a further study to find out the relationship between the rate of lesion development and pathogen population as determined by ELISA tests, it was shown that indirect ELISA measurements (at A405 nm) were positively correlated ($R^2 = 0.91$) with lesion length (development) on the main roots of peas infected by *A. euteichus.* Resistance to root rot disease appears to be associated with reduced oospore production, pathogen multiplication, zoospore germination, and slower lesion development (Kraft and Boge, 1996).

Resistance of hazelnut (*Corylus avellana*) to eastern filbert blight disease caused by *Anisogramma anomala* is tested conventionally by artificial inoculation and this procedure requires a period of 13–22 months for symptom expression. But infection of *A. anomala* could be detected in symptomless plants by indirect ELISA at 3–5 months after inoculation, reducing the period of observation for the development of external symptoms significantly. ELISA tests were found to be 100% efficient in detecting infection compared to conventional microscope examination, which could detect infection in only 36% of the samples. The use of ELISA in a breeding program resulted in the identification of a hazelnut progeny with a gene conferring high level of resistance derived from the cultivar Gasaway (Coyne et al., 1996). A competitive ELISA procedure was developed to assess different grades of resistance of narrow-leaved lupins (*Lupinus angustifolius*) to latent stem infection by *Diaporthe toxica*. Seedlings were inoculated with *D. toxica* and resistance of test entries (12) was assessed at 21 days after inoculation (DAI) by counting latent infection structures in epidermal tissues under the microscope. Infected stem pieces, after a 6-day incubation in moist chamber, were tested by ELISA. Positive signals of ELISA were highly correlated ($r = 0.92$) with the frequency of latent infection structures detected by microscopic examination. The susceptible, resistant, and highly resistant lines could be differentiated by ELISA tests (Shankar et al., 1998).

8.4.2 Dip Stick Immunoassay

A recent significant development in diagnostic assays are the visible immunodiagnostic assay methods in which the samples can be handled easily under field conditions. The 96-well titer plate is replaced by dip-stick formats (Dewey et al., 1989; Cahill and Hardham, 1994a). The dipstick immunoassay is based on the phenomena of chemotaxis and electrotaxis to attract the zoospores to a membrane on which they encyst and are then detected by immunoasasy. The zoospores are attracted by a variety of chemicals, such as amino acids, alcohols, phenols and isovaleraldehyde, pectin, and the phytohormone abscissic acid. Positively charged nylon membranes strongly attract the zoospores. The dipstick assay can detect as few as 40 zoospores/ml within 45 min. Immunolabeled cysts attached to the membrane can be seen with the naked eye or observed under low-power magnification after silver enhancement of a gold-labeled secondary probe. By using MABs *Phytophthora cinnamomi* can be rapidly detected (Cahill and Hardham, 1994b).

8.4.3 Dot Immunobinding Assay

Immunoblotting assays have been found to be useful to overcome problems with nonspecific interference in ELISA procedures. Gleason et al. (1987) developed the seed immunoblot assay (SIBA) to detect *Phomopsis longicolla* in infected soybean seeds (Appendix 8[xix]). The mycelium of *P. longicolla* growing onto a nitrocellulose sheet from infected soybean seeds forms a conspicuous colored blotch when assayed by the SIBA test. As SIBA detects only viable *P. longicolla*, this technique may be preferable to ELISA, which does not differentiate live and dead fungus (Fig. 8.10). Infection of wheat seeds of Karnal bunt disease caused by *Tilletia indica* can be detected by SIBA tests. A colored imprint is produced on the nitrocellulose sheet on which the infected wheat seeds are placed, following immunoprocessing (Anil Kumar et al., 1998).

Dot immunobinding assay (DIBA) was used to detect *Phomopsis phaseoli*, cause of pod and stem blight, and *P. longicolla*, cause of seed decay of soybean in asymptomatic soybean tissues. Polyclonal antiserum was raised against *P. longicolla* and the PABs reacted strongly with all *Phomopsis* spp. and *Colletotrichum truncatum*. The culture filtrates and mycelial extracts of *Phomopsis* spp. showed similar reactions with PABs. A method of quantification of antigen based on the "antigen unit," instead of absorption values, was developed (Velicheti et al., 1993).

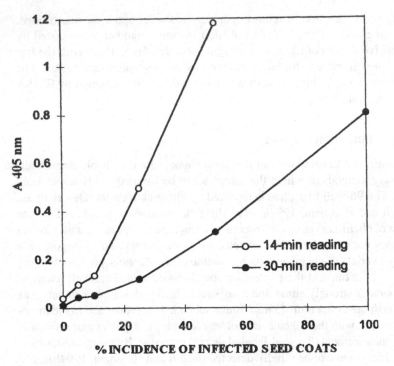

Figure 8.10 Effect of dilution of seed coats infected by *P. longicolla* with uninfected seed coats on detection of *P. longicolla* by indirect ELISA. (British Society for Plant Pathology, U.K.) (Courtesy of Gleason et al., 1987.)

8.4.4 Tissue Blot Immunoassay

The involvement of the enzyme endopolygalacturonase (EndoPG) in the development of diseases caused by *Fusarium oxysporum* has been suggested. A polyclonal antibody APG1 was raised against the purified preparation of PG from *Fusarium oxysporum* f.sp. *lycopersici* race 2. The production of endo PG in the stem tissue of inoculated plants was detected by using direct tissue blot immunobinding assay by Arie et al. (1998).

8.4.5 Immunofluorescence Assay

Aldehyde-fixed zoospores and cysts of *Phytophthora cinnamomi* were used to generate 24 MAB clones. Using the immunofluorescence assay (IFA), 11 MABs were found to be species-specific, reacting specifically with zoospores and cysts of *P. cinnamomi* only, but not with other species of fungi, such as *Phytophthora* spp., *Pythium* spp., *Saprolegnia* spp., *Fusarium* spp., *Verticillium* spp., *Rhizocto-*

nia spp., and *Schizophyllum* spp. One MAB was genus-specific, reacting with other *Phytophthora* spp. One MAB capable of reacting only with *P. cinnamomi* in both IFA and ELISA was identified (Gabor et al., 1993). For the detection of *Botrytis cinerea*, whole conidia, their extracellular material, and a putative cut in esterase isolated from conidia were used as antigens to prepare MABs. Three MABs capable of recognizing conidia of 43 isolates of *B. cinerea* from different hosts representing six countries were identified by Salinas and Schots (1994). These MABs could be used to detect *B. cinerea* in flowers of gerbera by immunofluorescence test. The presence of *Thielaviopsis basicola* in cotton roots was detected by immunofluorescence assay by Holtz et al. (1994). By using a PAB raised against whole ascospores of *Mycosphaerella brassicola*, causing ringspot disease of vegetable crucifers, the pathogen was detected by immunofluorescence (IF) assay. Antirabbit IgG-FITC conjugate reacted with ascospores and mycelial wall of *M. brassicola* under field conditions. The ascospores trapped by Melinex spore trap coated with bovine serum albumin (BSA) as a support medium and blocking agent were detected by the IF assay. Autofluorescence of spores and mycelial components of other fungal species was eliminated by the application of counter stain Evan's blue (0.2%) and eriochrome black (0.5%) in phosphate buffered saline, pH 7.2. The procedure has the potential for use in epidemiology and forecasting programs (Kennedy et al., 1999).

8.5 DETECTION OF VIROIDS

It has been demonstrated that viroids do not possess any messenger activity and that the viroid is not translated in the *Xenopus larvis* oocyte system (Davies et al., 1974; Semancik et al., 1977), leading to the absence of any viroid-specific protein in the infected plants. However, production of enhanced levels of host-specific proteins has been reported in tomato infected by potato spindle tuber viroid (PSTVd) and *Gynura aurantica* infected by citrus exocortis viroid (CEVd) (Zaitlin and Hariharasubramanian, 1972; Conejero et al., 1979).

Diener et al. (1985) reported that a host protein with a molecular weight of 70,000 accumulated in tomatoes after inoculation with planta macho viroid and this protein was isolated after density gradient centrifugation. The antiserum raised against the host-specific protein designated PM antigen could be used for the detection of tomato planta macho viroid (TPMVd) infection, using the double-diffusion test. However, stimulation of PM antigen synthesis does not appear to be specific to TPMVd infection. The presence of PM antigen in tomato plants infected by other viroids, such as PSTVd, chrysanthemum stunt viroid (CSVd), tomato apical stunt viroid (TASVd), columnea viroid (CVd), and the cucumber mosaic virus, along with CARNA5, has been detected (Diener et al., 1985). This reveals that use of serological tests for viroid detection and identification may not be very specific.

SUMMARY

Serological techniques have been employed for the detection of viruses, phytoplasmas, bacteria, and fungi present in infected seeds, seed materials, plants, soil, and water. The sensitivity and reliability of these techniques are greater than those of infectivity tests and chemodiagnostic methods in the case of many pathogens. Bacterial, phytoplasmal, and fungal pathogens are complex antigens and much difficulty is experienced in producing antisera against these pathogens. Polyclonal and monoclonal antisera have been prepared against these pathogens, and by using panels of monoclonal antibodies, the sensitivity and reliability of detection have been markedly enhanced. Production of antisera using bacterially expressed viral coat proteins and phage-displayed recombinant antibodies has resulted in significant increase in the effectiveness, sensitivity, specificity, and reliability of serological techniques. Serological techniques have been extensively used for the detection, differentiation, and quantification of microbial pathogens. The ELISA test and its variants, the dot immuno-binding assay (DIBA), tissue blot immuno-binding assay (TIBA), rapid immuno-filter paper assay (RIPA), enzyme-linked fluorescent assay (ELFA), and time-resolved fluoroimmunoassay (TR-FIA), have been the preferred methods. The development of cytofluorimetric method and immuno-capillary zone electrophoresis is the result of attempts to enhance the sensitivity of detection, and reduce the time required for detection. Only very low concentrations/populations of microbial pathogens and very small volumes of antigen and antisera are required for these tests and results are obtained in a few hours.

APPENDIX 8(i): PRODUCTION OF ANTISERUM TO VIRAL PROTEIN EXPRESSED IN BACTERIAL CELLS (HELGUERA ET AL., 1997)

A. Viral protein expression and purification
   i.   Introduce plasmid pRSET/C6CP into *Escherichia coli* cells by standard transformation protocol.
   ii. Add 0.3 mM isopropyl-beta-D-thiogalacto pyranoside (IPTG) to the bacterial culture grown to OD_{600} 0.5 to induce expression of C6CP polypeptide; incubate for 2 hr at 37°C for the formation of intracellular aggregates of the C6CP protein.
   iii. Resuspend the bacterial cells in STE buffer containing 100 mM NaCl, 50 mM Tris-HCl, pH 8.0, and 1 mM EDTA.
   iv. Disrupt the bacterial cells by sonication and centrifuge at 9000 g for 10 min.

    v.   Resuspend the pellet in 2 ml of STE buffer supplemented with 8 mM MgCl$_2$, DNase I (10 μg/ml); incubate at 4°C for 10 min and centrifuge at 9000 g for 10 min.

    vi.   Resuspend the pellet in STE buffer (1 ml)

    vii.   Analyze the expression products in Western blots using antibodies specific for the virus concerned.

B.   Antiserum production

    i.   Electrophorese the bacterially expressed viral CP in a preparative SDS-polyacrylamide gel; excise the band containing the viral CP from the gel, crush, and lyophilize.

    ii.   Immunize the rabbits by intradermal injection of fusion protein (0.6 mg) emulsified in Freund's complete adjuvant (1 ml).

    iii.   Administer a booster dose after 2 weeks by intramuscular injection of an equal amount of protein (0.6 mg) in Freund's incomplete adjuvant (1 ml).

    iv.   Bleed the rabbit at 15 and 21 days after the last injection.

    v.   Dilute the antiserum to the required level before use.

APPENDIX 8(ii): PRODUCTION OF PHAGE-DISPLAYED PEPTIDES (ZIEGLER ET AL., 1998)

A.   Peptide library

    i.   Use the Cys 1 library comprising of 13-mer random peptides displayed at the N-terminus of the major phage coat protein p VIII.

    ii.   Amplify the library by culturing infected *Escherichia coli* TG-1 cells overnight at 37°C in Luria-Bertani (LB) broth (100 ml) supplemented with tetracycline (12.5 μg/ml).

    iii.   Harvest phage particles by precipitation using polyethylene glycol (PEG)/NaCl

B.   Selection of phage clones

    i.   Prepare MAB immunoglobulins (IgG) by precipitation with ammonium sulfate from ascitic fluid.

    ii.   Coat the immuno tubes with 5 μg of MAB-IgG in 1 ml of 0.1 M carbonate buffer pH 9.6 (coating buffer).

    iii.   Add phage preparation to the immunoglobulin coated tube (= input); allow reaction between the phage and IgG at room temperature (22°C).

    iv.   Elute the bound phage particles with 100 mM triethylamine; add them to TG-1 cells for initiating infection.

    v.   Remove a small volume of infected TG-1 cells; plate on LB-tetracycline (LB-tet) plates to determine the number of eluted phage (= output).

    vi.   Culture the infected TG-1 cells by growing overnight at 37°C; precipitate the phage particles by adding PEG/NaCl.

    vii.   Determine the titer of the phage by infecting TG-1 cells and recording the number of tetracycline colonies capable of growing on LB-tet plates; repeat the above steps three times.

C.   Identification of reactive phage clones

    i.   Inoculate selected single tetracycline-resistant colonies from the third round of selection to LB-tet (5 ml) and grow at 37°C in a shaker for 18 hr.

    ii.   Centrifuge to remove the bacterial cells; assay the supernatant fluid (SN) by ELISA.

    iii.   Coat the microplates with MAB (100 µl at 2.5 µg/ml in coating buffer); remove excess MAB by washing.

    iv.   Block the wells by adding PBS (200 µl) containing 2% nonfat dried milk (MPBS) to each well at 30°C for 30 min and wash the plates three times with PBS containing Tween 20 (0.05%) followed by washing once with PBS.

    v.   Dispense MPBS (25 µl) into each well along with SN-containing phage particles (75 µl); incubate the plates at 30°C for 4 hr and wash the plates as in step (iv).

    vi.   Add the mixture (100 µl) of a rabbit anti-M13 PAB antiserum (1:500) and anti-rabbit IgG–alkaline phosphatase (AP) conjugate (1:10,000) to each well; incubate at 30°C for 1 hr and wash the wells as done before.

    vii.   Add p-nitrophenylphosphate (0.5 mg/ml) in 0.1 M diethanolamine buffer, pH 9.8, as substrate; stop reaction by adding 3 M NaOH; record the color intensity at 405 nm.

APPENDIX 8(iii): FORMATS OF ENZYME-LINKED IMMUNOSORBENT ASSAY (ELISA)

A.   Double-antibody sandwich (DAS) ELISA (Clark and Adams, 1977)

    i.   Precipitate globulins from the antiserum using 36% sodium sulfate; wash the precipitate with 18% sodium sulfate and store at −70°C. Conjugate a portion of globulin with alkaline phosphatase using glutaraldehyde as the coupling agent.

    ii.   Dilute the globulin fraction (unlabeled) with 0.05 M carbonate buffer at pH 9.6 to yield a concentration of 10 µg protein/ml. Add 200 µl of antibody solution to each well in polystyrene ELISA plates and incubate at 37°C for 3–5 hr. Empty the wells; wash thrice with 0.15 M phosphate-buffered saline solution at pH 7.2 containing 0.05% Tween 20 (PBS-Tween) and dry.

    iii.   Add samples (purified antigen or extracts of infected tissues) in 200-

μl quantities in PBS-Tween; incubate at 4°C overnight or for 18 hr and wash the wells as before.

iv. Add aliquots of 200 μl of enzyme-labeled antibody conjugate to each well; incubate for 4 hr at 37°C and wash the wells as before.

v. Add enzyme substrate p-nitrophenyl phosphate at a concentration of 1 mg/ml in diethanolamine buffer at pH 9.8 at room temperature. Stop the reaction after 30 min by adding 3 *M* NaOH at 50 μl/well.

vi. Determine the color intensity (OD) at 405 nm in an ELISA reader.

B. Direct antigen coating (DAC)-ELISA

i. Add samples at 200 μl to each well in the ELISA plate; incubate at 37°C for 1 hr and wash the wells with PBS-Tween.

ii. Add antiserum at suitable dilution at 200 μl/well; incubate for 1 hr at 37°C and wash the wells with PBS-Tween.

iii. Add enzyme-labeled antirabbit IgG at 200 μl to each well; incubate for 1 hr at 37°C and wash the wells with PBS-Tween.

iv. Follow steps (v) and (vi) as in DAS-ELISA.

C. Protein A-coating (PAC)-ELISA

i. Dissolve protein A (1–10 mg/ml) in carbonate buffer; dispense 200 μl/well in ELISA plate; incubate for 1 hr at 37°C and wash with PBS-Tween.

ii. Dispense antiserum (at suitable dilution) at 200 μl/well; incubate for 1 hr at 37°C and wash with PBS-Tween.

iii. Dispense 200 μl of samples (purified antigen/extracts of tissues at suitable dilution); incubate at 37°C for 1 hr and wash the wells with PBS-Tween.

iv. Dispense 200 μl of antiserum and proceed as in step (ii).

v. Dispense enzyme-labeled antirabbit IgG or Fc at 200 μl/well; incubate for 1 hr at 37°C and wash with PBS-Tween.

vi. Follow steps (v) and (vi) as in DAS-ELISA.

D. Indirect DAS-ELISA

i. Dispense goat or chicken antivirus globulins (1–10 μ/ml) at 200 μl/well; incubate at 37°C for 1 hr and wash with PBS-Tween.

ii. Dispense 200 μl of suitably diluted samples in each well; incubate at 37°C for 1–3 hr and wash with PBS-Tween.

iii. Dispense 200 μl of antivirus rabbit globulin/well; incubate at 37°C for 1–3 hr and wash with PBS-Tween.

iv. Dispense 200 μl of antirabbit globulin conjugate/well; incubate for 1 hr at 37°C.

v. Follow steps (v) and (vi) as in DAS-ELISA.

APPENDIX 8(iv): DOT IMMUNOBINDING ASSAY (DIBA)

i. Dip a nitrocellulose membrane grid (1.0 or 2.5 cm squares) in Tris buffer (0.02 *M* Tris-CL, 0.5 *M* NaCl, pH 7.5) (TBS) and dry it on filter paper for 5 min.

   ii.   Spot the samples (1 μl) prepared in TBS in the center of the grid and dry. Place the grid in blocking solution (3% gelatin, 2% Triton X-100 in TBS) in a petri dish and agitate for 1 hr.

   iii.   Dip the grid into distilled water; transfer to 50 ml antiserum (1 mg protein/ml) and 1% gelatin in a petri dish and agitate for 1 hr.

   iv.   Dip the grid into distilled water; then wash twice by agitation for 10 min in TBS containing 0.05% Tween 20 (TTBS).

   v.   Dip the grid into distilled water; transfer to horseradish peroxide conjugated IgG (1/1000 dilution) and 1% gelatin kept in a petri dish; agitate for 1 hr.

   vi.   Dip the grid into distilled water; wash in TTBS and TBS successively for 10 min.

   vii.   Transfer to substrate solution (dissolve 0.06 g 4-chloro-1-naphthol in 20 ml of 4-C-methyl alcohol, then add 100 ml of TBS and 0.06 ml of 30% hydrogen peroxide); incubate for 10–30 min in darkness.

APPENDIX 8(v): TISSUE BLOT ASSAYS

A.   Tissue blot immunoassay (TBIA) (Hu et al., 1997)

   i.   Make sharp cut across the leaf or plant organ using a razor blade; press the freshly cut edge firmly onto mitrocellulose membrane (0.45 μm) for 3–5 sec so that the impression of the vascular bundles remains on the membrane and store the membranes dry at room temperature.

   ii.   Place the membranes in a plastic container; block the blots with 2% powdered milk dissolved in TBS buffer containing Tris-HCl (50 mM), NaCl (50 mM), pH 7.5 for 30 min at room temperature.

   iii.   Incubate in another plastic container or sealable plastic bag with primary antibodies against the test virus (1 μg/ml) in TBS with gentle agitation for 1 hr at room temperature, followed by an overnight soak at 4°C.

   iv.   Wash the membranes with three changes of TBS + 0.5% Tween 20 (TBST) for 10 min each with constant agitation.

   v.   Incubate the membranes with goat antimouse (GAM) alkaline phosphatase conjugate at a dilution of 1:1000 in TBS for 2–3 hr at room temperature

   vi.   Wash the membrane as done before (step iv); place them in hybridization bottles with enzyme substrate and rotate in a hybridization oven for 1 hr at room temperature; rinse the membranes in distilled water and air dry.

   vii.   Examine under a dissecting microscope.

B.   Direct tissue blotting technique (Lawson, 1991)

   i.   Prepare tissue blots by pressing a newly cut leaf, flower, or insects on a nitrocellulose membrane (0.45 μm pore size); immerse in PBS containing 1% bovine serum albumin (BSA) for 60 min with gentle agitation.

ii. Incubate with alkaline phosphatase-labeled antibodies in PBS for direct detection; wash the blots thrice in PBS-Tween.

iii. For indirect detection, incubate blots with virus-specific primary antibodies in PBS for 60 min; wash thrice in PBS-Tween; incubate with enzyme-labeled species-specific secondary antibodies for 60 min; wash three times in PBS-Tween.

iv. Immerse in solution containing nitroblue tetrazolium (14 mg) and 5-bromo-4-chloro-3-indolyl phosphate (7 mg) in 40 ml substrate buffer consisting of 0.1 M Tris, 0.1 M NaCl, and 5 mM MgCl$_2$, pH 9.5). Purple color in blots indicates a positive result.

APPENDIX 8(vi): IMMUNOBLOT ANALYSIS (DAMSTEEGT ET AL., 1999)

i. Resuspend extracts from sucrose gradient–purified virus preparation representing 2 μg of the virus in Laemmli sample buffer; electrophorese on 4–20% acrylamide gradient sodium dodecyl sulfate–polyacrylamide gel electrophoresis (SDS-PAGE) minigels using Laemmli buffers.

ii. Transfer proteins to nitrocellulose membranes (pore size, 0.2 μM) in transfer buffer at 10 V for 15 hr.

iii. Block the blots in gelatin (3% w/v) in Tris buffered saline (TBS; 25 mm Tris-HCl, pH 7.2, and 150 mm NaCl).

iv. Probe with purified polyclonal IgG raised against the test virus (SbDV-D) at 25°C for 3.5 hr on a shaker at 50 rpm and wash the blots with TBS.

v. Probe with goat antirabbit IgG conjugated to alkaline phosphatase at 1:3000 dilution.

vi. Use avidin conjugated to alkaline phosphatase at 1:3000 dilution, if biotinylated molecular weight standards are to be detected, as per the protocols of manufacturer of detection kits.

APPENDIX 8(vii): FILTER PAPER SEROASSAY (FiPSA)

i. Heat filter paper disks (Whatman No. 1) (7 mm) at 150°C for 1 hr; preblock the binding sites with rabbit serum albumin; suction dry and store the disks in a desiccator.

ii. Extract the antigen in a few drops of spotting buffer containing 0.05 M Tris-HCl, pH 7.4; 0.2 M NaCl; and 20% sucrose with 0.02% phenol red as spotting aid. Transfer to plastic microvial (400–1500 μl size); centrifuge at 5000 rpm for 10 min; dilute the supernatant suitably.

iii. Spot a 2 μl sample in the center of the disk and dry; transfer to individual cylindrical vials (10–15 mm diameter) and dry the disks at 40°C for 10 min.

iv.   Incubate the disks in vials containing 150 μl of antiserum/normal serum at suitable dilutions for 30–60 min; remove the sera by aspiration from the wells; wash with 1 ml of TBS (spotting buffer without sucrose) for 10 min by shaking the vials; remove the excess buffer.

v.   Immerse the disks in protein A-peroxidase conjugate (2 μg/ml) in TBS and incubate with gentle shaking for 15–30 min; wash the disks.

vi.   Incubate the disks in substrate solution containing 5 parts of TBS + 1 part of 4-chloro-1-naphthol at 3 mg/ml in methanol + 0.018 part of 3% hydrogen peroxide for 1 min in darkness; remove the disks and immerse in water for 10 min; dry at 40°C. Development of violet-blue color indicates a positive reaction.

APPENDIX 8(viii): RAPID IMMUNO-FILTER PAPER ASSAY (RIPA)

i.   White latex beads, used as solid phase, and pink latex beads, used as tracer, in TBS are mixed with antibody (100 μg/ml) at pH 7.2 for coating the beads; incubate for 2 hr at room temperature; centrifuge at 15,000 rpm for 10–15 min with TBS-BSA; suspend the sediments in TBS-BSA.

ii.   Apply 5 μl of coated white latex beads at 1.5 cm from the lower end of a glass filter paper GF/A (8 × 0.5 cm) (Whatman) strip, and air-dry; store the strips in a desiccator till use; keep sensitized pink latex in suspension at 4°C.

iii. Dilute pink latex to 0.025% (v/v) with TBS and mix with an equal volume of purified virus solution or tissue extract in flat-bottomed tubes (Eppendorf). Add Tween 20 to the mixture to yield a final concentration of 0.3% (v/v).

iv.   Dip the filter paper strip with immobilized white latex into the solution containing coated pink latex to immerse 0.5 cm of the filter paper.

v.   Observe the appearance of a pink band on the immobilized white latex. Use a chromatoscanner to determine color intensity at 700 nm (visible light).

APPENDIX 8(ix): ENZYME-LINKED FLUORESCENT ASSAY (ELFA)

i.   Sensitize microtiter plates or polystyrene beads (6 mm diameter) with IgG in 0.05 M sodium carbonate, pH 9.6; block nonspecific reactions with 2% ovalbumin or 0.2% ovalbumin and 2% polyvinyl pyrrolidone in 0.02 M NaPO$_4$, pH 7.2, containing Tween 20 and 0.85% NaCl (PBS-Tween); wash with PBS-Tween thrice.

ii. Use aliquots of 200 μl for microtiter plates; 0.5 ml of antibody and 1.0 ml of antigen samples for beads.

iii. Apply the virus antigen to the solid phase; incubate for 12 hr at 4°C.

iv. Apply the antibody-labeled biotin, diluted with PBS-Tween or labeled with alkaline phosphatase diluted in PBS-Tween containing 2% polyethylene glycol; wash with PBS-Tween.

v.   Add 4-methyl-umbelliferyl phosphate (MUP) (0.1 mM) in 1.0 mM di-ethanolamine, pH 9.8, containing 0.01 mM MgCl$_2$; terminate the reactions by adding NaOH (to a final concentration of 0.27 M) for colorimetric assay or Na$_2$HFO$_4$ adjusted to pH 10.4 with KOH to a final concentration of 0.2 M for fluorescent assays.

vi. Determine color intensity/fluorescence at 405 nm spectrphotmetrically or in a fluorometer.

APPENDIX 8(x): FLUORESCENT ANTIBODY TECHNIQUES

A. Preparation of conjugated antibodies

i.   All steps are at 4°C. Precipitate globulins by using equal volumes of neutralized saturated ammonium sulfate; dialyze against phosphate-buffered saline solution at pH 7.0, using a 0.025 M sodium carbonate buffer, pH 9.8.

ii. Adjust the protein concentration of antibody to 1%; dialyze against 10 volumes of 0.01% fluorescein isothiocyanate in the same buffer for 24 hr.

iii. Dialyze the globulin solution against phosphate-buffered saline solution, pH 7.0, until unconjugated dye is washed out.

iv. Centrifuge the conjugated antibody at low speed and store at −40°C.

B. Fluorescence test with leaf tissues

i.   Place leaf segments (2 × 5 mm) in small aluminum foil boats containing 20% gelatin in 5% glycerol; freeze the samples with dry ice at −20°C; remove the aluminum foil.

ii. Cut sections (10–20 μm) at room temperature; mount the sections on slides smeared with gelatin-glycerol adhesive.

iii. Place drops of conjugated antibody solution on sections and incubate in a moist chamber 45–60 min in the dark.

iv. Wash the sections for 10–15 min with phosphate-buffered saline solution, pH 7.2, to remove excess dye; mount in a glycerol-phosphate mixture (1 ml acid-free glycerol + 9 ml phosphate buffer, pH 7.2).

v.   Observe under a fluorescent microscope.

C. Indirect immunofluorescent technique (Geminiviruses) (Hunter et al., 1997).

i.   Allow acquisition access period for the whiteflies on source plants [tomato infected by tomato mottle begomovirus (ToMoV)] for 5–6 days; transfer the whiteflies to a nonhost for ToMoV and maintain them for 1 or 2 days prior to testing.

ii. Collect the whiteflies in glass tubes; immobilize them by placing them at $-20°C$ for several min.

iii. Make a circle on each glass slide using a hydrophobic solution that, when dried, will confine solution within the circle; place a small drop of clear nail polish in the middle of the circle made on the slide.

iv. Place the immobilized whiteflies either ventral side down or on their sides in the nail polish with the wings spread out gently; air dry the nail polish.

v. Cover the whiteflies with 10 mM phosphate buffered saline (PBS), pH 7.4; dissect the whiteflies using a razor and fine-tip forceps to expose the intact digestive tract and other organelles.

vi. Block nonviral antigens (proteins) with protein blocking agent (PBA) (containing Trisbuffer, bovine immunoglobulin, albumin gelatin, and sodium azide) and goat normal serum (1:1,000); blot off excess PBA.

vii. Apply primary antibody raised against the test virus, diluted to 1:1000 concentration in 10 mM PBS, pH 7.4; incubate for 30–60 min.

viii. Wash the insect tissues with three changes of PBS at 15-min intervals.

ix. Apply the secondary antibody conjugated with FITC (either goat anti-rabbit or goat anti-mouse) at a dilution of 1:1000 in PBS; incubate for 20–30 min and wash the insect tissues as done before (step viii).

x. Mount the tissues in Aqua–Mount (Polysciences, Inc., Warrington, PA); examine under fluorescent microscope.

D. Indirect immunofluorescent technique (Nepoviruses) (Wang and Gergerich, 1998).

i. Allow acquisition access period of 10 days for *Xiphinema americanum* on cucumber plants infected by tobacco ringspot nepovirus (TRSV); select 200 active nematodes individually; transfer them to microcentrifuge tube containing tap water.

ii. Centrifuge at 14,000 g for 4 min; pour off the supernatant; add 2% formaldehyde (1 ml) to the pellet and incubate for 1 hr at 4°C.

iii. Centrifuge and pour off the supernatant; place the nematodes on a clean glass slide in a small amount of 2% formaldehyde; cut the nematodes into small pieces using a razor blade.

iv. Suspend the nematode body fragments in blocking buffer (2–4 drops) containing NaCl (0.14 M), phosphate buffer (0.01 M), BSA (3%), and Triton X-100 (0.2%), pH 7.2, kept in a clean microcentrifuge tube for 15 min at 4°C and then remove the blocking buffer.

v. Add 200 μl of purified primary antibodies (against TRSV) diluted in blocking buffer (1:50) to the centrifuge tube containing nematode body fragments; incubate for 18 hr at 28°C in an orbital shaker.

vi. Wash the nematode body fragments four times for 10 min each with blocking buffer at room temperature; incubate in 200 μl of a 1:50 dilution of FITC-conjugated goat anti-rabbit IgG in blocking buffer on an orbital shaker for 20 hr at 28°C.

vii. Wash the nematode body fragments as done before (step vi); vacuum-dry for 15 min.

viii. Mount the pellet in 10 μl of glycerol (50%) in PBS on a glass slide; tease the fragments gently to disperse them in glycerol.

ix. Examine under an epifluorescent microscope.

APPENDIX 8(xi): TIME-RESOLVED FLUOROIMMUNOASSAY (TR-FIA) (SIITARI ET AL., 1986)

A. Sample preparation
 i. Dilute the leaf or tuber extract to 10^{-6} with buffer.
B. Conjugation of antibodies with europium
 i. Purify IgG using a Sepharose column.
 ii. Conjugate IgG with Eu^{3+}-N^1 (p-iso-thiocyanatobenzyl)-diethylene triamine-N^1, N^2, N^3-tetraacetate for 18 hr at 0°C in 0.1 M carbonate buffer, pH 9.3; separate the Eu-IgG complex from free Eu reagent by gel filtration and store at 4°C in 0.05 M Tris HCl buffer, pH 7.7, containing 0.9% NaCl, 0.5% gelatin, 20 μM diethylenetriamine N^1, N^1, N^2, N^3-pentaacetic acid, 0.01% Tween 40, and 0.05% NaN_3 in 0.05 M Tris HCl buffer, pH 7.7.
C. One-step incubation procedure
 i. Wash the wells in polystyrene microtitration strips twice with TBS-T containing 20 mM Tris-HCl, 150 mM NaCl, pH 7.5, with 0.05% Tween 20; coat the wells with antibodies (IgG or MABs) for 18 hr at 37°C in carbonate buffer, pH 9.6, and block with BSA as in ELISA.
 ii. Incubate antigen (50 μl) and europium conjugate already coated with IgG or MABs for 1 hr at 37°C; wash with 20 mM Tris-HCl buffer, pH 7.5, containing 150 mM NaCl; add the enhancement solution containing 15 μM 2-naphtoyl trifluoroacetone, 50 μM tri-n-octylphosphine oxide, and 0.1% Triton X-100 in 0.1 M acetate phthalate buffer, pH 3.2; shake for 10 min.
 iii. Determine the fluorescence in a fluorometer.
D. Two-step incubation procedure
 i. Coat the wells with antibodies (IgG or MABs) as in the one-step incubation procedure.

ii.   Incubate the antigen (100 μl) for 1 hr at 37°C; wash the wells with TBS-T.

iii. Incubate the Eu conjugate for 1 hr at 37°C. Further steps as in the one-step incubation procedure.

APPENDIX 8(xii): CYTOFLUORIMETRIC METHOD (IANNELLI ET AL., 1996)

i.   Incubate latex particles (about 10^7) with leaf extracts (1 ml) overnight at 4°C or 2 hr at room temperature under agitation; the size of particles (1, 3, or 6 μ) and number of particles vary with experiments; centrifuge

ii.   Incubate the pellet with 1% bovine serum albumin (BSA) in borate buffer (0.1 M), pH 8.5 for 30 min at room temperature

iii. Wash the latex particles with 0.15M PBS, pH 7.2; incubate with primary antibodies diluted in PBS [PAB for CMV and MABs for PVY or plum pox virus (PPV)] for 4 hr at room temperature

iv.   Wash the sensitized particles with PBS; incubate with secondary antibodies labeled with FITC for 1 hr at room temperature.

v.   Wash the particles again with PBS; measure the fluorescence with flow cytometer (FACScan, Becton Dickinson Immunocytometry System, San Jose, CA).

vi.   Control samples are incubated with PBS in place of primary antibodies.

APPENDIX 8(xiii): VIROBACTERIAL AGGLUTINATION (VBA) TEST

A. Preparation of *Staphylococcus aureus* reagents (Lyons and Taylor, 1990)

i.   Cultivate the authentic culture of *S. aureus* (Cowan strain [NCTC, 8530; ATCC, 12598) on nutrient agar for 24–48 hr at 37°C; prepare the suspension of bacterial cells (10^8–10^9) in glycerol broth containing nutrient broth (Difco) (1.6 g), glycerol (30 ml), and distilled water (170 ml); transfer in 1–2 ml aliquots to sterile vials and equal volume of sterile 3 mm hollow glass beads; store at −80°C.

ii.   Thaw the container when required; transfer the contents to nutrient broth (10 ml); distribute the culture after shaking well to 40 nutrient agar plates; incubate overnight at 37°C.

iii. Flood the plates with phosphate buffered saline (PBS) solution that contains 0.02% sodium azide; scrape the surface of the plates gently with a glass streaker; centrifuge the bacterial suspension at 300 g for 30 min.

iv. Resuspend the pellet in 20 ml of 1.5% aqueous formaldehyde; shake the suspension well for 30 min; heat at 80°C for 30 min; rapidly cool to room temperature.

v. Wash twice by centrifugation and resuspension with PBS containing 0.05% sodium azide added at the rate of 9 volumes of PBS to 1 volume of pellet.

B. Virobacterial agglutination (Walkey et al., 1992)

i. Prepare appropriate dilutions of the antiserum against the test antigen in PBS containing $Na_2HPO_4 \cdot 12 H_2O$ (2.9 g), KH_2PO_4 (0.2 g), NaCl (8.0 g), and KCl (0.2 g) in distilled water (11), pH 7.2; add sodium azide (NaN_3) at 0.2 g/l.

ii. Mix *S. aureus* cells with diluted antiserum at a 1:5 ratio for conjugation; add ethanol saturated basic fuchsin to color the conjugate for easy recognition of bacterial agglutination; store at 4°C; stir the conjugate before use; check with a hand lens to detect autoagglutination, if any.

iii. Prepare the extracts from healthy and infected leaves using a small quantity of K_2HPO_4 (10 g/L) solution.

iv. Dispense a 4 μl aliquot of the bacterium-antiserum conjugate and 2 μl of extract on a multitest slide (Flow Laboratories Ltd., UK); mix the reactants well; maintain suitable controls in the same slide; shake the slide gently by hand as the reaction proceeds; observe the reaction with a hand lens.

v. Record agglutination, which indicates a positive reaction, after 30 sec to 5 min.

APPENDIX 8(xiv): IMMUNOCAPILLARY ZONE ELECTROPHORESIS (EUN AND WONG, 1999)

A. Preparation of plant extracts

i. Grind fresh plant tissue (0.5 g) in 0.01 *M* sodium borate buffer (1 ml), pH 7.5.

ii. Centrifuge the homogenate at 8000 g and store the supernatant at 4°C until required.

B. Conjugation with virus-specific antibodies

i. Prepare serial dilutions of antibodies in 50 m*M* sodium borate buffer, pH 9.7.

ii. Prepare similar dilutions of purified virus preparations/leaf extracts from infected plants.

iii. Mix virus-antibody solutions at appropriate dilutions and incubate for 10, 20, 30, and 50 min.

iv. Maintain healthy plant extracts as negative controls and purified virion samples as positive controls.

C. Capillary zone electrophoresis (CZE) analysis
    i. Perform CZE analyses with Beckman P/ACE System 2100 under normal polarity conditions, the anode and cathode being positioned at the capillary inlet and outlet, respectively.
    ii. Add the test samples into separate holders and place them in the inlet tray; capillary rinsing, sample injection, sample separation and detection, and data processing and generation of electropherograms are carried out by the fully automated analytical process system.
    iii. Use an uncoated 57-cm fused-silica capillary with an internal diameter of 75 μm housed in a temperature-regulated catridge; maintain a temperature of 25°C within the capillary during the CZE analysis.
    iv. Condition the capillary by successively rinsing with 0.1 M HCl for 5 min, 0.1 M NaOH for 10 min, and deionized water for 5 min before commencement of analysis for the day.
    v. Carry out rerun rinsing before each run, with 50 mM sodium borate buffer, pH 9.7, for 2 min and postrun rinsing successively with 0.1 M NaOH for 2 min and deionized water for 2 min.
    vi. Inject the samples pneumatically at 3.45 kPa for 5 sec; use 50 mM sodium borate buffer, pH 9.7, as running buffer for all runs performed.
    vii. The separated components, as they pass a glass window situated close to the outlet of the capillary, are detected by a wavelength-selectable UV detector system consisting of a deuterium lamp, mirrors, wavelength-selectable UV filters, and a photomultiplier tube.
    viii. Set the detection absorbance at 280 nm for all runs; link the UV detector system to a computer for plotting the signal graphically in the form of an electropherogram (relative absorbance against elution time).
    ix. Identify the major peaks in the electropherogram and record elution times after each CZE separation; perform 50 successive CZE runs to elute each identified peak.
    x. Store all pooled fractions from each peak at 4°C for examination under electron microscope to identify the virus in the fractions.

APPENDIX 8(xv): INDIRECT IMMUNOFLUORESCENCE MICROSCOPY FOR DETECTION OF BACTERIAL PATHOGEN (LI ET AL., 1993)

    i. Cut infected leaves and healthy parts of diseased leaves and leaves artificially inoculated by wounding into 2–3 mm pieces and immerse in 1 ml of sterile distilled water (SDW) in Eppendorf tubes for 10 min.

ii. Remove leaf pieces; centrifuge at 13,000 rpm for 5 min; resuspend the pellets in 0.1 ml SDW.

iii. Heat-fix drops of suspensions on glass slides; expose to a few drops of 95% ethanol for 10 min.

iv. Cover the slide with a drop of specific MAB supernatant for 15 min and wash five times with SDW.

v. Keep the slides in darkness and cover with a drop of goat antimouse IgG fluorescent conjugate (1:700 dilution) for 15 min; wash five times with SDW.

vi. Air-dry the slides in the dark; place a drop of 50% glycerol; cover with a cover slip; examine under fluorescent microscope fitted with a UG-1 ultraviolet excitation filter.

vii. Determine the number of bacterial cells by the following formula:

$$\text{Number of bacterial cells} = \frac{N \times A}{a}$$

where N = average number of cells/field, A = circle area (mm^2) on the glass slide, and a = area of microscope field = $(9.1 \times 0.01)^2$ mm^2.

APPENDIX 8(xvi): DETECTION OF BACTERIAL PATHOGENS (FROMMEL AND PAZOS, 1994)

A. Antigen preparation

i. Grow xanthomonads on yeast-dextrose-chalk (YDC) agar, pseudomonads on King's B medium, and other bacteria on Clark's medium or nutrient yeast dextrose agar for 48 hr at 25°C.

ii. Suspend the cells in 10 ml of 0.85% sterile saline solution (SSS); pour the suspension into a flask containing 200 ml of YDC broth and incubate for 48 hr at 26°C on a horizontal shaker (350 rpm).

iii. Centrifuge at 10,000 rpm for 15 min; wash the pellets once in SSS and then in 0.1% sarcosyl solution for 10 min; resuspend in SSS.

iv. Adjust the concentration of cells to 3×10^7 cells/ml; expose the cells to 100°C for 20 min to eliminate flagellar antigens.

B. Preparation of polyclonal antibodies (PABs)

i. Mix the immunogen prepared as above in equal quantities with Freund's incomplete adjuvant.

ii. Inject a New Zealand male rabbit intradermally, at 12 sites along the back of the rabbit, at the rate of 100 μl/site; repeat injections after 7, 21, and 28 days and inject 1 ml intramuscularly as a final dose at 60 days after the primary injection.

iii. Bleed the rabbit at 0, 30, 50, and 70 days after the first injection; assess the antibody titer of different fractions by indirect ELISA; preserve the fractions with titers above 1/5120 with 0.01% sodium azide and store at $-20°C$.

iv. Precipitate globulins in the antiserum by using saturated ammonium sulfate; dialyse against three changes of 0.1 M phosphate buffer, pH 7.4 (PB).

v. Fractionate gamma globulins on a DEAE-Sephadex A 50 column; place globulins to be used immediately at 4°C and store the rest at $-70°C$.

vi. Adjust the protein concentration to 1 mg/ml by using a spectrophotometer; conjugate with alkaline phosphatase; store unconjugated IgG at $-70°C$ in PB with bovine serum albumin (BSA).

C. Production of monoclonal antibodies (MABs) (Alvarez et al., 1985)

i. Streak the culture on yeast-glycerol agar (YGA); incubate for 48–72 hours at 28°C; collect the cells in 0.01 M phosphate-buffered saline (PBS) solution, pH 7.4, for immediate use or store in PBS containing 0.5% formalin at 6°C.

ii. Inject BALB/C mice intraperitoneally with 10^8 living cells of bacterium and repeat after 14 days; give a booster injection 2–3 days before hybridization.

iii. Remove the spleen and collect 5×10^7 cells; mix them with 5×10^7 myeloma cells; centrifuge and fuse, using 45% polyethylene glycol [molecular weight (MW) 1300–1600].

iv. Wash the cells; resuspend in Dublecco's modified Eagle's medium with 10% fetal calf serum, 1 mM sodium pyruvate, 0.1 mM hypoxanthine, 0.0004 mM aminopterin, 0.016 mM thymidine, and 1% nonessential amino acid solution (Gibco Laboratories, Grand Island, NY), and 10% NCTC 109 lymphocyte growth medium, dispense in culture plates.

v. Coat the culture plates with peritoneal macrophage "feeder" cells 1 day prior to transfer of fused cells.

vi. Screen the supernatant fluids from the well with healthy hybridomas for antibody titer by radioimmunoassay or ELISA; select stable and efficient clones.

vii. Inject selected stable clones into pristane-primed BALB/C mice intraperitoneally with about 10^6 hybridoma cells; collect the ascitic fluid; centrifuge; store the MABs at $-20°C$.

D. Double-antibody sandwich enzyme-linked immunosorbent assay (DAS-ELISA)

i. Adjust the concentration of primary antibody at 8 μg/ml in coating buffer consisting of 0.13 M $Na_2CO_3H_2O$ + 0.035 M $NaHCO_3$, pH 9.6.

ii. Dispense the sample at 200 μl/well and incubate overnight at 4°C;

wash three times with PBS, pH 7.2; block with 1% BSA for 1 hr at 37°C.

iii. Add IgG-alkaline phosphatase conjugate at a concentration of 3.2 μg/ml in PBS containing 1% BSA; incubate.

iv. Add *p*-nitrophenyl phosphate as substrate at a concentration of 1 mg/ml.

v. Perform other steps as in Appendix 8[i].

E. Semiselective enrichment procedure (Frommel and Pazos, 1994)

i. Prepare the semiselective enrichment broth (SSEB) consisting of gentamycin (5 μg/ml) cephalexin (6 μg/ml), tyrothricin (150 μg/ml), ampicillin (5.6 μg/ml), cycloheximide (200 μg/ml), benomyl (80 μg/ml), and Tween 20 (0.02%), pH 7.0, added to enrichment broth (EB) (basal medium of Schaad and Forster, 1985) consisting of nutrient broth (8 g/l), glucose (5 g/l, pH 7.0) or saline-Tween buffer (STB) (containing NaCl 8.5 g/l, Tween 20, 2 ml/l, pH 7.0).

ii. Dispense 120 ml of either SSEB or EB to groups of four Erlenmeyer flasks; seed with cells of bacteria at different concentrations for assay standardization.

iii. Place 400 wheat seeds-into each flask; incubate at 26°C in a shaker at 350 rpm.

iv. Draw 2 ml samples from flasks at 2, 4, 6, and 8 hr of incubation; assay by DAS-ELISA as in section D.

F. Semiselective enrichment broth (SSEB)-ELISA (Frommel and Pazos, 1994)

i. Place individual seeds in each well of acid-washed 96 well microtiter plates.

ii. Dispense SSEB, STB, or NB to 80 wells at 300 μl/well; remaining 16 wells serve as negative controls and bacterial standards; seal the plates with cellulose tape and incubate for 4 hr at 26°C in a wrist action shaker at 300 rpm.

iii. Transfer aliquots of 200 μl from each well to microtiter plate wells precoated with specific antiserum at 8 μg/ml in carbonate buffer.

iv. Follow the DAS-ELISA procedure as in section D.

APPENDIX 8(xvii): PRODUCTION OF RECOMBINANT SINGLE-CHAIN ANTIBODIES (SCFVS) AGAINST BACTERIAL LIPOPOLYSACCHARIDE (LPS) (GRIEP ET AL., 1998)

A. Production of LPS-phage antibodies (PhAbs) complex

i. Coat immunosorbent (IS) tubes with 500 μg of purified bacterial (*Ralstonia solanacearum*) lipopolysaccharide (LPS) (125 μg/ml in 50 m*M* NaHCO₃, pH 9.8) at 4°C for 18 hr; wash the tubes with phosphate-buffered saline (PBS) twice.

   ii.   Block with 2% PBM (PBS containing skimmed milk powder 2%) for 30 min at room temperature; mix simultaneously 2 ml of PhAbs stock derived from the Vaughan library (containing 2.5×10^{13} PhAbs in 2 ml of 4% PBM); preincubate for 30 min.

   iii.   Remove the blocking solution, wash with PBS; add 4 ml of PhAbs to the IS tubes.

   iv.   Incubate on a roller bench for 30 min and for another 90 min without rotation for the PhAbs by washing the IS tubes with PBS containing 0.1% Tween 20 (PBST) 10 times and wash further with PBS 10 times to remove the detergent.

   v.   Elute the bound PhAbs by adding 1 ml of 0.1 M triethylamine (TEA); incubate for 10 min on a roller bench and neutralize the pH with 0.5 ml of 1 M Tris-HCl, pH 7.4.

B.   Expression of LPS-PhAb complex in *E. coli*

   i.   Grow *E. coli* TG1 bacteria to an optical density of 0.5 at 600 nm (O.D.$_{600}$) in 2 × tryptone yeast (2TY) broth.

   ii.   Mix 1 ml of eluted PhAbs with 5 ml of *E. coli* TG1 bacteria freshly grown as above and incubate for infection for 30 min in a water bath at 37°C without shaking.

   iii.   Centrifuge the mixture, after infection, at 3000 g for 10 min; resuspend the pellet in 1 ml of 2TY broth amended with 100 μg of ampicillin and 2% glucose (2TYB-AMP-2% Glu) per ml.

   iv.   Assess the number of eluted PhAbs by plating 50 μl from the suspension and serial dilutions on 2TY agar plates containing 100 μg of ampicillin and 2% glucose (2TYA-AMP-2Glu) per ml; plate the remaining suspension (950 μl) separately on 225 × 225 mm 2TYA-AMP-2% Glu plates and grow the bacteria at 30°C for 18 hr.

   v.   Repeat the panning procedures (A and B) four times in succession.

APPENDIX 8(xviii): PRODUCTION ANTISERUM AGAINST FUNGAL PATHOGENS (BANKS ET AL., 1992)

A.   Antigen preparation

   i.   Prepare spore suspensions in 0.01% Tween 80; wash thrice by centrifugation; inoculate 1 ml of spore suspension (10^6 spores/ml) into 100 ml liquid medium supplemented with NaCl (100 g/l); incubate at 25°C for 7 days in the dark by placing the flask on a rotary shaker.

   ii.   Transfer the mycelium by filtering into a sintered glass filter; wash with sterile water and then with sterile phosphate-buffered saline (PBS); freeze overnight at −20°C; thaw and transfer to centrifuges; dry in a vacuum dryer.

iii. Collect the mycelium and add 50 ml of liquid nitrogen; mince the mycelium in a blender for 1 min and grind in a mortar with pestle to have a fine powder.

iv. Suspend the mycelial powder in PBS (200 mg in 10 ml); centrifuge at 4500 rpm (3000 G) for 10 min at 4°C; divide the supernatant containing soluble antigen into 0.5 ml aliquots and store at −20°C.

v. Estimate the total protein content of antigen preparation.

vi. Add 10 ml of warm serum-free RPMI 1640 medium (Gibco) over the next 60 sec with gentle stirring. Add another 20 ml of warm RPMI; centrifuge at 400 g for 3 min at room temperature.

vii. Suspend the pellet of cells in 50 ml of growth medium (RPMI 1640) with 20% (v/v) of Myoclone fetal calf serum (FCS); dispense cell suspension into five 96 well microplates at 100 μl/well.

viii. After 24 hr, add 100 μl of hypoxanthine-aminopterin-thymidine (HAT) medium (diluted to 1:50 in growth medium) to each well in the fusion plates.

ix. Add growth medium + HAT on 2, 4, 7, and 10 days by removing 100 μl of the medium and replacing with 100 μl of fresh medium.

x. Screen the hybridoma cells for antibody production by indirect ELISA.

xi. Clone healthy growing hybridomas twice by limiting dilution in nonselective medium; preserve by freezing slowly in 7.5% (v/v) dimethyl sulfoxide (DMSO) and store in liquid nitrogen.

B. Production of polyclonal antiserum

i. Mix soluble antigen preparation with equal quantities of Freund's complete adjuvant (Difco) to produce a final protein concentration of the mixture at 1 mg/ml.

ii. Inject rabbits intramuscularly with 1 ml of the mixture at predetermined intervals.

iii. Bleed the animal at 4 weeks after the first injection and subsequently at 14, 16, and 18 weeks.

iv. Take the serum after clotting and centrifugation.

C. Production of monoclonal antibodies

i. Mix soluble antigen preparation with an equal quantity of Freund's complete adjuvant to yield a final protein concentration of 1 mg/ml.

ii. Inject a BALB/C mouse, after anaesthetization, with 0.1 ml of the immunogen intraperitoneally and subsequently at 2, 4, and 6 weeks and at 8 weeks after the first injection with PBS; remove the spleen after sacrificing the animal by cervical dislocation.

iii. Carry out fusion of splenocytes with myeloma cell line P3-NS 1-Ag4 at a ratio of $1 \times 10^8 : 5 \times 10^7$ by gentle addition of 2 ml of 30% (w/v) of polyethylene glycol (PEG) over 60 sec.

D.   Solid-phase attachment of fungal hyphae in ELISA plates (Banks and Cox, 1992)

    i.   Coat 96 well microplates with 0.005% poly-L-lysine (PLL) at 50 µl/well; incubate for 45 min at 25°C.

    ii. Wash four times with PBS with 0.05% Tween 20 (PBST) at 250 µl/well; blot dry; treat with 2% glutaraldehyde at 50 µl/well; incubate for 15 min at 25°C and wash as before.

    iii. Add antigen (at a concentration of 3.4 µg/ml) at 50 µl/well pretreated with PLL and glutaraldehyde or untreated wells (to serve as control); incubate at 25°C overnight to allow the antigen to dry onto the bottom of the wells.

    iv. Wash the wells with PBST four times; add 3% bovine serum albumin (BSA) at 250 µl/well; incubate at 25°C for 60 min; wash and use immediately or store at −20°C after washing thrice before storage and once before using.

E.   Indirect enzyme-linked immunosorbent assay (ELISA) to screen antifungal antibodies (Banks and Cox, 1992)

    i.   Add antiserum at appropriate dilution to microplate wells prepared by the procedure in Appendix 8 (×) D, at 50 µl/well; incubate for 60 min at 25°C; wash four times with PBST.

    ii. Add horseradish peroxidase conjugated goat antimouse immunoglobulins at 200 µl/well; incubate for 60 min at 25°C; wash as before.

    iii. Add O-phenylenediamine dihydrochloride (0.4 g in 100 ml phosphate citrate buffer [PCB], diluted to 1:10 in PCB with 0.012% hydrogen peroxide) as substrate at 200 µl/well; allow the reaction in darkness for 10 min at 25°C.

    iv. Add 2.5 M sulfuric acid at 50 µl/well to stop the reaction; determine color intensity at 490 nm in a microplate reader.

APPENDIX 8(xix): SEED IMMUNOBLOT ASSAY (SIBA) (GLEASON ET AL., 1987)

    i.   Surface-sterilize the seeds in 0.5% sodium hypochlorite for 30 sec; rinse in deionized water three times; blot-dry the seeds with paper towels.

    ii. Place the seeds 2 cm apart on a nitrocellulose sheet; transfer the nitrocellulose sheet to a plastic tray (18 × 29 × 5 cm) containing three layers of moist germination towels; place three additional layers of moist paper towels above the seeds; cover the tray with aluminum foil to maintain high humidity and incubate for 2–3 days at 25°C.

    iii. Remove the seeds carefully; assay the antigens adsorbed to the nitrocellulose sheet; agitate the nitrocellulose sheet in a blocking solution containing Tris-buffered saline solution (50 mM Tris-HCl, 200 mM NaCl, pH 7.5) and 5% nonfat dry milk (TBS-milk) for 1 hr at room temperature.

iv. Replace the blocking solution with IgG preparation (10 μg/ml) in TBS-milk and agitate for 1 hr.

v. Rinse nitrocellulose sheet three times for 10 min each in TBS-milk; incubate with goat antirabbit IgG conjugated to horseradish peroxidase in TBS-milk in a shaker for 1 hr.

vi. Rinse the nitrocellulose sheets twice for 10 min each with TBS-milk and then with TBS.

vii. Add chloro-1-naphthol in hydrogen peroxide as substrate; observe for 15–20 min for the development of blue color, indicating a positive reaction; rinse in deionized water; dry under a heat lamp; store the blots.

9
Nucleic Acid–Based Techniques

It is generally considered that closely related organisms share a greater nucleotide sequence similarity than those that are distantly related. A highly specific nucleotide sequence present in an isolate or strain of virus or other pathogen, but absent from other strains or species, may be used to test for the presence of that virus or pathogen. Detection of plant pathogens by hybridization is based on the production of nucleic acids by specific hybridization between the single-stranded target nucleic acid sequence (denatured DNA or RNA) and a complementary single-stranded nucleic acid probe. Probes for plant viruses are mostly cDNA, as the genomes of most plant viruses are RNA. Either RNA or DNA sequences may be used as probes. Transcription vectors to produce RNA probes in vitro can be developed to yield RNA:RNA or RNA:DNA hybrids, which are more stable than DNA:DNA hybrids.

Nucleic acid hybridization has several advantages over immunological assays. The antigenic determinants of the viral coat proteins represent only about 2% to 5% of the nucleic acid of viral genomes, and hence the differences in characteristics of the virus governed by major portions of the viral genome cannot be determined by serological assays. The bacteria and fungi are complex antigens, the nature of which may vary, depending on the stage of their development, and so the antisera produced against one type of spore or mycelium at one stage may not react positively with spores or mycelium produced at all stages in the life cycle. But the nature of genomic elements is constant in all stages of the pathogen, making it possible to detect, differentiate, and establish relationships between strains or related species. For the detection of viroids, serological techniques cannot be employed as they do not have any protein, and necessarily hybridization methods have to be employed for their detection and characterization. The cloned probes with varying specifications can be produced to suit requirements for different assays and in unlimited quantities. Moreover, the sensitivity can be increased by amplification of desired sequences by using polymerase chain reaction (PCR).

9.1 HYBRIDIZATION METHODS

9.1.1 Probe Preparation and Labeling

DNA probes have been commonly used in experiments with DNA or RNA viruses, although labeled viral RNA itself has been used as a probe in certain cases. The probes may be labeled with either radioactive markers, such as ^{32}P or ^{3}H, or nonradioactive markers, such as biotin. For differentiation of viruses in a group or strains of a virus, the cDNA probes with appropriate common or specific sequences of nucleotides can be prepared and labeled in different ways.

9.1.1.1 Reverse transcription

The cDNA copies of the viral RNA are formed by using a retrovirus reverse transcriptase and labeled radioactively. These probes can be used to identify the virus or its strains.

9.1.1.2 Cloned probes

Double-stranded cDNA, after cloning in the bacterium, is labeled by nick translation (some nucleotides are replaced after excision with enzymes and insertion with labeled nucleotides). In another procedure, the double-stranded cDNA is separated by heating and randomly primed with short synthetic oligomers. Complete new double-stranded (ds) molecules are produced by using DNA polymerase and then labeled. Cloned cDNA probes have been used to detect viruses such as plum pox virus (Variveri et al., 1988), peanut mottle and stripe viruses (Bijaisoradat and Kuhn, 1988), and potato leaf roll virus (Robinson and Romerio, 1991, Smith et al., 1993). The probe representing a portion of the potato leaf roll virus particle protein gene could also detect beet western yellow luteovirus (BWYV) and the English strain of the RPV form of barley yellow dwarf luteovirus, but it reacted weakly with extracts from plants infected by groundnut rosette assistor luteovirus and carrot red leaf luteovirus (Robinson and Romerio, 1991). An increase in sensitivity of virus detection, as compared to ELISA, up to 250 times, has been reported (Variveri et al., 1988; Bijaisoradat and Kuhn, 1988).

9.1.1.3 Synthetic probes

Oligonucleotides (15–20 bases) representing the desired segment of viral genes can be chemically synthesized, if the nucleotide sequence of part or all of the viral genome is known. This method is useful because a) it is possible to produce required amounts of single-stranded probes which can be end-labeled with a ^{32}P-labeled nucleotide by using a polynucleotide kinase; b) a library of probes specifically designed to detect different segments of viral genome may be prepared; c) several oligonucleotides may be ligated in tandem and cloned; and d) probes specific to one strand of the viral genome may also be produced. Synthetic oligonucleotide probes have been used for the detection of viruses such as potato virus X (Rouhiainen et al., 1991), tulip breaking potyvirus (Langeveld et al., 1991), beet western yellows luteovirus (Jones et al., 1991), bean yellow mosaic geminivirus (Castro et al., 1993), and cymbidium mosaic potexvirus (Lim et al., 1993).

9.1.1.4 Nonradioactive probes

Nonradioactive probes are preferred by many researchers because of the short half-life of the isotope ^{32}P and the difficulty in handling it. Biotin has been shown to be a highly effective nonradioactive label, and the tests using biotin chemically linked to UTP and introduced into the probe by nick translation have revealed its usefulness as an alternative to radioactive markers. The biotin has a strong affinity for the bacterial protein streptavidin, which is conjugated with an enzyme. Photobiotin, an analog of biotin, has also been found to be a useful label, but it requires highly purified probe nucleic acid, since it can react with any organic material. Cloned photobiotin-labeled cDNA was used for the routine diagnosis of barley yellow dwarf luteovirus (BYDV) in nucleic acid extracts from field samples using the dot-blot method (Habili et al., 1987) and the sensitivity was comparable to that of tests employing ^{32}P-labeled probes. Similar sensitivity levels were reported in the case of biotin-labeled probe for papaya mosaic potyvirus (Roy et al., 1988). Welnicki and Hiruki (1992) reported that digoxigenin-labeled DNA probes were highly sensitive in detecting potato spindle tuber viroid, and that as little as 25 pg of the viroid RNA could be detected.

The hybridization reaction may be of three types: a) solution hybridization performed in solutions (Young and Anderson, 1985), b) in situ hybridization performed in cells or tissues (Pardue, 1985), and c) filter hybridization conducted on solid filter supports (Anderson and Young, 1985). For the detection of plant pathogens, filter hybridization and in situ hybridization methods have been most commonly used.

9.1.2 Filter Hybridization Methods

In filter hybridization denatured DNA or RNA is immobilized on an inert support such as nitrocellulose- or nylon-based membrane. In the case of plant samples, a small amount of sap is placed on nitrocellulose sheet and baked to bind the nucleic acid to the support. Prehybridization solution containing bovine serum albumin and small single-stranded fragments of an unrelated DNA, and salt is used to block nonspecific binding sites on the nitrocellulose membrane. Hybridization with labeled phosphorus (usually ^{32}P) is carried out, followed by extensive washing of the membrane to remove unreacted probe. The amount of probe bound to the target nucleic acid is estimated by autoradiography. Variants of filter hybridization are colony and dot-blot hybridization, and Southern and Northern blotting.

9.1.3 In Situ Hybridization Methods

With the in situ hybridization technique, it is possible to observe precise localization of target sequences at the organelle, cellular, or tissue level. Exposure of fixed tomato protoplasts to 0.2 N HCl at room temperature followed by heating at 70°C for 30 min and digestion with 10 μg/ml proteinase K resulted in localization of PSTVd-RNA and TMV-RNA by in situ hybridization without alteration in cellular structures (Yokoyama et al., 1990; Uehara and Hosokawa, 1994). Nucleic acids of pathogens present in only a small number of cells or at low concentrations in cells can be detected. This technique is most useful for detecting the latent infections, tissue- or organ-restricted diseases, and seed-borne pathogens (Chu et al., 1989). The hybridizations may be analyzed by either light microscope or electron microscope. Nonradioactive probes are increasingly preferred for in situ hybridization methods (Lewis et al., 1987).

9.1.4 Solution Hybridization Methods

Hybridization between target nucleic acid and probe is performed in solution, and the hybrids are analyzed by gel electrophoresis or by liquid scintillation counting. This procedure is useful for quantitative estimation of potato spindle tuber viroid concentration in purified RNA preparations (Owens et al., 1978) and for indexing of avocados for the presence of avocado sunblotch viroid (Palukaitis et al., 1981). This procedure has been used to detect potato virus X in crude leaf extracts by using both radioactive and nonradioactive labeled probes (Rouhiainen et al., 1991).

   The nucleic acid hybridization methods are highly sensitive and require only very small samples containing about 40–100 pg per gram of leaf tissue. They are particularly useful for viroids which do not have any protein. All genome sequences can be cloned simultaneously, and cloned probes can be prepared in unlimited supply. Detection and differentiation of viruses and other pathogens are possible with probes prepared for sequences which do not code for proteins. This provides a distinct advantage over serological methods, which can detect the presence of proteins which are coded by segments of viral genomes accounting for only about 10% of genomic content. Thus the serological methods cannot recognize the differences in the major portions of viral genomes. Hybridization methods can have greater applications and increased sensitivity when amplification techniques such as polymerase chain reaction (PCR) are combined (Chu et al., 1989).

9.2 DETECTION OF PLANT VIRUSES

Dot-blot hybridizations are extensively used for the detection of plant pathogens, especially viroids. These tests generally do not distinguish among different types and sizes of nucleic acids hybridizing to the probes. However, they can be very useful for qualitative detection, since this method can discriminate between closely related but different target sequences (Beltz et al., 1983). Diagnosis of diseases caused by RNA viruses using the dot-blot technique has been suggested by Palukaitis (1984). Banana bunchy top virus (BBTV) was detected by using BBTV-specific clones and radioactive or nonradioactive probes in a dot-blot hybridization assay, and this assay was as sensitive as ELISA (Xie and Hu, 1995). Tobacco mosaic tobamovirus was detected in tobacco and parsley by using radioactively labeled probe (Ogras et al., 1994). A variant of dot-blot hybridization termed squash blot was used to assay maize streak geminivirus in an individual leafhopper vector at different periods after acquisition feeding (Boulton and Markham, 1986). Navot et al. (1989) detected the tomato yellow leaf curl virus in squash blots of tomato leaves, roots, stems, flowers, and fruits, and in single whiteflies fed on infected tomato plants. Other viruses detected by this method are tobacco mosaic virus and potato virus Y. The presence of potato leaf roll virus in leaf and aphid vector extracts could be detected by dot-blot hybridization, which was equal to ELISA in sensitivity. Use of formaldehyde, instead of formamide, for denaturing leaf tissue extracts increased the effectiveness 32-fold (Smith et al., 1993). Using the digoxigenin (DIG)-labeled probes and colorimetric visualization in a dot-blot hybridization system, beet curly top intergeminivirus, beet yellows closterovirus, lettuce infectious yellows virus, squash leafcurl virus, and zucchini yellow mosaic potyvirus were detected and the test has the potential for routine diagnosis of these virus diseases (Harper and Creamer, 1995). The satellite RNA associated with groundnut rosette virus (GRV) in groundnuts infected with isolates of GRV from East and West Africa and isolates of GRV causing different kinds of symptoms could be detected by using a cloned cDNA copy of satellite RNA with ^{32}P or DIG and this test provides a reliable indicator of GRV infection in groundnuts (Blok et al., 1995). Likewise, the bamboo mosaic virus (BaMV) and its associated satellite RNA (sat BaMV) could be detected by using ^{32}P or DIG-labeled probes synthesized from cDNA clones of BaMV genomic (L probe) and sat BaMV (S probe) RNA. ^{32}P-labeled L and S probes were found to be more sensitive (25-fold) than DIG-labeled probes in detecting the virus in infected leaf extracts (Hsu et al., 2000).

The use of nonradioactive digoxigenin labeled probes has found wide application for the detection of several viruses. Dot-blot assays using DIG labeled probes for the detection of Indian peanut clump virus had sensitivity similar to that of ^{32}P-labeled probes (Wesley et al., 1996). DIG-labeled probes were found to be effective for diagnosis of infection by alfalfa mosaic alfamovirus (AMV), cucumber mosaic cucumovirus (CMV), potato Y potyvirus (PVY), tomato mosaic tobamovirus (TMV), tomato spotted wilt tospovirus (TSWV), and tomato yellow leafcurl geminivirus (TYLCV). With this assay system, 400–500 samples representing approximately 1.15 million tomato seedlings could be tested per day indicating its suitability for routine diagnosis of virus infection (Saldarelli et al., 1996). DIG-labeled cDNA probes have been successfully employed for the detection of peanut chlorotic streak virus (Satyanarayana et al., 1997), banana bract mosaic virus (Rodoni et al., 1997), beet necrotic yellow vein virus (Saito et al., 1997), and cucumber mosaic virus (Kiranmai, et al., 1998). Random primed cDNA probes labeled with ^{32}P or DIG were employed for the detection of pea seed–borne mosaic potyvirus (PSbMV) in the total nucleic acid extracts from infected pea plants at a dilution of 1:3125 (Ali et al., 1998).

The sensitivity of detection has been significantly improved by using DIG-labeled cRNA probes. A closterovirus associated with Nigerian sweet potato disease (SPVD) complex was detected. The SPVD complex was found to be due to the simultaneous infection of sweet potato by the closterovirus and sweet potato feathery mottle potyvirus (SPFMV) (Pio-Ribero et al., 1996). Labeled cRNA probe and ELISA were found to be equally sensitive for the detection of pelaragonium flower break carmovirus (Frank et al., 1996). The sensitivity of detection of PLRV in potato leaf extracts was enhanced by 2000-fold compared with ELISA, when DIG-labeled cRNA probe of approximately 2100 bp was employed. The probe detected PLRV readily in dormant potato tuber tissues and there was no cross-reaction between the probe and PVX or PVY (Loebenstein et al., 1997). DIG-labeled sense and antisense cRNA probes were employed for the detection of cymbidium mosaic potexvirus (CyMV) and odontoglossum ringspot tobamovirus (ORSV) in crude leaf extracts or total RNA from infected leaves. The probes remained stable for more than 1 year and provided sensitive and reliable detection for routine diagnosis of these viruses (Hu and Wong, 1998). A new whitefly-transmitted tomato chlorosis closterovirus (ToCV) was detected using DIG-11-UTP-labeled riboprobes derived from cDNA clones representing positions of RNAs 1 and 2 in infected tomato and it was differentiated from tomato infectious chlorosis virus (TICV) by the absence of cross-reaction in dot-blot hybridization tests (Wisler et al., 1998).

The assessment of sensitivity of cRNA probe labeled with DIG was made using dot-blot, Northern blot, and microplate hybridization techniques. In dot-blots, the probe could detect up to 10 fg of CMV-RNA, whereas Northern blot analysis revealed the riboprobe hybridizations to all CMV genomic RNAs in the total RNAs extracted from inoculated leaves. These probes can also be used to monitor the accumulation of CMV-RNA in the inoculated leaves of bottle gourd (*Lagenaria siceraria*) (Takeshita et al., 1999). The presence of prunus necrotic ringspot virus in the peach shoots cultured at 4°C for long periods could be detected by a cRNA probe (Heuss et al., 1999).

The microplate hybridization developed by Sano and Ishiguro (1996) is a simple and sensitive nucleic acid hybridization assay method performed in multi-well microplates instead of nitrocellulose or nylon membrane. The denatured nucleic acids are directly immobilized on polystyrene microplate and then hybridized with a cRNA probe labeled with DIG. Then the target nucleic acid is detected via hydrolysis of p-nitrophenyl phosphate by alkaline phosphatase conjugated with an anti-DIG antibody. This direct microplate hybridization technique can detect very low concentrations (below picogram level) of viral RNA or viroid cDNA. The important advantages of using this technique are the possibility of quantifying the results using a microplate reader and applying it for routine diagnosis of virus and viroid diseases. DIG-labeled synthetic oligonucleotide probes were used for subgroup-specific and quantitative detection of genomic RNA of CMV. The RNAs of CMV were adsorbed to the microplate wells and then hybridized with probes specific for respective subgroups of CMV. It was possible to detect about 1 ng of CMV-RNA under optimal conditions without any nonspecific reaction (Uchiba et al., 1999).

Using DIG-labeled probes, quantitative assessment of tomato yellow leafcurl geminivirus (TYLCV) in infected tomato plants was achieved (Caciagli and Bosco, 1996). Southern blot analysis is reported to be useful for the detection of tomato yellow leaf curl geminivirus (TYLCV) in whitefly vectors at different periods after the acquisition access period in 15% of whitelies tested TYLCV-DNA could be detected after a period of 30 min and in all whiteflies tested after an 8-hr access period. The virus could be detected earlier (within 2 hr) in insects fed on young leaves which contained higher concentrations of TYLCV. It was estimated that a single whitefly could acquire no more than 600 million viral genomes (1 mg viral DNA) (Zeidan and Czosnek, 1991).

Dot-blot hybridization technique has been used for quantitative assay of geminiviruses. DIG-labeled probes were employed for the detection and quantification of tomato yellow leafcurl geminivirus (TYLCV) in the whitefly (Bemisia tabaci) vectors that had different periods of acquisition access on infected tomato plants. The maximum amount of TYLCV-DNA (0.5–1.6 ng/insect) was present at the end of acquisition period and the viral DNA could be detected up to 20 days after the acquisition period. It appeared that TYLCV-DNA remained in the insect tissues for periods longer than the period for which the insect could transmit the virus (Caciagli and Bosco, 1997). A new virus infecting sweet potato causing leaf curl (SPLCV-US) transmitted by B. tabaci showed sequence homology with other geminiviruses. The DNA probes prepared with component A of pepper Huasteco geminivirus, bean golden mosaic geminivirus, and tomato mottle geminivirus hybridized with a 2.6-kb DNA band present in DNA extracts from plants infected by SPLCV-US isolate. Probes prepared with the B component of these geminiviruses, however, did not hybridize with the DNA extracts from sweet potato (Lotrakul et al., 1998).

The distribution and tissue specificity of maize streak virus (MSV) in maize were studied by Lucy et al. (1996). Using the DIG-labeled probes in in situ hybridization technique, the presence of MSV in the vascular tissues of the shoot apex was visualized, whereas in mature leaves MSV was located only in areas of lamina exhibiting disease symptoms. The mesophyll, vascular-associated parenchyma and bundle sheath cells of the leaf revealed the presence of positive and negative strands of MSV DNA (Lucy et al., 1996). The dot-blot hybridization technique was employed to identify two tomato lines showing highest level of resistance to TYLCV based on the determination of the contents of viral DNA in infected plants (Lapidot et al., 1997).

The tissue-print hybridization technique involves the transfer of the viral nucleic acid from the plant tissue directly on nitrocellulose or nylon membrane followed by hybridization of the printed membrane with nucleic acid probe labeled with radioactive or nonradioactive chemiluminescent markers. As the membrane has high binding affinity for nucleic acids, the viral RNA or DNA becomes affixed onto the solid matrix and immobilized. In the case of dsRNA or DNA that is not readily bound to the solid matrix, the infected plant tissues have to be denatured with denaturants like glyoxal or alkali prior to printing. Tissue-print hybridization technique was used for the detection of viruses infecting orchids (Chia et al., 1992). This technique can also be employed to demonstrate the localization of the virus in specific tissues of the infected plants. The local and systemic spread of tobacco mosaic virus in transgenic tobacco was monitored by this procedure (Wisniewski et al., 1990). A whole-plant printing protocol, developed to localize distribution of cymbidium mosaic potexvirus (CyMV) in cattleya, revealed the profile of CyMV-infected plant (Chia et al., 1995). Tissue print hybridization using DIG-labeled probes was shown to be effective for the differentiation of citrus tristeza virus (CTV) isolates from greenhouse or field, making it possible for monitoring spatiotemporal movement of specific CTV strains in epidemiological studies (Narváez et al., 2000).

9.3 DETECTION OF VIROIDS

The nucleic acid spot hybridization (NASH) technique was first developed for the detection of potato spindle tuber viroid (PSTVd) by Owens and Diener (1981), using ^{32}P-labeled cDNA probes. This method is based on hybridization of highly radioactive cDNA probe with the viroid RNA, which is bound to a solid matrix nitrocellulose; this method permits the detection of PSTVd in a large number of potato tubers (Fig. 9.1). Later, with the development of plasmid transcription vectors containing promoters for SP6, T3, and T7 bacterial polymerases, it was possible to prepare riboprobes (Mc Innes and Symons, 1989; Melton et al., 1984). The single-stranded cRNA probes are more sensitive than similar cDNA probes and

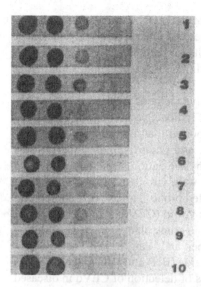

Figure 9.1 Sequential dot-blot hybridization of serially diluted sap extracts from PSTVd-infected potato leaves and stem with a digoxigenin-labeled PSTVd cRNA probe. 1–10, sequence in which membranes were hybridized (Elsevier Science B.V., Netherlands). (Courtesy of Podleckis et al., 1993.)

can be prepared more easily and uniformly labeled (Lakshmanan et al., 1986; Varveri et al., 1988; Candresse et al., 1990).

The usefulness of spot (dot-blot) hybridization for the early detection and assay of several viroids, which cause potato spindle tuber, avocado sunblotch, and coconut cadang-cadang diseases, has been reported (Owens and Diener, 1984). Latent infection of avocado by potato spindle tuber viroid was revealed by dot-blot hybridization (Querci et al., 1995). Employing a cDNA probe in a liquid-liquid system, coconut cadang-cadang RNA up to 1 ng/ml could be detected after a 96 hr hybridization time (Randles, 1985). A new viroid that causes fruit crinkle in apple could be differentiated from apple scar skin viroid by the hybridization technique (Ito et al., 1993). A spot hybridization test using purified cucumber pale fruit viroid (CPFVd) and hop stunt viroid (HSVd) showed that the sequence homology between CPFVd and HSVd was very high (Shikata, 1985), indicating the extent of relatedness. DIG-labeled cDNA probes were prepared from a full-length cDNA clone of chrysanthemum stunt viroid (CSVd). These probes detected CSVd specifically in dot-blot hybridization. The sensitivity of detection of CSVd in 2 M LiCl solution of nucleic acids corresponding to 17 μg of fresh weight, in dot-blot hybridization with DIG-labeled probes, was 100 times greater than the return-PAGE technique (Li et al., 1997). The potato spindle tuber viroid (PSTVd) was detected in seven potato cultivars using DIG-labeled cDNA probes. This test was suitable for diagnosis of PSTVd in potato leaves and tubers directly (Nakahara et al., 1997). The dot-blot hybridization test results formed the basis for preparing a calendar for diagnosis of apple scar-skin viroid (ASSVd) and dapple apple viroid (DAVd) (Paduch-Cichal et al., 1998).

The need for the development of alternative methods for detection arose because of the safety risks involved in the use of radioactive probes and short half-life of such probes. McInnes et al. (1989) reported that avocado sunblotch, coconut cadang-cadang, chrysanthemum stunt, and potato spindle tuber viroids could be detected in plant extracts by dot-blot hybridization, using nonradioactive photobiotin-labeled nucleic acid probes which were viroid-specific. The sensitivity of detection was similar to that of ^{32}P-labeled probe. The use of a photobiotin labeled F20 probe specific for citrus exocortis viroid (CEVd) sequence provided more sensitive and reliable detection of CEVd in crude plant extracts at a dilution of 1:100 and 1 ng in purified preparations (Chen and Wan, 1995). The sensitivities of cDNA probes labeled with biotin hydrazide (BH) and DIG in the detection of CEVd were compared. The minimum limits of detection of CEVd in diseased plant total nucleic acids were ca. 400 ng/spot and ca. 80 ng/spot for cDNA probes labeled with BH and DIG, respectively. A CEVd detection kit has been developed using the labeled probes for routine diagnosis of CEVd infection; the kit is rapid, specific, and easy to adopt (Hu et al., 1997).

The sensitivity of five types of synthetic oligonucleotide probes (PS-1, 2, 3, 4, and 5) labeled with biotin (BIO) was compared with that of the cDNA probes labeled with DIG or BIO for the detection of potato spindle tuber viroid (PSTVd). The BIO-labeled oligonucleotide probes, when tested individually, were less sensitive than DIG-labeled cDNA probe. However, the sensitivity was enhanced by using a mixture of two or all five oligonucleotide probes and the sensitivity was equal to that of DIG-labeled cDNA probes when the mixture of all five oligonucleotide probes (PS mix 1–5) was employed. The principal advantage of using PS mix 1–5 probe is that hybridization time can be shortened to 2 hr without any loss of sensitivity, whereas shortening of hybridization time decreases the sensitivity of detection by DIG-labeled cDNA probe (Nakahara et al., 1998).

Later chemiluminescent probes such as cRNA probe labeled with the steroid hapten digoxigenin were found to be suitable alternatives to the radioactive probes. In this system, the hybridized probe is detected by an antidigoxigenin polyclonal antibody conjugated to alkaline phosphatase. The conjugate probe complex can then be visualized by adding an enzyme substrate that, on dephosphorylation by alkaline phosphatase, yields sufficient light to expose film (Podleckis et al., 1993). The sensitivities of chemiluminescent digoxigenin-labeled cRNA probe with ^{32}P-labeled cDNA probe for the detection of PSTVd and apple scar skin group viroids (ASSVd) were compared. Dot-blot hybridization of purified viroids and sap extracts from infected plants showed that digoxigenin-labeled probes were as sensitive as ^{32}P-labeled probes. Positive detection was possible with 2.0–2.5 pg of purified viroid or 0.4 ng of total nucleic acid extract from infected tissue or in sap extracts diluted to 10^{-3} with healthy leaf extracts (Podleckis et al., 1993) (Fig. 9.1). DIG-labeled RNA probes have been employed for the rapid and sensitive detection of several viroids. In the case of citrus exocortis viroid (CEVd), a denaturation step with formaldehyde was included for the preparation of RNA extracts, resulting in stronger hybridization signals, when compared with extraction without formaldehyde treatment. CEVd RNA could be detected in as little as 0.2 mg of symptomatic *Citrus medica* leaves (Noronha et al., 1996). The hop stunt viroid (HSVd) was detected in the field samples of apricots by using specific DIG-labeled RNA probes. In addition to apricots, the infection of HSVd in almonds and pomegranates was also detected establishing them as new hosts for HSVd (Astruc et al., 1996).

Detection sensitivity was improved by using labeled cRNA probes for the detection of peach latent mosaic viroid (PLMVd) in field and greenhouse-grown peach and nectarine cultivars. PLMVd was detected in different parts such as leaf petioles, stem pieces, and dormant buds (Skrzeczkowski et al., 1996). This technique was applied for the detection of PLMVd in peach samples along with indexing on GF 305 seedlings as biological indicator of infection (Turturo et al., 1998). The cRNA probes were employed for the detection of apple scar-skin viroid (ASSVd), dapple apple viroid (DAVd), and peach rusty skin viroid (PRSVd) infection in Polish orchards (Paduch-Cichal et al., 1996).

The method of extracting nucleic acids from citrus tissues for diagnosis of citrus viroids was improved by using a differential 2-butoxyethanol precipitation of contaminating polysaccharides and phenolic compounds that interfere with nucleic acid extraction, when the conventional combination of 2-methyoxyethanol extraction and cetyltrimethyl ammonium bromide precipitation was used. This modified procedure was effective for the extraction of citrus viroids such as citrus exocortis viroid (CEVd), group I citrus viroid (CVd-I), and hop stunt viroid–citrus isolate, which is a variant of group II citrus viroid (CVd-II). These viroids were detected using DIG-labeled cRNA probes in dot-blot hybridization. Additionally infection by citrus viroid group III (CVd-III) in several citrus samples was detected for the first time in Japan using DIG-labeled cRNA probe prepared from the cloned cDNA (Nakahara et al., 1998).

The tissue print or imprint hybridization procedure was developed by Romero-Durban et al. (1995) for the detection of several viroids using labeled cRNA probes. This method is simple, requiring minimal sample manipulation. The sap from cells that have been mechanically ruptured is transferred directly onto the nitrocellulose or nylon membrane, which has high binding affinity for nucleic acids. The citrus exocortis viroid (CEVd), chrysanthemum stunt viroid (CSVd), hop stunt viroid (HSVd), and avocado sun blotch viroid (ASBVd) were detected in their respective host plants. This method, in addition to being rapid and sensitive, can be used to determine the sites of viroid accumulation. The potato spindle tuber viroid (PSTVd) could be detected in stem rachis of two tomato cultivars using a ^{35}S-labeled PSTVd RNA probe. PSTVd was detected earlier in susceptible cultivar Rutgers than in tolerant cultivar Gold Kugel, which remained symptomless. The distribution and tissue-specific localization of PSTVd could be studied by this technique (Stack-Lorenzen et al., 1997). The sensitivity and reliability of detection of citrus viroids by tissue imprint hybridization can be enhanced by first inoculating *Citrus medica* (Etrog citron) and employing DIG-labeled RNA or DNA probes (Palacio-Bielsa et al., 1999).

9.4 POLYMERASE CHAIN REACTION

The polymerase chain reaction (PCR) is a simple, sensitive, and versatile technique that can provide results rapidly. The procedure essentially involves heat denaturation of the target ds DNA and hybridization of a pair of synthetic oligonucleotide primers to both strands of the target DNA, one to the 5' end of the sense strand and one to the 5' end of the antisense strand by an annealing step. Then by using a thermostable Taq DNA polymerase enzyme from *Thermus aquaticus* (*Taq*), new DNA is synthesized on templates to produce twice the number of target DNAs. Newly synthesized DNA strands are used as targets for subsequent DNA synthesis, and the steps discussed are repeated up to 50 times (Fig. 9.2). There is an exponential increase in the number of target DNA molecules. In this procedure DNA sequences between the primers are reproduced with high fidelity and an efficiency of up to 85% per cycle (Weier and Gray, 1988); the procedure can be automated if required. *Taq* DNA polymerase is heat-stable and can be used at temperatures between 60°C and 85°C.

The PCR products can be used a) as a target for hybridization; b) for direct sequencing of the DNA to determine strain variations, and c) as a specific probe. The PCR has many advantages over traditional methods of disease diagnosis. The pathogens need not be cultured before detection; is enough if the pathogen DNA is extracted. With PCR, it is possible to detect a single pathogen or many members of a group of related pathogens, as in serological methods, but serological methods are more expensive and time-consuming. Hundreds of different PCR primers may be synthesized at costs comparable to those of methods for developing only a few monoclonal antibodies (Henson and French, 1993). Levy et al. (1994) reported that the commercial product Gene Releaser could be used to produce plant extracts suitable for PCR amplification without use of organic solvents, alcohol precipitation, or additional nucleic acid purification techniques. This procedure will be useful for the detection of viruses, viroids, and phytoplasmas that infect woody hosts, and the samples can be prepared in 1–2 hr, as against the

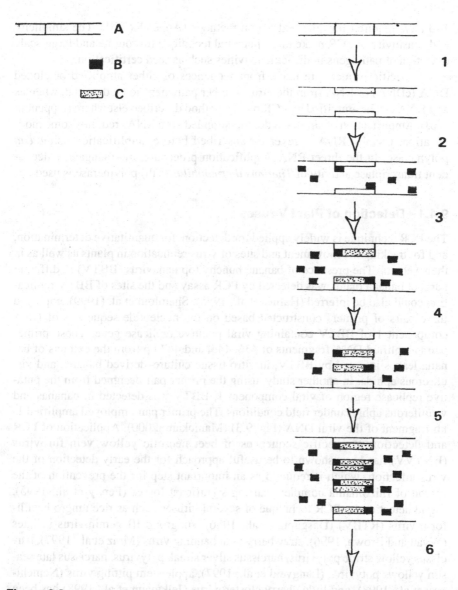

Figure 9.2 Amplification of pathogen DNA by polymerase chain reaction (A, target DNA; B, primer; C, new DNA): 1, Denaturation of DNA; 2, addition of primer and annealment; 3, synthesis; 4, denaturation; 5, synthesis; 6, repetition of denaturation and synthesis cycles (CAB International, U.K.). (Adapted from Leach and White, 1991.)

1–3 days required for other extraction methods (Appendix 9[i]). The simplicity and sensitivity of PCR make this a potential technique for routine and large-scale detection of pathogens in difficult activities such as seed certification.

Specific primers are made from sequences of either amplified or cloned DNA (cDNA) or RNA from the virus or other pathogens to be detected, whereas any DNA can be amplified by PCR by the method described elsewhere (Appendix 9[ii]). Amplification of single- or double-stranded viral RNAs requires some modification. Usually RNA is reverse-transcribed before amplification using *Taq* polymerase. In the direct RNA amplification procedure, the manganese-dependent transcriptase activity of *Thermus thermophilus* (Tth) polymerase is used.

9.4.1 Detection of Plant Viruses

The PCR technique is widely applied for detection, for quantitative determination, and for tracking the movement and sites of virus replication in plants as well as in their vectors. The presence of banana bunchy top nanavirus (BBTV) in different parts of banana plants was detected by PCR assay and the sites of BBTV replication could also be inferred (Hafner et al., 1995). Shamloul et al. (1999) employed three pairs of primers constructed based on the nucleotide sequences of DNA component 1 of BBTV containing viral putative replicase gene. These primer pairs amplified DNA fragments of 436, 446, and 447 bp from the extracts of banana leaves infected by BBTV, in vitro tissue culture–derived banana, and viruliferous aphids. In another study, using the primer pair designed from the putative replicase region of viral component 1, BBTV was detected in bananas and viruliferous aphids under field conditions. The primer pair employed amplified 1-kb fragment of the viral DNA (Fig. 9.3) (Manickam, 2000). Application of PCR and detection of specific sequences of beet necrotic yellow vein furovirus (BNYVV) has been shown to be useful approach for the early detection of the virus infection, which is required as an important step for the prevention of the spread of rhizomania complex causing significant losses (Fenby et al., 1995). Rapid detection by PCR technique of several viruses such as rice tungro bacilliform virus (RTBV) (Dasgupta et al., 1996), subgroup III geminivirus isolates (Wyatt and Brown, 1996), strawberry vein banding virus (Mráz et al., 1997), narcissus yellow stripe potyvirus, narcissus silver streak potyvirus, narcissus late season yellows potyvirus (Langveld et al., 1997), apple stem pitting virus (Nemchinov et al., 1998), and little cherry closterovirus (Jelkmann et al., 1998) has been useful for eliminating the infected plants or planting materials, a mandatory requirement for quarantine and certification programs.

The identity and relationship of new viruses have been established by using PCR technique. Pineapple bacilliform virus (PBV) was found to be related to, but distinct from, other badnaviruses. Specific primers designed from PBV sequences yielded a 403-bp fragment when used in PCR with extracts from infected pineap-

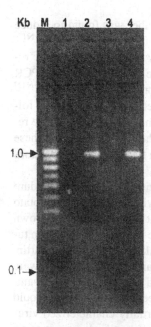

Figure 9.3 Detection of banana bunchy top virus (BBTV) by PCR. Lane: M, 100 bp DNA ladder, molecular markers; 1, healthy banana; 2, BBTV infected banana; 3, healthy aphid; 4, BBTV infected aphid. (Courtesy of Manickam, 2000.)

ple plants only, whereas no amplification occurred with extracts from plants infected with sugarcane bacilliform or banana streak badnaviruses (Thomson et al., 1996). Likewise, the citrus mosaic disease has been shown to be caused by a badnavirus whose genome is a dsDNA as indicated by PCR amplification of degenerate oligonucleotide primers based on conserved badnavirus genomic sequences (Ahlawat et al., 1996)

The North American necrosis-inducing isolates of potato Y potyvirus (PVY^{N}) were screened by PCR for the European-type members of the tuber necrosis inducing PVY^{NTN}. One isolate from garden potato tuber was found to induce potato tuber necrotic ringspot disease (PTNRD) (Singh et al., 1998). The PVY^{NTN} strain could be differentiated from the PVY^N strain by using strain-specific restriction endonucleases that may cause unique cleavage of their respective PCR products. Single cleavages of PCR products derived from the 5' end of PVY^{NTN} genome by *NcoI* and that of the PVY^N by *BglII* were effected and this was followed by polyacrylamide gel electrophoresis (PAGE) analysis of digests. The resultant digestion patterns provided a reliable basis for the differentiation of these strains. Single and mixed infections could be detected in field samples of potatoes (Rosner and Maslenin, 1999).

The reverse-transcription-polymerase chain reaction (RT-PCR) procedure has been used for the detection and assay of many plant viruses (Table 9.1). Potato leaf roll virus (PLRV) was detected by PCR in dormant tubers from field-grown plants and in vitro-propagated microtubers, whereas the presence of PLRV in tubers of field-grown plants could not be detected by ELISA (Spiegel and Martin, 1993). The RNA1 and RNA2 of wheat soil-borne mosaic virus (WSBMV) were individually detected in *Triticum aestivum;* RNA2 could be observed in root samples of both susceptible and resistant cultivars up to 7 weeks before the virus could be detected by ELISA (Pennington et al., 1993). A partially characterized virus isolated from *Gloriosa rothschidiana* could be positively identified as a potyvirus by specific amplification and subsequent sequence analysis of an amplified DNA fragment (Langeveld et al., 1991). Zerbini et al. (1995), using PCR, detected as much as 12% divergence in the coat protein-hypervariable region of the genome of lettuce mosaic potyvirus isolates; this variability does not lead to any change in the biological properties. The luteoviruses potato leaf roll virus, beet western yellows virus, and New York barley yellow dwarf virus (BYDV) were easily distinguished by restriction enzyme analysis of the amplified DNA products; this test could simultaneously detect all five BYDV serotypes (Robertson et al., 1991).

Two subgroups of cucumber mosaic cucumovirus (CMV) were distinguished by using two CMV specific primers that the flank CMV capsid protein gene and amplifying the cDNA fragment. Restriction enzyme analysis of this DNA fragment offers distinct restriction patterns that help to classify the CMV isolates accurately in the respective subgroups. Xie and Hu (1995) developed a PCR assay for the detection of banana bunchy top virus in banana plants and aphid vectors, and the assay was 1000 times more sensitive than dot-blot hybridization and ELISA tests.

The transmission of banana bunchy top virus (BBTV) by the aphid *Pentalonia nigronervosa* was studied by using PCR assay. The aphids acquired BBTV within 4 hr after acquisition access feeding on infected banana plants and retained the virus for 15 days. There was no evidence for the transmission of BBTV from the adults to the offspring (Hu et al., 1996). The PCR assay was employed for indexing the aphids collected from an endemic area in Tamil Nadu

Table 9.1 Target Sequences of Plant Viruses Amplified by PCR

Virus	Target sequence/primers	Reference
Abutilon mosaic bigemini virus (AbMV)	Virus-specific sequence; plant extract	Wu and Hu, 1996
Apple chlorotic leaf spot virus (ACLSV)	Virus-specific fragment; total nucleic acid from apple leaves	Kinard et al., 1996
Apple stem grooving virus (ASGV)	Primer pair amplifying 345-bp coat protein (CP) gene	Crossley et al., 1998
Apple stem pitting virus (ASPV)	Sequences in the 3' terminal region	Schwarz et al., 1998
Banana bunchy top virus (BBTV)	Virus specific sequences; plant extracts	Xie and Hu, 1995
	Aphids extract	Hu et al., 1996
Banana streak badnavirus (BSV)	Episomal and integrated sequences of virus	Harper et al., 1999
Barley mild mosaic virus (BaMMV)	Virus fragment of 9.8 kb	Lee, 1998
Barley yellow mosaic virus (BaYMV)	Virus fragment of 0.9 kb	Lee, 1998
Bean common mosaic virus (BCMV) and Bean common mosaic necrosis virus	Virus-specific sequences	Xu and Hampton, 1996
Bean golden mosaic virus (BGMV)	BGMV-DNA sequence from infected plant	Gilbertson, et al., 1991
Bean yellow mosaic virus (BYMV)	BYMV-DNA sequence; plant extracts containing viral DNA	Vunish et al., 1990, 1991
Beet pseudo-yellows virus (BPYV)	BPYV sequence; plant extract	Coffin and Coutts, 1992
Beet western yellows (BWYV) and beet mild yellowing viruses	Virus-specific sequences; plant extracts	Jones et al., 1991
Beet yellows virus (BYV)	Cp gene sequences from plants and aphids	Stevens et al., 1997
Blackcurrant reversion associated virus (BRAV)	Virus-specific sequence	Lemmetty et al., 1998
Blueberry scorch virus (BBScV)	Virus-specific sequences	Halpern and Hillman, 1996

Table 9.1 Continued

Virus	Target sequence/primers	Reference
Cacao swollen shoot virus (CSSV)	CP gene sequences	Hoffmann, et al., 1997
Cauliflower mosaic virus (CaMV)	CaMV-specific sequence; aphid extracts with viral DNA	Lopez-Moya et al., 1992
Cherry leaf roll virus (ChLRV)	Virus-specific sequences; plant extract	Borja and Ponz, 1992; Rowhani et al., 1997
Citrus mosaic badnavirus (CMBV)	Degenerate oligonucleotide primers	Ahlawat et al., 1996
Citrus psorosis ringspot virus (CtRSV)	Fragment of bottom component RNA	Garcia et al., 1997
Citrus tristeza virus (CTV)	CP gene sequence; plant extract	Nolasco et al., 1997
	Aphid extract	Mehta et al., 1997
	Sequences in both 5′ and 3′ regions	Hilf et al., 1999
Cowpea mild mottle virus (CMMV)	Carlavirus specific primer	Badge et al., 1996
Cucumber mosaic virus (CMV)	CP gene sequence; plant extracts	Yang et al., 1997; Lee et al., 1998
	Conserved sequence CMV, RNA-3	Raj et al., 1998
Cucurbit yellow stunting disorder virus (CYSDV)	Fragments of RNA-2	Livieratos et al., 1999
Cymbidium mosaic virus (CyMV) and *Odontoglossum* ringspot virus (ORSV)	A pair of common primers	Seok et al., 1998
Fiji disease virus (FDV)	Virus-specific sequences; plant extracts	Smith et al., 1992
Garlic virus 1 carla virus (GV1); garlic virus 2 potyvirus (GV2); garlic mite-borne mosaic virus (GMbMV) and onion yellow dwarf poty-virus (OYDV)	CP gene sequences of respective viruses	Takaichi et al., 1998

Table 9.1 Continued

Virus	Target sequence/primers	Reference
Grapevine A trichovirus (GAV), grapevine B trichovirus (GBV)	Genomic fragments of respective viruses	Notte et al., 1997
Grapevine fanleaf virus (GFLV)	Virus-specific sequences; nematode extracts	Esmenjaud et al., 1994
Grapevine leafroll-associated closterovirus 3 (GLRa V3)	Genomic fragments of the virus	Notte et al., 1997
Grapevine rupestris stem-pitting associated virus (GRSPaV)	Virus-specific sequences	Zhang et al., 1998; Meng et al., 1999
Japanese yam mosaic virus	Virus-specific sequences	Fuji and Nakamae, 1999
Lettuce-mosaic virus (LMV)	Virus-specific sequences	Revers et al., 1997; van der Vlugt et al., 1997
Lily mottle virus (LMoV) and lily symptomless virus (LSV)	Virus-specific sequences	Lee et al., 1998
Lily cherry closterovirus (LchV)	Virus-specific sequences	Vitushkina et al., 1997; Jelkmann et al., 1998
Luteoviruses	Virus-specific sequences; plant extracts	Robertson et al., 1991
Maize streak virus (MSV)	Conserved virus sequences; plant extracts	Rybicki and Hughes, 1990
Melon necrotic spot virus (MNSV)	CP gene sequences	Matsuo et al., 1998
Narcissus late season yellows (NLSYV); narcissus silver streak virus (NSSV) and narcissus yellow stripe virus (NYSV)	Virus-specific sequences, plant extracts	Langaveld et al., 1997
Odontoglossum ringspot virus (ORSV) and Cymbidium mosaic virus (CyMV)	A pair of common primers	Seok et al., 1998
Peanut stripe virus (PStV)	CP gene sequences	Pappu et al., 1998
Pea seed-borne mosaic virus (PSbMV)	Virus-specific sequences; plant or seed extracts	Kohnen et al., 1992; Phan et al., 1997

Table 9.1 Continued

Virus	Target sequence/primers	Reference
Plum pox virus (PPV)	Virus-specific sequences; plant extracts	Korschineck et al., 1991; Wetzel et al., 1992; Cambra et al., 1998
	Primers for the noncoding region (NCR)	Levy et al., 1995
	CP gene sequences	Bitóová et al., 1997
Potato leaf roll virus (PLRV)	Virus-specific sequences	Nolasco et al., 1993; Leon, et al., 1997
	Aphid extracts	Singh et al., 1997
Potato moptop virus (PMTV)	Virus-specific sequences, tuber extracts	Sokmen et al., 1998
Potato virus A (PVR)	Primer from P1 gene	Singh and Singh, 1997
Potato virus Y (PVY) and potyviruses	Conserved virus sequences; plant extracts	Langeveld et al., 1991
	Oligonucleotide primers	Singh et al., 1998
	Specific primers for 835-bp product	Tomassoli et al., 1998
	Common primers for 4 strains	Nielsen et al., 1998
	Virus-specific sequence; tuber extracts	Singh et al., 1999
Prunus necrotic ringspot virus (PNRSV)	Primers for 300-bp product	Spiegel et al., 1996
	Short and long primers for 200 bp and 785 bp, respectively	Rosner et al., 1997
	Sequences of RNA4	Aparicio et al., 1999
	CP gene sequence	Candresse et al., 1998
Raspberry bushy dwarf virus (RBDV)	Virus-specific sequence	Barbara et al., 1995
	Sequences of RNA3	Kokko et al., 1996
Rice tungro baciliform virus (RTBV)	Virus-specific sequences; leafhopper extracts	Venkitesh and Kogenezawa, 1995
	Viral DNA sequences in leaf extracts	Dasgupta et al., 1996
Rice tungro spherical virus (TRSV)	CP gene sequences	Yambao et al., 1998
Squash leafcurl virus (SLCV)	Viral DNA in whitefly extracts	Rosell et al., 1999
Strawberry vein banding virus (SVBV)	Virus-specific sequence; plant extracts	Mráz et al., 1997

Table 9.1 Continued

Virus	Target sequence/primers	Reference
Sugarcane mosaic virus (SCMV)	CP gene sequences	Huckett and Botha, 1996
	Strain group–specific primers	Yang and Mirkov, 1997
Sugarcane yellow leaf syndrome virus (SYLSV)	Virus-specific sequences	Comstock et al., 1998
Sweet potato feathery mottle virus (SPFMV), Sweet potato latent virus (SPLV) and Sweet potato virus G	CP gene–specific sequences	Colinet et al., 1998
Tobacco mosaic virus (TMV)	MP gene sequences; plant extracts	Drygin et al., 1992
	Primer pairs RS1 and RS2	Chen et al., 1996
Tobacco rattle virus (TRV)	Virus-specific sequence; potato tuber extracts	Muchalski, 1997; Yamaji et al., 1998
	Sequences in RNA1; potato tuber extracts	Kawchuk et al., 1997
Tomato mosaic virus (ToMV)	Virus-specific sequences; cloud and fog samples	Castello et al., 1995
Tomato ringspot virus (TRSV)	Virus-specific sequences; plant extracts	Chen et al., 1996
Tomato spotted wilt virus (TSWV)	Virus-specific sequences; insect extracts	Tsuda et al., 1994; Mumford et al., 1994, 1996; Jain et al., 1998
Tomato yellow leaf curl (TYLCV)	Viral DNA sequence, whitefly extracts	Navot et al., 1992; Navar-Castino et al., 1995
Turnip mosaic virus (TuMV)	Specific primers for 876 bp	Park et al., 1998
Yam mild mosaic virus (TMMV) and yam mosaic virus (YMV)	Virus-specific sequences; plant extracts	Mumford and Seal, 1997
Zucchini yellow mosaic virus (ZYMV)	Virus-specific sequences; plant extracts	Barbara et al., 1995

State in India. The aphids were tested singly or in groups of 5, 10, and 20. The presence of BBTV even in single aphids could be inferred by observing the 1-kb fragment of BBTV-DNA amplified by the primer pair designed from the putative replicase region of viral component 1 (Fig. 9.4) (Manickam, 2000). The PCR-based procedure provides a simple alternative to the serological assays used for

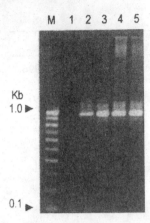

Figure 9.4 Detection of banana bunchy top virus (BBTV) by PCR in field-collected aphids. Lane: M, 100 base pair markers; 1, healthy aphid; 2, single viruliferous aphid; 3, group of five viruliferous aphids; 4, group of 10 viruliferous aphids; 5, group of 20 viruliferous aphids. (Courtesy of Manickam, 2000.)

classifying CMV isolates (Rizos et al., 1992). The RT-PCR procedure has been employed for the detection of several other viruses, such as cymbidium mosaic virus (Lim et al., 1993), strawberry mild yellow edge virus (Hadidi et al., 1993), sweet potato feathery mottle virus (Colinet et al., 1994), tomato spotted wilt virus (Mumford et al., 1994), and viruses that infect woody plants (Rowhani et al., 1995). Using PCR, the cauliflower mosaic virus (CaMV) was detected in a single viruliferous aphid, and this method could detect about 10 pg/ml of purified CaMV and was more sensitive than serological methods (Lopez-Moya et al., 1992). Takahashi et al. (1993) detected the presence of rice tungro bacilliform badnavirus (RTBV) in tungro disease-affected rice leaves, but not in the leafhopper vector, by PCR, which was found to be 1000 to 10,000 times more sensitive than ELISA. However, Venkitesh and Koganezawa (1995) showed that a small fragment (569 bp) of RTBV-DNA could be amplified by PCR from the total nucleic acid extract of a single viruliferous leafhopper and detected on agarose gel.

The pathway of plant virus movement in the insect vector was studied by using a membrane feeding system and PCR technique. The squash leafcurl geminivirus (SLCV) DNA was detected in whole whitefly body extracts and in saliva, honey dew, and hemolymph of the vector *Bemisia tabaci* and nonvector *Trialeurodes vaporariorum* after an acquisition access period of 0.5–96 hr on the infected plants. Detection of SLCV-DNA in the honeydew of both *B. tabaci* and *T. vaporariorum* indicated that SLCV viroins, viral DNA, or both were able to pass through the digestive system of both the vector and nonvector insects. On the other hand, SLCV DNA was detected in the saliva and hemolymph of *B. tabaci* only, providing evidence that SLCV could pass through the gut barrier, enter hemolymph, and reach salivary glands of the vector species only. The digestive epithelia of nonvector species appeared to be impermeable to the passage of SLCV (Rosell et al., 1999).

The RT-PCR method was performed directly in crude extracts of CMV-infected plants using primers complementary to conserved sequences of cucumber mosaic virus RNA 3 for broad-spectrum detection of isolates belonging to subgroups I and II from different geographical locations (Blasde et al., 1994). The stone fruit anthers are sent from one country to another for use in breeding programs. Hence it is essential to ensure that anthers are free of viruses. For this purpose RT-PCR was used for detection of plum pox potyvirus (PPV) in the desiccated anthers stored at 4°C for 1–2 years. After preparing the anther tissue with Gene Releaser polymeric matrix, RT-PCR protocol was employed using DNA primers from the 3′ noncoding region (NCR) of PPV. It was established that stone fruit anthers could be a source of PPV dissemination during international movement of *Prunus* germplasm (Levy et al., 1995). Using RT-PCR technique, PPV could be detected in both seed coat and cotyledon of apricots, whereas the virus was detected by ELISA only in seed coat, demonstrating the high sensitivity of RT-PCR in detecting PPV in plant tissues with low virus titer (Pasquini et al., 1998). The presence of plant viruses can be detected not only in plant materials but also in as diverse sources as clouds and fog. Tomato mosaic tobamovirus (ToMV) was detected in cloud samples collected from the summit of Whiteface Mountains, New York, and fog samples from two collection sites along the coast of Maine by using RT-PCR technique (Castello et al., 1995).

The citrus tristeza virus (CTV) isolates can be grouped based on the genomic nucleic acid sequence divergence assessed by PCR technique. The cDNA sequences of the Florida CTV isolates T3 and T30 showed a relatively consistent or symmetrical distribution of nucleotide sequence identity in both the 5′ and 3′ regions of the 19.2-kb genome. Selective amplification of these isolates occurred when primers designed from cDNA sequence for PCR were used, indicating that they can be considered as similar to T3, T30, or T36. PCR primers designed from the T36-CP gene sequence amplified successfully from all isolates. Based on the results of PCR, two groups may be recognized. The T36 group can be differentiated from the VT group from Israel by the highly divergent 5′ genomic sequence. It is suggested that the 5′ region of the viral nucleic acid may serve as a measure of the extent of sequence divergence and this parameter can be used to define new groups or group members of the CTV complex (Hilf et al., 1999).

The sequences of geminivirus DNA fragments comprising part of the *rep* gene, the common region, and part of the *CP* gene were compared to assess the extent of variability in geminivirus isolates associated with *Phaseolus* spp. in Brazil. The geminivirus infecting bean had sequences nearly identical to that of bean golden mosaic geminivirus (BGMV-B2), whereas lima bean was found to be infected by a new species of geminivirus that induces symptoms similar to those caused by BGMV-B2. Thus the new species could be differentiated and it was named lima bean golden mosaic virus (LBGMV-BR). All sequences from bean isolates clustered with BGMV-B2, while the sequences of lima bean isolate were distinctly different. Another new virus infecting the weed *Leonurus sibiricus* was also identified based on the differences in the DNA sequences (Faria and Maxwell, 1999).

The study for assessing the sensitivity of ELISA, dot-blot, and PCR assays showed that PCR technique was 10,000-fold more sensitive than dot-blot and ELISA for the detection of abutilon mosaic bigeminivirus (AbMV) (Wu and Hu, 1996). The efficiency of two ELISA formats DAS-ELISA and TAS-ELISA, nucleic acid hybridization with chemiluminescent or chromogenic labeled cDNA probes, and PCR amplification for the detection of barley yellow dwarf virus (PAV-IL) was compared. The PCR technique was found to the most sensitive with a detection limit of 0.1 pg of RNA extracted from purified virus and viral RNA from 0.5 pg of infected leaf tissue. The detection limits of ELISA format were much higher (1 ng of purified virus or 78 ng of infected tissue) and nucleic hybridization technique was also sensitive with similar higher detection limits (Figueira et al., 1997). The complementary nature of ELISA and PCR assays was brought out by the study on the serotypes of plum pox potyvirus (PPV). The results obtained with indirect DAS-ELISA using specific MABs for PPV-D and PPV-M serotypes had excellent correlation with the results of PCR assays or random fragment length polymorphism (RFLP) analysis of PCR products. All isolates of PPV reacting positively with the PPV-M-specific MAB were assigned to M-serotype by PCR based assays also. Likewise, 51 of 53 isolates showing positive reaction to D-specific MAB were placed under D-serotype by PCR typing also (Candresse, 1998). The plant viruses detected by the RT-PCR procedure have been listed by Henson and French (1993).

Plant viruses that have been subsequently detected by adopting RT-PCR protocol include garlic viruses (Tsuneyoshi and Sumi, 1996; Takaichi et al., 1998), blueberry scorch virus (Helpern and Hillman, 1996), citrus psorosis ringspot virus (Garcia et al., 1997), plum pox virus (Bittóová et al., 1997), little cherry virus (Vitushkina et al., 1997), potato leaf roll virus (Leone and Shoen, 1997), grapevine virus (Notte et al., 1997; Habili et al., 1997), prunus necrotic ringspot virus (Spiegel et al., 1996; Rosner 1997), cucumbr mosaic virus (Yang et al., 1997; Raj et al., 1998), tobacco rattle virus (Muchalski, 1997; Yamaji et al., 1998), sugarcane yellow leaf syndrome (Comstock et al., 1997), tomato infectious chlorosis virus (Li et al., 1998), potato moptop virus (Sokmen et al., 1998), barley yellow mosaic virus and barley mild mosaic virus (Lee KuiJae, 1998), *Cymbidium* mosaic virus and *Odontoglossum* ringspot virus (Seoh et al., 1998), lily viruses (Lee et al., 1998), turnip mosaic virus (Park et al., 1998), apple stem grooving virus and apple stem pitting virus (Kummert et al., 1998; Malinowski et al., 1998), potato virus A (Singh and Singh, 1998), potato virus Y (Nielsen et al., 1998; Singh et al., 1999), and cherry mottle leaf virus 1 (James et al., 1999). These examples show the effectiveness of RT-PCR technique for the detection and differentiation of plant viruses infecting a wide range of crop plants. This technique requires only very small amounts of infected leaf tissue (1 mm^2) or viral nucleic acid (10 fg), and the results are obtained in about 5 hr (Lin et al., 1993).

The RT-PCR technique can also be used to detect specific virus sequences or components of viral nucleic acid. The pathotypes P1 and P4 of pea seed–borne mosaic virus could be differentiated by employing sequence-specific RT-PCR (Kohnen et al., 1995). Closely related furoviruses, beet necrotic yellow vein virus, and beet soil–borne mosaic virus could also be differentiated (Rush et al., 1994). Further, studies by Kruse et al. (1994) showed that by determining restriction fragment length polymorphism (RFLP) patterns of the RT-PCR products of different regions of viral genome, two major strain groups, designated A and B, of BNYVV could be differentiated. Hataya et al. (1994) developed the PCR-microplate hybridization method for detection of plant viruses. In this method a cDNA fragment from coat protein region of the PVY-RNA genome amplified by RT-PCR was detected by a digoxigenin-labeled cDNA probe. Detection of 10 fg of PVY genomic RNA was possible, and it was 10,000 times more sensitive than ELISA (Figs. 9.5 and 9.6) (Appendix 9[iii]).

Prunus necrotic ringspot ilarvirus (PNRSV) could be detected in dormant peach trees by adopting a modified RT-PCR technique. The total RNA was extracted from bark tissue using the lithium chloride–based method and used for reverse transcription and subsequent amplification of viral sequences. The PCR product (about 300 bp) was analyzed by gel electrophoresis and stained with ethidium bromide for visualization of PCR product. This RT-PCR protocol was quite sensitive for the detection of PNRSV in dormant woody bark tissues and it can be used for screening imported buldwood materials in postentry quarantine programs and for production virus-free planting materials (Spiegel et al., 1996).

A combination of immunocapture and a gel-free RT-PCR–based fluorogenic detection procedure for PLRV in dormant potato tubers was developed by

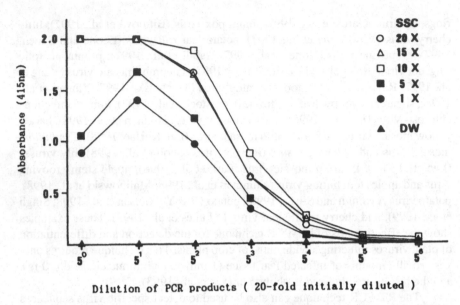

Dilution of PCR products (20-fold initially diluted)

Figure 9.5   Effect of standard saline citrate (SSC) concentrations on adsorption of DNA to microplate wells. Absorbance values are determined after hydrolysis of substrate for 1 hr (Elsevier Science B.V., Netherlands). (Courtesy of Hataya et al., 1994.)

Shoen et al. (1996). A PLRV-specific oligonucleotide containing a 5'-terminal "reporter" fluorescein, and a 3'-terminal rhodamine "quencher" was specifically degraded during amplification and this led to a relative increase in reporter-associated fluorescence. Immunocapture of PLRV from tuber extracts was carried out using paramagnetic beads coated with an antiserum raised against PLRV. The cell wall–degrading enzymes, cellulase, and macrozyme were added to the tuber extracts to improve the detection sensitivity. The presence of PLRV in primarily infected dormant tubers of four potato cultivars was detected by this procedure performed in microplates, which is rapid, reproducible, semiquantitative, and amenable for automation. The inspection time of seed tubers for PLRV infection can be reduced to 1 day from 5 days required for conventional testing.

The PCR primers designed from fragments of cDNA from the bottom component RNA of citrus psorosis ringspot virus (CtRSV-4) were employed for the detection of this virus in citrus leaves at a tissue dilution of 1:12,800 representing as little as 2 μg of tissue indicating high level of sensitivity of RT-PCR technique (Garcia et al., 1997). A modified RT-PCR technique was developed for the detection of prunus necrotic ringspot ilarvirus (PNRSV) in dormant peach and almond trees. Two different pairs of primers yielding short (70 bp) and long (200 bp) products were used. The relative amount of short product was higher than the

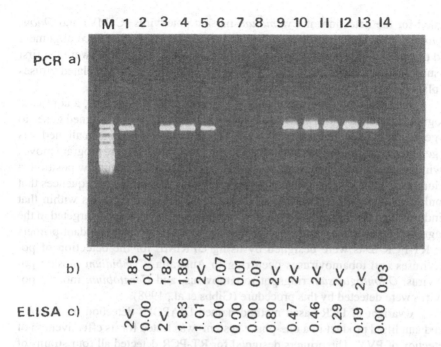

Figure 9.6 Detection of PVY in samples of potatoes from the field: a, by RT-PCR; b), by microplate hybridization; c, by ELISA, b and c, absorbance values at 415 nm after substrate hydrolysis for 2 hr (Elsevier Science B.V., Netherlands). (Courtesy of Hataya et al., 1994.)

longer product. Amplification of short PCR product was effective for the detection of PNRSV in plant tissues with low virus titer as in dormant trees, whereas the long product was amplified in tissues with high virus concentrations (Rosner et al., 1997). Another modification of PCR, designated spot-PCR, was developed for the detection of grapevine A trichovirus (GAV), grapevine B trichovirus (GBV), and gravevine leaf roll–associated closterovirus 3 (GLRaV3). Specific amplification of genomic fragments of these viruses was achieved by RT-PCR on total nucleic acid solubilized from small pieces of charged nylon membrane on which a drop of crude sap of infected grapevine had already been spotted. The ease of release of viral template from the spot on nylon membrane was enhanced by a heat treatment at 95°C for 10 min. The detection limit of spot-PCR was comparable with standard PCR technique with the additional advantage of storing the blots up to 1 month after spotting (Notte et al., 1997).

It is possible to detect two viruses infecting the plants simultaneously by using a pair of common primers in a single PCR. A pair of common primers was designed for the detection of *Cymbidium* mosaic potexvirus (CyMV) and *Odontoglossum* ringspot tobamovirus (ORSV) infecting orchids. Sequence alignment and primer analysis were used for primer design. Seoh et al. (1998) were the first to successfully apply a single pair of PCR primers to detect two unrelated viruses in plants infected by them simultaneously.

This approach was further enlarged by the GPRIME package, a computer program developed with a view to identifying the best regions of aligned genes to target in nucleic acid hybridization tests. The homologous regions in aligned sets of gene sequences of viruses to be targeted are identified. The core program moves a window over the aligned sequences and determines, at each window position, a redundancy value. The redundancy value represents the number of sequences that would represent all permutations of the variable sequence positions within that window. Then the regions with minimal redundancy values may be targeted in the diagnostic tests, based on oligonucleotide hybridization. The redundant primers for RT-PCR tests were designed by using GPRIME for the detection of potexviruses and tobamoviruses occurring in Australia. *Cymbidium* mosaic potexvirus, *Odontoglossum* ringspot tobamovirus, and *Ceratobium* mosaic potyvirus were detected by this procedure (Gibbs et al., 1998).

A variant of PCR-based method named DIAPOPS-detection of immobilized amplified product in a one-phase system- was tested for its effectiveness of detection of PVY. The primers designed for RT-PCR detected all four strains of PVY, viz., O, N, NTN, and C, but not potato V and potato A potyviruses in dormant potato tubers. However, ELISA tests detected these strains in some samples that gave negative results with DIAPOPS. Possibly this variation may be due to larger sample size (10 times more) used for ELISA compared with DIAPOPOS (Nielsen et al., 1998).

The direct RNA amplification procedure was used to amplify the transport protein gene of TMV in addition to the RT-PCR method. Both amplified cDNA products were restricted with NCo I or Hae III endonucleases and identical restriction fragments were formed. Hence Drygin et al. (1992) suggested that direct RNA-PCR may be adopted for the detection of RNA viruses. Direct RNA-PCR can be used to amplify the target fragment in total RNA extracts from leaves infected by viruses such as beet western yellows virus. The PCR products are then detected by staining with ethidium bromide after agarose gel electrophoresis. The limit of detection may be further increased by Southern blotting. This procedure is far more sensitive than ELISA and dot-blot hybridization and is able to distinguish beet western yellows virus and beet mild yellowing virus (Jones et al., 1991).

Rapid detection, identification, and differentiation of viruses in crop plants is a basic requirement for development of effective strategies for disease management. RT-PCR technique has been widely used as a tool to provide reliable data to achieve the goal of efficient management of crop diseases. Grapevine rupestris stem pitting (RSP) is a widerspread disease of unknown etiology. The graft transmissible agent associated with RSP was found to be a virus, very similar to apple stem pitting virus, and it may be present along with carla and potex viruses causing the RSP complex in grapevine. A specific RT-PCR–based protocol detected the virus in 60 of 62 sources known to be infected with RSP disease. This virus was designated grapevine rupestris stem pitting–associated virus (RSPaV-1) (Zhang et al., 1998). Further studies by Meng et al. (1999) showed that this virus (RSPaV-1) could be detected in from 85% to 100% of grapevine that had indexed positive for RSP, by using a specific set of primers. Thus RT-PCR has provided rapid and reliable results compared to biological indexing for the diagnosis of RSP disease complex. Groundnut (peanut) rosette disease is a complex caused by groundnut rosette assistor luteovirus (GRAV), groundnut rosette umbravirus (GRV) and its satellite RNA. These causative agents could be detected in plants and aphid vectors by using the total RNA extraction kit supplied by Qiagen and RT-PCR technique. Both GRV and GRAV were detected in single aphids (*Aphis craccivora*) exposed to either green or chlorotic rosette-infected groundnut plants. On the other hand, the satellite RNA could be amplified only when extracts from two or more aphids were used for testing (Naidu et al., 1998).

The identity of cucurbit yellow stunting disorder virus (CYSDV) and its relationship with lettuce infectious yellows virus (LIYV) were established by using a modified RT-PCR protocol with gel-extracted dsRNA templates. The sequence analysis of the amplified and cloned fragments of CYSDV RNA2 showed that the coat protein (CP) was in a contiguous gene arrangement similar to that of LIYV. These two viruses have been proposed as members of a new genus *Crinivirus* in Closteroviridae (family) (Livieratos et al., 1999). By using specific primers designed from RNA replicase cDNA sequence, the cowpea mottle carmovirus (CPMoV) was detected in newly acquired germplasm of *Vigna* spp. The RT-PCR was 10^5 times more sensitive than DAC-ELISA. Furthermore, no false negative reaction was observed in RT-PCR, as sometimes seen with ELISA tests (Gillaspie et al., 1999).

Differentiation of strains of viruses and classifying them into pathogroups have been possible based on the results of RT-PCR assays. Use of a specific primer pair (RS1 and RS2) resulted in the precise detection and identification of tobacco mosaic tobamoviruses rakkyo strain (TMV-R) infecting *Allium chinense* (Chen et al., 1996). RT-PCR–based assay helped to differentiate tospoviruses, tomato spotted wilt virus (TSWV), impatiens necrotic spot virus (INSV), tomato chlorotic spot virus (TCSV), and groundnut ringspot virus (GRSV) (Mumford et al., 1996; Dewey et al., 1996). The corky ringspot disease was shown to be a distinct isolate of tobacco rattle virus (TRV). Amplified cDNA was cloned and sequenced for, confirming the identity of this TRV isolate and determining the relationship of this isolate to other sequenced TRV isolates (Kawchuk et al., 1997).

Differentiation of strains of sugarcane mosaic virus (A, B, D, and E) and sorghum mosaic virus strains (SCH, SCI, and SCM) was possible by using group-specific primers for RT-PCR–based random fragment-length polymorphism (RFLP) analysis. The use of differential hosts producing characteristic symptom was dispensed with for the first time (Yang and Mirkov, 1997). For the differentiation of two variants (A and Vt6) of rice tungro spherical waikavirus (RTSV) specific oligonucleotides were used to amplify coat protein gene fragments of RTSV by RT-PCR. Variation between the two viral strains was confirmed at nucleotide positions 2556 and 3032 based on published sequence data. The RT-PCR technique may be useful to study the RTSV coat protein variations in natural field populations also (Yambao et al., 1998). The designed specific synthetic oligonucleotide primers were used to differentiate melon necrotic spot virus (MNSV) strains NK, NH, and S in the purified preparations as well as in melon plants inoculated with these strains. Among these three strains, MNSV-NK was widely distributed under natural conditions in Japan (Matsuo et al., 1998). The degenerate genus specific primers were designed to amplify the variable 5' terminal region of the potyvirus CP gene. Based on the molecular analyses of the amplified fragments, the Chinese strains of sweet potato feathery mottle virus, sweet potato latent virus, and sweet potato virus G were identified (Colnet et al., 1998).

A RT-PCR protocol in combination with restriction endonuclease analysis, based on the nucleotide sequences, was developed for the detection and differentiation of bean common mosaic potyvirus (BCMV) and bean common mosaic necrosis potyvirus (BCMNV) and their pathogroups (PGs). Two virus-specific primer pairs that could amplify a PCR product specific for each virus were designed. With RT-PCR procedure, four BCMV-PG-V isolates were differentiated from isolates of BCMV pathogroups PGI, II, IV, and VII. By using restriction enzyme *XbaI* digestion of BCMNV-PCR products, two BCMNV pathogroups, PGIII and PGIV, were distinguished. Thus, the combination of RT-PCR and restriction enzyme analyses, can be used to differentiate both BCMV and BCMNV, two pathotypes of BCMNV, and one pathogroup of BCMV from other potyviruses and pathogroups (Xu and Hampton, 1996). Similar approach was made by Marie-Jeanne et al. (2000) for the differentiation of four potyviruses, viz., maize dwarf mosaic virus (MDMV), sugarcane mosaic virus (SCMV), johnsongrass mosaic virus (JGMV), and sorghum mosaic virus (SrMV). Extraction of total RNA followed by RT-PCR steps yielded a 327-nucleotide fragment of capsid protein gene. Virus-specific patterns were obtained after enzymatic restriction of this fragment with two nucleases *AluI* and *DdeI*. It was possible to reliably identify well-characterized strains and field-collected isolates of potyviruses infecting maize.

The strain-differentiating oligonucleotides were designed by cloning and sequencing the 3' region of a necrotic strain of peanut stripe potyvirus (PStV-T) including a part of the N1b region, the complete coat protein (CP) gene, and the 3'-untranslated region. Nucleotide sequence differences unique to the necrotic strains were identified by comparing with the sequences of nonnecrotic isolates of

PStV. The necrotic strain was differentiated from nonnecrotic isolates, based on the nucleotide polymorphism existing in the CP gene sequences. The 3' end mismatch in nucleotides was the basis of differentiation of the strains of PStV and it can be utilized for the rapid, sensitive, and reliable detection of PStV strains (Pappu et al., 1998).

An attempt was made to correlate the molecular variability of prunus necrotic ringspot virus (PNRSV) isolates with their biological diversity. Differences in the electrophoretic mobility of the RNA4 transcripts of PNRSV synthesized from the corresponding polymerase chain reaction (PCR) products from six different PNRSV isolates were observed (Rosner et al., 1998). By using restriction-fragment-length polymorphisms (RFLPs), viral sequences amplified by PCR from 25 isolates of PNRSV, differing in the type of symptoms induced in six different *Prunus* spp., were analyzed. Most of the isolates could be differentiated by employing three different restriction enzymes, *EcoRI, Taq1*, or *RsaI*. All isolates clustered into three groups based on sequence comparison and phylogenetic analyses of RNA4 and coat proteins (CPs). However, no clear relationship between the type of symptom caused or host specificity and the molecular variability could be established (Aparicio et al., 1999). On the other hand, a relatively rare isolate of barley yellow dwarf virus (BYDV-PAV-DK1) could be differentiated based on the unique restriction enzyme profile obtained after digestion of the PCR products with *HaeIII* from the coat protein region. The RFLP profile of this isolate was significantly different from that of a laboratory isolate (BYDV-PAV-IL). Significant differences in the rates of transmission of these isolates by two of three biotypes of *Rhopalosiphum padi* were discernible, although no differences in symptom expression was observed. The unique restriction enzyme profile of BYDV-PAV-DK1 isolate may be used as a trackable characteristic in the epidemiological studies (Moon et al., 2000). RFLP analysis of DNA of rice tungro bacilliform virus (RTBV) was used to differentiate four strains G1, G2, Ic, and L. The strains G1 and Ic had identical restriction patterns, when the viral DNA was digested with endonucleases *PstI, Bam* HI, *Eco* RI, and *Eco* RV. However, they could be differentiated from strains G2 and L by digestion with *Eco* RI and *Eco* RV endonucleases, which could also be used for differentiating strain G2 from strain L. Identical restriction patterns for strains G2 and L were formed from the extracts of roots, leaves, and leaf sheaths of infected rice plants. The RFLP analysis may be useful to determine the variability of a large number of field samples (Cabauatan et al., 1998).

The mechanism of cross-protection of a plant species by one strain against another strain of a virus or closely related virus has been elucidated by employing RT-PCR assay. Sugarbeet plants are protected by beet soil–borne mosaic virus (BSBMV) against the closely related, but serologically distinct beet necrotic yellow vein mosaic virus (BNYVV). The presence of RNAs of both protecting and challenging viruses could be detected by RT-PCR assay, indicating that the RNA of challenging virus also was present in the protected beet plants. But the coat protein of the challenging virus was not detected by ELISA in the protected plants (Mahmood and Rush, 1999).

Rice tungro disease (RTD) is a complex caused by rice tungro bacilliform virus (RTBV), which induces the symptoms, and rice tungro spherical virus (RTSV), which assists the transmission of both viruses by green leafhoppers. Transgenic *japonica* rice plants expressing the full-length RTSV replicase (Rep) gene in the sense orientation were 100% resistant to RTSV even when challenged with high levels of virus inoculum. The accumulation of RTSV-RNA was low as determined by RT-PCR assay and the resistance was exhibited to geographically distinct RTSV isolates. The RTSV-resistant transgenic rice plants can be used to restrict the spread of RTD, since they are unable to assist the transmission of RTBV (Huet et al., 1999).

Another important application of RT-PCR is the determination of virus contents of vectors of plant viruses. A region of CP gene of grapevine fanleaf virus was detected by RT-PCR in nematode vectors by Esmenjand et al. (1994). The virus-specific SRNA of tomato spotted wilt virus could be detected in thrips vectors by RT-PCR technique (Tsuda et al., 1994). The RT-PCR protocol was used to quantify the potato leaf roll luteovirus (PLRV) in three aphid species, *Myzux persicae, Aphis nasturtii,* and *Macrosiphum euphorbiae,* collected from yellowpan traps. The RT-PCR bands specific for PLRV could be observed in all three vector species (Singh et al., 1997). A rapid and simple RT-PCR procedure was followed for the detection of citrus tristeza virus (CTV) in two vector aphid species, *Toxoptera citricida, Aphis gossypii,* and one nonvector aphid *Myzus persicae.* After acquisition access (24–48 hr), nucleic acid extracts from the aphids were reverse-transcribed and amplified using primers for CP gene of CTV [Florida B3 (T-36) isolate]. The amplified product from the aphids fed on citrus infected with B3 isolate was found to be the CTV-CP gene by digestion with restriction enzymes (Mehta et al., 1997).

The comparative efficacy of RT-PCR and ELISA tests for the detection of some plant viruses has been assessed. When the plate-trapped antigen (PTA) ELISA was used to determine the relationship between banana bract mosaic potyvirus (BBrMV) and other members of the family Potyviridae, weak serological signals were observed with many potyviruses. On the other hand, the examination of RT-PCR products from BBrMV and abaca mosaic virus (AbaMV) by Southern blot hybridization using virus-specific DIG-labeled DNA probes showed no cross-reaction (Thomas et al., 1997). For the detection of citrus tristeza virus (CTV), phloem-rich tissues such as petioles or midribes were the suitable tissues for ELISA tests, whereas the RT-PCR methods could detect CTV in all tissue types. Furthermore, RT-PCR methods can be employed even during the months when CTV titer drops below the level that is required for detection by ELISA (Mathews et al., 1997).

The sensitivity of RT-PCR assay is generally higher than that of ELISA tests, in addition to a higher level of reliability and rapidity. The detection limit of RT-PCR for the detection by *Cymbidium* mosaic potexvirus (CyMV) and *Odontoglossum* ringspot tobamovirus (ORSV) was 1000 times lower than that of ELISA tests indicating its greater sensitivity. RT-PCR was found to be more rapid, time-saving, and more reliable (Park et al., 1998). The detection limits of prunus necrotic ringspot ilarvirus (PNRSV) by RT-PCR, nonisotopic dot-blot hybridization, and DAS-ELISA were 1.28 pg/ml, 0.8 ng/ml, and 4 ng/ml, respectively, showing that sensitivity of the RT-PCR assay was much higher than that of the other two methods (Sanchez-Navarro et al., 1998). The prunus necrotic ringspot virus (PNRSV) was detected by RT-PCR directly in plant extracts from infected peach trees. When undiluted plant sap was used, PCR amplification did not occur, presumably owing to the presence of higher concentration of inhibitors of PCR. This situation could be overcome by diluting the sap (at least 1:50). This sap-dilution PCR can be easily adopted for large-scale testing. However, immunocapture RT-PCR, which is not affected by inhibitors, is more sensitive and particularly preferable when the virus titre of the plant tissues is low (Rosner et al., 1998).

Detection of viral pathogens becomes more sensitive when antibody binding and PCR are combined. These methods not only detect the presence of viruses but also may indicate their viability. In immunocapture PCR, as applied to plum pox virus, the virus is concentrated by using specific antibody; then the specific RNA sequence of the "captured" virus particles is amplified by PCR. The sensitivity of detection is 250 times that of direct PCR (Wetzel et al., 1992). A similar procedure was employed for the detection of bean yellow mosaic, cherry leaf roll, cucumber mosaic, citrus tristeza, grapevine fanleaf, potato leaf roll, pepper mild mottle, and tomato spotted wilt viruses and satellite RNA of CMV and PSTVd by Nolasco et al. (1993). Immuno-PCR is another highly sensitive procedure in which a DNA fragment is linked to an antigen-antibody complex, using protein A (linking to antibody) and streptavidin (attached to DNA). Protein A and streptavidin have strong affinity. This complex is then bound to a biotin-labeled DNA sequence, which is subsequently amplified by PCR (Appendix 9[iv]). The immuno-PCR technique is 10^5 times more sensitive than ELISA (Sano et al., 1992). Only antigen-specific antibody is required for immuno-PCR, whereas nucleic acid sequence information is also needed for the immunocapture technique in addition to specific antibody. These techniques have great potential for the rapid and reliable detection of pathogens.

Adoption of IC-RT-PCR technique has been reported to provide sensitive, reliable, and rapid detection of viruses such as raspberry bushy dwarf idaeovirus (RBDV) and zhucchini yellow mosaic potyvirus (ZYMV) (Barbara et al., 1995; Kokko et al., 1996), cacao swollen shoot virus 1A (Hoffmann et al., 1997), citrus tristeza virus (Nolasco et al., 1997), lettuce mosaic virus (Vlugt et al., 1997), yam mosaic potyvirus (Mumford and Seal, 1997; Revers et al., 1997), potato leafroll virus (Leone and Schoen, 1997), apple stem pitting virus (Jelkmann and Keim-Konard, 1997; Schwaz and Jelkmann, 1998), beet yellows virus (Stevens et al., 1997), tomato spotted wilt virus (Jain et al., 1998), apple stem grooving virus (Crossley et al., 1998), plum pox virus (Varveri and Boutsika, 1998; Adams et al., 1999), black currant reversion-associated virus (Lemmeitty et al., 1998), prunus necrotic ringspot virus (Rosner et al., 1998), potato tuber necrotic ringspot isolate of PVY (Tomassoli et al., 1998), and banana streak virus (Harper et al., 1999).

The IC-RT-PCR procedure followed for the detection of lettuce mosaic virus (LMV) consists of trapping the virions on the wall of an antiserum (specific for LMV)-coated tube followed by washing to remove inhibitory substances of plant origin. Using the viral RNA as template, reverse transcriptase was employed for the synthesis of cDNA, which was then amplified in PCR with virus-specific primers. The PCR product was analyzed by electrophoresis in agarose gel. The IC-RT-PCR assay was about 1000 times more sensitive than ELISA and it could detect small concentrations of LMV in leaf extracts (Vlugt et al., 1997). An IC-RT-PCR capable of detecting two yam potyviruses in the same format was developed by Mumford and Seal (1997). The test was performed in a single tube using immunocapture and a single buffer RT-PCR for the detection of yam mosaic potyvirus and yam mild mosaic virus infecting yams. This IC-RT-PCR format was 100-fold more sensitive than ELISA tests.

Another format developed for the detection of potato leafroll virus (PLRV) in potato tubers consists of an immunocapture of PLRV coupled to a one-tube RT-PCR using *Thermus thermophilus* (Tth) instead of *Taq* DNA polymerase. The important advantage of this method is the reduction of inspection time of seed potatoes for infection of PLRV to just 1 day from 5 weeks required for conventional testing (Leone and Schoen, 1997). The IC-RT-PCR protocol performed in the same tube was used for the detection of tomato spotted wilt tospovirus in a wide range of host plants including tomato, pepper, tobacco, and impatients (Jain et al., 1998).

The IC-RT-PCR assay can be performed in microplate wells also. For the detection of raspberry bushy dwarf idaeovirus (RDBV), the virus particles are captured by antibodies coated in the PCR microplate wells for enrichment, followed by lysis of virus particles and RT-PCR of the viral RNA. The reaction mixtures containing reverse transcriptase and DNA polymerase are used for lysis and amplification in a single step of four fragments of RNA3 of RBDV with different combinations of four primers. This procedure was found to be highly sensitive in detecting RBDV in in vitro–cultured plants in which detection of RBDV by conventional immunological procedures is difficult or even impossible (Kokko et al., 1996). Oligonucleotide primers derived from the flanking regions of the putative CP gene of cacao swollen shoot badnavirus (CSSV) isolate 1A were employed to amplify products directly from extracts of leaves infected by CSSV using IC-RT-PCR. This technique was at least 100 times more sensitive than DAS-ELISA and specifically detected the CSSV isolate 1A (Hoffmann et al., 1997).

The quick-decline strain of citrus triteza virus (CTV) from Florida was detected by the IC-RT-PCR format that combines the simplicity of ELISA. The sequence of the CP gene of the CTV strain was used for designing the primers and fluorescent probes. After PCR amplification, the electrophoretic analysis of the results was substituted for measurements of the fluorescence. This assay procedure detected the CTV strains not only in all ELISA positive samples, but also in an additional 20% samples that were ELISA negative, revealing the higher sensitivity level of this technique (Nolasco et al., 1997). The fusion protein-specific antiserum raised against viral coat protein expressed in *Escherichia coli* was employed to enhance the sensitivity of detection of apple stem pitting virus in woody plant tissue by IC-RT-PCR assay. The apple stem pitting virus and pear vein yellows virus could be detected by this procedure (Jelkmann and Keim-Konard, 1997) (Fig. 9.7).

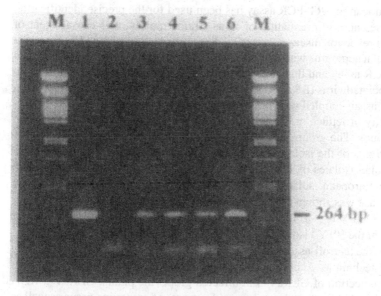

Figure 9.7 Agarose gel electrophoresis of IC-RT-PCR products to detect a 264-bp product specific for apple stem pitting virus (ASPV). Lane: M, lambda *Pst* 1 cut DNA marker; 1, p. 206 plasmid control; 2, water control; 3, ASPV from *Nicotiana occidentalis;* 4, PVY isolate from *N. occidentalis;* 5, ASPV from "Golden Delicious" apple; 6, PVY isolate from "Williams" pear. (Blackwell Wissenchafts-Verlag, Berlin, Germany.) (Courtesy of Jelkmann and Keim-Konrad, 1997.)

The usefulness of IC-RT-PCR technique for large-scale application for the detection of plum pox potyvirus (PPV) under field conditions was demonstrated by Varveri and Boutsika (1998). The virus was detected in 31% of ELISA-negative leaf samples and in 23% of ELISA-doubtful trees. Additionally, by analyzing PCR products by RFLP technique, the majority of the isolates were found to be M serotype of PPV. Although IC-PCR assay is, generally, more sensitive (about 1000 times) than ELISA, its superiority could not be observed when aerial plant tissues were tested for PPV in plum trees, possibly owing to irregular distribution of PPV in shoots. However, when root samples from infected trees were tested, IC-PCR detected PPV in 92–100% of samples as against 38–65% by ELISA, indicating that bulked samples from fibrous roots were the suitable tissues for PPV detection (Table 9.2) (Adams et al., 1999). For the detection of prunus necrotic ringspot virus (PNRSV), use of avian myeloblastosis virus reverse transcriptase and incubation at 46°C for RTase reaction resulted in higher levels of amplified products compared to incubation at 37°C. Further improvement in PCR yields was obtained by preheating the reaction mixture at 55°C for 5 min. These modifications in the procedure increased the sensitivity of the IC-RT-PCR assay (Rosner et al., 1998).

Immunocapture-RT-PCR assay has been used for the precise identification of new viruses and differentiation of virus strains rapidly. The causal agent of black current reversion disease is yet to be determined conclusively. However, the association of a nepovirus with the black currant reversion disease was established by IC-RT-PCR assay and this virus has been tentatively named black currant reversion-associated virus (BRAV) (Lemmetty et al., 1998). The data obtained from IC-RT-PCR assay coupled with direct sequencing were used to study the molecular variability of lettuce mosaic virus (LMV) isolates and to classify them into different groups. The groups of LMV isolates seemed to correlate with the geographical origins of the isolates rather than with their virulence (pathogenicity). The Californian isolates of LMV showed significant sequence similarities with the Western European isolates (Revers et al., 1997). The potato tuber necrotic ringspot disease was shown to be caused by a tuber necrotic isolate of potato potyvirus (PVY) (PVY[NTN]). IC-RT-PCR technique was used to differentiate the PVY[NTN] from the PVY[N] isolate (Tomassoli et al., 1998).

Grapevine leafroll-associated viruses (GLRaV) have been detected by using IC-RT-PCR technique. Acheche et al. (1999) developed the IC-RT-PCR procedure for the detection of GLRaV-3 in infected grapevine plants. GLRaV-3-RNA could be detected in total crude nucleic acid extracts of grapevine tissue as well as in the extracts of the mealy bug vector (*Planococcus ficus*). This technique was found to be specific, sensitive, and rapid with no background interference. Sefc et al. (2000) designed specific primers based on the sequence from the conserved HSP70 region of closteroviruses for the detection of GLRaV-1. The specificity of the primers was indicated by the failure of amplification of any PCR product from

GLRaV-2, 3, and 4 viruses in infected plants. This procedure was about 125 times more sensitive than ELISA test. The effectiveness of IC-RT-PCR assay for the detection of viruses in peanut seed lot was reported by Gillaspie et al. (2000). This technique could be applied to detect peanut stripe virus (PStV) and peanut mottle virus (PeMV) in the extracts of small slices removed from each seed distal to the radicle. The IC-RT-PCR method was more sensitive than ELISA currently used for virus detection and it has the potential for testing large number of seed lots of peanut germplasm.

The RT-PCR assay, though sensitive, requires extensive manipulation of each sample prior to the RT-PCR reactions and electrophoretic or blot hybridization analysis of PCR products. The immunocapture technique reduced the complexity of sample preparation for RT-PCR to some extent. The development of a colorimetric RT-PCR, which combines the sensitivity of PCR and convenience of ELISA, marked the progress toward the simplification of detection procedure without losing the sensitivity. PCR-ELISA test was applied for the simultaneous detection and typing of two serotypes D and M of plum pox potyvirus (PPV) more efficiently when compared to IC-PCR (Olmos et al., 1997). By using strain-specific capture probes, the two major serotypes could be differentiated without subsequent random-fragment-length polymorphism (RFLP) analysis of PCR products (Pollini et al., 1997). The apple stem grooving capillovirus (ASGV) was detected by directly labeling the RT-PCR-amplified product with DIG. Alternatively, the unlabeled amplification products were hybridized to an internal DIG-labeled detection probe. A biotin-labeled probe was employed for both systems for trapping the amplification products on streptavidin-coated microplate strips. The alkaline phosphatase–labeled anti-DIG conjugate was employed for visualization and determination of color intensity following the reaction between substrate and enzyme conjugate. The detection of ASGV in apple tissues by colorimetric PCR assay showed high level of sensitivity (Daniels et al., 1998).

The effectiveness of colorimetric-RT-PCR assay was further demonstrated for the detection of viruses infecting woody plants. A simplified colorimetric assay for virus-specific RT-PCR products, which eliminates the requirement of electrophoretic analysis, was developed by Rowhani et al. (1998). The colorimetric detection, when combined with immunocapture of virions directly from plant extracts, enlarges the scope of applying PCR technique to a large number of samples and also provides the quantitative determination of virus titer in plant samples. The colorimetric PCR was employed for the detection and quantitation of a walnut isolate of cherry leaf roll virus (CLRV-W), citrus tristeza virus (CTV), prune dwarf virus (PDV), prunus necrotic ringspot virus (PNRSV), and tomato ringspot virus (ToRSV) in woody and herbaceous plants [Appendix 9(v)].

A PCR-ELISA procedure was developed for the simultaneous detection and identification of prunus necrotic ringspot virus (PNRSV) and apple mosaic virus

(ApMV). Multiple alignments of PNRSV and ApMV CP gene sequences were made to select PCR primers in the regions conserved between two viruses. Virus-specific oligonucleotides were used as capture probes in the PCR-ELISA assay. The selected primer pairs allowed detection of both PNRSV and ApMV as revealed by the analysis of a range of isolates of both viruses. Simultaneous use of the specific capture probes of both viruses resulted in the enhancement of sensitivity of detection of both viruses by the PCR-ELISA protocol developed by Candresse et al. (1998).

In the attempt to further simplify the RT-PCR assay, the print or spot-capture PCR (PC-PCR) has been developed for the detection of viruses in plant hosts and their vectors. The print capture RT-PCR, analogous to tissue blot immunoassay (TBIA), was found to be simple, rapid, and sensitive for the detection of plum pox virus (PPV) (Olmos et al., 1996). However, Varveri and Boutsika (1998) found this simplified version to be effective for detection of PPV in glasshouse samples, but less effective with field samples. The PC-PCR assay was employed for efficient detection of PPV, apple chlorotic leaf spot trichovirus (ACLSV), prunus necrotic ringspot ilarvirus (PNRSV), and apple mosaic ilarvirus (ApMV) in tissue imprints of different stone fruit species. It is possible to store the immobilized target nucleic acids on imprinted papers that can be mailed to testing centers for amplification by PCR. The PC-PCR, in conjunction with heminested print-PCR and PCR-ELISA, could be used for simultaneous detection and typing of PPV isolates to serotypes M or D in plant tissues as well as in individual aphid (*Aphis gossypii*) vectors (Cambra et al., 1998).

The print capture-PCR (PC-PCR) has been applied successfully for the detection of geminiviruses in plants and whitefly vectors. This technique allows direct amplification of viral DNA from infected plant or whitefly tissues printed directly on Whatman 3MM paper. This format thus eliminates the need for grinding the tissues and incubation or washing steps prior to the amplification step, in addition to the absence of cross-contamination between samples that may be possible with standard PCR procedure. The PC-PCR assay was used for the detection and differentiation of tomato yellow leafcurl bigeminivirus (TYLCV) isolates TYLCV-Sv and TYLCV-Is (Navas-Castillo et al., 1998). In another study by Atzmon et al. (1998), squashes of tissues of tomato plants infected by TYLCV and viruliferous whiteflies (*Bemia tabaci*) were applied onto a nylon membrane stripe (1 × 2 mm) that was introduced into PCR reaction mixture. The PCR products were electrophoresed in gel, blotted, and hybridized with radiolabeled virus-specific DNA probe. TYLCV was detected in leaves, roots, and stems of infected tomato plants and individual viruliferous whitefly. TYLCV could be detected at the site of inoculation just 5 min after inoculation feeding by whitefly in some plants and all inoculated tomato plants had detectable concentrations of TYLCV at 30 min after inoculation feeding. TYLCV could be detected in the head, thorax, and abdomen of *B. tabaci* at 5, 10, and 25 min after acquisition feeding on infected tomato plants (Atzmon et al., 1998). This technique has been used to monitor the movement of TYLCV in the whitefly vector.

The multiplex RT-PCR technique can be applied to amplify multiple viruses or viroids from a single sample in single reactions. In this technique, multiple primer sets, instead of one primer pair in standard PCR or RT-PCR assay, at 100 pmol of complementary primer for cDNA and 20 pmol each of the primer sets, are used. It is possible that formation of PCR product by one template or another may be favoured in the multiplex RT-PCR. Use of dimethylsulfoxide (DMSO) may prevent this condition. This technique was employed for the detection of grapevine leafroll-associated virus III and grapevine virus B (Minafra et al., 1993; Hadidi et al., 1995). The conventional RT-PCR and multiplex RT-PCR techniques were tested for their efficacy in detecting prunus necrotic ringspot virus and plumpox virus. Both tests gave similar results for the detection of single or multiple infection by these viruses. Multiplex RT-PCR may be used for testing a limited number of samples to verify the health status of plant materials, especially when ELISA tests do not provide reliable results (Kölber et al., 1998). Detection and discrimination of closely related viruses such as wheat spindle streak mosaic virus (WSSMV) and wheat yellow mosaic virus (WYMV) could be achieved by employing a multiplex RT-PCR protocol. By using specific primers designed from coat protein gene sequence (834–837 nucleotides), WSSMV and WYMV could be detected and identified. No PCR product was produced with soil-borne wheat mosaic virus, which is commonly associated with both diseases infecting wheat (Clover and Henry, 1999).

9.4.1.1 Single-strand Conformation Polymorphism Analysis

Orita et al. (1989) reported that single-strand conformation polymorphisms (SSCP) could be used as the reliable alternative for the detection of differences in the genomic DNA. The + and − strands of a ds DNA, if separated, become metastable sequence-specific folded structures with distinct electrophoretic mobilities in nondenaturating polyacrylamide gels. Under such conditions, it is possible to detect even single nucleotide exchanges. Koening et al. (1995) employed SSCP analysis as a tool for rapidly assigning large numbers of beet necrotic yellow vein virus (BNYVV) isolates to strain groups A, B, or P and for detecting mixed infections, minor variants, or new strain groups (Table 9.3). The SSCP analyses are much less time-consuming than RFLP analysis and may be particularly useful for differentiating serologically indistinguishable strains of viruses.

Application of SSCP analysis for the differentiation of citrus tristeza closterovirus (CTV) isolates has been reported to be a rapid and inexpensive method, as it is not necessary to have full sequence data for comparison. The CTV isolates from diverse geographical areas were compared for differences in their coat protein (CP) gene. The CP genes of 17 isolates of CTV were reverse-transcribed, am-

Table 9.2 Comparative Efficacy of Detection of Plumpox
Potyvirus (PPV) in Different Tissues of Dormant Plum Trees
by ELISA and IC-PCR Technique

	Percent detection	
Nature of Samples	ELISA	IC-PCR
Bark	71–80	85–86
Shoots (1 year old)	66–81	81–87
Fibrous roots	38–65	92–100%

Source: Adams et al., 1999.

plified by PCR, and cloned. The sequences of the clones showed between 91.7%
and 99.8% sequence homology. After amplificaton of clones and denaturation of
PCR products, they were compared by SSCP analysis in 8% polyacrylamide gels.
The patterns of 16 or 17 clones showed variations under two different electrophoretic conditions. The SSCP analysis with combination of two electrophoretic conditions and restriction of eight clones with *Eco* 9II allowed discrimination between 21 of 22 CP gene clones selected for comparison. This study
shows that SSCP analysis may provide a sound basis for identification and differentiation of several genes or gene regions (Rubio et al., 1996).

The SSCP analysis of a nested asymmetrical PCR product was applied to
characterize the population of sequence variants of apple stem grooving
capillovirus (ASGV). Two to four bands in the PCR products from ASGV-in-
fected apple, Japanese pear, or European pear trees were detected, indicating that
ASGV existed in the form of a mixture of sequence variants. Each sequence variant seems to be distributed irregularly within a tree as reflected by the composition of sequence variants with differing number of bands and their relative quantity as assessed by SSCP analysis. After a serial passage of ASGV isolates in
Chenopodium quinoa plants, the composition of sequence variants was altered as
revealed by SSCP analysis. Some sequence variants dominated, while others were
reduced to undetectable levels, possibly owing to selection pressure applied by the
host plant, *C. quinoa* (Magome et al., 1999).

9.4.2   Detection of Viroids

9.4.2.1   Reverse transcription–polymerase chain reaction

Reverse transcription–polymerase chain reaction (RT-PCR) assay has been successfully applied for detection, differentiation, and characterization of viroids that
cause serious crop diseases. By using the RT-PCR technique, citrus exocortis,
cachexia and citrus viroid IIA (Yang et al., 1992; Saito et al., 1995; Turturo et al.,

1998), potato spindle tuber viroid (Shamloul et al., 1997), peach latent mosaic viroid (Shamloul et al., 1995; El-Dougdoug, 1998), blueberry mosaic viroid (Zhu et al., 1995), chrysanthemum stunt viroid (Shiwaku et al., 1997; Weidemann and Buchta, 1998), avocado sunblotch viroid (Mathews et al., 1997; Schnell et al., 1997), hop stunt viroid (Puchta and Sanger, 1989; Nakahara et al., 1999), apple scar skin and pear rusty skin viroids (Hadidi and Yang, 1990; Osaki et al., 1996), grapevine viroids (Rezaian et al., 1992; Wah and Symons, 1997), coconut tinangaja viroid and coconut cadang-cagang viroid (Hodgson et al., 1998), and hop latent viroid (Nakahara et al., 1999) have been detected in their respective plant hosts.

Peach latent mosaic viroid (PLMVd) was detected by using DNA primers for cDNA synthesis and PCR amplification of a full-length viroid DNA product from extracts of PLMVd-infected peach tissue. Amplified viroid cDNA hybridized to ^{32}P-labeled PLMVd cRNA probe. PLMVd could be detected from extracts of infected peach fruits, leaf, and bark tissues. Assessment of PLMVd infection of peach germplasm revealed the world wide distribution of PLMVd. (Shamloul et al., 1995). The viroid causing disease in apple shared similarities in sequences with PLMVd as revealed by RT-PCR and dot-blot hybridization analysis of the amplified product with PLMVd-cRNA probe. (El-Dougdoug, 1998).

The total nucleic acids (TNA) were extracted from citrus leaves, and using CEVd-specific primers, TNA was amplified by RT-PCR. The DIG-labeled probe prepared by PCR was employed for detecting CEVd in all CEVd-infected samples as well as in two citrus samples that were found CEVd negative by indexing on Etrogcitron Arizona 861-S1 as indicator plants. The biological assays and electrophoresis fail to differentiate viroid species and also need long periods for viroid detection (Saito et al., 1995). Potato spindle tuber viroid (PSTVd) was detected using RT-PCR from TNA or Gene Releaser™–treated extracts of tree potato seeds and pollen. It was possible to obtain an amplified full-length PSTVd cDNA by treating as few as five pollen grains with Gene Releaser™ (Shamloul et al., 1997). Detection of avocado by sunblotch viroid (ASBVd) by RT-PCR assay in avocado trees that biologically indexed either positive or negative provided reliable results rapidly (Mathews et al., 1997). Furthermore, the RT-PCR assay requires only small amounts of tissue, and does not require either highly purified ASBVd or molecular hybridization (Schnell et al., 1997).

The detection of five viroids infecting grapevines was possible by applying a modified RNA extraction procedure and a high-sensitivity RT-PCR assay. The DNA primers were carefully selected for optimization for viroids in low copy number. This protocol was particularly effective for the detection of grapevine viroids in vines regenerated by shoot apical meristem culture (SAMC) and fragmented shoot apex culture (FSAC) (Wah and Symons, 1997). A rapid and simple procedure for preparing RNA templates for single-tube RT-PCR amplification of citrus exocortis viroid (CEVd) and citrus cachexia viroid (CCaVd) was applied.

The total nucleic acids (TNA) extracts from infected plants or viroidal template released by directly boiling small tissue pieces were used for conducting RT-PCR assay. The TNA (0.15 ng) from infected plants spotted on nylon N+ membrane and released in glycine-NaCl buffer was successfully used in single-tube amplification of these viroids. The amplification efficiency was enhanced by wetting nylon membrane in NaOH-EDTA solution prior to amplification (Turturo et al., 1998).

The extraction of nucleic acids from coconut leaves for viroid detection was performed by a special method [Appendix 9(vi)]. The identity of coconut tianangaja viroid (CtiVd) established by two-dimensional polyacrylamide gel electrophoresis was confirmed by RT-PCR assay with the oligonucleotide probe as one of the two PCR primers or by diagnostic oligonucleotide probe (DOP) hybridization assay. RT-PCR was not found to be significantly more sensitive than DOP hybridization assay for the detection of CTiVd in coconut leaf extracts. Coconut cadang-cadang viroid (CCCVd) and CtiVd could be differentiated by employing viroid-specific oligoncleotide probes (Hodgson et al., 1998). Another simple and rapid method of nucleic acid extraction developed consists of liberation of nucleic acids from citrus plant tissues by incubation in a buffer containing potassium ethylxanthogenate (PEX) without tissue homogenization followed by precipitation with ethanol (NA-PEX). Hop stunt viroid (HSVd), hop latent viroid (HLVd), and PSTVd were detected in NA-PEX by RT-PCR, whereas four citrus viroids, HSVd, HLVd, and PSTVd could be detected by hybridization (Nakahara et al., 1999).

A method of tissue printing followed by RT-PCR involving immobilization of plant extract onto filter paper was developed for the detection of PSTVd in primarily and secondarily infected potato plants, primarily infected in vitro plants, and potato tubers. This print-PCR protocol simplifies the processing of samples and it is suitable for testing large number potato plants. The dot-PCR, involving dotting of plant sap on filter paper, can be used for testing in vitro potato plantlets and tuber tissues for indexing. Reliable results can be obtained even when one infected plant is bulked with a maximum of nine healthy plants for the detection of secondary infection. The dotted or printed filter paper squares can be stored for at least 2 weeks in TritonX-100 at 4°C under dry conditions (Weidemann and Buchta, 1998).

9.4.2.2   Single-Strand Conformation Polymorphism (SSCP) Analysis

The variations in the sequences of field isolates of citrus exocortis viroid (CEVd) were determined by single-strand conformation polymorphism (SSCP) analysis. Shifts in the migration of the cDNA and/or hDNA strands of 311 cloned full-length CEVd DNA inserts were observed when electrophoresis in nondenaturing

Table 9.3 Differentiation of Beet Necrotic Yellow Vein Virus Strains by SSCP Analysis

BNYVV RNA	Amplified region (nt)	Size of PCR product	No. nt exchanges between published sequences for A- and B- types	Differentiation between A- and B- types
1	6150–6651	501	Unknown	Excellent[a]
2 (Triple gene	2480–3241	761	30	Excellent
block region)	2711–3398	687	23	Excellent
	2811–3108	297	12	Excellent
	2811–3241	430	19	Excellent[a]
	2811–3398	587	21	Excellent
	2950–3398	448	15	Excellent
3	409–1268	859	Unknown	Excellent[a]
	911–1268	357	Unknown	Excellent
4	699–1301	602	Unknown	Excellent[a]

[a] Recommended for screening tests.
Source: Koenig et al. (1995).

14% polyacrylamide gels was performed. Seven different groups of variants showing one to six changes that did not reflect the overall variability among the CEVd clones were recognized. The sequence analysis confirmed the relationship between the different SSCP profiles exhibited by CEVd clones and the variation in their nucleotide changes. The SSCP analysis could detect additional single nucleotide variations among clones that initially clustered together. Furthermore, the viroid region affected by specific changes can also be revealed by SSCP analysis of partial-viroid-length DNA (Palacio and Duran-Vila, 1999).

9.5 DETECTION OF PHYTOPLASMAS

9.5.1 Dot-Blot Hybridization Assay

Phytoplasma specific DNA probes have been prepared from chromosomal or plasmid (extrachromosomal) DNA of the phytoplasma pathogens. These probes are usually labeled with either radioactive ^{32}P or nonradioactive biotin. Recently another nonradioactive substance, digoxigenin, has been used to label probes specific for sweet potato witches'-broom phytoplasma (Ko and Lin, 1994). Specific tissues such as phloem sieve tube elements or hemolymph or salivary glands of vector insects may be used to extract the phytoplasma DNA, since the concentration of phytoplasmas is higher in such tissues (Kirkpatrick et al., 1987; Davis et al., 1988; Kirkpatrick, 1989). The phytoplasmas are usually detected by extracting the DNA from infected plants or inoculative insects and using a specific DNA probe in a dot-blot hybridization.

Dot-blot hybridization is useful to detect, differentiate, and quantify nonculturable phytoplasmas infecting plants (Appendix 9[vii]). Kirkpatrick et al. (1987) employed cloned DNA probes successfully in dot-blot assays to detect western X phytoplasma in infected plants and leaf hoppers (Appendix 9[viii]). The phytoplasmas that cause maize bushy stunt disease (Davis et al., 1988), aster yellows disease (Lee and Davis, 1988; Kulske and Kirkpatrick, 1992), apple proliferation disease (Bonnet et al., 1990), periwinkle little leaf disease (Davis et al., 1990), chrysanthemum yellows disease (Bertaccini et al., 1990), clover proliferation disease (Deng and Hiruki, 1990), and palms lethal yellowing disease (Harrison et al., 1992) were also detected by this technique (Fig. 9.8). Nakashima et al., (1993) reported that rice yellow dwarf phytoplasma could be differentiated from rice orange leaf phytoplasma by using specific chromosomal and extra chromosomal DNA probes. Davis et al. (1988) reported that biotinylated DNA probes could be used to detect alfalfa witches'-broom, clover yellow edge, X-disease, clover phyllody, eastern and western aster yellows phytoplasmas in different host

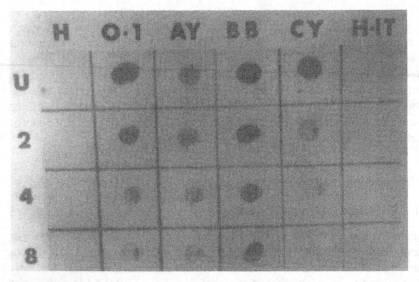

Figure 9.8 Dot hybridization of biotinylated cloned DNA probe to nucleic acid preparations extracted from healthy and phytoplasma-infected *Catharanthus roseus* plants: H, healthy; H-IT, healthy Italy; CY, chrysanthemum yellows; 0–1, periwinkle little leaf; AY, aster yellows; BB, tomato big bud; U, undiluted; 2, 4, and 8, reciprocals of dilution (American Phytopathological Society, Minnesota, USA). (Courtesy of Bertaccini et al., 1990.)

plants, and aster yellows in the leafhopper vector *Macrosteles fascifrons*. By the dot-blot hybridization procedure, the presence of ash yellows phytoplasma was detected in the innermost phloem at the trunk base, roots, twigs, and leaves of white ash trees. The DNA probe detected phytoplasmas as consistently as the fluorescence dye DAPI (4,6-diamidino-2, phenyl indole 2 HCl) fluorescence test (Sinclair et al., 1992; Davis et al., 1992). The presence of phytoplasmas in plant tissues micropropagated in vitro could be detected by dot hybridization (Bertaccini et al., 1992).

Phytoplasma-specific probes may be used to determine the host range of the phytoplasma and latent infections may also be recognized. The presence of phytoplasma causing lethal yellowing (LY) was detected in true date (*Phoenix dactylifera*), cliff date (*P. rupicola*), chinese fan (*Livistona chinensis*), and five coconut palm cultivars. *Caryota rumphiana* and *L. rotundifolia*, two palm species, which were not known to be infected by lethal yellowing, also revealed the presence of LY phytoplasma (Harrison et al., 1992).

The distribution and multiplication of the phytoplasma in the infected plant can be studied by using labeled phytoplasma-specific DNA probes, as in the case of aster yellows phytoplasma in periwinkle plants. The phytoplasma moved from grafted shoots into ungrafted shoots and then systemically throughout the plant. Distribution and concentration of the phytoplasma were directly correlated with expression of virescence and proliferation symptoms, and the concentration was maximum in symptomatic, actively growing shoots (Kulske and Kirkpatrick, 1992). Clover proliferation (CP) and potato witches'-broom (PWB) phytoplasmas were first detected in the external primary pholem tissues of periwinkle and then in the secondary phloem elements. The phytoplasmas later spread into the internal phloem tissue (Hiruki and Deng, 1992).

The extent of relationship between phytoplasmas can be reliably determined by dot-blot hybridization based on the amount of probes hybridizing with the target DNA. Probes specific to sweet potato witches'-broom phytoplasma hybridized with serologically related peanut witches'-broom phytoplasma. On the basis of differences in the band patterns formed in Southern blots, sweet potato witches'-broom phytoplasma could be differentiated from peanut witches'-broom phytoplasma (Ko and Lin, 1994). The relationship between the phytoplasmas affecting six species of ash (*Fraxinus*) and lilac (*Syringa*) was studied by Griffiths et al. (1994). Dot-blot hybridization with clover proliferation phytoplasma DNA probes showed that there was cross-hybridization with nucleic acid from potato witches'-broom phytoplasma infected periwinkle, but not with the nucleic acid from plants infected by either western aster yellows phytoplasma or clover phyllody phytoplasma. This result indicates that phytoplasmas that cause clover phyllody and proliferation are distinct and unrelated (Deng and Hiruki, 1990a).

Patterns of hybridization using DNA probes labeled with ^{32}P or biotin indicated that aster yellows phytoplasma, orchard phytoplasma, tomato big bud phytoplasma, and blueberry stunt disease phytoplasma belonged to the cluster of strains that share greater nucleotide sequence homology with one another than with other phytoplasmas tested. Kulske et al. (1991a) observed that the native ^{32}P-labeled plasmid isolated from the severe western aster yellows (SAY) strain phytoplasma hybridized with small extrachromosomal DNA molecules present in many virescence-inducing phytoplasmas and the maize bushy stunt (MBS) phytoplasma but not with DNA from decline-inducing phytoplasmas or spiroplasmas.

Lee and Davis (1988) constructed a ^{32}P-labeled single-stranded RNA probe (riboprobe) with plasmid vector pS64. This riboprobe was more sensitive and reliable than cDNA probe in detecting western X phytoplasma. At higher concentrations of cDNA probe, a nonspecific hybridization signal was observed with nucleic acid from healthy plants and from plants infected by other phytoplasmas. On the other hand, sensitivity of detection with complementary riboprobe was increased at higher concentration. Lee and Davis (1988) and Davis et al. (1988, 1990a), using ^{32}P-labeled riboprobes, showed that aster yellows (AY)–related phytoplasma strain cluster could be recognized and that these probes distinguished the strains of this cluster.

As the phytoplasmas are usually found in low concentrations in infected plants, the sensitivity of detection has to be improved. Many phytoplasmas contain extrachromosomal DNA (Denes and Sinha, 1991), and the probes for extrachromosomal DNA usually give stronger hybridization signals than chromosomal DNA probes. It is possible that the higher sensitivity may be due to the presence of multiple copies of extrachromosomal DNA in phytoplasma cells. The amount of plant tissue required for detecting maize bushy stunt phytoplasma using extrachromosomal DNA probe was only 0.02 g of plant tissue, whereas 0.3 g of tissue was necessary for detection when chromosomal DNA probe was used (Davis et al., 1988). Depending on the type of host tissue, about 15–30 ng of phytoplasma DNA may be required for detection of phytoplasmas by dot-blot hybridization (Goodwin and Nassuth, 1993).

Use of nonradioactive labels has distinct advantages, as in serological assays (Chapter 8). Davis et al. (1990a, 1990b) employed biotinylated cloned DNA probes for detecting aster yellows (AY) phytoplasma in infected plants and in the leafhopper vector *Macrosteles fascifrons*. Sinclair et al. (1992) found the hybridization signals, when biotin-labeled cloned DNA probe for ash yellows phytoplasma was used, to be most consistent and intense with samples from innermost phloem at the trunk base of infected plants, indicating the possibly high concentration of the phytoplasma in that tissue. Davis et al. (1992) showed that ash yellows phytoplasma could be detected in leaves, twigs, trunk phloem, and roots of white ash trees and that this represented a distinct strain cluster. Using digoxigenin-labeled DNA probes, the sweet potato witches'-broom phytoplasma was detected in infected sweet potato and periwinkle plants. The majority of the probes tested hybridized with serologically related peanut witches'-broom phytoplasma.

The probes could detect sweet potato witches'-broom phytoplasma DNA at 10 ng and 0.39 ng of DNA from periwinkle and sweet potato, respectively (Ko and Lin, 1994), indicating that the phytoplasma may reach a higher concentration in sweet potato than in periwinkle. The restriction enzyme digests of the extra chromosomal DNAs of rice yellow dwarf phytoplasma and sugarcane white leaf phytoplasma exhibited polymorphism among the isolates collected within a single field. Hybridization tests revealed significant homology in the sequences of the extra chromosomal DNAs of these phytoplasmas indicating the relatedness of rice yellow dwarf and sugarcane white leaf phytoplasmas. In contrast, little homology existed with those of sesame phyllody phytoplasma and aster yellows type phytoplasma (Nakashima and Hayashi, 1995).

9.5.2 Polymerase Chain Reaction

Polymerase chain reaction (PCR) can be successfully used to detect and differentiate phytoplasmas even when they are in very low concentrations. It can be employed to develop a specific assay for phytoplasma detection. A cloned fragment of a plasmid from the phytoplasma is sequenced to identify oligonucleotide primers for PCR. Amplified DNA fragments of the predicted size are then obtained from the DNA extracted from plants or insects infected by the phytoplasma, whereas no amplification occurs in healthy plant or insect DNA. The PCR-based assays are over 500 times more sensitive than a hybridization-based assay, as observed in the case of aster yellows phytoplasma (Goodwin et al., 1994). Deng and Hiruki (1990a) synthesized two PCR primer pairs to amplify specifically the two clover proliferation (CP) phytoplasma DNA fragments from crude nucleic acids containing CP phytoplasma DNA and host plant DNA. Using 5' end-labeled sequence-specific internal probes, PCR products were identified by liquid hybridization. To detect CP phytoplasma only about 2.5×10^{-8} to 2×10^{-5} ng nucleic acid was required, whereas 2.5 ng nucleic acid was necessary when PCR was not used.

A simple and efficient procedure for extraction of high-quality DNA from phytoplasma-infected woody and herbaceous plants for PCR assays was developed by Green et al. (1999). Commercially available microspin-column matrices are employed in place of phenol, chloroform, or alcohol conventionally used for precipitation of nucleic acids. The total DNA could be extracted in less than 1 hr. The method was effectively employed for the purification of rDNA from different types of tissues such as whole leaves, petioles, midribs, roots, and dormant buds from various plants. The PCR assay was employed for the detection of phytoplasmas causing pear decline, Western X-disease, peach yellow leaf roll, peach rosette, apple proliferation, Australian grapevine yellows, and *Vaccinium* witches' broom diseases. This procedure may be particularly effective in the case of phytoplasmas infecting woody plants (Green et al., 1999).

Application of PCR assay for the detection of several phytoplasmas causing diseases such as palm yellowing (Harrison et al., 1994), paulownia witches' broom (Nakamura et al., 1996), plum leptonecrosis (Malisano et al., 1996), maize bushy stunt (Harrison et al., 1996), papaya yellow leaf crinkle and mosaic (Gibb et al., 1996; Guthrie et al., 1998), eucalyptus little leaf (Marcone et al., 1996), pear decline (Schneider and Gibb, 1997), peanut witches' broom (Chen and Lin, 1997), sesame phyllody (Han et al., 1997), Bermuda grass white leaf (Marcone et al., 1997), European stone fruit yellows (Jarausch et al., 1998), parsley yellows (Khadhair et al., 1998), sugarcane yellow leaf (Cronjé et al., 1998) *Rubus* stunt and yellows (Davies, 1998), and cotton phyllody (Marcone et al., 1999) has been successful.

The sequence analysis of the 16S rRNA gene has been used as a basis for the selection of primer pairs specific for phytoplasmas and to establish the extent of relatedness between phytoplasmas. Specific oligonucleotide primers were synthesized by using sequence data of 16S rRNA gene from three plant pathogenic phytoplasma groups. PCR technique was found to be effective for detecting the phytoplasma 16S rRNA genes from diseased plants and inoculative vectors (Namba et al., 1993). The phytoplasmas causing pear quick decline, apple proliferation and the sweet potato little leaf phytoplasma were detected by amplifying the 16S rRNA gene sequences using PCR technique (Giunchadi et al., 1993; Firrao et al., 1994; Gibb et al., 1995). Using the oligonucleotide primer pair amplifying ribosomal protein (rp3/rp4), a paulownia witches' broom phytoplasma–specific DNA fragment from as little as 150 pg of nucleic acid samples was amplified. This phytoplasma could be detected in 95% of asymptomatic infected paulownia trees (Nakamura et al., 1995).

By comparing the 16S rRNA sequences of plum lepotnecrosis and apple proliferation phytoplasmas, two oligonucletoides were selected, differing by two nucleotides specific for apple proliferation and plum leptonecrosis phytoplasma, respectively. The specific oligonucleotides labeled with DIG hybridized to the 16S rDNA fragments amplified from apple and plum leaf samples. This procedure was employed to differentiate these two phytoplasmas (Malisano et al., 1996). The phytoplasmas associated with maize bushy stunt (MBS) disease was detected by the amplification of a DNA product of about 740 bp from symptomatic corn plants infected with MBS phytoplasma. The presence of a Florida isolate of MBS phytoplasma was detected in all leaf and stalk samples from presymptomatic plants at 12 days after inoculation by the leafhoppers (*Dalbulus maidis*) (Harrison et al., 1996). The newly identified eucalyptus little leaf elm yellows phytoplasma was reported to belong to the elm yellows phytoplasma group by determining the restriction patterns of the PCR-amplified rDNA (Marcone et al., 1996).

The infection of Bermuda grass by white leaf phytoplasma was detected by employing specific primers and PCR amplification in all symptomatic plants and no amplification product was present in nonsymptomatic plants, indicating the specificity of the primers used (Marcone et al., 1997). A PCR-based assay was employed for the detection of European stone fruit yellows phytoplasma, by amplifying a 237-bp DNA fragment from total DNA extracts derived from over 300 stone fruit samples. Both specific and universal primers were employed and there was high correlation (97%) between the results obtained with these primers. The results obtained have the potential for application in epidemiological studies (Jarausch et al., 1998). The DNA fragments of phytoplasmas causing papaya dieback, yellow crinkle, and mosaic diseases were detected by PCR with primers specific in general and for the stolbur group of phytoplasmas (Guthrie et al., 1998).

The distribution of the phytoplasma associated with papaya dieback (PDB) disease occurring in Australia was studied using electron microscopy and PCR assay. By employing primers based on the 16S rRNA gene and 16S–23S intergenic spacer region, the PDB phytoplasma DNA was detected in leaves, stem, and roots, but not in mature leaves of infected plants. Electron microscopy did not reveal the presence of PDB phytoplasma in mature sieve elements of PDB-affected leaf, stem, or fruit tissues from plants at different stages of symptom expression, though PCR assays indicated the presence of the phytoplasma DNA in these tissues. The variations in the results may be possibly due to the presence of phytoplasma cells in low titer and irregular distribution of the phytoplasma cells in diseased tissues (Siddique et al., 1998).

The identity of the phytoplasma isolates obtained from different geographical locations or even in the same location can be established by performing PCR assays with certainty. *Rubus* plants showing stunting symptoms were shown to contain two distinct isolates by using universal primers to sequences in the 16S rRNA gene and group-specific primers to sequences in internal transcribed spacers (ITS) region. One isolate associated with *Rubus* stunt disease was similar to the phytoplasmas included in the group V (elm yellows), whereas the other isolate seemed to have similarity with the members of group III (X-disease) (Davies, 1998). The phytoplasma causing cotton phyllody disease could be differentiated from faba bean phyllody phytoplasma by PCR amplification of rDNA followed by RFLP analysis (Marcone et al., 1999).

The PCR is very useful in determining the phylogenetic relationships among phytoplasmas that cause various plant diseases (Lee et al., 1992b). The faba bean phyllody phytoplasma could be detected by employing a specific primer pair and DNA amplification by PCR (Saeed et al., 1994). Lee et al. (1994) reported that a phytoplasma group-specific primer pair allowed sensitive detection and simultaneous classification of phytoplasma strains and that the sensitivity of detection may be further increased by employing nested-PCR assays using the universal primer pair R 16 F2/R2 and a group-specific primer pair. The recycled PCR (R-PCR), a modified method, may be used for both diagnosing phytoplasma diseases and developing a phylogenetically based phytoplasma taxonomy. In this method, several different PCR reactions may occur simultaneously in one tube by addition of a second primer to the tube containing the first PCR products. Recycled PCR helps in detection of mollicute-specific DNA fragments and phytoplasma specific or group-specific DNA fragments as multiple bands, and each phytoplasma can be reliably identified (Namba et al., 1993).

The phytoplasma causing yellows symptoms in parsley (*Petroselinum hortense*) was shown to be a member of aster yellows (AY) phytoplasma group by amplifying the DNA extracted from the infected leaves with a 16S rDNA universal primer pair P1/P6. The PCR amplification resulted in the expected PCR product of 1.5 kb. The identity was confirmed by amplification with the specific primer pair R16 R1/F1, which was designed on the basis of AY phytoplasma 16S rDNA sequences. The expected DNA fragment of 1.1 kb was amplified with the specific primer set in the direct and nested PCR assays [Appendix 9(ix)]. By using universal phytoplasma primers from 16S rDNA in a nested PCR procedure, sugarcane Ramu stunt (SCRS) phytopalsma–specific products were consistently amplified from the leaves of field-grown sugarcane, greenhouse-grown sugarcane plants with SCRS symptoms, and the delphacid planthoppers (*Eumetopina flavipes*). Digestion of the amplimers with restriction enzymes *Rsa*I and *Hae* III resulted in profiles that exhibited similarity with those of the members of sugarcane white leaf (SCWL) phytoplasma group. This newly recognized phytoplasma showed a 95.98% homology in the DNA sequences of intergenic spacer region with SCWL phytoplasma, indicating the close relationship between these phytoplasmas (Cronjé et al., 1999).

The importance of assessing the population of inoculative leafhoppers as one of the critical factors that may influence the incidence of phytoplasmal diseases has been stressed by many workers. Direct detection of phytoplasmas in the vectors of PCR assay has provided reliable data that may be used in disease-forecasting systems. By employing three different primer pairs—two universal primer pairs designed on conserved sequences of the 16S rRNA gene and one designed on extrachromosomal DNA of a severe strain of Western aster yellows (AY) phytoplasma—the chrysanthamum yellows (CY) phytoplasma was detected in *Macrosteles quadripunctulatus* and *Euscelis incisus*. The CY phytoplasmal DNA could be amplified from the total DNA (at a dilution of 1/10,000) extracted from a single leafhopper-carrying phytoplasma. The direct PCR protocol developed by Marzachi et al. (1988) seems to be very specific, sensitive, and rapid for the detection of CY phytoplasma in the leafhoppers. For the detection of AY phytoplasma, two PCR-based assays were employed for the detection and quantitative determination of AY phytoplasma in the aster leafhoppers. The phytoplasma was detected in individual leafhoppers to monitor the number of leafhoppers carrying the AY phytoplasma. Additionally, the total amount of phytoplasma DNA in a leafhopper population was also estimated. The study demonstrated that the peaks in phytoplasma in the aster leafhopper populations present in lettuce plots in Ontario, Canada, preceded or coincided with the development of AY symptoms in the crops. The number of leafhoppers with detectable AY phytoplasma appeared to be a better predictor of AY incidence than the total number of aster leafhoppers (Goodwin et al., 1999).

A highly sensitive system PCR-ELISA has been developed for the determination of PCR-amplified products and this method eliminates the need for electrophoresis and associated procedures. The effectiveness of this approach for the detection of plant viruses has already been discussed. The immunoenzymatic detection of PCR products for identification of phytoplasmas causing apricot chlorotic leafroll (ACLR), plum leptonecrosis (PLN), and pear decline (PD) has been successfully performed. The PCR-ELISA was found to be highly reliable as the infected samples always showed very high absorption, visible even to the naked eye (Fig. 9.9). The sensitivity of PCR-ELISA was greater than elec-

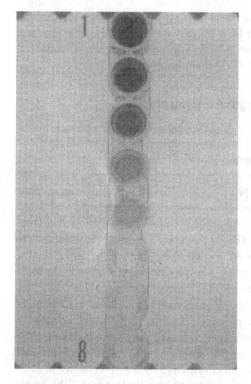

Figure 9.9 Colorimetric detection of PCR products hybridized with biotin rPDS probe. Samples: 1–5, from pear decline phytoplasma-infected pear tree serially diluted (1/10, 1/20, 1/40, 1/80, and 1/160); 6, undiluted from Japanese plum with plum leptonecrosis (PLN) phytoplasma; 7, from healthy pear tree; 8, sample without probe. (Blackwell Wissenchafts-Verlag, Berlin, Germany.) (Courtesy of Pollini et al., 1997.)

Table 9.4 Comparison of Sensitivities of PCR-ELISA and Electrophoresis

	Dilution limits	
Phytoplasma tested	PCR-ELISA	Electrophoresis (amplification with P1/P7)
Olive (OY)	1:160	1:10
Rubus (RS)	1:160	1:10
Pear (PD)	1:160	1:20
Japanese plum (PLN)	1:80	1:10
Apricot (ACLR)	1:80	1:80

Source: Pollini et al., 1997.

trophoresis after amplification with universal primers of P1/r P7 or with specific primers (Table 9.4). The specificity of PCR-ELISA was revealed by the absence of cross-reaction that may be possible with PCR. This method has the potential for large-scale screening (Pollini et al., 1997) [Appendix 9(x)].

9.5.3   Restriction Fragment Length Polymorphism

The phytoplasmal relatedness may be established by digesting the phytoplasmal DNA with restriction endonucleases followed by restriction fragment length polymorphism (RFLP) analysis. The restriction profile of the phytoplasma associated with apricot decline, obtained by using *AluI* restriction endonulcease, was similar to that of phytoplasma causing apple proliferation disease revealing the possible relationship between the phytoplasma causing these disease in apricot and apple (Marcone et al., 1995). The papayas in Australia are affected by yellow leaf crinkle (PYC), mosaic (PM), and dieback (PDB) diseases. The phytoplasmas associated with yellow crinkle and mosaic were shown to be identical by RFLP analysis using restriction enzymes, whereas the phytoplasma causing dieback symptoms showed distinct differences in RFLP patterns compared with the other two phytoplasmas (Gibb et al., 1996). RFLP analysis was employed to differentiate dieback, yellow crinkle, or mosaic phytoplasma in papayas showing dual infection (Guthrie et al., 1998). The genetic relatedness of the phytoplasmas associated with PYC, PDB, and PM diseases was examined by RFLP analysis of the 16S rRNA gene and 16S rRNA/23S rRNA spacer region (SR). The PYC and PM phytoplasmas were found to be identical, confirming the findings of Gibb et al. (1996), and they were more closely related to members of the faba bean phyllody strain cluster based on the RFLP patterns and SR sequence comparison. The PDB phytoplasma showed similarities to phormium yellow leaf (PYL) phytoplasma from New Zealand and the Australian grapevine yellows (AGY) phytoplasma

(Gibb et al., 1998). Analysis of the sequences in the intergenic region between 16S and 23S rDNA genes showed that the sugarcane yellow leaf (SYL) phytoplasma belonged to the Western X group of phytoplasmas (Cronjé et al., 1998).

RFLP analyses of PCR-amplified rDNA have been useful for the characterization of newly recognized phytoplasmas and for establishing their genetic relatedness with other known phytoplasmas. The Bermuda grass white leaf (BGWL) phytoplasma occurring in Italy was shown to have similar RFLP patterns as BGWL phytoplasma present in Thailand, which is known to be a member of the sugarcane white leaf phytoplasma group (Marcone et al., 1997). The RFLP patterns obtained after digestion with restriction enzymes *MseI* and *AluI* of the PCR fragments amplified with the primer pair fU5/rU3 were shown to be identical to those from the sweet potato little leaf phytoplasma (Schneider and Gibb, 1997). The phytoplasma associated with lethal decline (CLD) disease of coconuts occurring in west and east Africa could be detected by using primers designed based on conserved sequences of 16S rRNA gene followed by PCR amplification. The phytoplasmas in coconuts showing decline syndrome were detected by dot-blot hybridization by employing two probes from palm lethal yellowing (PLY) phytoplasma from Florida. Polymorphisms were detected in rDNA fragments by RFLP analysis suggesting that CLD and PLY phytoplasmas are not identical, although the genetic relationship between them is indicated by dot-blot hybridization (Tymon et al., 1997).

9.5.4 Heteroduplex Mobility Analysis

The heteroduplex mobility analysis (HMA) is based on the delay in the rate of migration of a DNA heteroduplex in comparison with a DNA homoduplex to identify mismatches or deletions in DNA sequences. This approach was first used to determine the variability of human immunodeficiency virus type 1 (HIVs) surface envelope (*env*) glycoprotein-coding sequences (Delwart et al., 1994). The HMA was employed to establish genetic relatedness among different field isolates of aster yellows phytoplasmas by Ceranic-Zagorac and Hiruki (1996). As many phytoplasmas induce similar symptoms such as witches' broom, a simple diagnostic method is required for reliable conclusions especially for certification and quarantine programs. The HMA procedure requires only universal primers and a standard DNA. Differences in the migration distance on gel electrophoresis between hetero- and homoduplex reveal the nucleotide heterogeneity of the test phytoplasma samples and standard phytoplasma DNA.

The French and German isolates of *Populus nigra* cv. Italica witches' broom (WB) were differentiated by HMA. The French isolate was found to be identical to the aster yellows type strain, whereas the German isolate was different, though it is closely related to the French isolate. The phytoplasmas causing stolbur and big bud symptoms in tomato could also be distinguished by HMA technique. A high degree of nucleotide sequence heterogeneity existed between the DNA strands of reference stolbur and aster yellows and hence they have to be placed in different groups (Cousin et al., 1998) [Appendix 9(xi)]. The HMA in conjunction with PCR showed a shift in the mobility for a heteroduplex formed in combination with the chestnut little leaf (CLL) and jujube witches' broom (JWB) phytoplasmas. No change in the mobility for the heteroduplex was observed for the combination with CLL and each of paulownia witches' broom and mulberry dwarf phytoplasma DNAs. The combination of HMA and PCR was demonstrated to be a useful technique for the detection and differentiation of phytoplasmas (Han et al., 1998).

9.6 DETECTION OF FUNGAL PATHOGENS

Detection of fungal pathogens by using nucleic acid-based techniques has several advantages. Serological methods have been found to be more difficult, since the fungal antigens are complex and variable at different growth stages. The presence or absence of spore-bearing structures and slow-growing nature of certain fungal pathogens will not affect their detection by nucleic acid-based techniques, since only fungal cells containing DNA are needed. Using appropriate DNA probes, fungi which are generally not amenable for rapid identification can be detected and identified. Soil-borne fungal pathogens such as *Phytophthora, Gaeumanno-myces, Pythium,* and *Leptosphaeria* that cause relatively nonspecific symptoms, such as generalized rotting and death of plants, may be detected by using specific DNA probes. Adoption of PCR-based assays allows enhancement of sensitivity and specificity of detection and quantification of fungal pathogens in plant tissues and assessment of relatedness of pathogens.

9.6.1 Dot-Blot Hybridization Assay

The soil-borne pathogens have to be detected in both the soil and plant materials rapidly and reliably to monitor the buildup of population of fungal pathogens that will have significant influence on the incidence and subsequent spread of diseases caused by them. *Rhizoctonia solani* AG-8, which induces root rot and damping-off diseases in several crops, was detected in soil samples by using a specific DNA probe pRAG12. The specificity and high copy number of AG-8 probe provide the means of a sensitive diagnostic assay for *R. solani* in infested soils (Whisson et al., 1995). A plasmid DNA fragment designated PE-42 hybridized to DNA of all 22 isolates of *R. solani* AG-2-2 IV causing large patch disease of Zoysia grass, but not to the DNA of other pathogens infecting *Zoysia,* revealing the specificity of hybridization assay with PE-42 plasmid DNA. When the PE-42 plasmid DNA fragment was employed as a probe in the Southern hybridization, it was demonstrated that the PE-42 plasmid DNA fragment could be employed as a marker to distinguish *R. solani* AG-2-2-IV from other intraspecific groups of *R. solani* and for diagnosis of large patch disease of *Zoysia* grass (Takamatsu et al., 1998).

A slot-blot hybridization procedure employing a specific DNA probe (pG158) showed that pG158 hybridized strongly to pathogenic isolates of *Gaeumannomyces graminis* var. *tritici*, moderately to *G. graminis* var. *avenae*, but not at all to nonpathogenic isolates of *G. graminis* var. *tritici* and other soil fungi. It is essential to differentiate pathogenic isolates from the morphologically similar nonpathogenic isolates to relate the soil population to the incidence of wheat take-all disease caused by *G. graminis* var. *tritici*. It is possible to use pG158 to detect the pathogenic isolates both in the soil and wheat roots and also for intraspecific classification of *G. graminis* isolates (Harvey and Ophel-Keller, 1996).

The effectiveness of detection and quantification of *Phytophthora cinnamomi* in avocado roots was demonstrated by using a probe selected from a library of genomic DNA. This specific probe detected as little as 5 pg of *P. cinnamomi* DNA in dot-blot and slot-blot assays. The extent of colonization of avocado roots by *P. cinnamomi* was assessed by determining relative amounts of pathogen and host DNA, over a period of time (Judelson and Messenger-Routh, 1996). A reverse dot-blot protocol developed for the identification of oomycetes is based on assays of oligonucleotides labeled with DIG. This procedure showed far fewer cross-hybridization than the one based on entire amplified internal transcribed spacer (ITS) fragments. By just observing the positive hybridization reaction between the DNA labeled directly from the sample and the specific oligonucleotides immobilized on nylon membrane, the unknown species of oomycetes could be identified. This assay can be used for the identification of *Pythium aphanidermatum, P. ultimum, P. acanthium,* and *Phytophthora cinnamomi* (Lévesque et al., 1998). *Sporisorium reiliana* causing head smut disease and *Ustilago maydis* causing common smut disease in maize were detected by employing DIG-labeled inserts in dot-blot hybridization assay (Xu et al., 1999).

9.6.2 Restriction Fragment Length Polymorphism

Application of restriction fragment length polymorphism (RFLP) analyses for fungal plant disease diagnosis has been found to be useful. The RFLP technique is based on the natural variations in the genomes of different groups or strains of organisms. Loss or gain of restriction endonuclease recognition sites or other events such as deletions or insertion in the DNA sequences may result in variations (polymorphisms) in fragment sizes. The DNA of the test organism is digested with restriction enzymes, and the fragments are separated by electrophoresis in agarose or polyacrylamide gel to detect the differences in the size of the DNA fragments. The number and size of the fragments formed after digestion are determined by the distribution of restriction sites in the DNA. Hence, depending on the combination of each restriction enzyme and target DNA, a specific set of fragments, that can be considered a fingerprint for a given strain is formed. The specific sites of fragments are usually identified by Southern blot analysis (Hamer et al., 1989; Leach et al., 1990), but they can also be directly observed by staining the gels with ethidium bromide for observation under ultraviolet light (Klich and Mullaney, 1987; Jones et al., 1989). The DNA fragments are then transferred to a nitrocellulose or nylon membrane and hybridized with an appropriate probe (Appendix 9[xii]).

By using an appropriate probe, detection and identification of pathogenic fungi may be achieved. Nicholson et al. (1994) developed a pathotype-specific DNA probe for the identification of the R type of *Pseudocercosporella herpotrichoides*. They isolated a 6.7-kb DNA fragment from an R-type isolate of the pathogen which showed specific hybridization to R-type isolates and not to N, C, or S pathotypes or to *P. anguioides*. Infection of rye seedlings by R type was detected by hybridization of this probe to DNA extracted from infected plants.

The RFLP data may be used for assessing the genetic diversity of the pathogen population as well as for determining the extent of relatedness of the pathogen groups on the basis of the numerical analysis of the data (Lynch, 1988; Nei and Li, 1979). Restriction fragment length polymorphisms were used to estimate the genetic divergence and relationship among isolates of *Fusarium oxysporum* f. sp. *gladioli* (Me et al., 1994). The usefulness of RFLP analysis has been increasingly recognized in fungal taxonomy (Garber and Yoder, 1984; Anderson et al., 1987, Coddington et al., 1987; Foster et al., 1987; Kistler et al., 1987; Manicom et al., 1987; Klich et al., 1993). Genomic DNA RFLPs combined with random probes can be used for differentiating species, formae speciales, races, and isolates of *Fusarium* (Coddington et al., 1987; Manicom et al., 1987; Kim et al., 1993). Ko et al. (1993) reported that RFLPs in nuclear DNA were correlated to some extent with the prevailing races of *Magnaporthe grisea* in Korea. However, the relationship between RFLPs in nuclear DNA and virulence of *M. grisea* was inconclusive.

Both total DNA (Manicom et al., 1987; Coddington et al., 1987) and mitochondrial (mt) DNA (Anderson et al., 1987; Foster et al., 1987) have been subjected to RFLP analyses for taxonomic studies. But in many cases RFLP analyses have been performed with mitochondrial DNA (mt DNA) rather than with genomic DNA, since the mt DNA is much smaller. When cut with a single restriction enzyme, mt DNA produces about 10–20 fragments, forming distinct patterns on electrophoresis in agarose gels. The RFLP analysis of ribosomal DNA (rDNA) and mt DNA of isolates of *Cylindrocarpon heteronema,* which causes European canker in apples, by using rDNA from *Saccharomyces carlsbergensis* and mt DNA of *C. heteronema,* revealed intraspecific heterogenicity. Four rDNA and six mt DNA restriction pattern categories were noted among the isolates tested (Brown et al., 1994) (Fig. 9.10). The isolates of *Phytophthora parasitica* var. *nico-*

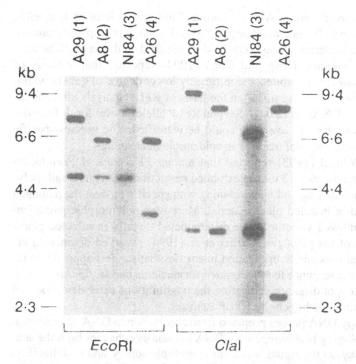

Figure 9.10 RFLP patterns of DNA from four isolates of *Cylindrocarpon heteronema* digested with *Eco* RI and *Cla* I and hybridized with the yeast rDNA probe pMY 60; *ClaI* restriction patterns of all isolates show a third faint band of approximately 1.05 kb (Blackwell Science Ltd., and British Society for Plant Patholgy, U. K.). (Courtesy of Brown et al., 1994.)

tianae, causing tobacco black shank disease do not produce elicitin (TE⁻) whereas avirulent isolates and nontobacco isolates of *P. parasitica* produce elicitin (TE), which is known to induce the initiation of resistance reaction in tobacco. Elicitin production (TE⁺) was generally associated with low virulence on tobacco and frequent pathogenicity on tomato, whereas TE⁻ isolates generally are highly virulent and specialized to tobacco. RFLP analysis of both mitochondrial and nuclear DNA could be used to differentiate isolates infecting tobacco (TE⁺) from other *P. parasitica* isolates (TE⁻). It is suggested that monitoring the loss of elicitin production (leading to virulence) may be an important factor in disease management programs (Colas et al., 1998).

Okoli et al. (1994) examined the relationship of two host-adapted pathotypes of *Verticillium dahliae* by RFLP analysis. They found that isolates obtained from and adapted to peppermint formed a subgroup (M) distinct from the nonhost-adapted subspecific group A of *V. dahliae*. Similarly isolates of *V. dahliae* from cruciferous hosts formed another group (D). By using two specific probes, the isolates from cruciferous plants could be distinguished on the basis of the variation in polymorphisms. Ueng and Chen (1994) reported that isolates of *Phaeosphaeria nodorum* exhibited a significantly lower degree of genetic variation than the isolates of *P. avenaria,* on the basis of RFLP analysis after the digestion of genomic DNAs by Eco RI. Several RFLP alleles useful for differentiation of *P. nodorum* from *P. avenaria* could be identified. The species-specific probes may be used as natural markers in epidemiological studies.

Frei and Wenzel (1993) reported that among 21 clones of *Pseudocercosporella herpotrichoides,* 13 clones exhibited restriction fragment length polymorphisms among isolates, and by combining with specific probes, the pathogen could be detected in infected plant material. Moreover, polymorphic pathogen-specific probes allowed varieties to be differentiated directly in infected plants without isolation of the pathogen. Carlier et al. (1994) observed distinct differences in RFLP patterns and hybridization intensities that suggest appreciable interspecific genetic divergence in *Mycosphaerella musicola* and *M. fijiensis,* which cause banana leaf spot diseases, indicating the possibility of early detection and identification of these pathogens by RFLP analyses.

Muliple-copy DNA probes prepared from chromosomal DNA have several advantages. Employing highly repetitive DNA sequences enhances both the sensitivity of the assay, as the signal is present in multiple copies, and its reliability, as a result of the lack of influence of variation in one copy in the genome on the total signal observed in a hybridization-based assay. Moreover, repetitive DNA has a very high probability of being species-specific. Repetitive DNA fragments of 12 species of *Phytophthora* tested appeared as continuous discrete bands over a faint smear in agarose gels when stained with ethidium bromide. Similar digestion patterns were observed for the isolates belonging to the same species; different species of *Phytophthora* exhibited different patterns. Very similar *P. cryptogea* and *P. drechsleri* could be differentiated by the repetitive DNA profiles. The heterogeneous status of *P. megasperma* and complete homogenicity of 12 isolates of *P. parasitica* could be established by examining the DNA profiles. As this method is relatively simple, it may be useful for investigation of taxonomic problems and identification of different species of *Phytophthora* (Panabieres et al., 1989) (Appendix 9[xii]). Cloned DNA probes prepared from chromosomal DNA of *Phytophthora parasitica* hybridized to *P. parasitica* DNA only, but not to DNA of other *Phytophthora* spp. and *Pythium* spp. DNA from all isolates of *P. parasitica,* including *P. parasitica* var. *nicotianae,* hybridized strongly with the probes, indicating their species-specific nature (Goodwin et al., 1989, 1990). Development of such species-specific DNA probes for the detection of *P. citrophthora* (Goodwin et al., 1990), *Gaeumannomyces graminis* (Henson et al. 1993), *Phoma tracheiphila* (Rollo et al., 1987), and *Leptosphaeria korrae* (Tisserat et al., 1991) has been reported. One of the cloned DNA sequences had as many as 50 to 100 copies of the *L. korrae* genome and can be employed for the detection of the pathogen in the root tissues of the turfgrass (Tisserat et al., 1991) (Fig. 9.11) (Appendix 9[xiii]).

As mitochondrial DNA is smaller than chromosomal DNA, restriction maps can be developed to determine conserved and nonconserved regions. Cloned DNA probes generated from mt DNA of *Gaeumannomyces graminis* (Henson et al., 1993) and *Peronosclerospora sorghi* (Yao et al., 1991) were highly specific and did not hybridize with DNA of other fungi. However, in the case of *Pythium* sp., some probes hybridized to a subset of isolates sharing the same mitochondrial restriction map, whereas many probes hybridized to the DNA of more than one *Pythium* sp. (Martin, 1991).

The probe specific for a dispersed repeated DNA sequence (called MGR) was employed to construct genotype-specific *Eco* RI restriction fragment length

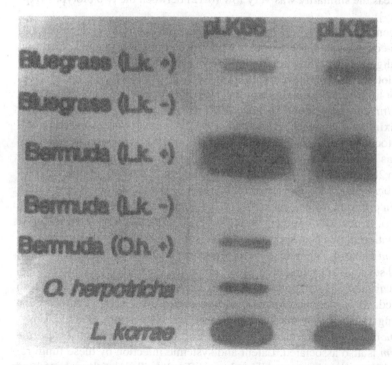

Figure 9.11   Slot hybridization of two probes (pLK 66 and pLK 88) to fungal and plant DNA extracted from Kentucky bluegrass and bermuda grass root samples: LK +, roots colonized by *L. korrae;* LK-, noninfected roots (American Phytopathological Society, Minnesota, USA). (Courtesy of Tisserat et al., 1991.)

profiles (MGR-DNA fingerprints) from field isolates of rice blast pathogen (*Magnaporthe grisea*) in the United States of America. The MGR-DNA fingerprints could be used as the basis for distinguishing major pathotypes of *M. grisea,* identifying the pathotypes accurately, and defining the organization of clonal lineages within and among pathotype groups (Levy et al., 1991). The genetic relationships among isolates of *Pyricularia grisea* from rice and other hosts were analysed by RFLP analysis by using the repetitive probe MGR 586. Rice blast isolates representing four distinct races differentiated by inoculation on Korean differential rice varieties showed multiple bands hybridizing to the probe MGR 586. The study indicated that *P. grisea* populations from nonrice hosts, such as *Digitaria sanguinalis, Eleusine indica, Lolium boucheanum, L. multiflorum,* and *Festuca elatior,* could be sources of inoculum for the rice crop (Han et al., 1995). Ueng et al. (1995) found a correlation between molecular and biological characters of the biotypes of *Stagonospora nodorum* from barley and wheat. The genetic similarity, as determined by RFLP analysis, was very high (70.82) within each of the two biotypes, whereas the similarity was very low (0.12) between the two biotypes (Appendix 9[xiv]).

The genetic similarity in 39 isolates of *Fusarium oxysporum* encompassing 5 formae speciales, that cause vascular wilts in cucurbits was studied. The total DNA was digested with three restriction enzymes (*Pst* I, *Hind* III, and *Eco* RI), Southern-blotted, and hybridized with a mt DNA polyprobe from *F. o. niveum.* Within each forma speciale unique RFLPs were present. *Fusarium oxysporum/niveum* was found to have least divergence; *F. o. cucumerianum* showed maximum divergence (Kim et al., 1993). A probe (P449) consisting of a 3.38-kb mitochondrial DNA fragment obtained from *Fusarium oxysporum* f.sp. *cubense* was employed to determine RFLPs in the restriction digests of total DNAs from 28 isolates of *F. oxysproum* that could infect a variety of plant species in various locations. The existence of mitochondrial DNA polymorphisms within and between different *formae speciales* of *F. oxysporum* was reported by Bridge et al., (1995). A new DNA fingerprinting probe, (Cat)5, was employed to detect genetic variation in two host specialized groups (on pine and fir) of *Heterobasidium annosum* in North America. Several fingerprint bands were specific to these groups, and several bands unique to isolates of either of the host specialized varieties (var. *wageneri* and var. *ponderoseum*) of *Leptographium wageneri* were also distinguished. More polymorphisms could be detected by using (CAT)5 probe than by using 21 isoenzyme markers (DeScenzo and Harrington, 1994).

Phomopsis longicolla causes soybean seed decay, with which *Diaporthe phaseolorum* is also associated. Latent and systemic infection by these fungi results in significant yield losses and it has been difficult to distinguish such infected seeds by conventional methods. RFLP analysis of PCR amplification products was used to differentiate these two pathogens. Specific primers Phom. I and Phom. II were designed from the polymorphic regions of *P. longicolla* and *D.*

phaseolorum isolates from soybean. The presence of specific bands in the PCR products from 10 pooled seed samples and also from individual infected seeds was observed. DNA extracts of tissues from symptomless plants inoculated with *P. longicolla* and *D. phaseolorum* var. *sojae* exhibited the specific band indicating the specificity of the RFLP analysis (Zhang et al., 1997). The isolates of *Ascochyta pisi* were genotyped by rDNA-RFLP and UP (universal primed)-PCR using eight UP-primers and two arbitrary primers individually or in pairwise combination. The isolates were differentiated by the polymorphic UP-PCR products that may be used as markers for developing isolate- or pathotype-specific PCR-based diagnostic assays (Lübeck et al. 1998).

9.6.3 Polymerase Chain Reaction

Enhancement of the sensitivity of the diagnostic assay to a high level becomes necessary for the detection of fungal pathogens especially those causing vascular wilts and root rots, since they have to be detected when their populations in soil or in infected plant tissue are low and also very rapidly. Polymerase chain reaction (PCR)-based assays allow detection of the fungal pathogens as well as their quantitation very effectively. Among the molecular methods, PCR-based assays have been demonstrated to be more sensitive, reliable, and rapid for the detection, differentiation, and quantitation of fungal pathogens (Table 9.5). The presence of *Phytophthora parasitica* in infected tomato roots and soil (Goodwin et al. 1990), *Leptosphaeria korrae* in turfgrass (Tisserat et al., 1991), *Phytophthora citrophthora* in citrus roots (Goodwin et al., 1990), *Gaeumannomyces graminis* in wheat (Schesser et al., 1991; Ward and Gray, 1992; Henson et al., 1993; Elliott et al., 1993), *Phomopsis tracheiphila* in citrus (Rollo et al., 1990), *Verticillium* spp. in potato (Moukhamedov et al., 1994), and *Monosporascus* spp. in muskmelon (Lovic et al., 1995) has been detected by employing PCR-based assays. The primer pair K1 and K3 designed on portions of the sequences of the internal transcribed spacer (ITS) region of rDNA amplified the isolates of *Pythium ultimum* in the PCR assay. The pathogen could be detected in single diseased seedling after diluting the extracts (10-fold) with Tris-EDTA (TE) buffer (Kageyama et al., 1997).

PCR assays have been applied for the detection of different species of *Phytophthora* causing important diseases using the repetitive sequences. *Phytophthora infestans,* causing potato late blight disease, was detected in the potato leaves 1 day after inoculation (Niepold and Schöber-Butin, 1995). A region in the ITS specific to *P. infestans* was used to construct a PCR primer (PINF) which could detect the pathogen in infected tomato and potato field samples (Trout et al., 1997). A one-tube PCR (nested or multiplex PCR) procedure was developed using the sequences of a repetitive satellite DNA fragment of *P. infestans* for designing specific primers. These primers were able to amplify the target DNA from

Table 9.5 Target Sequences of Fungal Pathogens Amplified by PCR

Pathogen	Target sequence/primers	Reference
Alternaria linicola	rDNA ITS 1 and 2	Mc Key et al., 1999
Colletorichum gloeosporioides	ITS1 and conserved rDNA	Mills et al., 1992
C. lindemuthianum	Bulked DNA	Mesquita et al., 1998
Diaporthe phaseolorum	ITS sequence of rDNA	Zhang et al., 1999
Fusarium avenaceum; F. culmorum and *F. graminearum*	20-mer oligonucleotides	Schilling et al., 1996
F. oxysporum f.sp. *ciceris*	DNA sequences	Garcia Pedrajas et al., 1999
F. oxysporum f.sp. *vasinfectum*	ITS sequences in rDNA	Moricca et al., 1998
F. moniliforme	Nucleotide sequences of pUCF2 genomic clone	Murillo et al., 1998
Gaeumannomyces graminis	Mitochondrial (mt) DNA rDNA of mitochondria	Eliot et al., 1993 Ward and Gray, 1992
Helminthosporium solani	ITS region of nuclear ribosome	Olivier and Loria, 1998
Leptosphaeria maculans	ITS1 and conserved region of nuclear rDNA	Xue et al., 1992
	ITS size polymorphism	Balesdent et al., 1998
Magnaporthe grisea	Primer sequences from *Pot2*	George et al., 1998
Mycosphaerella fijiensis and *M. musicola*	ITS1 and conserved region of rDNA	Johanson and Jeger, 1993
Peronospora sparsa	ITS sequences	Lindquist et al., 1998; Hellqvist et al., 1998
Phaeoisariopsis griseola	DNA sequences	Guzman et al., 1999
Phialaphora gregatum	ITS sequences of rDNA	Chen et al., 1999
Phoma tracheiphilla	DNA from infected plants	Rollo et al., 1990
Phomopsis longicolla	ITS sequences of rDNA	Zhang et al., 1999
Phytophthora sp.	ITS sequences of rDNA	Lee et al., 1993; Ristains et al., 1998
Phytophthora erythrosepitca	ITS region 2 sequences of rDNA	Tooley et al., 1997
P. fragariae	ITS sequences of rDNA	Bonants et al., 1997
	P. FRAGINT primers; ITS region of rDNA	Hughes et al., 1998
	rDNA sequences	Bundry 1999
P. infestans	DNA from mycelia; inoculated potato tuber slices	Niepold and Schöber-Bulin, 1995; 1997
	ITS 4 and 5 sequences of rDNA	Trout et al., 1997
	ITS 2 sequences of rDNA	Tooley et al., 1997; 1998

Table 9.5 Continued

Pathogen	Target sequence/primers	Reference
P. medicaginis	Intergenic spacer region rDNA	Liew et al., 1998 Tooley et al., 1997; 1998
P. nicotianae	Sequences of elicitin gene *Par A1*	Lacourt and Duncan, 1997
	ITS 2 sequences of rDNA	Tooley et al., 1998
Plasmodiophora brassicae	Primers PBTZS-2, PBTZS-3, and PBTZS-4	Ito et al., 1999
	rDNA repeat section	Faggian et al., 1999
Plectosporium tabacinum	ITS sequences of rDNA	Chen et al., 1999
Pseudocercosporella herpotrichoides	DNA sequences specific for R and W types	Beck et al., 1996
Pythium ultimum	ITS sequences	Kageyama et al., 1997
Rhizoctonia oryzae	ITS sequences of rDNA	Mazzola et al., 1996
R. solani AG-8; *R. solani* AG1-IA	ITS sequences of rDNA	Mazzola et al., 1996; Matsumoto and Matsuyama, 1998
Septora tritici	ITS of nuclear DNA	Beck and Ligon, 1995
	ß-tubulin gene sequence	Fraaije et al., 1999
Spongospora subterranea	ITS sequences of rDNA	Bulman and Marshall, 1998
Sporisorium reiliana	Fungal DNA sequences; primers SR1 and SR3	Xu et al., 1999
Tapesia acuformis *T. yallundae* *(Pseudocercosporella herpotrichoides)*	DNA sequences	Bradsley et al., 1998
Tilletia indica	DraI fragments of mitochondrial DNA	Smith et al., 1996
T. barclayana, T. controversa, T. indica, T. laevis, and T. tritici	ITS region of rDNA or mitochondrial DNA	McDonald et al., 1999
Ustilago hordei	ITS sequences of rDNA	Willits and Sherwood, 1999
U. maydis	Fungal DNA sequences; primers SR1 and SR3	Xu et al., 1999
U. scitaminea	Fungal DNA sequences	Schenck, 1998
Venturia inaequalis	ITS 1 and 2 of rDNA	Schnabel et al., 1999
Verticillium albo-atrum	ITS 1 and 2 of nuclear rDNA	Nazar et al., 1991; Hu et al., 1993; Mahuku et al., 1999
V. dahilae	DNA from spores; ribosomal DNA	Moukhmedov et al., 1994

all known A1 mating types of *P. infestans* races 1, 3, 4, and 1–11 occurring in Germany and A2 mating types (Niepold and Schöber-Butin, 1997). By using specific primers designed based on sequences of ITS 2 region of DNA, *P. infestans* and *P. erythroseptica* (causing pinkrot of potato tubers) were detected in potato tubers as early as 72 hr after inoculation well before any visible symptoms appeared on the tubers (Tooley et al., 1998).

The sequences of the flanking and coding regions of the elicitin gene *ParA1* of *P. nicotianae* were used for designing primers. Using a combination of IL7 (flanking region)/IL8 (coding region) primers, a diverse collection of isolates of *P. nicotinae* (some causing tobacco black shank disease) could be detected in their host plants. As little as 100 zoospores trapped onto a nylon membrane could be detected after two rounds of PCR (Lacourt and Duncan, 1997). By employing the primers developed based on sequences of ITS region of ribosomal gene repeat (rDNA), *P. fragariae* was detected more efficiently when compared to ELISA technique (Bonants et al., 1997). The primers P-FRAGINT and the universal primer ITS 4 were used for the detection of *P. fragariae* var. *fragariae* and *P. fargariae* var. *rubi* in the roots of strawberry and raspberry, respectively. The detection efficiency was maximum between 1 and 5 days after inoculation, since degradation of the coenocytic mycelium and development of thick-walled oospores occurred later (Hughes et al., 1998). A PCR procedure based on the amplification of 5.8S rDNA gene and ITS 4 and ITS 5 primers was developed for the rapid identification of economically important *Phytophthora* spp. belonging to six taxonomic groups. The pathogens include *P. cactorum, P. capsici, P. cinnamomi, P. citricola, P. citrophthora, P. erythorseptica, P. fragariae, P. infestans, P. megasperma, P. mirabilis,* and *P. palmivora.* A highly sensitive diagnostic procedure based on a pair of oligonucleotide primers (PPED 04 and PPED 05) that can amplify a specific fragment within the intergenic spacer (IGS) 2 region was employed for the detection of *P. medicaginins* in stems and roots of lucerne at a dilution of 1:1,000,000 of pathogen DNA (Liew et al., 1998).

The fungal pathogens causing wilts and root rots have been detected and differentiated by employing PCR-based assays. The primers Fov1 and Fov2 designed using the nucleotide difference in the ITS sequences between 18S, 5.8S, and 28S rDNAs unambiguously amplified a 500-bp DNA fragment of all isolates of *Fusarium oxysporum* f.sp. *vasinfectum* causing cotton wilt disease. This assay system has the potential for use in disease diagnosis as well as in disease monitoring and forecasting programs (Morica et al., 1998). An efficient method of extracting DNA of *F. oxysporum* f.sp. *ciceris* involving disruption of fungal tissues by grinding dry soil using its abrasive properties in the presence of skimmed milk powder, which prevented the loss of DNA by adsorption to soil particles, was developed. The skimmed milk powder also reduces the coextraction of PCR inhibitors along with pathogen DNA. Specific detection of *F. oxysporum* f.sp. *ciceris* was possible in both artificial and natural soils (Garcia Pedrajas et al., 1999).

Rhizoctonia solani has many anastomosis groups (AGs) within the morphological species. *R. solani* AG1 IA causing rice sheath rot disease can be detected and identified rapidly by using primers designed from unique regions within the ITS regions of rDNA. The pathogen could be accurately identified in rice tissue and paddy field soils by this PCR protocol (Matsumoto and Matsuyama, 1998). Likewise, *R. solani* AG2 was detected by the PCR protocol involving amplification of 5.8S rDNA and part of the ITS region. The designed primers in combination with the general fungal primers ITS IF and ITS 4B were used. Six primers specifically amplified *R. solani* AG2, the subgroups AG2-1, AG2-2, and AG2-3, and the ecological type AG2-t. The DNAs from *R. solani* AG2 and AG4 in infected radish plants were amplified as in the case of DNA from axenic cultures (Salazar et al., 2000). These studies have clearly indicated that PCR-based detection techniques provide a powerful tool for rapid and reliable detection and identification of different anastomosis groups of *R. solani.*

Leptosphaeria maculans causing blackleg disease of oilseed rape could be identified and differentiated from other components involved in the species complex. The ITS region was directly amplified using the intact conidia as a substrate (Balesdent et al., 1998). Primers were designed based on DNA sequences of cloned random amplified polymorphic DNA (RAPD) fragments. Two major groups of *Phaeoisariopsis griseola* infecting common bean (*Phaeolus vulgaris*) were detected and identified by amplification of different-sized DNA fragments with PCR using group-specific primer pairs. The mycelia or synnemata or conidia collected from angular leaf spots on bean leaves after sonication could be used for PCR detection of *P. griseola* (Guzman et al., 1999). For the detection of *Septoria tritici* (*Mycosphaerella graminis*), specific primers were designed by aligning and comparing β-tubulin sequences from other fungi to target the β-tubulin gene. This procedure was highly sensitive, capable of detecting as little as 10 pg DNA of *S. tritici* in the presence of 200 ng of wheat leaf DNA (Fraaije et al., 1999).

The nested PCR detection protocol was developed for the detection of *Plasmodiophora brassicae* causing clubroot disease in cruciferous crops. Primers targeted to ribosomal RNA genes and ITS regions were employed to detect *P. brassicae* in the soil and water. The detection limits of the assay were 0.1 fentogram (fg = 10^{-15} g) for pure template and as low as 1000 spores/g of potting mix. The pathogen could be detected in all soils containing inoculum that may result in disease incidence (Faggian et al., 1999). In another study using a single-tube nested PCR (STN-PCR) for the detection of *P. brassicae,* the outer primer PBTZS-2 for amplifying a 1457-bp fragment from *P. brassicae* DNA and nested primers PBTZ-3 and PBTZ-4 for amplifying a 398-bp fragment internal of the 1457-bp fragment were employed. The assay could detect even a single resting spore present in 1 g of soil. The sensitivity of detection could be further improved by subjecting the STN-PCR product to second PCR amplification (double PCR) using the nested primers PBTZS-3 and PBTZS-4 (Ito et al., 1999). Another soilborne fungal pathogen, *Spongospora subterranea*, was detected in potato peel, tuber washings, and soil by using specific primers (Sps1 and Sps2) constructed from the sequences of ITS region of rDNA of the pathogen. These primers amplified a 391-bp product from *S. subterranea* but not from a range of other soilborne microbes. The detection threshold was *S. subterranea* DNA equivalent to 25×10^{-5} cystosori or one zoospore per PCR. The PCR assay was considerably more rapid and sensitive than the immunoassays or conventional bait-plant bioassays employed earlier. Furthermore, it may be useful for the development of disease risk assessments for field soils and seed potato seed stocks (Bell et al., 1999).

In the case of smut diseases, the symptoms of infection are not generally seen until heading, hampering the development of breeding programs for resistance to these diseases. The results of PCR assays were positively correlated with eventual production of smut whips by *Ustilago scitaminea* in sugarcane (Schenek, 1998). The direct amplification of DNA from single ungerminated teliospores of *Tilletia indica* and *T. tritici* was effective for the detection and identification of the smut pathogens. This technique does not require extraction of DNA from mycelia (McDonald et al., 1999). The maize pathogens *Sporisorium reiliana* causing head smut and *Ustilago maydis* causing common smut were detected by employing primer pairs SR1 and SR3 specific for *S. reiliana* and UM11 specific for *U. maydis (U. zeae)*. *S. reiliana* could be detected in the extracts of pith, node, and shank, but not in leaves of infected plants, the detection limit being 1–6 pg of fungal DNA irrespective of the maize DNA (Xu et al., 1999). Likewise, *U. hordei* could be detected using the primer pair designed based on the sequences of ITS region of rDNA. The pathogen DNA was amplified from leaf tissues of inoculated susceptible and resistant plants at different stages of plant development. The resistant plants had detectable amounts of *U. hordei* DNA in the first or three or four leaves but not in leaves produced later. The possibility of detecting *U. hordei* prior to heading would greatly help breeding for resistance to barley smut (Willits and Sherwood, 1999).

Polymerase chain reaction was used to amplify a ribosomal DNA fragment from *G. graminis*. This fragment, after labeling, was used as a probe which hybridized to Eco RI digests of target DNA. Consistent differences in the band pattern among three varieties of *G. graminis* (*G. tritici, avenae,* and *graminis*) were observed, indicating that such probes have considerable potential for use in the identification of these pathogens (Ward and Gray, 1992). Elliott et al. (1993) used a 188 bp DNA fragment derived from boiled mycelium of *G. graminis* for amplification by PCR as a probe for detection and identification of the pathogen from different grass hosts. Amplification of the specific DNA and presence of lobed hyphopodia in culture may be used for the identification of this pathogen. Using PCR assay, on the basis of sequence differences in their ribosomal RNA genes, *Verticillium dahliae, V. albo-atrum,* and *V. tricorpus* can be detected reliably, and a diagnostic set is now available for the investigation and monitoring of the *Ver-*

ticillium-potato pathosystem (Moukhamedov et al., 1994). The study to determine the relative efficacy and rapidity of detecting *Verticillium albo-atrum* in the soil and potato stem tissues by PCR assay using specific primers and media-plating method showed that the PCR assay was faster and more efficient requiring only 2 days for species identification, whereas the media-plating method needed more than 4 weeks. Both methods, however, detected *V. albo-atrum* and *V. tricorpus* readily in the soil samples indicating the ability of both fungi for survival and proliferation in the soil (Mahuku et al., 1999). Henson et al. (1993) showed that by using nested primers to amplify a fragment of mt DNA, it is possible to detect *G. graminis* in development and to diagnose resistance in pathogens to fungicides. The biomass of *Verticillium albo-atrum* or *V. dahliae* could be determined by adopting PCR assay (Hu et al., 1993). The assay was used to study the colonization of lucerne by *V. albo-atrum* and sunflower by *V. dahliae* comparatively. The study accurately showed the substantial differences between the two pathogens, and the results could be obtained more rapidly and accurately than by conventional cytological or maceration and plating techniques (Hu et al., 1993). Many fungal pathogens have been reported to have developed resistance/tolerance to methyl benzimidazole carbamate (MBC), as a result of point mutation at amino acid 198 in the β-tubulin subunit, causing a change from glutamic acid to alanine. The resistant and sensitive strains of *Botrytis cinerea* could be successfully diagnosed by a PCR-based assay (Martin et al., 1992).

Sequences of specific regions, such as the internal transcribed spacers (ITS), of ribosomal DNA may be amplified by PCR with universal primers and used to differentiate *Pythium* spp. that are difficult to identify on the basis of morphological characters. The restriction fragment probes from ITS 1 showed a high degree of species specificity to *P. ultimum* when tested by dot-blot hybridization against 24 other *Pythium* spp. There was no difference among 13 isolates of *P. ultimum* var. *ultimum* and var. *sporangiferum* from eight countries and two isolates of *Pythium* group G recently classified as *P. ultimum* (Levesque et al., 1994). The tandem arrays of 5S genes unlinked to the ribosomal DNA repeat unit present in some *Pythium* spp. have been used to prepare probes to target the genomic DNA of 92 species of *Pythium*. Probes specific for *P. ultimum* var. *ultimum* and *P. ultimum* var. *sporangiferum* were found to be species- or variety-specific. It was shown that 5S rRNA gene spacer sequences may be useful in defining species boundaries in the genus *Pythium,* as these sequences diverged rapidly after speciation (Klassen et al., 1996). Probes generated from the ITS region of ribosomal DNA of *Colletotrichum gloeosporioides* (*Glomerella cingulata*) amplified by PCR were employed to study 39 different isolates. These isolates were divided into 12 groups linked to host plant species and geographical origin (Mills et al., 1992). By using species-specific PCR primers for ITS regions, *Cylindrocarpon heteronema, Stagonospora nodorum, Septoria tritici,* and *Monosporascus* spp. were detected by Brown et al. (1993), Beck and Ligon (1995), and Lovic et al. (1995).

Phialophora gregata isolates from soybean, mungbean, and adzuki bean obtained from diverse geographical locations were differentiated based on the intraspecific genetic variation in the nuclear ribosomal DNA. A unique banding pattern, after digestion of PCR-amplified ITS and the 5' end of the large subunit rDNA with restriction enzymes, was exhibited by all 79 isolates. The isolates from soybean in the midwestern states of the United States and Brazil contained identical ITS sequences, whereas the ITS sequence of the isolates from adzuki bean from Japan showed 98% homology with the soybean isolates (Chen et al., 1996). *Monilinia fructicola* infecting plum fruit could be differentiated from two other related species, *M. fructigena* and *M. laxa,* by using the PCR primers specific to the 3' regions of the intron (present only in *M. fructicola*) together with the small subunit (SSU) rDNA primer NS 5. These primers amplified a 444-bp product only from *M. fructicola* and plum fruit tissue infected with *M. fructicola,* but not from *M. fructigena* and *M. laxa* (Fulton and Brown, 1997). *Fusarium oxysporum* f.sp. *canariensis* causes the wilt disease of Canary Island date palm (*Phoenix canariensis*), a highly prized ornamental palm. Two oligonucleotide primers designed to amplify a 567-bp fragment of *F. oxysporum* f.sp. *canariensis* were specific for the isolates of this pathogen. The DNA from 61 outgroup isolates were not amplified by these primers indicating that they can be used for the reliable detection of the pathogen and establishing *F. oxysporum* lineage (Plyler et al., 1999).

The genetic variations in different species of *Venturia* were determined by PCR amplification of a portion of the 18S rDNA gene, the internal transcribed spacers (ITS1 and ITS2), and the 5.8S rDNA gene. The optional group I intron in the 18S rDNA gene of *V. inaequalis* was detected in 75% of 92 strains collected worldwide. Four intron alleles were identified using sequence and restriction analysis of rDNA and three of these intron alleles were present in worldwide strains. The *Venturia* spp. strains were classified into three monophyletic groups based on the ITS 1–5.8S–ITS 2 sequences. The intron and ITS 1 alleles in *V. inaequalis* can be used as markers for subdividing populations of *V. inaequalis* and the relationship between species of *Venturia* can be established based on the ITS 1–5.8S–ITS 2 sequences (Schnabel et al., 1999)

Another application of PCR that has practical utility is for the evaluation and optimization of use of fungicides applied against fungal pathogens. By using two sets of oligonucleotides that could amplify a PCR product from DNAs of *Pseudocercosporella herpotrichoides* R and W types causing eyespot disease on cereals, the other common cereal pathogens such as *Fusarium* spp. and *Rhizoctonia solani* were discriminated by the PCR assay. The distinct PCR products amplified by specific oligonucleotides from R and W types could be quantified by incorporating PCR into a colorimetric microplate format using ELISA protocol. This assay system may be employed as an effective tool for optimizing application of fungicides to support the introduction of Unix for eyespot disease management (Beck et al., 1996).

The effectiveness of treatment of wheat seeds with bitertanol, fuberidazole, and fludioxonil against *Fusarium culmorum* and *Microdochium nivale* was assessed by employing PCR tests to identify these pathogens and quantify pathogen inoculum. Observations on seedling emergence and disease assessment by PCR tests were also made under field conditions (Edwards et al., 1998). The efficacy of four fungicides, prochloraz, cyprodinil, azoxystrobin, and flusilazole, was tested against wheat stem-base diseases caused by *Tapesia* spp., *Fusarium* spp., *Microdochium nivale,* and *Rhizoctonia cerealis*. PCR assay was employed to rapidly differentiate the components of the disease complex at early growth stages of wheat crop before fungicides were applied and to confirm fungicidal effects on disease development. It was concluded that fungicide application did not result in significant disease-related yield increases (Morgan et al., 1998), Wheat ear blight is caused by two fungal pathogens, *Fusarium culmorum* and *F. poae*. The disease severity was assessed by quantitative PCR assay and visual disease assessment (VDA) method to determine the effects of demethylase-inhibiting (DMI) fungicides perchloraz, and tebuconazole. The assessment of disease severity after fungicide application showed that PCR assay indicated higher levels of disease control by the fungicides than by VDA method. Yield analysis based on 1000-grain weight indicated that pathogen DNA content as determined by PCR more accurately predicted yield loss than VDA could indicate. Furthermore, the importance of identifying the causal organism up to the species level by PCR is underscored in this study, since yield reduction by *F. culmorum* is significant, whereas *F. poae* does not affect field significantly (Dooham et al., 1999).

It is essential to develop a rapid and sensitive method of detecting and differentiating fungicide-tolerant strains of fungal pathogens that develop following continued and indiscriminate application of fungicides. McRay and Cooke (1997) developed a PCR technique to identify the strain tolerant to thiabendazole (TBZ). Differentiation of TBZ-tolerant strains by conventional methods requires several weeks. Primers designed from the conserved region of the fungal β-tubulin gene were employed for amplification by PCR and the amplified product was sequenced. In the TBZ-tolerant strains, a change in the amino acid sequence from glutamic acid to alanine or glutamine was observed due to point mutation at codon position 198. It is possible to detect fungicide-tolerant strains by using species-specific PCR primers to amplify the desired regions in conjunction with restriction endonuclease to cause cleavage in sensitive isolates, providing reliable data for taking actions for fungicide resistance management. However, in some pathogenic fungi such as *Fusarium graminearum* molecular mechanism of resistance to fungicides like carbendazim may not depend on a mutation in the amino acid sequence. By using fungal β-tubulin consensus oligonucleotide primers B1 and B3, a 821-bp fragment of β-tubulin gene of *F. graminearum* was amplified by PCR. The DNA sequences of PCR products from strains sensitive or resistant to carbendazim were compared. No change in codon 165, 198, 200, and 257 was observed, suggesting the operation of a different mechanism that does not depend on amino acid mutation at these positions (Lu et al., 2000).

9.6.4   TaqMan PCR Detection

Conventional PCR-based assays need the use of ethidium bromide–stained agarose gel electrophoresis in conjunction with either hybridization/blotting procedures or different types of capture techniques. Detection of PCR products is time-consuming and labor-intensive with little possibility of automation. By using PCR-RFLP, the species and varieties of *Diaporthe* and *Phomopsis* have been identified (Zhang et al., 1997). This protocol, however, requires incubation of soybean seeds on PDA plates and extraction of DNA from mycelia prior to PCR amplification. The need for post-PCR processing is eliminated by the recent advances in PCR technology leading to real-time detection and automated product analysis.

The TaqMan PCR detection system utilizes TaqMan chemistry and the 7700 Sequence Detection System (SDS) (Perkin-Elmer Applied Biosystems, CA), which has been demonstrated to be extremely sensitive, capable of detecting even a single copy of DNA in some systems. A fluorogenic probe labeled on the 5' end with a reporter dye and on the 3' end with a quencher dye anneals between the PCR primers during PCR assay. *Taq* polymerase displaces a small portion at the 5' end of the probe, during the extension phase, creating a fork-like structure. The endogenous 5' to 3' nucleolytic activity of *Taq* polymerase is then initiated, resulting in cleaving of the probe and freeing of the reporter dye, which emits an intense fluorescent signal. Exponential synthesis of the amplicon occurs following repeated cycles of PCR, followed by a concomitant increase in reporter fluorescence automatically quantified by the SDS software. The 7700 SDS is an integrated hardware and software system that can be used for real-time quantification of nucleic acids. This system consists of a built-in 96-well thermal cycler, an argon ion laser that excites the samples at 488 nm, a charge-coupled device, a spectograph that records emissons ranging from 500 to 650 nm, and a complete software package for analyzing the data for each run automatically.

TaqMan primer/probe sets PL3, PL-5, and DPC-3 were designed based on the sequence data of ITS regions of rDNA of *Phomopsis longicolla, Diaporthe phaseolorum* var. *caulivora, D. phaseolorum* var. *meridionalis,* and *D. phaseolorum* var. *sojae* using Primer Express (Perkin-Elmer Applied Biosystems) for differentiation of variants of *Phomopsis* and *Diaporthe.* The set PL-3 amplified a 86-bp DNA fragment within the ITS 2 region of *P. longicolla,* whereas set PL-5 amplified a 96-bp fragment within the ITS 1 region of *P. longicolla* and *D. phaseolorum* varieties. The set DPC-3 amplified a 151-bp DNA fragment within the ITS 2 region of *D. phaseolorum* var. *caulivora.* The detection limit of TaqMan primer/probe sets was 0.15 fg (four copies) of plasmid DNA, whereas the PCR-

Table 9.6 Relative Cost and Time Required for Detection of *D. phaseolorum* and *P. longicolla* in Soybean Seeds per Reaction or Plate

Detection Technique	Time (hr)	Cost[a] (US $)
TaqMan single	<2	2.29
TaqMan two	<2	2.45
PCR-RFLP	5	6.70
PCR with Phom[b]	3	2.41
Plating assay	>280	0.90

[a] Cost includes cost of primers, reagents, buffers, etc.
[b] Specific primers designed for total infection of *D. phaseolorum* and *P. longicolla* as described by Zhang et al. (1997).
Source: Zhang et al., 1999.

RFLP technique could detect 100 pg of pure DNA. The frequency of detection of *P. longicolla* in soybean seed lots was similar in TaqMan and PCR-RFLP systems. However, results can be obtained more rapidly when compared to other detection techniques tested (Table 9.6) (Zhang et al., 1999).

9.6.5 Random Amplified Polymorphic Deoxyribonucleic Acid Technique

The random amplified polymorphic DNA (RAPD) method is a PCR technique which uses arbitrary primers and can be employed to distinguish races, strains, and pathogenic or nonpathogenic isolates of fungi. The primers used in this method are very short (10 or fewer bases) pieces of DNA from a desired source. It is highly probable that these primers find some complements in target DNA, producing a mixture of DNA fragments of various sizes. When the products from such a reaction are run on an electrophoresis gel, distinct banding patterns are produced, and somes of these patterns may prove to be specific to certain species or varieties or strains. The patterns themselves may be useful for detection and diagnosis of some pathogenic fungi, but some of the bands, in certain cases, may be cut out of a gel and sequenced to produce specific primers for more precise PCR analysis or probes for dot hybridization and other detection protocols.

The RAPD technique has been demonstrated to be simple, sensitive, and rapid for the detection, differentiation, and establishment of phylogenetic relationship between isolates of a morphological species of several fungal pathogens such as *Verticillium* spp. (Koike et al., 1996), *Gaeumannomyces graminis* (Fouly et al., 1996; Bryan et al., 1999), *Puccinia hordei* (Jennings et al., 1997), *Mycosphaerella fijiensis* (Müller et al., 1997), *Plasmodiophora brassicae* (Buhariwalla et al., 1995; Möller and Harling, 1996), *Tapesia* spp. (Nicholson et al., 1997), *Elsinoe* spp. (Tan et al., 1996), *Colletotrichum lindemuthianum* (Sicard et al., 1997), *Uncinula necator* (Délye et al., 1997), *Phytophthora* spp. (Zheng and Ward, 1998), *Fusarium oxysporum* f.sp. *phaseoli* (Woo et al., 1996), *F. oxysproum* f.sp. *erythroxyli* (Nelson et al., 1997), *F. oxysporum* f.sp. *ciceris* (Kelly et al., 1998), *Fusarium culmorum* and *F. graminearum* (Nicholson et al., 1998), *F. oyxporum* (Paavanen-Huhtala et al., 1999), *Rhizoctonia cerealis* (Nicholson and Parry, 1996), *R. solani* (Bounou et al., 1999), and *Pyrenophora teres* f.sp. *teres* (Weiland et al., 1999).

The DNA sequence with 232 bp obtained from random amplified polymorphic DNA of *Peronospora tabacina* had homology to *P. tabacina* DNA only, and this sequence was amplified by PCR by using required oligonucleotides. Using this DNA fragment, *P. tabacina* was detected in local lesions, systemic vascular infections, and other infected parts of tobacco plants. Prediction of a disease epidemic may be possible by the use of spore traps, followed by amplification of the specific DNA fragment. This procedure may be valuable to regulatory agencies and in epidemiological and ecological studies (Wiglesworth et al., 1994). Johanson et al. (1994) reported that the RAPD technique could be used to differentiate *Mycosphaerella fijiensis* and *M. musicola,* which cause Sigatoka disease of banana, *M. mussae* and *M. minima,* the two other species commonly found on banana. The DNA from these *Mycosphaerella* species produced distinct RAPD banding patterns with all PCR primers tested.

Dobrowolski and O'Brien (1993) reported that fragments obtained from the products of RAPD-PCR amplification of *Phytophthora cinnamomi* DNA were tested for specific hybridization to *P. cinnamomi* DNA. The DNA fragments that hybridized specifically to *P. cinnamomi* were cloned and could be used for detecting the fungal pathogen. Nicholson et al. (1993) showed that by using specific primers that produced distinct profiles of three races, the races of *Bipolaris maydis* could be differentiated. Genetic variation in *Magnaporthe poae* and *Erysiphe graminis* isolates was determined by using RAPD markers by Huff et al. (1994) and Mc Dermott et al. (1994). Kolmer et al. (1995) reported that there was a correlation of 0.58 between virulence and molecular similarity in 64 single uredinal isolates of *Puccinia recondita* f.sp. *tritici.* Virulence and molecular polymorphisms were determined by using near-isogenic wheat differential lines and RAPD technique employing 10 arbitrary decamer primers.

The pathogenic and nonpathogenic isolates of *Fusarium oxysporum* f.sp. *dianthi* from carnation were distinguished clearly by the RAPD method. The genetic markers useful for the detection of *F. oxysporum* f.sp. *dianthi,* causing carnation wilt disease, were identified by employing RAPD technique. Four amplification groups among the isolates of *F. oxysporum* f.sp. *dianthi* were identified based on the RAPD banding patterns generated by the primer OPA17. This primer was also useful for discriminating isolates of *F. oxysporum* f.sp. *dianthi* from nondianthi isolates. However, no direct correlation could be seen between the RAPD patterns and races of the pathogen (Hernandez et al., 1999). Isolates of *Crinipellis perniciosa,* which causes witches'-broom disease of cacao, and members of Sterculiaceae, Solanaceae, and Bixaceae were analyzed by RAPD, which revealed distinct RAPD banding patterns. Banding patterns were found to be similar among basidiocarps on the same broom in *Theobrama cacao.* However, differences among monospore cultures from the same basidiocarp could be detected (Andebrhan and Furtek, 1994). This procedure was found to be simple, rapid, and reproducible when compared to other methods of identification (Manulis et al., 1994).

For the identification of strains of *Trichoderma,* increasingly used as biocontrol agents, use of serological methods or isoenzyme patterns has not offered reliable results. By using RAPD with 10 arbitrary oligonucleotide primers, the strains of *Trichoderma* could be consistently distinguished, especially the isolate T-39 (the strain of *T. harzianum* used commercially as a biocontrol agent against *Botrytis cinerea*). The procedure developed by Zimand et al. (1994) requires only a small amount of DNA and less time and does not involve the use of radioisotopes (Appendix 9[xv]). *Colletotrichum gloeosporioides* f.sp. *malvae* is used for the control of the weed *Abutilon theophrasti.* RAPD technique was used to identify isolates of *C. gloeosporiodes* f.sp. *malvae* with greater virulence to enhance of the effectiveness of the biocontrol agent. All isolates tested were highly pathogenic on *A. theophrasti* and they did not show significant differences in the RAPD patterns. However, differences in the RAPD patterns could be used as the basis for the differentiation of isolates of *C. gloeosporioides* f.sp. *malvae* from other *Colletorichum* spp. (Kutcher and Mortensen, 1999).

The scab disease affecting citrus is caused by *Elsinoe fawcetti* (citrus scab), *Sphaceloma fawcetti* var. *scabiosa* (Tryon's scab in Australia), and *E. australis* (sweet orange scab), which do not show significant morphological differences. But their pathogenicity to different *Citrus* spp. varies. The sequence analysis of ITS region of rDNA and restriction analysis of amplified ITS region with several endonucleases could reveal the differences that may be used to differentiate *E. australis* from *E. fawcettii* and *S. fawcetti* var. *scabiosa.* RAPD analysis further demonstrated that *E. fawcettii* isolates from Australia and Florida were more closely related to each other than to *E. australis* isolates from Argentina. RAPD profiles further differentiated the Australian isolates from all Florida isolates. The two pathotypes identified in Australia could be differentiated by RAPD profiles indicating the correlation between the results of RAPD analysis and pathogenicity tests (Tan et al., 1996). By using random 10-base primers for amplification of the DNA of *Plasmodiophora brassicae,* causing club root disease, RAPD profiles were prepared for three single-spore isolates of the pathogen. The RAPD profiles were compared with race classification based on inoculation of the European club root differentials (ECD) series of *Brassica* hosts. Among the 40 primers tested, these primers provided isolate-specific profiles and one primer gave profiles that corresponded with ECD race classification. This study thus clearly indicates that RAPD profiling offers a faster means of race classification in *P. brassicae* (Möller and Harling, 1996).

Genetic variation in the isolates of *Uncinula necator,* causing grapevine powdery mildew disease, from Europe and India were studied by RAPD analysis and three main groups could be differentiated. The majority of the European isolates (53) clustered together in one group, whereas others (nine) formed a second group. Of the Indian isolates, 15 isolates formed one group that was recognized as a subgroup of one of the group of European isolates, while 13 other isolates formed a distinct group. A PCR primer derived from a RAPD fragment was found to be specific for Indian isolates and it has the potential for use for rapid identification of field isolates existing in India (Délye et al., 1997). The name *Alternaria alternata* has been applied to a variety of morphologically distinct taxa, because of the morphological plasticity of *Alternaria* spp. under nonstandard cultural conditions. RAPD-PCR analysis of 216 isolates of *Alternaria* was carried out using total genomic DNA and three different primers. Morphological groups or species were resolved as distinct branches of the dendrogram: *Alternaria gaisen* (= *A. kikuchiana, A. alternata* Japanese pear pathotype group 2), *A. longipes* (*A. alternata* tobacco pathotype group 5), the *"tenuissima"* group (group 5), the *"arborescens"* group (group 3), and the *"infectoria"* group (group 6). Roberts et al. (2000) observed that the "pathotype" system should be abandoned, as there is no predictive value relative to observable morphological and genetic characters.

The usefulness of RAPD analysis for classification of *Phytophthora* spp. infecting a diverse range of host plants has been brought out by Zheng and Ward (1998). Ten randomly chosen 10-mer primers were employed to study the variations among 39 isolates of *P. citrophthora, P. parasitica, P. capsici, P. palmivora,* and *P. meadii* from rubber and citrus trees and *P. colocasiae* from taro (yams). RAPD profiles generally were similar within each species of *Phytophthora* and were different between species. None of the primers could differentite all the six species studied. However, the pooled data from all primers established that the isolates of each species clustered together forming six groups corresponding to the six morphological species. Moreover, the group corresponding to *P. citrophthora* could be subdivided into groups that were related to host plant species infected and geographical locations. The RAPD analysis can be used for confirmation and validation of classical taxonomic classification.

The pathogenic and nonpathogenic isolates of *Fusarium oxysporum* f.sp. *phaseoli* (FOP) were tested by RFLP technique using five different restriction enzymes and RAPD analysis employing four different primers. The banding patterns exhibited correspondence to the vegetative compatibility groups (VCGs), but not to pathogenic races. However, RFLP and RAPD markers could differentiate nonpathogenic and self-incompatible isolates of FOP (Woo et al., 1996). The genetic fingerprinting and RAPD technique were used to a differentiate the pathotypes of

Fusarium oxysporum f.sp. *ciceris* (FOC) by amplification of DNA by RAPD. Yellowing and wilt-inducing pathotypes of FOC could be recognized. Identification of these two pathotypes was confirmed by inoculating chickpea differentials (Kelley et al., 1994). Later a combination of RAPD and PCR was employed to differentiate the wilt-inducing and yellowing-inducing isolates of FOC infecting chickpea. A 1.6-kb fragment of RAPD was present in wilt-inducing isolate, but not in yellowing-inducing isolate of FOC or in other fungi tested. Specific primers designed from the sequence data of this DNA fragment were employed in PCR to detect the pathogen genomic DNA in symtomless chickpea plants at 16 days after inoculation. PCR products were detected in the DNA extracted from roots and stem tissues, but not in the leaves of the same chickpea plant (Kelly et al., 1998). The RAPD analysis of DNA from isolates of *Pyrenophora teres* f. *teres* (causing barley net blotch disease) with high virulence and low virulence and the progeny isolates of the cross between the parental isolates was carried out, with a view to selecting molecular genetic markers associated with virulence phenotype. Five RAPD markers associated with low virulence were identified. RAPD technique can be utilized to tag genetic determinants of virulence in fungal pathogens (Weiland et al., 1999).

RAPD analysis was employed to identify amplification products that could differentiate *Fusarium culmorum* causing stem base disease of winter wheat and *F. graminearum* infecting wheat grains. Competitive PCR assay with primer specific to *F. culmorum* was used to assess the extent of fungal colonization by determining the fungal DNA content. The trichothecene-producing isolates of *F. graminearum* were predominantly present on wheat grains as determined by the competitive PCR, suggesting that trichothecene may act as a virulence factor in colonizing wheat grains (Nicholson et al., 1998). *Rhizoctonia solani* AG-3 causes significant losses in potato in eastern Canada and the United States, necessitating rapid and reliable identification of this pathogen. By employing RAPD technique, specific genetic markers of AG-3 isolates were identified. RAPD amplification revealed the presence of a specific DNA fragment (2.6 kbp) in all AG-3 isolates. By using the restriction enzyme *Xho I*, a PCR-based restriction mapping method was developed for the identification the AG-3 isolates. These virulent isolates could be rapidly detected in both plant tissues and infested soil samples (Bounou et al., 1999).

9.6.6 Single-Strand Conformation Polymorphism Analysis

Single strand conformation polymorphism (SSCP) analysis has been employed to determine genetic variations in plant viruses as discussed earlier (Chapter 9.4.1.1). SSCP analysis can detect small nucleotide changes in the fungal DNA fragments by electrophoresis and it can be a tool for distinguishing different species of a fungal genus. This technique was used to detect DNA polymorphism in the ITS 1 region of DNA of nine species of *Melampsora* causing the willow rust diseases in Japan.

SSCP analysis is preceded by the amplification of the ITS 1 region of rDNA fragments using the primer pair ITS 1-F and ITS 4-B. The amplified DNA fragments are partially purified and dissolved in TE buffer consisting of Tris (45 mM) and EDTA (2 mM). Some DNA fragments are digested with restriction enzyme *Rsa1*. The preparations (2 μl) containing the fungal DNA fragments (about 50 μg) are mixed with 8 μl of loading buffer containing formamide (95%), xylene cyanol (0.5%), and bromophenol blue (0.5%). This mixture is then heated at 95°C for 5 min and cooled on ice for 10 min. Then the samples (2 μl) are applied onto a poly-acrylamide gel (6.5%) containing acrylamide:bisacrylamide (49:1), the gel size being 13 × 12 × 0.1 cm, without glycerol in TBE buffer consisting of Tris (45

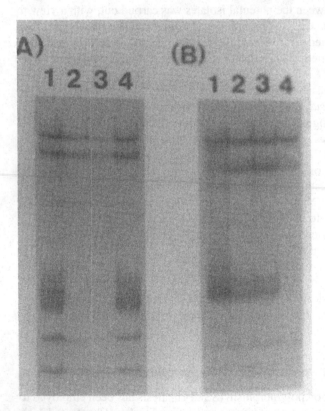

Figure 9.12 Improved single-strand conformation polymorphism (SSCP) patterns of the ITS 2 regions from *Melampsora cheliodonii-pierotii* and *M. coleosporioides* (A) and *M. epitea* and *M. humilis* (B) by electrophoresis on a gel without glycerol at 5°C. (Phytopathological Society, Japan.) (Courtesy of Nakamura et al., 1998.)

mM), boric acid (45 mM), and EDTA (2 mM) to perform electrophoresis at 200 V for 2–8 hr in a chamber at 5°C or 25°C followed by silver staining.

The SSCP patterns in the amplified ITS 1 region of *M. capracearum, M. epiphylla, M. larici-urbaniana, M. microsora,* and *M. yezoensis* were found to be species-specific. In contrast, the SSCP patterns between *M. chelidonii-pierotii* and *M. coleosporioides* and between *M. epitea* and *M. humilis* did not show significant variations indicating their close relationship. The PCR-SSCP analysis to detect genetic variations in *Melampsora* spp. was found to be more sensitive than PCR-RFLP analysis (Fig. 9.12). The SSCP technique has great potential in the case of fungal pathogens that cannot be differentiated by conventional methods (Nakamura et al., 1998).

9.7 DETECTION OF BACTERIAL PATHOGENS

9.7.1 Nucleic Acid Hybridization Technique

Detection and identification of plant bacterial pathogens may be possible by following conventional isolation and by employing physiological and pathogenicity tests which are time-consuming. But they yield doubtful results. Development of diagnostic DNA probes specific for the pathogen(s) concerned has helped in detecting, differentiating, and quantifying the population of bacteria very rapidly, especially the pathogens known to be transmitted through seeds and other planting materials. Specific DNA probes are available for detecting *Erwinia carotovora* subsp. *atroseptica* (Ward and DeBoer, 1984), *Pseudomonas syringae* pv. *phaseolicola* in beans (Schaad et al., 1989; Prossen et al., 1991), *Xanthomonas campestris* pv. *phaseoli* in beans (Gilbertson et al., 1989), *Clavibacter michiganense* subsp. *michiganense* (Thompson et al., 1989) and *P. syringae tomato* (Cuppels et al., 1990) in tomato, *Erwinia carotovora* (Ward and De Boer, 1990) and *C. michiganense* subsp. *sepedonicum* (Verreault et al., 1988) in potato tubers, *X. oryzae* pv. *oryzae* and *X. oryzicola* in rice (Cottyn et al., 1994), and *Pseudomonas glumae* in rice seeds infected by bacterial grain rot (Tsushima et al., 1994) (Appendix 9[xvi]).

X. axonopodis pv. *vesicatoria* and *X. vesicatoria* causing bacterial leafspots in tomato and peppers were detected by employing a 1.75-kb fragment (KK 1750) that preferentially hybridized to both pathogens (Kuflu and Cuppels, 1997). Fluorescent in situ hybridization (FISH) technique was used to detect *Ralstonia solanacearum* race 3 biovar 2, causing brown rot disease of potato. The probe RSOLB can be used for specific detection of *R. solanacearum* in pure cultures and potato tissues, which show strong fluorescent signal (Wullings et al., 1998). For the detection of *Clavibacter xyli* subsp. *xyli,* causing sugarcane ratoon stunting disease, the tissue blot hybridization assay was developed by using 560-bp PCR-amplified product as a probe. This probe was amplified from the intergenic region of the 16S/23S rDNA of the pathogen DNA with two universal primers. This assay procedure has the potential for large-scale application along with other diagnostic methods (Pan et al., 1998).

The diagnostic DNA probes may show specificity at the genus, species, pathovar, or race level. A probe that can be used for taxonomic comparisons is prepared by identifying a DNA fragment that is present only in the bacterial species to be identified, but not in other closely related species. The DNA sequences of 16S ribosomal RNA of different bacterial species have been compared for selecting probes specific at genus level. De Parasis and Roth (1990) developed a DNA probe for comparison of partial sequences of 16S rRNA from 52 strains of bacteria including *X. oryzae* pv. *oryzae*. The rRNA molecule is present in large numbers (>10,000 copies/cell) in actively growing bacterial cells and has diverse sequence regions that can be correlated with phylogenetic relatedness. The sensitivity of the probe derived from unique 16S rRNA sequences is greatly increased, because of the large number of copies present in a bacterial cell. A DNA sequence based on the 16S rRNA which hybridized only with plant pathogenic pathovars of *X. oryzae* pv. *oryzae* has been identified. By screening cloned DNA fragments at random, species- and subspecies-specific probes for *Erwinia carotovora* (Ward and De Boer, 1990) and *C. michiganense* subsp. *sepedonicum* (Verreault et al., 1988) and *C. m. michiganense* (Thompson et al., 1989) have been developed. Tsushima et al. (1994) prepared a probe specific for *Pseudomonas glumae*, causative agent of bacterial grain rot of rice for the detection of the pathogen in rice seeds by using two restriction enzymes, Eco RI and Kpn 1. The resultant fragment (PG2Ia) hybridized to all strains of *P. glumae*, but not to *X. campestris* pv. *oryzae*, *X. campestris* pv. *campestris*, other species of *Pseudomonas, Agrobacterium, Erwinia*, and gram-positive *Clavibacter, Arthrobacter*, and *Carobacterium* tested. A probe for plasmid DNA present in *Xanthomonas campestris* pv. *citri* was employed to distinguish the pathotypes of this bacterial pathogen (Pruvost et al., 1992).

Dreier et al. (1995) reported that by employing Southern hybridization with DNA probes derived from plasmid-borne genes *cel* A (encoding an endocellulose) and *pat*-1 (involved in pathogenicity), *Clavibacter michiganensis* subsp. *michiganensis*, causative agent of tomato wilt and canker, could be detected. The *cel* A probe differentiated the subspecies of *C. michiganensis*, while *pat*-1 could distinguish virulent and avirulent strains of *C. michiganensis* subsp. *michiganensis*. *C. michiganensis* subsp. *sepedonicus* causing potato ring rot disease was detected by radiolabeled probes obtained from unique three single-copy DNA fragments designated Cms50, Cms72, and Cms 85 isolated from CS 3 strain of *C. michiganensis* subsp. *sepedonicus* by subtraction hybridization. These probes specifically hybridized to all North American strains of the pathogen tested by Southern hybridization. All strains including plasmidless and nonmucoid strains of *C. michiganensis* subsp. *sepedonicus* were specifically detected. The detection limit of a PCR assay using a primer pair that amplified Cms 85 was found to be 100 colony forming units (cfu)/ml of the pathogen cells in potato core fluid (Mills et al., 1997).

Ryba-White et al. (1995) compared the strains of *Xanthomonas oryzae* pv. *oryzae* (Xoo) from Africa, North America, and Asia by RFLP analysis using three repetitive DNA sequences cloned into Bluescript as probes. Total genomic DNA from each strain was digested with Eco RI, separated by electrophoresis, and blotted to membranes on which hybridization with ^{32}P-labeled probes was carried out. They found that the DNA banding patterns of African and Asian Xoo strains were substantially different, representing distinct geographically isolated populations. The RFLP patterns for strains from North America had least similarity with any other group of strains. This study suggests that RFLP analysis, though useful for comparing genomes within a pathovar, cannot be used for reclassifying the strains on the basis of RFLP analysis only.

Pathovars are identified by their pathogenic potential, differentiated by a specific set of differential cultivars. Development of a specific probe for the DNA fragment related to the production of toxic metabolite involved in pathogenesis is possible. By using the probe for the DNA related to phaseolotoxin, pathovars of *P. s. phaseolicola* could be identified (Schaad et al., 1989; Prossen et al., 1991). Likewise probe for DNA involved in coronatine toxin production has been developed to detect *P. s. tomato* (Cuppels et al., 1990). The role of extracellular cell-wall-degrading enzyme pectin methyl esterase (Pme) in pathogen development and pathogenesis can be studied by identifying DNA fragment coding for enzymes. Primers designed from *pme* sequence generated an 800-bp DNA fragment of *Ralstonia solanacearum* and this fragment, when employed as probe, identified Pme-encoding plasmids from the *R. solanacearum* genomic library. This study demonstrated that Pme was required for the growth of *R. solanacearum* but not for virulence, since a *pme* mutant was just as virulent as the wild-type strain on tomatoes, aubergines (eggplant), and tobacco (Tans-Kersten et al., 1998). The DNA probes for T-DNA in Ti-plasmid can be used for the detection of *Agrobacterium tumefaciens* (Burr et al., 1990). Probes can also be developed to monitor the distribution of genes encoding for copper resistance in *X. c. vesicatoria* (Garde and Bender, 1991) and streptomycin resistance in *P. papulans* (Norelli et al., 1991). Development of probes specific for races has not been very successful, except in the case of a race-specific DNA probe for *P. solanacearum* race 3 (Cook and Sequira, 1991).

To perform colony hybridization, the bacterial suspension is prepared by macerating the infected tissue in liquid and incubated for diffusion of the bacteria into the liquid. The bacterial suspension is then spread onto the culture medium, permitting its growth, and covered with a nylon or nitrocellulose membrane. The filter membrane is removed after sufficient growth of the bacteria. Soil extracts or seed-soak washes may also be tested for the presence of the bacteria. By employing an appropriate DNA probe the bacteria can be detected (Cuppels et al., 1990; Ward and De Boer, 1990).

The sensitivity of dot-blot hybridization for detection of bacterial pathogens may vary, depending on the probe and bacterial species, and the minimum number of cells required for detection, which varies from 200 colony forming unit (cfu) to 10^6 cfu. As the number of bacterial cells in seed-soak washes is likely to be less, it is necessary to grow the bacteria first on a semiselective medium for 96 hr and to concentrate bacteria before spotting on the nitrocellulose membrane as in the case of *P. c. phaseolicola* (Schaad et al., 1989). Citrus greening disease caused by *Liberobacter* (tentatively named) was detected by employing greening fastidious bacteria (GFB)-specific DNA fragment (0.24 kb) labeled with biotinylated nucleotides by a polymerase chain reaction (PCR)-labeling technique. The GFB was detected using this probe in various citrus hosts including mandarins, tangerins, oranges, and pummelos. Detection of GFB in samples from several Asian countries, but not in those from South Africa, was possible indicating the feasibility of using this procedure in quarantine programs (Hung et al., 1999).

9.7.2 Restriction Fragment Length Polymorphism

Restriction fragment length polymorphism (RFLP) analysis employing specific restriction enzyme has been used to detect and differentiate bacterial pathogens and their strains. Hartung and Civerolo (1989) used the genomic DNA prepared from 21 strains of *Xanthomonas campestris* pv. *citri*, 14 strains of *X. campestris* isolated from Florida citrus nurseries, and 10 strains of five other pathovars of *X. campestris* for RFLP analysis. A significant separation between the strains of *X. campestris* and *X. c. pv. citri* was observed, indicating that *X. campestris* that was prevalent in citrus nurseries was not a form or strain of *X. c. pv. citri*. The cosmid clone PXCF 13–38 isolated from the genomic library of *Xanthomonas axonopodis (campestris)* pv. *citri* covers almost the entire *hrp* genes cluster (involved in hypersensitive response and pathogenicity). The clone was used as a probe for RFLP analysis of Xanthomonads. A dendrogram that was different from the ones based on biochemical characteristics was generated for RFLP analysis and it was found to be useful for the identification of various Xanthomonads (Kanamori et al., 1999).

A simplified method for RFLP analysis of *X. oryzae* pv. *oryzae* was developed by Raymundo and Nelson (as cited by Leach and White, 1991). Digestion of bacterial DNA with the restriction enzyme Pst 1 resulted in only a few high-molecular-weight DNA fragments which formed distinct patterns for different strains on electrophoresis. The RFLP patterns could be recognized without Southern blot analysis. Most of the RFLP patterns revealed by digestion with Pst 1 were found to be specific for a group of strains within a single race. The strains of race 2, commonly prevalent in the Philippines, could be distinguished by RFLP analysis. The strains of *X. oryzae* pv. *oryzae* found in the United States are considered to be not closely related to the Asian strains because of the differences in RFLP patterns, and hence it is suggested that the bacterium that causes rice leaf blight may be a distinct pathovar of *X. oryzae* (Leach and White, 1991).

The genetic diversity of *X. oryzae* pv. *oryzae* (Xoo) strains collected from 18 locations in India was assessed. It was observed that strains of Xoo belonging to a single lineage were distributed in 16 of the 18 locations sampled. All strains with this lineage belonged to pathotype 1b as shown by pathotyping analysis (Yashitola et al., 1997). The strains of *X. campestris* infecting cereals and grasses were classified into three RFLP groups, which corresponded to the groups recognized based on the results of biochemical and physiological tests and host range. RFLP group 1 enclosed strains pathogenic to barley but not to wheat, rye, *Bromus, Lolium, Agropyron,* and oat. Strains that could infect wheat, rye, and grasses mentioned above were placed in RFLP group 2. Only one strain with the same host range as group 2 had to be placed in a separate RFLP group 3. The strains in group 1 and 3 showed highly conserved RFLP patterns, whereas group 2 could be further subdivided. This study indicates that RFLP analysis can be used to distinguish strains of *X. campestris* causing bacterial leaf streak disease of cereals (Alizadeh et al., 1997). The different *Burkholderia* spp. infecting rice can be distinguished by RFLP analysis of ITS region (5S plus ITS 1 and ITS 2) of rDNA repetitive units amplified by PCR. This technique was used to differentiate *B. glumae, B. gladioli, B. plantarii,* and *B. vandii.* The DNA samples from rice leaf sheath tissues inoculated with *B. glumae* and *B. gladioli* and the cultured bacterial cells of the two species were subjected to digestion with *HhaI* and *Sau 3AI* restriction enzymes followed by PCR-RFLP, and they formed polymorphism identical to those formed by pure cultures. Thus the pathogens present in the naturally infected rice plant tissues can be directly distinguished (Ura et al., 1998).

9.7.3 Polymerase Chain Reaction

When the target bacterium is contaminated with saprophytic bacteria, it will be difficult to detect by dot-blot hybridization. Employing a PCR-based assay will be useful under such condition in addition to enhancing the sensitivity of the detection as observed in the case of several bacterial pathogens (Table 9.7). For the detection of *P. s. phaseolicola,* a PCR-based assay using the primers from DNA sequences of the phaseolotoxin gene was employed. The assay was not affected by the presence of a high population of nontarget bacteria, and it was highly sensitive in detecting the target bacteria at 1–5 cfu/ml of seed-soak wash (Prossen et al., 1991). *Xanthomonas campestris* pv. *phaseoli* was detected by using primers from plasmid DNA in PCR assay. As little as 10–100 fg of *X. c. phaseoli* DNA (1–10 CFU) was sufficient for detection of this bacterial pathogen (Audy et al., 1994). The PCR assay could detect 34 *Agrobacterium tumefaciens* strains of *Vitis* spp. re-

Table 9.7   Target Sequences of Bacterial Plant Pathogens Amplified by PCR

Pathogen	Target sequence/primers	Reference
Agrobacterium tumefacines	Vir D2 sequences	Haas et al., 1995
	T-DNA sequences	Sachadyn and Kur, 1997; Cubero et al., 1999
Burkholderia gladiolii	Primers PL-12f/PL11r and GL 13f/GL 14r	Takeuchi et al., 1997
B. glumae, B. plantarii, and B. vandii	ITS1 and 2 regions of rDNA	Ura et al., 1998
Clavibacter michiganensis subsp. *michiganensis*	Primers CM3 and CM4 Pathogen-specific DNA sequences	Sousa Santos et al., 1997 Niepold, 1999
C. michiganensis subsp. *sepedonicus*	Fragment (434 bp) pathogen DNA	Karjalainen et al., 1995; Lee et al., 1997
C. xyli subsp. xyli	Intergenic spacer region of 16S–23S rDNA sequences	Pan et al., 1998; Fegan et al., 1998
Erwinia amylovora	16S rDNA sequences	Bereswill and Geider, 1996
	DNA fragment of plasmid pEA 29	McManus and Jones, 1996
E. carotovora subsp. *atroseptica*	Pectate-lyase-encoding gene sequence	Hélias et al., 1998
	Conserved regions of 16S rRNA gene	Toth et al., 1999
E. carotovora subsp. *carotovora*	Pathogen DNA sequence (Probelia kit)	Fechon et al., 1998
	16S rRNA sequences	Toth et al., 1999
E. herbicola pv. *gysophilae*	Cytokinins or IAA-biosynthetic gene sequences	Manulis et al., 1998
Liberobacter sp. and *L. asiaticum*	16S rDNA sequence	Nakashima et al., 1996, 1998
Pseudomonas savastanoi (syringe) pv. *phaseolicola*	Gene sequence encoding phaseolotoxin-insensitive ornithyl carbamoyl transferase (arg K)	Mosqueda-Cano and Herrera-Esterella, 1997
P. syringae pv. *atropurpurea*	Plasmid pCOR1-associated with pathogenicity and coronatine synthesis	Takahashi et al., 1996
P. syringae pv. avellanae	Unique DNA fragment sequence	Scortichini et al., 1998
Ralstonia (Pseuduomonas) solanacearum	16S rDNA sequence	Seal et al., 1999
	Primers PS96H and PS96I	Hartung et al., 1998

Table 9.7   Continued

Pathogen	Target sequence/primers	Reference
Xanthomonas albilineans	Unique DNA sequence	Pan et al., 1997
	Primers PGBL1 and PBGL2	Pan et al., 1999
	Primers XAF1/XAR1	Wang et al., 1999
Xanthomonas axonopodis (*campestris*) pv. *citri*	Pathogen DNA sequences	Gillings et al., 1995
	Plasmid DNA sequence	Hartung et al., 1996
	Intergenic region between 16S and 23S rRNA	Miyoshi et al., 1998
X. axonopodis pv. *glycines*	DNA sequence encoding glycinecin	Oh et al., 1999
X. axonopodis pv. *manihotis*	Pathogenicity gene sequence (898 bp)	Verdier et al., 1998
X. axonopodis pv. *phaseoli and X. axonopodis* pv. *phaseoli* var. *fuscans*	Primers Xf1 and Xf2 with sequences conserved amplified region (SCAR)	Toth et al., 1998
X. fragariae	Repetitive DNA sequences	Opgenorth et al., 1996
	Pathogen DNA sequences	Hartung and Pooler, 1997
	Oligonucleotide primers	Mahuku and Goodwin, 1987
	Primers specific to *hrp* gene	Zhang and Goodwin, 1997
X. hortorum (*campestris*) pv. *pelargoni*	Pathogen-specific DNA sequences	Sulzinski et al., 1998
X. oryzae pv. *oryzae*	Pathogen-specific DNA sequences	George et al., 1997
X. translucens pv. *cereals*	Intergenic spacer region between 16S and 23S	Maes et al., 1996
Xylella fastidiosa	tRNA consensus primers	Beretta et al., 1997
	Pathogen-specific DNA sequence	Berisha et al., 1998

liably, and in most cases the PCR results confirmed the results of pathogenicity tests and DNA-slot blot hybridization (Dong et al., 1992). By employing two PCR primers based on the sequences of *vir* D2 and *ipt* genes, a wide variety of pathogenic strains of *A. tumefaciens* were detected. The T-DNA-borne cytokinin synthesis gene was detected by the primers corresponding to sequences of *ipt* gene in *A. tumefaciens* but not in *A. rhizogenes* (Haas et al., 1995). The *tms2* genes present in T-DNA coding for indole acetamide amidohydrolase is required for the pathogenicity of *A. tumefaciens*. Primers flanking a 220-bp fragment of one of the conserved regions of *tms2* gene were designed for PCR amplification, which revealed the presence of T-DNA in infected plants and infested soils (Sachadyn and Kur, 1997). PCR assay has been found to be the most efficient method for detecting *A. tumefaciens* in more than 200 samples including naturally infected almond, peach, apricot, rose, tobacco, tomato, raspberry, grapevine, and chrysanthemum. The pathogen was more effectively detected in crown and root tumors than in aerial tumors (Cubero et al., 1999).

The DNA probes generated from the fragment obtained after digestion of genomic DNA of *Xylella fastidiosa* (Grape Pierce's disease bacterium) and amplification by PCR were used to detect the bacterium in grape petiole and citrus stem tissues. Detection of the bacterial pathogen by PCR was 100 times more sensitive than by ELISA; the limits of detection were 1×10^2 cfu/ml for PCR and 2×10^4 cfu/ml for ELISA. The two pathotypes infecting grape and citrus could be differentiated by PCR (Minsavage et al., 1994). Further studies by Beretta et al. (1997) showed that three different fingerprint groups could be recognized when PCR amplification of DNA of strains of *X. fastidiosa* with tRNA consensus primers was carried out. The strains causing citrus variegated chlorosis (CVC) and mulberry leafscorch (MLS) were found to be different from other strains. Primers designed on the sequence of ITS region of rDNA of CVC strain detected the pathogen in trees with CVC symptoms in a survey conducted in Brazil. The occurrence of Pierce's disease (PD) in Europe was confirmed by PCR assay and this PD strain was found to be closely related to US-PD strains and several other strains of *X. fastidiosa* (Berisha et al., 1998). In a later study by Chen et al. (2000), the usefulness of 16S rDNA sequences as signature characters for the identification of *X. fastidiosa* was demonstrated. The 16S rDNA sequences of all *X. fastidiosa* strains studied were highly homologous and characteristically different from other bacteria, as revealed by the PCR amplification with two conserved primers.

By using PCR based on primers derived from the *pat-1* region of the plasmid (involved in pathogenicity), virulent strains of *Clavibacter michiganensis* subsp. *michiganensis* could be detected in homogenates of infected tomato plants and contaminated seeds (Dreier et al., 1995).

Li et al. (1995) reported that a pair of PCR primers (Sp1f and Sp5r) specifically amplified a 215 bp fragment of genomic DNA of *Clavibacter michiganensis* subsp. *sepedonicus,* causal agent of bacterial ring rot disease of potato, but did not amplify DNA from phenotypically and serologically related bacteria isolated from potato stem or tubers. Comparison of sensitivity of detection of PCR, ELISA, and immunofluorescence (IF) using monoclonal antibodies (MABs) shows that detection of the bacterial pathogen by PCR was more sensitive than that by the other two assay methods. Tubers from ring rot-infected plants which had negative results in ELISA and IF tests, gave positive results in PCR (Table 9.8). De Boer and Ward (1995) showed that by using primers capable of specifically amplifying the fragment of genomic DNA of *Erwinia carotovora* subsp. *atroseptica,* the pathogen could be reliably detected in potato stem and tuber tis-

Table 9.8 Comparative Sensitivity of Detection of *Clavibacter michiganensis* subsp. *sepedonicus* by ELISA, IF, and PCR Assays

Sample No.	Symptom rating[a]	ELISA (OD at 405 nm)	IF[c]	PCR[d]
1	1	0.059 (−)[b]	0.7 (−)	++
2	2	0.054 (−)	1.2 (−)	+
3	3	0.040 (−)	~40	+
4	1	0.047 (−)	~40	+
5	1	0.244	>50	+++
6	3	0.287	>50	+++
7	2	0.253	>50	+++
8	2	0.098 (−)	>50	+++
9	1	0.082 (−)	8.0	+++
10	2	0.172	−40	+++
11	2	0.261	>50	+++
12	2	0.275	>50	+++
13	1	0.103	7.0	+
14	2	0.052 (−)	8.0	+
15	1	0.040 (−)	9.0	+
16	4	0.272	>50	+++
17	3	0.097 (−)	7.0	++
18	3	0.198	>50	+++
19	1	0.080 (−)	6.0	++
20	3	0.318	>50	+++
21	1	0.093 (−)	1.8 (−)	+
22	1	0.066 (−)	1.0 (−)	+
23	1	0.063 (−)	0.6 (−)	+
24	1	0.045 (−)	1.0 (−)	+

[a] Symptom rating: 1—no symptom, 2—slight symptom; 3—well developed symptom; 4—rot.
[b] Tests negative.
[c] Average number of fluorescing cells/microscopic field.
[d] Intensity of bands in ethidium bromide stained agarose gel; + weak, ++ moderate and +++ strong band.
Source: Li et al. (1995).

sues. The PCR method was more sensitive than ELISA employing MABs (Table 9.9). Since infected seed potatoes are the primary sources of infection, detection of low levels of bacterial pathogens in tubers is essentially required for elimination of infected tubers. Hence PCR, with higher sensitivity of detection, can play an important role in disease management programs. This requirement was fulfilled by the development of a one-step PCR-based method for the detection of all five species of *E. carotovora* including subspecies *carotovora* and *atroseptica* and

Table 9.9 Comparison of ELISA and PCR for Detection of *Erwinia cartovora* subsp. *atroseptica* in Potato Tissues

Potato tissue	No. of samples tested	ELISA	No. of samples positive by PCR
Stem			
Symptomatic	25	+	25
Asymptomatic	8	+	6
Seed tubers	52	−	11
	6	+	2
	14	−	2
Progeny tubers			
Symptomatic	5	+	5
Asymptomatic	35	+	32
	25	−	3

Source: De Boer and Ward (1995).

all pathovars/biovars of *E. chrysanthemi* in micropropagated potato plants. The primers SR3F1 and SRIcR based on conserved region of the 16S rRNA gene amplified a DNA fragment of 119 bp from all strains (65) tested. When an enrichment step was used prior to PCR amplification, the sensitivity of detection was enhanced by about 200 fold (Toth et al., 1999).

Schaad et al. (1995) developed the BIO-PCR technique, which combines biological and enzymatic amplification of PCR targets. Bean seeds are soaked overnight, and the aqueous extract is plated onto a general agar medium and incubated for 45–48 hr. The plates are then washed with water to collect the bacterial cells, and aliquots of the washings are subjected to two consecutive cycles of PCR without prior DNA extraction using nested pairs of primers required to amplify the *tox* (phaseolotoxin) gene region of *Pseudomonas syringae* pv. *phaseolicola.* Another PCR protocol using two oligonucleotides designed based on the sequences of the gene encoding the phaseolotoxin-insensitive ornithyl carbamoyl transferase (arg K) of *P. savastanoi (syringae) phaseolicola* was developed for the detection of the pathogen in water extracts of soaked bean seeds. The sensitivity of the technique could be improved by allowing the bacteria present in seed extract to multiply in a semisolid medium for 18 hr prior to PCR amplification (BIO-PCR). The BIO-PCR technique can detect the pathogen even if one infected seed is present in seed lots of 400–600 seeds (Mosqueda-Cano and Herrera-Estrella, 1997). The pathovar-specific detection of *X. axonopodis (campestris) glycines* was achieved by employing two oligonucleotide primers heu 2 and heu 4 designed from the sequences of DNA region containing the gene encoding glycinecin A (bacteriocin) for amplification in PCR. The presence of a 0.86-kb DNA fragment in the gel was detected only when *X. axonopodis glycines* DNA was present in the PCR assay (Oh et al., 1999). The BIO-PCR has several advantages over other PCR techniques. There is no need for DNA extraction from a test organism before amplification, and false-positive results due to the presence of dead bacterial cells and false-negative results due to the presence of PCR inhibitor in seed can be eliminated.

The PCR-based assays have been demonstrated to be more rapid, sensitive, reliable, and economical for the detection, identification, and differentiation of a large number of bacterial pathogens infecting economically important crops (Table 9.7). A primer pair that amplified the 222-bp DNA fragment of *Xanthomons axonopodis* (*campestris*) pv. *citri,* was employed for the first time to diagnose a field outbreak of citrus canker (Gillings et al., 1995). A highly sensitive nested-PCR assay was developed based on a sequential nested amplification by PCR of a region of plasmid DNA that is highly conserved in *X. axonopodis* pv. *citri.* The amplified PCR products were detected colorimetrically by DIANA (detection of immobilized amplified nucleic acid) method, which was 50–100-fold more sensitive than conventional PCR assay. The sensitivity of the assay could be further improved by immunocapture PCR (IC-PCR) by concentrating the target bacteria from dilute plant extracts, by 100-fold over nested PCR assay. This procedure will be particularly suitable for plant quarantine and clean planting stock programs (Hartung et al., 1996). The PCR assay based on the use of primers designed from sequences in the intergenic regions between 16S and 23S rRNA genes had a detection limit of 30 cfu/ml and the results can be obtained within 6 hr of sample collection (Miyoshi et al., 1998).

The sensitivity of PCR detection of bacterial pathogen may be considerably improved by inclusion of immunological steps. An immunomagnetic separation (IMS) step prior to PCR was followed to concentrate *Acidovorax avenae* subsp. *citrulli* present in watermelon seeds. The sensitivity of detection by IMS-PCR protocol showed a 100-fold increase over direct PCR, the threshold of detection by IMS-PCR being 10 cfu/ml in watermelon seed wash. The presence of *A. avenae* subsp. *citrulli* could be consistently detected even in seed lots with 0.1% infection (Walcott and Gitaitis, 2000). The immunoenzymatic determination of PCR products provided a sensitive method of detection of *Erwinia amylovora*. The amplicons were labeled with 11-digoxigenin (DIG) dUTP during the PCR amplification and then captured by hybridization to a biotinylated oligonucleotide in streptavidin-coated ELISA microplates. The amplicons were detected using anti-DIG-Fab'-peroxidase conjugated antibodies. This PCR-ELISA protocol was shown to be more specific and sensitive in detecting and identifying strains of *E. amylovora* from different host plants and geographical origins (Merighi et al., 2000).

The repetitive sequence-based (rep)-PCR genomic finger printing technique is based on PCR-mediated amplification of DNA sequences located between specific interspersed repeated sequences in prokaryotic genomes. These repeated sequences are differently designed BOX, REP, and ERIC elements. A range of different-sized DNA fragments from the genomes of individual strains of a bacterial species is generated by amplification of the DNA sequences between primers based on these repeated elements. By separating the fragments on agarose gel, highly specific DNA fingerprints can be obtained. The rep-PCR genomic fingerprinting technique eliminates the need for DNA extraction and it can be applied directly to cell suspensions prepared from infected plant tissues. Another advantage is the possibility of differentiating isolates of the same pathovar as in the case of *X. campestris* pv. *vesicatoria* strains (Louws et al., 1995) and *X. oryzae* pv. *oryzae* (Vera Cruz et al., 1996).

The genomic DNA fingerprints were used to identify field isolates of *X. fragariae* collected in nurseries in California. The rep-PCR fingerprints of isolates were in agreement with pathogenicity tests, which required a long time. The rep-PCR analysis was more sensitive than indirect ELISA and provided results more rapidly and accurately (Opgenorth et al., 1996). By employing rep-PCR, genomic fingerprints of *Clavibacter michiganensis* subsp. *michiganensis, C. michiganensis* subsp. *sepedonicus, C. michiganensis* subsp. *nebraskensis, C. michiganensis* subsp. *tessellarius,* and *C. michiganensis* subsp. *insidiosum* were generated. The patterns of DNA fragments generated by rep-PCR, after agarose gel electrophoresis, corresponded to the five subspecies recognized by current classification. Furthermore, based on the limited DNA polymorphisms, at least four types, A, B, C, and D, within *C. michiganensis* could be identified using the rep-PCR fingerprints. Many naturally occurring avirulent strains of *C. michiganensis* subsp. *michiganensis* with identical fingerprints were recovered from tomato plants, indicating the potential use of this technique in studies on epidemiology and host-pathogen interactions (Louws et al., 1998). *Pseudomonas avellanae,* causing the destructive hazelnut decline disease, was rapidly and accurately identified by rep-PCR assay. Among the ERIC, REP, and BOX primer sets tested, ERIC primers yielded the most discriminating clustering of strains of *P. avellanae* that could be grouped according to their geographical origin. All of the 60 strains were accurately identified by rep-PCR using ERIC primers and also by traditional methods that require more than 6 months. This procedure was suggested by Scortichini et al. (2000) for sanitation of the infested area in central Italy.

A unique DNA fragment of plasmid pEA29 and conserved in *Erwinia amylovora,* causing fire blight disease, was employed to design two oligonucleotide primers that were used in nested-PCR assay. This assay could detect even single bacterial cell in pure cultures and had 1000-fold increased sensitivity compared to single-round PCR. *E. amylovora* was detected in a greater number of leaf, axillary bud, and mature fruit calyx samples from infected apple trees when compared with PCR and dot-blot hybridization assays (McManus et al., 1996). The oligonucleotide primers derived from the intergenic region between 16S and

23S rRNA genes amplified the specific DNA sequences in *Clavibacter xyli* subsp. *xyli* causing sugarcane ratoon stunting disease, but did not produce any amplificons from closely related *C. xyli* subsp. *cyanodontis* indicating the specificity of detection (Fegom et al., 1998). This pathogen was also successfully detected by employing two 20-mer oligonucleotide primers, which amplified a 438-bp DNA fragment from 21 strains of *C. xyli* subsp. *xyli* (Pan et al., 1998).

The pectate lyase enzyme produced by *Erwinia carotovora* subspecies has a vital role in the development of softrot diseases. Several *E. carotovora* subspecies could be detected by a PCR-RFLP test based on a pectate lyase-encoding gene. The isolates of *E. carotovora* subsp. *atroseptica,* infecting potatoes, exhibited wide molecular diversity extending across 19 RFLP groups detected among *E. carotovora* subspecies. The detection threshold of *E. carotovora* subsp. *atroseptica* was significantly improved by coupling PCR with a 48-hr enrichment step in a polypectate-rich medium. The bacterial pathogen could be detected in wash water, leaves, stem, and tuber peel extracts (Hélias et al., 1998). A PCR-based kit (Probelia) detects DNA-specific PCR amplification products by hybridization to a peroxidase-labeled DNA probe in a microplate. The PCR primers amplified the specific sequences of all serogroups of *E. carotovora* subsp. *carotovora* (Frechon et al., 1998). *E. herbicola* pv. *gysophilae* infecting *Gysophila paniculata* plants was detected by employing three primer pairs based on cytokinins *(etz)* or IAA-biosynthetic genes of the pathogen. With nested-PCR using *etz* primers even a single bacterial cell in pure culture could be detected, indicating an increase of sensitivity by 100-fold over single-round PCR assay. The BIO-PCR involving cultivation of bacteria on a semisolid medium was found to be suitable for use in programs to establish disease-free nuclear stock of mother plants of gysophila (Manulis et al., 1998).

Ralstonia (Pseudomonas) solanacearum infects a wide range of host plants and has a worldwide distribution. By employing the specific primers (PS 96H and PS 96I), 28 strains of *R. solanacearum* could be identified. Sequencing the region between the primers indicated the presence of six different sequence groups (Hartung et al., 1998). A distinct pathotype of *R. solanacearum* race 1, biovar 1 infecting pothos (*Epipremnum aureum*) was intercepted while entering Florida from Costa Rica by employing PCR amplification followed by electrophoretic analysis (Norman and Yuen, 1998). In another study, three 16S rDNA subgroups of *R. solanacearum* were differentiated by using primers designed based on the sequences of intergenic region between 16S and 23S rDNA genes. Different combination of forward and reverse primers allowed selective PCR amplification of DNA sequences of biovars 3, 4, and 5 included in Division I and biovars 1 and 2 in Division II (Seal et al., 1999). A rapid and efficient method of extracting DNA from *R. solanacearum* and *C. michiganensis* subsp. *michiganensis* using a mixture of lysozyme and proteases combined with minimized TRIS/HCL/BSA buffer volume was developed for these two important quarantine pathogens. The PCR detection threshold was between 10^4 and 10^5 cfu/ml of potato tuber extracts, providing reliable results rapidly (Niepol, 1999).

The occurrence of yellow vine disease of cucurbits was observed in the south-central United States. The presence of a bacterium-like organism (BLO) in the phloem of infected plants was observed. Based on the nucleotide sequence of the prokaryotic DNA isolated from symptomatic plants, three primers, YV1, YV2, and YV3, were designed. In the PCR assay the primer pair YV1/YV2 amplified a product with 640 bp, whereas YV1/YV3 primer pair amplified a 1.43-kbp product. The phylogenetic analysis showed that the prokaryote causing yellow vine disease was a gamma-3 proteobacterium. This agent may cause the yellow vine disease in cantaloupe, squash, and watermelon (Avila et al., 1998). The PCR assay has provided the reliable evidence to establish the identity of another new pathogenic agent. The bacterial pathogen causing a new bacterial blight disease of leek (*Allium porrum*) in California was identified as *Pseudomonas syringae* pv. *porri* by rep-PCR analysis. The DNA fingerprints of leek isolate were indistinguishable from those of known strains of *P. syringae* pv. *porri*. The rep-PCR analysis helped to identify the pathogen unambigously, while fatty acid analysis did not provide a clean pathogen designation (Koike et al., 1999).

9.7.4 Real-Time Polymerase Chain Reaction

The standard PCR assay requires considerable time for post-PCR manipulation and processing of the reaction with slabgel and capillary electrophoresis or hybridization to immobilized oligonucleotides. To overcome these limitations during field use, a modification of PCR designated real-time PCR was developed to obtain the results rapidly without losing the sensitivity. The real-time PCR consists of the fluorigenic 5′-nuclease assay known as TaqMan and a spectrofluorimetric thermal cycler.

TaqMan, representing a homogenous PCR, employs a fluorescence resonance energy transfer (FRET) probe typically consisting of a green fluorescent "reporter" dye at the 5′ end and an orange "quencher" dye at the 3′ end. During the PCR, the probe anneals to a complementary strand of an amplified product, whereas *Taq* polymerase cleaves the probe during extension of one of the primers and the dye molecules are displaced and separated. As the electronically excited reporter is no longer suppressed by the quencher dye, variations in green emissions can be monitored by a fluorescence detector. The measurements of green fluorescence intensity reflect the concentration of PCR amplicons in the reaction.

Silicon chip–based spectrofluorimetric thermocyclers are used for testing the samples. The Advanced Nucleic Acid Analyzer (ANAA) suitable for field use offers real-time monitoring and requires only low power supply. Ten reaction modules and a laptop computer are placed in a protective casing. Each reaction module consists of a silicon reaction chamber with highly efficient integral thin-film heaters and a dedicated low-power optical system. Using a blue-light-emitting diode as an exciting source and two photodiode detectors with bandpass filters centered at 530 and 590 nm, the TaqMan probe is monitored. The positive signal, as determined by a real-time analysis alogrithm, is informed to the user by the software automatically via an audible alert and green-to-red indicator. The minimum time required for the detection and characterization of *Erwinia herbicola* at a concentration of 500 cfu was 7 min after which strong positive signal was indicated. The real-time PCR assay provides both qualitative detection and quantitative determination of the bacterial pathogen (Belgrader et al., 1999). This technique has the potential for large-scale application for the detection of pathogens. The operational costs of tests and suitability for detection of other kinds of microbial pathogens have to be determined.

9.7.5 Random Amplified Polymorphic DNA

The random amplified polymorphic DNA (RAPD) technique, as in the case of other microbial pathogens, has been used to identify, differentiate, and establish the relationship between bacterial pathogens and their strains/pathovars. Unique RAPD profiles for each of five strains of *Xanthomonas albilineans* were revealed when four 10-mer arbitrary primers were used individually. However, most of the RAPD markers were common to all the strains. DNA fingerprint analysis showed that RAPD analysis would be useful to identify and differentiate the strains of *X. albilineans* and to monitor the appearance of any new strain(s) in a geographical location (Peramul et al., 1996). The races of *Pseudomonas syringae* pv. *pisi* were detected by using two unique DNA fragments generated by a RAPD-PCR protocol, as probes, as the DNA from each isolate hybridized to only one of the two probes. Two pairs of oligonucleotide primers were designed based on the cloned sequences of DNA fragments of race 7 and 2. *P. syringae* pv. *pisi* could be reliably identified by using these oligonucleotide primers and the isolates of *P. syringae* pv. *pisi* included in the study could be classified into two phylogenetic groups I and II (Arnold et al., 1996).

Erwinia spp. and pectolytic pseudomonads are commonly associated with softrot diseases, making it essential that the causative bacteria should be rapidly detected and precisely identified. Mäki-Valkama and Karjalainen (1994) reported that *Erwinia carotovora* subsp. *atroseptica* and *carotovora* could be differentiated by the RAPD-PCR technique. The use of two randomly chosen primers in combination was effective in differentiating *Erwinia carotovora* from pectolytic, fluorescent *Pseudomonas* spp. These primers could also be used to distinguish *E. carotovora* and *E. carotovora* subsp. *atroseptica* based on the RAPD analysis. The results of RAPD analysis for the identification of 49 additional softrot bacteria corresponded to those of biochemical tests (Parent et al., 1996). The strains of *E. amylovora* were distinguished by comparing RAPD banding profiles obtained by

PCR amplification using six different 10-mer primers. The number of RAPD fragments shared between the strains formed the basis of cluster analysis. The strains from the subfamily Pomoideae constituted a single group, whereas the second group consisted of two strains from *Rubus* sp. (subfamily Rosoideae). The third group included two strains from Asian pear in Hokkaido, Japan. The three strain groups of *E. amylovora* based on host range and geographical region can be reliably distinguished based on the sets of RAPD fragments identified for each group. Thus the RAPD analysis has been shown to be useful for differentiation of bacterial strains and for determining the relatedness between them (Momol et al., 1997).

A PCR-based restriction analysis and ligation-mediated PCR procedure revealed polymorphisms among individual strains of *X. oryzae* pv. *oryzae* (*Xoo*) and the genetic relationships could be deduced based on the RFLPs of *Xoo* strains. The strains from Indonesia and Philippines were found to be closely related (George et al., 1997). The usefulness of RAPD fingerprinting strains of *Xylella fastidiosa* causing Pierce's disease (PD) of grapevines was demonstrated for the differentiation of a large number of strains (158) isolated from a vineyard in North Florida. RAPD analysis could differentiate the PD strains from non-PD strains of *X. fastidiosa* causing citrus variegated chlorosis, mulberry large scorch, periwinkle wilt, plum leaf scald, and phony peach diseases (Albibi et al., 1998).

SUMMARY

Nucleic acid-based techniques are used to determine the nucleotide sequence similarities between related species, races, strains, or biotypes of pathogens. Closely related organisms show greater nucleotide sequence similarity. A highly specific nucleotide sequence present in a strain or isolate of a pathogen may be used to detect and differentiate that particular strain or isolate present in infected plant material and rapidly establish its identity. Interspecific and intraspecific variations can be assessed by using appropriate probes. Chromosomal and extrachromosomal nucleic acid sequences have been used for the preparation of probes. Radioactive and nonradioactive materials have been employed for labeling the probes. Nucleic acid hybridization methods are particularly useful for the detection of viroids. Dot-blot hybridization, restriction fragment length polymorphism, polymerase chain reaction, and random amplified polymorphic DNA are the techniques that are being employed for the detection, differentiation, and quantification of plant pathogens in seeds, seed materials, soil, and water. Attempts to avoid sample manipulations prior to PCR and time-consuming procedures for the detection of PCR products have been made to increase the speed of PCR-based assays without loss of sensitivity of detection. The development of print capture-PCR, PCR-ELISA, and TaqMan PCR detection techniques indicates the progress made toward these objectives. The usefulness of the nucleic acid–based techniques in seed certification, quarantine and disease resistance breeding programs, and evaluation of crop disease management strategies is discussed.

APPENDIX 9(i): METHOD OF RAPID PREPARATION OF INFECTED PLANT TISSUE EXTRACTS FOR PCR AMPLIFICATION OF VIRUS, VIROID, AND PHYTOPLASMAS NUCLEIC ACID (LEVY ET AL., 1994)

A. Sample preparation (per Gene Releaser [GR] manufacturer recommendations)

 i. Cut out leaf disks (about 30 mg) by using an inverted pipette tip (larger end) and transfer them to 1.5 ml microfuge tubes.

 ii. For testing of anthers (for the presence of viroid) place 10–20 anthers into a microfuge tube.

 iii. For testing of the phloem tissue for the presence of phytoplasmas/viruses, excise 2–4.5 mm long leaf midrib sections and place them into microfuge tubes.

 iv. Grind the tissues with a disposable pestle in 100 μl/sample of ice-cold TE buffer containing 10 mM Tris-HCl; 1 mM EDTA, pH 8.0, and 0.4 mg of 120 grit carborundum/μl of buffer.

 v. For phytoplasmas, alternatively, grind the tissues in phytoplasma grinding buffer consisting of 95 mM K_2HPO_4, 30 mM KH_2PO_4, 10% sucrose, 0.15% bovine serum albumin fraction V, 2% polyvinylpyrolidone (PVP-10), and 0.53% ascorbic acid, pH 7.6, with carborundum or liquid N; then grind in phytoplasma buffer without carborundum.

 vi. Centrifuge at 12,000 rpm at 4°C for 1–2 min for virus samples; centrifuge for 10–45 sec or gravity-settle for phytoplasma samples, and transfer the supernatant to sterile 1.5 ml microfuge tubes and place them on ice.

 vii. Dispense 1 μl (for viruses and viroids) and 2 μl (for phytoplasmas) aliquot of each sample in thin-walled PCR tubes containing 20 μl (for phytoplasmas) and 23 μl (for virus and viroid samples of freshly resuspended gene releaser GR).

 viii. Gently vortex GR-tissue extract mixtures at low speed for 30 sec and place them on ice till all samples are prepared.

 ix. Place the samples in a microwave safe rack, overlay with 50 μl of mineral oil, close the lid, and microwave at the high-power setting for the required period (heating time multiplied by the oven power rating [watts] to be equal to 4500 watt-minutes).

B. Reverse-transcription-polymerase chain reaction (RT-PCR) amplification

 i. Take out 10–20 μl aliquot of GR matrix containing the sample immediately after microwaving; add to a primer annealing reaction mixture containing 6 μl of 5 X reverse transcriptase buffer (consisting of 250 mM Tris-HCl, pH 8.3; 375 mM KCl; and 15 mM MgCl$_2$); 3 μl of 0.1 M dithiothreitol [DTT), 1 μl complementary primer, and sterile water to yield a final volume of 30 μl.

ii. Vortex the sample-primer mixture briefly and denature by heating at 100°C for 5 min; chilling on ice for 2 min; and annealing the primers at room temperature for 30–45 min.

iii. Add the annealed reaction mixture to 20 μl of a cDNA reaction mixture consisting of 4 μl 5 X reverse transcriptase buffer, 2 μl 0.1 M DTT, 1 μl RNasin (40 units), 5 μl of 0.3 M 2-mercaptoethanol, 2.5 μl 10 mM dNTPs (2.5 mM each dGTP, dATP, dTTP, dCTP), and either 1 or 2 μl of Maloney murine leukemia virus (MMLV) reverse transcriptase (200 U/μl) for virus and viroid, respectively.

iv. Vortex the reaction mixture and incubate at 42°C for 1–1 1/2 hr.

v. Carry out amplification in thin-walled PCR tubes containing the reaction mixture consisting of 5 μl of 10 X PCR buffer (10 mM Tris-HCl, pH 8.3; 50 mM KCl; and 0.001% gelatin), 3 μl of 25 mM MgCl$_2$ (1.5 mM final concentration), 1 μl of 10 mM dNTPs, 1 μl each of 6 μM complementary and homologous DNA primer, 2.5 units of DNA *Taq* polymerase, and sterile water to produce a volume of 45 μl. Overlay the mixture with 75 μl of mineral oil and place at 85°C in a DNA thermocycler for 5 min.

vi. Add 5 μl of GR-cDNA mixture and amplify through the following steps: denaturation at 94°C for 30 sec, primer annealing at 62°C for 30 sec, and extension at 72°C for 45 sec for 30 cycles with a final extension at 72°C for 7 min.

C. Analysis of PCR products by electrophoresis

i. Electrophorese PCR products in 5% native polyacrylamide gels of 150 V for 1.5 hr in IX TBE (containing 89 mM Tris, 89 mM borate, and 2.5 mM Na$_2$EDTA, pH 8.3) and stain the gels with silver nitrate in the case of viruses and viroids; electrophorese in 1.2% agarose gels and stain with ethidium bromide and observe under ultraviolet (UV) illumination.

APPENDIX 9(ii): ONE TUBE PCR AMPLIFICATION PROCEDURE (WETZEL ET AL., 1991)

i. Grind the infected leaf tissue in a plastic bag containing gauze along with sterile water added at the rate of 1:4 (w/v) in a rolling grinder;

centrifuge the sap for 10 min in a microfuge at full speed; dilute the supernatant with sterile water (10-fold).

ii. Disrupt the virus particles in 10 μl aliquots of diluted extract by treatment with 1% Triton X-100 for 10 min at 65°C.

iii. Incubate the target nucleic acid with methyl mercury hydroxide (10 mM final concentration) for denaturation at room temperature for 10 min; neutralize with 20 mM 2-mercapto-ethanol for 10 min at room temperature.

iv. Incubate the denatured target nucleic acid with reverse transcription mixture containing 50 mM Tris-HCl, pH 8.3; 50 mM KCl; 7.5 mM MgCl$_2$; 0.1 V Inhibit ACE (5'3' Inc); 20 μM dNTPs; 0.1 μM primer; and 0.5 U avian myeloblastoses virus reverse transcriptase in the tube to yield a final volume of 20 μl for 45 min at 42°C.

v. Add 80 μl PCR buffer (10 mM Tris-HCl, pH 8.3; 50 mM KCl; 1.5 mM MgCl$_2$; 0.1% gelatin) containing 1 μM of both primers, 200 μm dNTPs, and 2 V of Taq DNA polymerase; overlay with 100 μl of mineral oil; amplify for 40 cycles of template denaturation at 92°C (1 min), primer annealing at 62°C (2 min), and DNA synthesis at 72°C (2 min).

vi. Analyze reactions by electrophoresis of the reaction mixture; observe bands by either ethidium bromide or silver staining.

APPENDIX 9(iii): PCR-MICROPLATE HYBRIDIZATION METHOD (HATAYA ET AL., 1994)

A. Preparation of digoxigenin (DIG)-labeled cDNA probe for the CP coding region of PVY-RNA

i. Use the cDNA clone pUCYTCP1-19 as a template DNA; carry out PCR in 50 μl reaction mixture containing 10 mM Tris-HCl (pH 8.9); 80 mM KCl; 1.5 mM MgCl$_2$; 0.5 mg/ml BSA; 0.1% sodium cholate; 0.1% Triton X-100; 0.1 mM each dGTP, dATP, dCTP; 0.065 mM dTTP; 0.035 mM DIG-11-dUTP; 50 pmol each of PCR primers; 20 ng of template DNA, and 1U Tth DNA polymerase.

ii. Carry out PCR amplification as follows: 30 cycles of 30 sec denaturation at 94°C (5 min for the first cycle), 1 min annealing at 56°C, and 2 min primer extension at 72°C (10 min for the last cycle); place a drop of light mineral oil to prevent evaporation before the commencement of amplification; extract the reaction product with TE (10 mM Tris-HCl, 1 mM EDTA, pH 8.0) saturated phenol; chloroform (1:1) mixture to remove light mineral oil and BSA; mix the water phase with 1/4 volume of 10 M ammonium acetate and 2.5 vol of ethanol and keep at $-80°C$ for 30 min.

iii. Purify DIG-labeled probe by an Ultra free C3-TTk (Millipore Ltd.) spun column for removing unreacted dNTP, DIG-II-dUTP, and primers; dissolve the DIG-labeled probe in the upper filter cup on 50 μl of TE.

B. Nucleic acid extraction from potato samples

  i. Extract the nucleic acids from leaves (about 1 g) (symptomatic and asymptomatic) by using a metal rod pestle; transfer about 200 μl of sap to a microcentrifuge tube (1.5 ml); immediately add the extraction buffer containing 100 mM Tris-HCl (pH 9.0), 2 mM EDTA, 1 mM sodium diethyl dithiocarbamate (DIECA), 1% sodium dodecylsulfate (SDS), and 0.1 mg/ml bentonite; centrifuge.

  ii. Treat the aqueous phase with TE saturated phenol; chloroform (1:1) mixture; centrifuge.

  iii. Precipitate the nucelic acids in aqueous phase with 0.1 volume of 3 M sodium acetate, pH 5.2; and 0.6 volume of isopropyl alcohol; extract the precipitated nucleic acids with TE saturated phenol; chloroform (1:1) mixture; precipitate the total nucleic acids again with 0.1 volume of 3 M sodium acetate (pH 5.2) and 2.5 volume of ethanol.

  iv. Dissolve the nucleic acids in 50 μl of distilled water; mix with 50 μl of 4 M LiCl; keep on ice overnight.

  v. Dissolve the precipitated RNAs in 400 μl of distilled water; use samples of 4 μl of RNA for RT-PCR.

C. Amplification of cDNA to viral RNA sequences by RT-PCR

  i. Incubate the reverse transcription (RT) reaction mixture (20 μl) containing 0.5 μg of extracted viral RNA, 0.5 μg of oligo (dT)$_{12-18}$ primer, 50 mM Tris-HCl (pH 8.3), 75 mM KCl, 10 mM DTT, 3 mM MgCl$_2$, 0.5 mM each dNTP and 100 UM-MLV RTase, at 37°C for 60 min.

  ii. Add one half of the reaction mixture (10 μl) to 40 μl of a PCR premixture containing 75 mM Tris-HCl, pH 8.9; 81.25 mM KCl; 1.125 mM MgCl$_2$; 0.375 mg/ml bovine serum albumin (BSA); 0.075% sodium cholate; 0.075% Triton X-100; 50 pmol each of plus and minus sense primer; 187.5 μm each of dNTP and IU of Tth DNA polymerase.

  iii. Place a drop of light mineral oil on the PCR reaction mixture to prevent evaporation; proceed through 30 cycles of 30 sec denaturation at 94°C (5 min for the first cycle), 1 min annealing at 56°C, and 2 min primer extension at 72°C (10 min for the last cycle) by using a Temp Control System PC-700 program.

iv. Extract the reaction product with TE saturated phenol/chloroform (1:1) mixture to remove light mineral oil and BSA; mix the water phase with 1/4 volume of 10 mM ammonium acetate and 2.5 volume of ethanol; keep at −80°C for 30 min; centrifuge.

v. Dry the precipitate in a vacuum and dissolve in 50 μl of TE.

D. Microplate hybridization

i. Denature the amplified cDNA fragments at 100°C for 5 min and quickly chill in ice water; dilute 5-fold serially with different dilutions of standard saline citrate (SSC) (1xSSC:0.15 M NaCl, 0.015 M sodium citrate, pH 7.0) containing 10 mM EDTA.

ii. Dispense 100 μl of diluted DNA into wells of a 96-well polystyrene microplate; incubate at 37°C for 2 hr and wash three times with phosphate-buffered saline (PBS) solution containing 137 mM NaCl, 8.1 mM Na_2HPO_4, 1.47 mM KH_2PO_4, 2.7 mM KCl, pH 7.4, containing 0.05% Tween 20 (PBS-T).

iii. Transfer to each well 100 μl of hybridization solution containing the heat-denatured DIG-labeled probe (1 μl/ml: 1000-fold dilution); 50% formamide; 5X SSC; 10 mM EDTA, pH 7.0; 0.1% Tween 20; and 100 μg/ml yeast tRNA; seal the plates with adhesive tape; incubate at 42°C for 12 hr; wash three times with PBS-T.

iv. Dispense 100 μl of alkaline phosphatase-conjugated anti-DIG antibody diluted 5000-fold with PBS-T; incubate at 37°C for 1 hr; wash three times with PBS-T.

v. Dispense 200 μl of P-nitrophenyl phosphate (1 mg/ml in diethanolamine buffer, pH 9.8) as substrate; incubate at room temperature for 1–2 hr; determine absorbance values at 415 nm by using a microplate reader.

APPENDIX 9(iv): COMBINATION OF IMMUNOCAPTURE AND PCR AMPLIFICATION IN MICROTITER PLATE FOR DETECTION OF VIRUSES (NOLASCO ET AL., 1993)

A. Immunocapture of viruses from plant tissues

i. Grind infected plant tissues (1:10 w/v) in 500 mM Tris-HCl, pH 8.2, containing 2% PVP-40, 1% PEG 6000, 140 mM NaCl, 0.05% Tween 20, and 3 mM NaN_3; centrifuge at 5000 g for 5 min.

ii. Dispense 50 μl of the supernatant to each well of the microtiter plate, already coated with antibody specific to the virus to be detected by the method of Clark and Adams (1977); incubate overnight at 4°C.

    iii. Wash the wells three times by flooding with PBS-Tween, taking care to prevent cross-contamination between wells (see Appendix 8 [i]).

B.   RT-PCR amplification

    i. Perform an RT reaction in the microtiter plate by adding to each well 20 μl of the reverse transcription mixture containing 50 mM Tris-HCl, pH 8.3; 75 mM KCl; 3 mM MgCl$_2$; 1 mM each dNTP; 25 units of ribonuclease inhibitor; 1 μM downstream primer; 200 units of M-MLV reverse transcriptase (BRL); incubate at 37°C for 1 hr.

    ii. Carry out amplification of cDNA in the same wells (using a Techno PHC-3 thermocycler with a microtiter plate adaptor) by adding 80 μl of amplification mixture consisting of 60 mM Tris-HCl, pH 9.0; 15 mM KCl; 2.1 mM MgCl$_2$; 20 mM (NH$_4$)$_2$SO$_4$; 0.2 mM each dNTP; 0.2 μM each primer; 0.005% BSA; overlay with a drop of mineral oil.

    iii. Heat the mixture at 94°C for 2 min; cool to 72°C and add 1.6 units of thermostable DNA polymerase.

    iv. Proceed through 30–35 cycles, each cycle consisting of an annealing step of 1 min at 52°C, an elongation step of 1 min at 72°C, a denaturation step of 30 sec at 93°C, and an elongation step of 5 min for the final cycle.

C.   Detection of amplified DNA products by Southern hybridization

    i. Transfer the DNAs to nylon membranes (Hybond N$^+$, Amersham Inc.) using the alkaline transfer procedure; soak the membranes in sodium citrate (SSC) (20 × SSC = 3 M NaCl, 0.3 M sodium citrate, pH 7.0); hybridize to the probes labeled with digoxigenin.

APPENDIX 9(v): COLORIMETRIC PCR ASSAY (ROWHANI ET AL., 1998) FOR COLORIMETRIC DETECTION OF PCR PRODUCTS

    i. Coat microtiter plates with streptavidin (10 μg/μl, 200 μl/well) in carbonate buffer containing 50 mM sodium carbonate pH 9.6 overnight at 4°C.

    ii. Wash the plates with ELISA washing buffer three times [Appendix 8(1)] and incubate with blocking buffer containing PBS with 1% Tween-20 for 30 min at room temperature.

    iii. Add virus-specific 5′-biotin-labeled oligonucleotide detection probe (7 pmol/μl, 200 μl/well); incubate for 30 min at 37°C in hybridization buffer containing 5 × SS PE, 0.5 M NaCl, 0.1% n-lauroyl sarcosine, pH 6.5 and wash with ELISA washing buffer three times.

iv. Denature 20 μl of the PCR product by adding 40 μl of denaturing buffer containing 0.2 N NaOH, and 1.5 M NaCl; vortex the mixture briefly; incubate for 10 min at room temperature and neutralize the denatured PCR products by adding hybridization buffer to a final volume of 500 μl.

v. Dispense the denatured PCR products (200 μl) in duplicate wells and incubate for 1.5 hr at 46°C.

vi. Wash the plates with ELISA washing buffer three times and incubate with 200 μl/well of 0.75 U/μl anti-digoxigenin f(ab′)2 fragments conjugated to alkaline phosphatase in loading buffer containing 0.05% Tween-20, 2.0% PVP-40, bovine serum albumin fraction V for 30 min at 37°C.

vii. Wash the plates again; incubate with p-nitrophenyl phosphate (0.75 mg/ml in 9.5% diethanolamine, pH 9.8) for 1–2 hr for color development and record the absorbance at 405 nm with microplate reader.

APPENDIX 9(vi): REVERSE TRANSCRIPTION-POLYMERASE CHAIN REACTION ASSAY FOR VIROID DETECTION (HODGSON ET AL., 1998)

A. Nucleic acid extraction

i. Place the coconut leaf samples (0.5 g) in a thick-walled plastic bag (7.5 × 140 mm) containing 2 ml of sterile NETM buffer consisting of 2 M NaCl, 100 mM sodium acetate, 10 mM EDTA, 50 mM Tris-HCl, pH 7.5, and 0.25% 2-mercaptoethanol (v/v) and crush by hammering the leaf tissue.

ii. Add the extract (800 μl) to 20% sodium dodecylsulfate (SDS) (40 μl) kept in a 1.5-ml microfuge tube; thoroughly mix the solution for 30 min at 25°C and add the mixture to a mixture (750 μl) of phenol-chloroform-isoamylalcohol (25:24:1).

iii. Vortex the mixture for 20 sec and centrifuge at 10,000 g for 15 min for separation of aqueous phase (upper).

iv. Separate the aqueous phase containing nucleic acids; mix the nucleic acid solution (700 μl) with equal volume of isopropanol for precipitation of nucleic acids (NA).

v. Wash the pellet (precipitate) with 70% ethanol; air-dry and dissolve the pellet in sterile diethylpyrocarbonate-treated water (50 μl).

B. Reverse transcription-polymerase chain reaction

i. Mix nucleic acid extract (2 μl) with RT buffer containing 50 mM Tris (pH 8.8), 75 mM KCl, and 3 mM MgCl$_2$, 3.3 mM dithiothreitol, 200 mM antisense primer D1-2 or D3-2, and water to a final volume of 26 μl; heat the mixture to 94°C for 4 min and chill on ice.

ii. To this mixture, add RNasin (Bresatec) (10 units), dNTP mix (333 μM), and MMLV-RT (GIBCO/BRL, Gaithersburg, MD) (400 units) to a final volume of 30 μl and incubate at 50°C for 20 min.

iii. Amplify the resulting first-strand cDNA by adding the RT reaction mix (5 μl) to PCR buffer [final concentration of 10 mM KCl, 10 mM $(NH4)_2SO_4$, 2 mM $MgSO_4$, 20 mM Tris (pH 8.8), 0.1% Triton X-100] containing 333 μM dNATP mix, 2 units of Deep Vent polymerase (New England Biolabs Inc., Beverly, MA), 300 nM each antisense (D3-2 or D1-2), and sense primers (D4-2, residues 102–131) in a final volume of 30 μl, which is overlaid with mineral oil.

iv. Provide the following amplification conditions: 94°C for 2 min, 70°C for 1 min, and 94°C for 30 sec for 5 cycles, 70°C for 30 sec, and 94°C for 30 sec for 40 cycles, 72°C for 5 min, and 25°C for 1 min for one cycle, using a Corbett FTS 320 thermal sequencer.

APPENDIX 9(vii): DETECTION OF PHYTOPLASMA IN INFECTED PLANT WITH DNA PROBES (LEE AND DAVIS, 1988)

i. Excise 0.3 g of leaf midrib or young shoot and grind in liquid nitrogen; transfer to a microcentrifuge tube containing 0.4 ml of extraction buffer (0.1 M Tris, pH 8.0; 0.5 M EDTA; 0.5 M NaCl; 0.5% 2-mercaptoethanol; 0.5% SDS), crush the sample with a minipestle; centrifuge for 10 min at 2000 rpm.

ii. Transfer the supernatant to another clean microcentrifuge tube; centrifuge the sediment again at 8000 rpm for 10 min; combine the supernatant with the one obtained earlier.

iii. Heat the supernatant at 65°C for 5 min; centrifuge at 14,000 rpm for 5 min; transfer the supernatant to another clean microcentrifuge tube.

iv. Extract with 200 μl phenol and 200 μl chloroform isoamyl alcohol (24:1); centrifuge at 14,000 rpm for 5 min; retain the aqueous phase.

v. Denature DNA and transfer to nitrocellulose membrane in a dot-blot apparatus; bake the membrane for 2 hr at 80°C.

vi. Hybridize with DNA probe as for Southern hybridizations.

APPENDIX 9(viii): DETECTION OF PHYTOPLASMAS IN INOCULATIVE INSECTS WITH DNA PROBES (KIRKPATRICK ET AL., 1987)

i. Crush the leafhoppers (frozen at -20°C) on a moistened nitrocellulose membrane; place the membrane with crushed leafhoppers on filter paper moistened with 0.3 M NaOH; incubate for 3 min.

    ii. Transfer the nitrocellulose membrane to a second filter paper soaked in 0.3 M NaOH; incubate again for 3 min.

    iii. Transfer the membrane sequentially to a pair of filter papers soaked with 1 M Tris-HCl, pH 8.0; incubate for 3 min, then to pairs of filter papers soaked with 0.5 M Tris-HCl, pH 8.0, and 15 M NaCl; remove the insect debris.

    iv. Dry the nitrocellulose membrane; bake at 80°C for 2 hr.

    v. Hybridize with a DNA probe as for Southern hybridization.

APPENDIX 9(ix): DETECTION OF PHYTOPLASMAS BY PCR ASSAYS AND RFLP ANALYSIS (KHADHAIR ET AL., 1998)

A. PCR amplification

    i. Dilute the DNA samples to a final concentration of approximately 20 ng/µl.

    ii. Prepare the PCR mixture (100 µl) containing 200 µM of each dNTP, 0.4 µM of each primer, 1× PCR reaction buffer, 2.5 mM MgCl$_2$, and 2.5U Taq DNA polymerase; cover the rection mixture with 30 µl in mineral oil.

    iii. Perform amplification in a DNA-Thermocycler-480 (Perkin-Elmer Cetus, Norwalk, CT).

    iv. Set the thermal conditions as follows: 35 cycles of denaturation for primer set P1/P6 at 94°C for 30 sec (except 2 min for the first cycle); annealing at 65°C for 60 sec, extension at 72°C for 1.5 min; and extension of the last cycle for an additional 3 min at 72°C.

B. Nested PCR

    i. Use the PCR products of DNA amplification with primer pair P1/P6 as templates.

    ii. Prepare dilutions of 1/40 from each DNA template using deionized water and amplify with specific primer pair R16 F1/R1.

    iii. Prepare reaction mixture for nested PCR as in step A(ii) above.

    iv. Set the thermal conditions as follows: a total of 35 thermal cycles consisting of denaturation at 94°C for 1 min (4 min for the first cycle), annealing at 55°C for 1 min and extension at 72°C for 1.5 min; extension for 7.5 min in the last cycle.

C. Restriction fragment length polymorphism (RFLP) analysis

    i. Use the PCR products obtained by amplifying 16S rDNA sequences of the phytoplasma concerned by direct PCR with primer pair R16F1/R1.

    ii. Purify the PCR products and concentrate by using a PCR Clean Up Kit (Boehringer, Mannheim, Germany).

    iii. Digest an aliquot of 5 µl of each PCR product with each of the restriction nucleases *Alw* I, *Hha* I, *Rsa* I, and *Sau* 3A according to the manufacturer's instructions (Promega, Madison, WI).

APPENDIX 9(x): DETECTION OF PHYTOPLASMAS BY POLYMERASE CHAIN REACTION-ENZYME LINKED IMMUNOSORBENT ASSAY (PCR-ELISA) (POLLINI ET AL., 1997)

A. Extraction of DNA by phytoplasma enrichment (Ahrens and Seemüller, 1992)
    i.   Prepare the samples of midribs (0.5 g) or shoot tips (0.5 g) by cutting them into small pieces with scissors
    ii. Place the tissues in a mortar containing 6 ml of ice-cold grinding buffer consisting of 125 mM potassium phosphate, 30 mM ascorbic acid, 10% sucrose, 0.15% bovine serum albumin (BSA), 2% polyvinyl pyrrolidone, pH 7.6 (PVP-15); incubate for 10 min and grind.
    iii. Add another aliquot of 8 ml of fresh buffer and grind again; centrifuge at 1000 g for 3 min at 4°C.
    iv. Centrifuge the supernatant at 14,600 g for 25 min at 4°C; resuspend the phytoplasma enriched pellet in 1.5 ml of extraction buffer consisting of 2% CTAB, 1.4 M NaCl, 0.2% 2-mercaptoethanol, 20 mM EDTA, 100 ml mM Tris-HCl, pH 8.0, and incubate at 60°C for 30 min.
    v.   Extract the lysate with equal volumes of chloroform/isoamyl alcohol (24:1 v/v) and centrifuge.
    vi. Precipitate the aqueous layer with a 2/3 volume of −20°C isopropanol and centrifuge at 15000 g with a microfuge.
    vii. Wash the pellet with 70% ethanol, dry under vacuum, and dissolve in 100 μl of sterile water.
    viii. Digest the extracted DNA with 50 μg/ml RNAse A at 37°C for 30 min followed by two extractions with chloroform/isoamylalcohol, ethanol precipitation.
    ix. Wash the pellet with ethanol and estimate the DNA content at OD$_{260}$.
B. DNA amplification and incorporation of digoxigenin in amplified products
    i   Amplify each sample using 5 μl of extracted DNA (150 ng), 250 nM of the universal primers of P1 and P7 for phytoplasmas, 250 μM of PCR-DIG labeling mix (Boehringer) containing the four nucleotides, one of which is bound to DIG, 1 unit of the *Taq* enzyme polymerase (Boehringer) and 100 μl of mineral oil.
    ii. Set the amplification conditions as follows: at 94°C for 60 sec, at 55°C for 75 sec, and at 72°C for 90 sec for 35 cycles.
C. Hybridization of amplified product and specific probe
    i.   Add 10 μl of amplified product (at appropriate dilutions) in sterile wa-

ter to 20 μl of denaturation solution; incubate for 10 min at room temperature.

ii. Add, after denaturation, 220 μl of the hybridization solution containing different dilutions of the specific probes; transfer 200 μl of this mixture to a well in an ELISA plate coated with streptavidin (from the kit) and incubate for different periods.

iii. Use oligonucleotides probes consisting of 20 or 21 nucleotides corresponding to segments of intergenic spacer region; group-specific or phytoplasma-specific probes are to be used for specific diagnosis. All reagents for the two steps, except specific probes, are provided by the PCR-ELISA-DIG detection kit of Boehringer.

D. Immunoenzymatic reaction

i. Wash the wells in ELISA plates after incubation, with washing buffer three times (3 min each).

ii. Add anti-DIG-horseradish peroxidase conjugate at 1:100 dilution and incubate for 1 hr or 4 hr at 37°C with moderate shaking.

iii. Wash the wells again as done before; add 200 μl of ABTS substrate to each well.

iv. Record the readings at 405 nm using an ELISA reader.

APPENDIX 9(xi): HETERODUPLEX MOBILITY ANALYSIS (HMA) FOR DETECTION AND DIFFERENTIATION OF PHYTOPLASMAS (COUSIN ET AL., 1998)

A. DNA isolation and polymerase chain reaction

i. Ioslate phytoplasma DNA by enrichment method [Appendix 9(ix)A]

ii. Using the unviersal phytoplasma primer pair P1/P7 (P1 located at the beginning of 16S rRNA gene and P7 located at the beginning of the 23S rRNA), perform PCR with thermal conditions set as follows: 95°C for 1 min (denaturation)., 55°C for 45 sec (hybridization), and 72°C for 45 sec (extension) during 37 cycles; analyze PCR products in agarose gels (1%).

B. Heteroduplex mobility assay (HMA)

i. Prepare the PCR-amplified products (DNA fragments) from the standard and test phytoplasma DNA using the same primers as mentioned above; mix the amplified DNA fragments of standard and test samples in equal volumes (15 μl) and add 2 μl of 10× annealing buffer consisting of 1 M NaCl, 100 mM Tris-HCl, pH 7.8, and 20 mM EDTA.

ii. Heat the mixture at 95°C for 2 min for the formation of four DNA single strands originating from the two PCR-amplified DNA fragments present in the mixture.

    iii.   Hybridize the single strands of DNA rapidly on ice and subject them to polyacrylamide gel electrophoresis (PAGE) for 5hr at room temperature; using 8% polyacrylamide in TBE buffer consisting of 45 mM Tris-borate, 1 mM EDTA, pH 8.0 with a gel thickness of 0.7 mm, stain the gel with ethidium bromide for 20 min.

APPENDIX 9(xii): IDENTIFICATION OF FUNGAL PATHOGENS BY REPETITIVE DNA POLYMORPHISM (PANABIERES ET AL., 1989)

A. Preparation of fungal DNA

    i.   Grow the fungal pathogen in an appropriate medium; harvest the cultures by filtration on filter paper under a vacuum; rinse the mycelia in 250 ml of distilled water; store by freezing.

    ii.   Grind the frozen mycelium (250 mg) in liquid nitrogen; suspend the fungal powder in 0.5 ml of NIB buffer containing 100 mM NaCl; 30 mM Tris-HCl, pH 8.0; 10 mM EDTA; 10 mM β-mercaptoethanol; 0.5% (v/v) NP-40; centrifuge for 1 min at 12,000 g.

    iii.   Resuspend the pellet in NIB buffer; repeat the procedure in (ii) above; resuspend the pellet in 0.8 ml of homogenization buffer consisting of 0.1 M NaCl, 0.2 M sucrose, and 10 mM EDTA; add 0.2 ml of lysis buffer containing 0.25 M EDTA; 0.5 M Tris, pH 9.2; and 2.5% sodium dodecyl sulfate; incubate at 55°C for 30 min.

    iv.   Extract twice with 1 volume of phenol-chloroform-isoamyl alcohol (50:48:2); then twice with 1 volume of ether.

    v.   Add 1 volume of ethanol; centrifuge for 1 min in a microcentrifuge at room temperature; collect the DNA as a pellet.

    vi.   Wash the pellet with 70% ethanol; centrifuge again; resuspend in 50 μl of TE (10 mM Tris, pH 8.0; 1 mM EDTA); store at −20°C.

B. Digestion of DNA and electrophoresis analysis

    i.   Digest 5 μg of total DNA overnight with 20 units of restriction enzyme (per manufacturer's recommendations).

    ii.   Separate DNA fragments on 1% agarose gels at 5 v/cm in 90 mM Tris borate buffer, pH 8.3.

    iii.   Stain the gels with ethidium bromide; view under UV light.

APPENDIX 9(xiii): DETECTION OF FUNGAL PATHOGENS WITH DNA PROBES (TISSERAT ET AL., 1991)

    i.   Place 200–400 mg of infected plant tissue in a 1.5 ml microfuge tube; freeze by adding liquid nitrogen; grind with a smooth-lipped steel rod.

ii. Suspend the ground samples in 600 μl 2X CTAB buffer (2 × CTAB = 1.4 M NaCl, 2% hexadecyltriethylammonium bromide, 1% 2-mercaptoethanol, 10 mM Tris-HCl, pH 8.0); extract twice with chloroform.

iii. Precipitate by adding 0.8 volume of isopropanol; resuspend the pelleted DNA in 40 μl TE buffer (TE = 10 mM Tris, pH 7.6; 1 mM EDTA).

iv. Denature DNA at 95°C for 4 min; transfer 20 μl to a nylon membrane in a slot-blot apparatus; bake at 80°C for 2 hr.

v. Hybridize with DNA as for Southern hybridizations.

APPENDIX 9(xiv): IDENTIFICATION OF FUNGAL PATHOGENS BY RESTRICTION FRAGMENT LENGTH POLYMORPHISMS (RFLP) ANALYSIS (UENG ET AL., 1992, 1995)

A. Preparation of fungal genomic DNA

i. Cultivate the fungal pathogen in an appropriate medium for 7–10 days at room temperature (25°C–28°C) with constant shaking (150 rpm); harvest the mycelia by filtration and centrifugation.

ii. Resuspend the mycelial pellet in 150 ml of 0.1 M EDTA, pH 8.0; stir for 30 min; centrifuge; collect the pellet.

iii. Grind the mycelium in liquid nitrogen in a mortar with pestle; add 100 ml of lysis buffer containing 50 mM Tris-HCl, pH 8.0; 50 mM EDTA; pH 8.0; 3% sodium dodecyl sulfate (SDS); 1% 2-mercaptoethanol; heat in a water bath at 65°C for 60 min.

iv. Add 100 ml of phenol-chloroform mixture (1:1, v/v); stir for 30 min; centrifuge at 16,500 relative centrifugal force (RCF) for 10 min.

v. Retain the upper phase and reextract with phenol/chloroform again; precipitate the DNA by adding 0.1 volume of 3.0 M sodium acetate, pH 7.0, and 2.5 volume of absolute alcohol.

vi. Transfer the DNA by using a sterile pipette to a centrifuge tube; wash once with 70% ethanol; drain to dry for a brief period.

vii. Add 15 ml of 1X TE buffer containing 10 mM Tris-HCl, 1 mM EDTA, pH 8.0 to the pellet; shake gently, heat intermittently at 68°C; add 1X TE buffer, if required, to dissolve the pellet.

viii. Add 8.6 g of cesium chloride (CsCl) and 0.3 ml of ethidium bromide (5 mg/ml) at the rate of 7.5 ml of DNA solution; centrifuge at 157,516 RCF for 15 hr in a vertical rotor.

ix. Collect the DNA band (distinct from the brownish polysaccharide band); extract with CsCl-saturated 1-butanol.

x. Precipitate the genomic DNA with 3 volumes of water and 4 volumes of isopropanol; wash with 70% ethanol; vaccum dry; resuspend the DNA in 1X TE buffer.

B. DNA hybridization
    i.   Cleave the genomic DNA with *Eco* RI (15 units) at 37°C overnight; fractionate DNA fragments by 1% agarose gel electrophoresis at a constant voltage of 25 V, after pretreatment of gel with 0.25 *N* HCl for 15 min, 1.5 *M* NaCl/0.5 *N* NaOH for 25 min, and 1.5 *M* NaCl/0.5 *M* Tris-HCl, pH 7.4, for 25 min, successively.
    ii. Transfer DNA fragments to membranes (Nytran) (pore size 0.45 μ*M*) using X 10 SSC buffer consisting of 1.5 *M* NaCl, 0.15 *M* sodium citrate, pH 7.0.
    iii. Bake the blots in a vacuum at 80°C for 2 hr; use the appropriate probe for hybridization.

APPENDIX 9(xv): RANDOM AMPLIFIED POLYMORPHIC DNA PROCEDURE FOR THE DIFFERENTIATION OF STRAINS OF FUNGI (ZIMAND ET AL., 1994)

A. Preparation of genomic DNA
    i.   Grow the test fungus in appropriate medium for 5–7 days; harvest the mycelium; freeze-dry overnight.
    ii.   Grind the mycelium (50 mg) in an Eppendorf tube; suspend in 500 μl of extraction buffer containing 200 m*M* Tris-HCl, pH 8.5; 250 m*M* NaCl; 2.5 m*M* EDTA; 0.5% sodium dodecyl sulfate (SDS).
    iii. Mix the slurry in 350 μl of phenol (melted at 45°C and equilibrated with 1 volume of extraction buffer); then add 150 μl chloroform; mix well; centrifuge at 13,000 g for 1 hr.
    iv.   Transfer the upper aqueous phase to a tube containing 25 μl RNase A solution (20 mg/ml in 10 m*M* Tris-HCl, pH 7.5; 15 m*M* NaCl); boil for 10 min; incubate for 10 min at 37°C.
    v.   Add 0.7 *M* NaCl and 0.1 volume of cetyltrimethyl ethyl ammonium bromide (CTAB) to remove the polysaccharides; extract with chloroform/isoamyl alcohol (12:1); remove the upper phase; repeat CTAB extraction till no interface is seen.
    vi.   Transfer the clean upper phase to an Eppendorf tube; mix with 0.54 volume of isopropanol; DNA precipitates as a lump; remove as much of the supernatant as possible.
    vii. Centrifuge for 5 sec; remove the supernatant with a pipette; rinse the pellet with 70% ethanol; dry under a vacuum; resuspend in 100 μl of 10 m*M* Tris-HCl, pH 8.0, and EDTA.
B. PCR amplification
    i.   Perform amplification in 25 μl containing 1–1.5 units of *Taq* DNA

polymerase; 0.5 μl each of dCTP, dGTP, dATP, and dTTP (1.25 mM); in 10 mM Tris-HCl, pH 9.0; 50 mM KCl; and 0.1% Triton X-100; 1.9 μl of $MgCl_2$ (25 mM); 0.5 μl of (10 μM) primer and 25–50 ng of DNA; overlay the reaction mixture with mineral oil in a Thermo Cycler through one cycle of denaturation at 94°C for 7 min; perform low stringency annealing of the primer at 35°C for 1 min and extension at 72°C for 1 min.

ii. Run additional 40 cycles with following steps: 90°C for 1 min, 35°C for 1 min, and 72°C for 2 min, final cycle at 72°C for 5 min.

iii. Electrophorese the PCR products in 1.2% agarose gels; stain with ethidium bromide for detection; use size markers obtained by digesting lambda DNA with *Pst* 1.

APPENDIX 9(xvi): PCR AMPLIFICATION OF BACTERIAL DNA FOR IDENTIFICATION OF BACTERIAL PATHOGENS (COTTYN ET AL., 1994)

A. Preparation of bacterial DNA

i. Grow the bacteria in nutrient broth to the early log phase; harvest the cells by centrifugation at 12,000 rpm for 10 min; resuspend the pellet in 2 ml of Tris-HCl, pH 8.3, and 1.0 mM EDTA (1X TE).

ii. Add 250 μl of 10% sodium dodecyl sulfate (SDS) and 50 μl of 10 mg/ml of proteinase K; incubate at 37°C for 1 hr with gentle shaking.

iii. Add 0.45 ml of 5 M NaCl and mix thoroughly; add 0.4 ml of hexadecyl trimethyl ammonium bromide (CTAB) solution (10% CTAB in 0.7 M NaCl); incubate at 65°C for 20 min.

iv. Add an equal volume of chloroform:isoamyl alcohol (24:1); shake the mixture gently for 30 min; centrifuge at 15,000 rpm for 30 min.

v. Transfer the upper aqueous phase by using a bent Pasteur pipette to a fresh tube containing an equal volume of cold isopropanol; gently shake till completion of DNA precipitation.

vi. Take out the DNA with a Pasteur pipette; wash in 70% ethanol; dissolve in 1X TE; store at −20°C.

B. PCR amplification and analysis of PCR products

i. Prepare the reaction mixture consisting of 20 ng of DNA template in a 50 μl reaction volume containing 10 mM Tris-HCl, pH 8.3; 50 mM KCl, 200 M each of dNTP; 0.4 M each of primer; and 1.5 units of *Taq* polymerase.

ii. Carry out amplification on a DNA Thermocycler (Perkin Elmer Cetus)

with an initial denaturation step at 94°C for 2 min followed by 40 cycles of the denaturation step at 94°C for 1.5 min, a primer annealing
step at 62°C for 2 min, and an extension step at 72°C for 2 min with an
additional extension step of 72°C for 5 min as the last cycle.

iii. Perform gel electrophoresis on 2% agarose in Tris-borate EDTA at 2
      V/cm; stain with ethidium bromide.

10
Detection of Double-Stranded RNAs

Infected plants may have double-stranded-(ds)-RNAs when a) the infecting viral genome is ds-RNA, as in phytoreoviruses and cryptoviruses, or b) ds-RNA is produced as a replicating form during the process of replication of single-stranded (ss)-RNA viruses. In the case of certain viruses, such as velvet tobacco mottle virus, ds-RNA may accumulate during replication. This technique of detecting the ds-RNA is useful for the early and rapid recognition of virus infection. The presence of ds-RNA can be detected by either polyacrylamide gel electrophoresis (PAGE) or antiserum reaction with ds-RNA. The quantity of ds-RNA obtained may vary with host-virus combinations (Valverde et al., 1986). The ds-RNA may be isolated and labeled or cloned as cDNA for developing nucleic acid probes.

Labeled ds-RNA or cDNA probes may be used to detect RNA viruses (Bar-Joseph et al., 1983; Jordan, 1986). It is possible to detect infection by viruses which reach only very low concentrations or are difficult to purify (Chu et al., 1989) and before symptom expression (Lejour and Kummert, 1986). Groundnut rosette virus (GRV) infection was detected by the presence of the ds-RNA even when the assistor virus was absent. As the antiserum against GRV is not available, detection of ds-RNA will be helpful for diagnosing GRV infection (Breyel et al., 1988). Latent infection by plant viruses can also be recognized by detecting the ds-RNA of the viruses. Habili (1993) reported that widespread latent infection of citrus with tristeza virus in Australia could be detected by ds-RNA analysis. Saldarelli et al. (1994) employed two ^{32}P-labeled cDNA clones specific for ds-RNA sequences as probes for detecting grapevine leaf roll-associated closterovirus III (GLRa V III) in grapevine extracts from leaves, petioles, or cortical tissues infected by the virus. The probes were virus-specific and did not hybridize with total RNA from healthy controls or from vines infected by other viruses. Tomato infectious chlorosis virus (TICV) has a bipartite genome consisting of ss-RNA1 (7880 nt) and ss-RNA2 (7400 nt). Ten clones containing cDNAs of viral genomic RNAs were used to generate DIG-UTP-labeled transcripts. These transcripts hybridized with TICV ssRNAs in Northern blot hybridization analysis. The labeled transcripts could be employed as probes in dot-blot analysis to detect infection of tomatoes by TICV (Wisler et al., 1996). Tomato chlorosis virus (ToCV), another distinct whitefly-transmitted, phloem-limited bipartite clostervirus, has two prominent ds RNAs of approximately 7,800 and 8,200 bp with several small ds RNAs. DIG-11-UTP-labeled probes derived from cDNA clones representing portions of RNAs 1 and 2 were used in dot-blot hybridizations to detect ToCV in infected tomato (Fig. 10.1) (Wisler et al., 1998).

The ds-RNA profile may help in the classification of an unknown virus or virus strain. On the basis of the characteristics of double stranded forms of subgenomic mRNAs, strains of the viruses may be identified (Valverde et al., 1986). By determining ds-RNA profiles of 26 cucumber mosaic cucumovirus isolates, 7 distinct ds-RNA profile types can be identified. The differences between the profiles are stable and hence may be used as the basis of classifying CMV isolates. They reflect a greater range of biological characteristics than other methods currently available (Pares et al., 1992). Additionally, ds-RNA analysis may help to detect

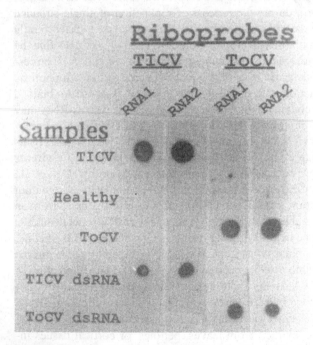

Figure 10.1 Reciprocal dot-blot hybridization analysis using DIG-11-UTP-labeled RNA transcripts representing portions of RNA 1 and RNA 2 from tomato infectious chlorosis virus (TICV) and tomato chlorosis virus (ToCV) (American Phytopathological Society, Minnesota, USA). (Courtesy of Wisler et al., 1998.)

the presence of infection complexes such as satellite RNA, multiple infections, and cryptic virus (Lejour and Kummert, 1986). The ds-RNA analysis of tomato aspermy cucumovirus (ch-TAV) infecting chrysanthemum revealed the presence of four distinct viral dsRNAs, whereas only one dsRNA was detected in plants infected by chrysanthemum B carlavirus (Ch-CVB) indicating the possibility of using dsRNA profiles for the identification and differentiation of these viruses (Chung et al., 1999).

The possibility of utilizing dsRNA detection for virus disease diagnosis for plant quarantine purposes was demonstrated by Yamashita et al. (1996). The ds-RNAs of CMV and potato X potexvirus (PVX) could be differentially detected in infected plants. Likewise, *Chenopodium quinoa*, when infected individually by different viruses such as CMV, tobacco rattle virus, tobacco ringspot virus, arabis mosaic virus, grapevine fan leaf virus, carnation mottle virus, and citrus tatter leaf virus, could be detected by ds-RNA analysis. Each virus exhibited a characteristic pattern of ds-RNAs, indicating that ds-RNA analysis can provide a useful procedure for first screening of virus infection in plant materials to be examined by quarantine personnel. (Yamashita et al., 1996). An RNA probe corresponding to the 3'-terminal region of apple stem grooving capillovirus (ASGV) was employed to detect ds-RNAs in infected plant tissues using Northern hybridization analysis. All infected plant samples contained three ASGV-specific ds-RNAs consisting of 6.5, 2.0, and 1.0 kbp. When additional RNA probes corresponding to different parts of the viral genome were employed, five virus-specific ds RNAs with 6.5 kbp (G-ds), 5.5 kbp (ds1), 4.5 kbp (ds2), 2.0 kbp (SG-ds1), and 1.0 kbp (SG-ds2) were detected. The G-ds is suggested to be the replicative form (RF) of the ASGV genome. The ds-RNA profile may be useful for the detection and identification of ASGV (Magome et al., 1997).

Virus strain differentiation may be possible by studying the polymorphism of heterologous duplexes of RNA transcripts. Complementary RNA (cRNA) transcripts were prepared from PCR products of prunus necrotic ringspot ilarvirus (PNRSV) isolates. The electrophoretic mobilities of the heterologous duplexes made from the cRNA transcripts originating from different isolates of PNRSV varied considerably. This polymorphism, designated double-stranded (ds)-transcript polymorphism (ds-TCP), appears to be conferred by conformational variations among the transcript duplexes. The ds-TCP may serve as an additional tool for the identification and differentiation of strains of PNRSV (Rosner et al., 1999).

The antiserum prepared against polyinosinic-polycytidylic acid (In-Cn) may be used to detect the ds-RNA of plant viruses by indirect enzyme-linked immunosorbent assay (ELISA-I). The ds-RNA from cucumber mosaic cucumovirus (CMV) and plum pox potyvirus (PPV) from infected plants were obtained by heating the leaf extracts at 80°C for 2 min and maintaining a pH of 6. The ds-RNA of PPV from *Nicotiana benthamiana* plants could be detected readily at 50 days after inculation (Aramburu and Moreno, 1994).

There are, however, certain limitations that prevent the procedure from being adopted widely. The presence of non-pathogenesis-related ds-RNA in healthy controls, lower level of sensitivity than bioassay or electron microscopy, and laborious nucleic acid purification methods prevent the wider use of ds-RNAs as an aid for virus detection and classification (Chu et al., 1989).

SUMMARY

Plants infected by plant viruses containing RNA as a genome may have ds-RNA, because some viruses have ds-RNA as their genome, or a ds-RNA may be produced as a replicating form during the process of viral nucleic acid synthesis. Detection of this ds-RNA by ds-RNA or cDNA probes is possible. The ds-RNA profile is useful for the classification of unknown viruses or virus strains.

11
Diagnosis and Monitoring of Plant Diseases

In the past, diagnosis of different diseases affecting crop plants was considered a form of art, and disease was diagnosed by an intuitive judgment as to its nature after examination of the visible symptoms of the disease concerned. Later, using determinations based on biochemical and physiological studies of the mechanisms of infection or the metabolic characteristics of pathogens, the diseases were diagnosed (McIntyre and Sands, 1977). It is now well recognized that the techniques for pathogen detection and diagnosis of diseases caused by them are critical factors to be considered for developing strategies for effective crop management and regulatory programs. The techniques are also required for determining the cause, epidemiological characteristics, and distribution of diseases in a geographical location.

During the past two decades remarkable advances in the fields of molecular biology and immunology have been responsible for significant improvements in the accuracy, rapidity, and sensitivity of diagnostic techniques. The diagnostic methods also enhance the effectiveness of quarantine and control measures, resulting in saving of millions of dollars (Leach and White, 1992).

Crop disease diagnosis is required essentially to recognize the primary disease-causing factor(s). The general procedure followed involves examination of symptoms, microscopic examination of diseased tissue, isolation and purification of the pathogen, and inoculation on the appropriate host to induce the disease for comparison of the symptoms with those observed on the plant submitted for diagnosis. Pathogens may then be grouped taxonomically on the basis of their morphological, physiological, or biochemical characteristics. If the pathogen occurs in the form of physiological races, a set of differential varieties of plant species has to be inoculated to identify the pathogen race or biotype. But this procedure is frequently found to be tedious and impractical when early diagnosis is desired.

The diagnosis of the disease without the requirement of isolation of the pathogen and subsequent adoption of Koch's postulates became the imperative need. The molecular diagnosis of plant diseases based on the immunological properties and genomic characteristics of plant pathogens is being increasingly preferred by diagnosticians, since the pathogens can be reliably identified very rapidly and suitable management decisions can be made at the appropriate time. Immunochemical assays employing polyclonal antibodies were the first molecular techniques tested for the detection of pathogens. Polyclonal antibodies have been found to be very effective for detection, identification, and taxonomic classification of viruses by employing immunosorbent electron microscopy or immunosorbent assays (Chapter 7). The use of polyclonal antibodies for bacterial or fungal pathogens is somewhat limited because of lack of specificity at the species, pathovar, or race level. However, the development of hybridoma technology leading to production of monoconal antibodies which react specifically with different epitopes helped to overcome some of the problems associated with polyclonal antibodies. Further improvements could be obtained by using nucleic acid-based techniques which can resolve the differences in the nucleic acid sequences which are not involved in the immunological properties of the pathogen. The immunoassays and nucleic acid-based methods provide the level of sensitivity and specificity required by diagnosticians.

The precision, sensitivity, and rapidity of the diagnosis may determine the quality and utility of the diagnostic tool employed. Molecular probes can appreciably reduce the time required for assay and increase the sensitivity of assay, allowing detection of the pathogen before symptom expression. Detection of viruses in presymptomatic plants or in single-vector insects can be achieved by ELISA (Hibino et al., 1988). As the viruses have much simpler structure than bacterial and fungal pathogens, immunological assays are more useful in detection of viral infections. Use of monoclonal antibodies, however, has enhanced the usefulness of immunological assays for the detection of bacterial and fungal pathogens.

Of the several requirements to be considered for selecting a diagnostic method for adoption, speed of obtaining the results may override other requirements to be satisfied. The molecular diagnostic methods have clear edge over the traditional methods. The real-time PCR assay has the potential of providing the test results most rapidly (within 7 min) (Belgrader et al., 1999) while the traditional methods may take several days or even weeks. Considerable experience on the part of an investigator is needed to perform traditional tests, whereas experience/knowledge is increasingly becoming replaced by expensive equipment and reagents for conducting modern molecular assays. The drawback of nonmobility associated with molecular techniques has to be overcome by developing less expensive kits for performing the tests under field conditions. Unless such developments occur soon, the advantages of inventing precise and sensitive techniques will not reach the field worker or cultivator for realizing the ultimate aim of preventing losses caused by the microbial plant pathogens.

Several additional advantages of employing molecular methods have been demonstrated. Mycotoxins produced by fungal pathogens are present in food and feed materials posing great danger to human beings and animals. Diagnostic methods are based on hapten technology in which the mycotoxin is bound to a known antigen. Detection of aflatoxin by immunological techniques in cereals and peanuts has been very effective in eliminating contaminated food materials (Chen and Chen, 1998). Development of immunological assays such as ELISA has been useful to detect fungicide residues in food materials eliminating the need to have expensive equipment (Cairoli et al., 1996; Viviani-Nauer et al., 1997; Watanabe et al., 1998). Furthermore, rapid detection of isolates of fungal pathogens showing resistance to fungicides such as carbendazim by using PCR-based assays greatly helped to plan planting and fungicide use optimization strategies well in advance (Mc Kay and Cooke, 1997). Equally important is the usefulness of PCR and RFLP techniques in identifying genes and genetic markers that can be effectively utilized for developing cultivars resistant to economically important diseases (Shen et al., 1998).

11.1 PLANT DISEASE DIAGNOSTIC CENTERS

Detection and identification of microbial plant pathogens rapidly, reliably, and accurately will be required in various contexts. Accurate identification is the cornerstone for success of all crop disease management strategies, whether advisory or statutory aimed at prevention, control, or eradication and irrespective of use of chemical, biological, or administrative means.

11.1.1 Prevention of Spread of Pathogens in International Trade (exclusion)

Every national government has a sovereign right and duty to protect both natural and managed agricultural, horticultural, and forest environments by preventing the introduction of destructive pathogens and insect pests that are not present in the native environment. Specific detection of pathogens listed as quarantine organisms by respective countries has to be undertaken rapidly and reliably. As per international regulations, specific detection of pathogens is required to be performed by exporting countries. All plants and propagative materials are subjected to different protocols of inspection and laboratory tests. At the point of entry by air, sea, or surface transport, a plant health inspection is carried out by personnel of importing countries and detailed detection methods are employed in postentry quarantine (PEQ) facilities.

11.1.2 Disease Eradication Programs

If a plant pathogen has reached the level of breaching the regulatory barrier, its further spread has to be contained by eradication of all kinds of host plants. For the eradication program to be effective, early detection and diagnosis of the disease-causing agent, especially in latent infections, have to be carried out. A suitable method that can detect low levels of pathogen population at a stage prior to widespread dispersal has to be adopted. The beet rhizomania disease caused by beet necrotic yellow vein virus could be contained largely by employing rapid and relatively cheap diagnostic techniques such as ELISA, coupled with statutory controls for the restriction of import and movement of soil and beets. Likewise, eradication campaign against potato brown rot disease due to *Ralstonia solanacearum* was successful in the United Kingdom because of the application of sensitive and specific diagnostic tests. It is interesting to note that the source of *R. solanacearum* was traced to *Solanum dulcamara,* a symptomless carrier, growing on the banks of an adjacent river, providing water for irrigation of crops. The roots of this weed host immersed in river water were found to be infected, by ELISA and PCR assays (Stead et al., 1996).

11.1.3 Production of Healthy Stocks

Use of disease-free seeds and propagative materials is one of the effective principles of crop disease management. Certification programs are being operated in several countries for the production of disease-free nuclear stocks. These programs rely on backup diagnostic tests to different degrees. The ELISA test amenable for automation and PCR assay are being used extensively for detection and identification of pathogens in seeds and propagative materials. Micropropagation has been shown to be an effective approach to eliminate microbial pathogens and to provide a disease-free source of plant materials for rapid multiplication. By using different ELISA formats or PCR-based assays, the pathogen-free condition of source plants is ensured.

11.1.4 Advisory Service

Reliable and rapid diagnosis of microbial pathogens is a necessity for providing effective advisory service to the growers. The diagnostic tests employed have to be specific, sensitive, reproducible, and frequently quantitative to plan effective strategies for crop disease management. The complex diseases due to two or more pathogenic causes have to be resolved using suitable diagnostic assays.

The need for the establishment of disease diagnostic centers (DDCs) was realized because an organized, systematic, and professional effort is required to assist in the rapid detection of pathogens and accurate identification of plant disease problems to implement suitable solutions. It is also rightly acknowledged that establishment of DDCs will ultimately reduce plant disease losses, both quantitative and qualitative since there can be no effective disease management system without proper diagnosis. Most of the departments of plant pathology in the agricultural or traditional universities at the state or province level or advanced research centers at the national level perform crop disease diagnostic services. In some countries, DDCs form part of the state's department of agriculture or are private or commercial services.

The responsibilities and motivation of DDCs may vary, depending on the nature of the agency/institution offering such a service. An extension-university clinic may offer disease diagnosis as a service, usually along with various aspects of the education programs of the cooperative extension service. The state departments of agriculture in the United States have diagnostic laboratories primarily to perform regulatory functions with the aim of identifying pathogens so as to limit or prevent their interstate or intrastate dissemination. Commercial or private laboratories also are engaged in disease diagnosis, mainly on a customized basis, as observed in many North American states (Barnes, 1994).

The capacity to handle diseased specimens may vary with DDCs, depending on the facilities available. They may receive several hundreds of plant disease samples in a year, and the number may go up progressively in future. The increase of over 900% in the number of specimens submitted to the Texas Plant Disease Diagnostic Laboratory over the past 11 years clearly indicates both the need for and acceptance of diagnostic service by the clientale group. Yet, in developing countries adequate facilities and expertise may not be available, and such a deficiency has to be overcome, wherever necessary, for the efficient and useful functioning of DDCs.

Adoption of modern methods will lead to dramatic changes in the sensitivity and reliability of disease diagnosis, increasing the credibility of the agency offering diagnostic service. The enzyme-linked immunosorbent assay (ELISA) and nucleic acid hybridization-based diagnostic procedures have been widely adopted, and these techniques have provided the modern clinics with a continued conduit of state-of-the-art diagnostic techniques. Many clinics in North American states routinely employ highly technical diagnostic procedures, which have attained the level of sophistication that prevails in medical diagnostic facilities, which require careful maintenance of equipment and instruments and professional expertise. Modern clinics also possess powerful computers for maintenance and retrieval of information. Traditional diagnostic methods involving isolating pathogens and culturing them on differential or selective media, inducing fungal sporulation by moist chamber incubation, performing examinations under the light microscope, proving pathogenicity to satisfy Koch's postulates, and inoculating onto differential diagnostic hosts or cultivars for race, strain, or biotype

identification, etc., have to be carried out under certain conditions for the interpretation of some disease problems.

The disease diagnostic centers have also been involved in activities associated with disease diagnosis, including disease surveys on a regional or national basis, host-pathogen indexing, and phytosanitary certificate services. These activities also demand accurate and rapid detection of plant pathogens for diagnosing
the various crop diseases without loss of time in taking required plant protection
measures. With the realization of the importance and need for effective functioning of DDCs, it was decided to bring out the *Plant Diagnostics Quarterly* (PDQ),
which is published by and for diagnosticians and those interested in disease diagnostics.

11.2 PLANT QUARANTINES

After the liberalization of regulations for import of plants and plant materials that
followed General Agreement on Tariffs and Trade (GATT) ratification by partic-
ipating countries, there is a potential risk of introducing several destructive
pathogens and pests, especially viruses, viroids, and physoplasmas. Implementa-
tion of adequate quarantine safeguards has become imperative to prevent the introduction of new pathogens. Establishment of a postentry quarantine facility
along with expertise in propagation of plants by tissue culture techniques will lead
to large-scale production of pathogen-free plants within short periods.

Various methods are used for the detection of viruses, viroids, and phytoplasmas in vegetatively propagated planting materials. The grow-out test, indica-
tor-inoculation test, indexing method, histopathological test, electron microscopy,
and serological tests are employed for the early detection of infection and elimination of infected materials or plants. Effective functioning of postentry quarantines will greatly help to prevent the introduction and subsequent spread of plant
pathogens.

SUMMARY

The history of development of plant disease diagnosis and the need for disease
monitoring to prevent introduction/incidence and spread of various crop diseases
are described. Establishment of disease diagnostic centers (DDCs) will help to detect and differentiate plant pathogens, and these centers can help farmers under-
take effective disease management programs.

References

Abbott, E.V. 1953. Redrot of sugarcane. *Yearbook Agriculture,* U.S. Department of Agriculture, pp. 536–539.

Acheche, H., Fattouch, S., M'Hirsi, S., Marzouki, N. and Marrakchi, M. 1999. Use of optimised PCR methods for the detection of GLRaV3: a closterovirus associated with grapevine in Tunisian grapevine plants. *Plant Molecular Biology Reporter, 17:* 31–42.

Adam, G., Lesemann, D.E., and Vetten, H.J. 1991. Monoclonal antibodies against tomato spotted wilt virus: Characterization and application. *Annals of Applied Biology, 118:* 87.

Adams, A.N. 1978. The detection of plum pox virus in *Prunus* species by enzyme-linked immunosorbent assay (ELISA). *Annals of Applied Biology, 90:* 215–221.

Adams, A.N., Barbara, D.J., Morton, A., Darby, P. and Green, C.P. 1995. The control of hop latent viroid in hops. *Acta Horticulturae, No. 385,* 91–97.

Adams, A.N., Guise, C.M. and Crossley, S.J. 1999. Plum pox virus detection in dormant plum trees by PCR and ELISA. *Plant Pathology, 48:* 240–244.

Adsuar, J. 1950. On the physical properties of sugarcane mosaic virus. *Phytopathology, 40:* 2.

Agrios, G.N. 1969, 1988. *Plant Pathology.* Academic Press, San Diego.

Ahlawat, Y.S., Pant, R.P., Lockhart, B.E.L., Srivastava, M., Chakraborty, N.K. and Varma A. 1996. Association of a badnavirus with citrus mosaic disease in India. *Plant Disease, 80:* 590–592.

Ahn, J.H. and Walton, J.D. 1997. A fatty acid synthase gene in *Cochliobolus carbonum* required for production of HC-toxin, cyclo (D-prolyl-L-alanyl D-alanyl-L-2-amino-9,10-epoxi-8-oxodecanoyl). *Molecular Plant-Microbe Interactions, 10:* 207–214.

Ahn, J.H. and Walton, J.D. 1998. Regulation of cyclic peptide biosynthesis and pathogenicity in *Cochliobolus carbonum* by TOXEp, a novel protein with a ZIP basic DNA binding motif and four ankyrin repeats. *Molecular and General Genetics, 260:* 462–469.

Ahoonmanesh, A., Hajimorad, M.R., Ingham, B.L., and Francki, R.I.B. 1990. Indirect double antibody sandwich ELISA for detecting alfalfa mosaic virus in aphids after short probes on infected plants. *Journal of Virological Methods, 30:* 271–282.

Ahrens, V. and Seemuller, E. 1992. Detection of DNA of plant pathogenic mycoplasma-like organisms by a polymerase chain reaction that amplifies a sequence of the 16Sr-RNA gene. *Phytopathology, 82:* 828–832.

Al Ani, R., Pfeiffer, P., Whitechurch, D., Lesot, A., Lebewier, G., and Hirth, L. 1980. *Ann. Virol (Institut Pasteur) 131E:* 33–55.

Albibi, R., Chen, J., Lamikanra, O., Banks, D., Jarret, R. and Smith, B.J. 1998. RAPD finger printing *Xylella fastidiosa* Pierce's disease strains isolated from a vineyard in North Florida. *FEMS Microbiology Letters, 165:* 347–352.

Aldon, D., Brito, B., Boucher, C. and Genin, S. 2000. A bacterial sensor of plant cell contact controls the transcriptional induction of *Ralstonia solanacearum* pathogenicity genes. *EMBO Journal, 19:* 2304–2314.

Alexopoulos, C.J., and Mims, C.W. 1979. *Introductory Mycology*, 3rd Edition, Wiley, New York.

Ali, A., Randles, J.W. and Hodgson, R.A.J. 1998. Sensitive detection of pea seed-borne mosaic potyvirus by dot-blot and tissue print hybridization assays. *Australian Journal of Agricultural Research, 49:* 191–197.

Alizadeh, A., Arlat, M., Sarrafi, A., Boucher, C.A. and Barrault, G. 1997. Restriction fragment length polymorphism analyses of Iranian strains of *Xanthomonas campestris* from cereals and grasses. *Plant Disesase, 81:* 31–35.

Allen, C., Gay, J. and Simon-Buela, L. 1997. A regulatory locus *pehSR*, controls polygalacturonase production and other virulence functions in *Ralstonia solanacearum*. *Molecular Plant-Microbe Interactions, 10:* 1054–1064.

Allen, T.C. and Riker, A.J. 1932. A rot of apple fruit caused by *Phytomonas melophthora* n. sp following invasion by the apple maggot. *Phytopathology, 22:* 557–571.

Allison, R.M. 1953. Effect of leaf roll virus infection on the soluble nitrogen composition of potato tubers. *Nature, 171:* 573.

Alvarez, A.M., Benedict, A.A., and Mizumoto, C.Y. 1985. Identification of xanthomonads and grouping of strains of *Xanthomonas campestris* pv. *campestris* with monoclonal antibodies. *Phytopathology, 75:* 722–728.

Alvarez, A.M., Benedict, A.A., Mizumoto, C.Y., Hunter, J.E., and Gabriel, D.W. 1994. Serological, pathological and genetic diversity among strains of *Xanthomonas campestris* infecting crucifers. *Phytopathology, 84:* 1449–1457.

Alvarez, A.M., Teng, P.S., and Benedict, A.A. 1989. Methods for epidemiological research on bacterial blight of rice. *Proceedings of International Workshop on Bacterial Blight of Rice*, pp. 99–110. International Rice Research Institute, Philippines.

Ambrós, S., Desvignes, J.C., Llácer, G. and Flores, R. 1995. Pear blister canker viroid: sequence variability and causal role in pear blister canker disease. *Journal of General Virology, 76:* 2625–2629.

Amelunxen, F. 1958. Die Virus-Eiweisspindeln der kakteen Darstellung elektronenmikroskopische und biochemische Analyse des Virus. *Protoplasma, 49:* 140–178.

Ammar, E.D., Jarlfors, U. and Pirone, T.P. 1994. Association of potyvirus helper component protein with virions and the cuticle lining the maxillary food canal and foregut of an aphid vector. *Phytopathology, 84:* 1054–1060.

Ammar, E.D., Rodriguez-Cerezo, E., Shaw, J.G., and Pirone, T.P. 1994. Association of virions and coat proteins of tobacco vein mottling potyvirus with cylindrical inclusions in tobacco cells. *Phytopathology, 84:* 520–524.

Andebrhan, T. and Furtek, D.B. 1994. Random amplified polymorphic DNA (RAPD) analysis of *Crinipellis perniciosa* isolates from different hosts. *Plant Pathology, 43:* 1020–1027.

Anderson, C.W. 1954. Two muskmelon mosaic virus strains from central Florida. *Phytopathology, 44:* 371–374.

Anderson, J.B., Petsche, D.M., and Smith, M.I. 1987. Restriction fragment polymorphisms in biological species of *Armillaria melea. Mycologia, 79:* 69–76.

Anderson, M.L.M. and Young, B.D., 1985. Quantitative filter hybridization. In: *Nucleic Acid Hybridization—a Practical Approach* (Eds.) B.D. Hames and S.J. Higgins, pp. 73–111, IRL Press, Oxford.

Anil Kumar, Singh, A. and Garg, G.K. 1998. Development of seed immunoblot binding assay for the detection of Karnal bunt *(Tilletia indica)* of wheat. *Journal of Plant Biochemistry and Biotechnology, 7:* 119–120.

Anonymous 1983. 2nd Edition. Commonwealth Mycological Institute, *Plant Pathologist's Pocketbook,* Kew, Surrey, England.

Aparicio, F., Myrta, A., Di Terlizzi, B., and Pallás, V. 1999. Molecular variability among isolates of *Prunus* necrotic ringspot virus from different *Prunus* spp. *Phytopathology, 89:* 991–999.

Arabatova, J., Lehto, K., Pehu, E. and Pehu, T. 1998. Localization of the P1 protein of potato associated with (1) cytoplasmic inclusion bodies and (2) cytoplasm of infected cells. *Journal of General Virology, 79:* 2319–2323.

Aramburu, J. and Moreno, P. 1994. Detection of double stranded RNA (dsRNA) in crude extracts of virus-infected plants by indirect ELISA. *Journal of Phytopathology, 141:* 375–385.

Archer, D.B. and Daniels, M.J. 1982. The biology of Mycoplasmas. In: *Plant and Insect Mycoplasma Techniques* (Eds.) M.J. Daniels and P.G. Markham, pp. 9–39. John Wiley & Sons, New York.

Arie, T., Gouthu, S., Shimagaki, S., Kamakura, J., Kimura, M., Inoue, M., Takio, K., Ozaki, A., Yoneyama, K., and Yamaguchi, I. 1998. Immunological detection of endopolygalacturonase secretion by *Fusarium oxysporum* in plant tissue and sequencing of its encoding gene. *Annals of Phytopathological Society, Japan, 64:* 7–15.

Ariovich, D. and Garnett, H.M. 1989. The use of immunogold staining technique for detection of a bacterium associated with greening diseased citrus. *Phytopathology, 79:* 382–384.

Ark, P.A. and Thomas, H.E. 1936. Persistence of *Erwinia amylovora* in certain insects. *Phytopathology, 26:* 375–381.

Arnold, D.L., Athey-Pollard, A., Gibbon, M.J., Taylor, J.D. and Vibian, A. 1996. Specific oligonucleotide primers for the identification of *Pseudomonas syringae* pv. *pisi* yield one of possible DNA fragments by PCR amplification: evidence for phylogenetic divergence. *Physiological and Molecular Plant Pathology, 49:* 233–245.

Arnott, H.J. and Smith, K.M. 1967. Electron microscopy of virus infected sunflower leaves. *Journal of Ultrastructure Research 19:* 173–195.

Arwiyanto, T., Goto, M., Tsuyumu, S., and Takikawa, Y., 1996. Biological control of bacterial wilt of tomato by an avirulent strain of *Pseudomonas solanacearum* isolated from *Strelitzia reginae. Annals of Phytopathological Society, Japan, 60:* 421–430.

Asai, M., Ohara, T., Takahashi, T., Saito, S. and Tanaka, K., 1998. Detection of viroids in fruit trees by return electrophoresis. *Research Bulletin of Plant Protection Service, Japan, No. 34,* 99–102.

Astruc, N., Macros, J.F., Macquaire, G., Candresse, T. and Pallás, V. 1996. Studies on the diagnosis of hop stunt viriod in fruit trees: identification of new hosts and application of a nucleic acid extraction procedure based on nonorganic solvents. *European Journal of Plant Pathology, 102:* 837–846.

Atzmon, G., Oss, H. van, and Czosnek, H., 1998. PCR-amplification of tomato yellow leafcurl virus (TYLCV) DNA from squashes of plants and whiteflies: application to the study of TYLCV acquisition and transmission. *European Journal of Plant Pathology, 104:* 189–194.

Audy, P., Laroche, A., Saindon, G., Huang, H.C., and Gilbertson, R.C. 1994. Detection of bean common blight bacteria, *Xanthomonas campestris* pv. *phaseoli* and *X. c. phaseoli* var. *fuscans* using polymerase chain reaction. *Phytopathology, 84:* 1185–1192.

Aveling, T.A.S. and Rong, I.H., 1994. Scanning electron microscopy of conidium formation of *Stemphylium vesicarium* on onion leaves. *Journal of Phytopathology, 140:* 77–81.

Avila, F.J., Burton, B.D., Fletcher, J., Sherwood, J.L., Pair, S.D. and Melcher, U. 1998. Polymerase chain reaction of an agent associated with yellow vine disease of cucurbits. *Phytopathology, 88:* 428–436.

Badami, R.S. and Kassanis, B. 1959. Some properties of three viruses isolated from a diseased plant of *Solanum jasminoides* Paxt. from India. *Annals of Applied Biology 47:* 90–97.

Badge, J., Brunt, A., Carson, R., Dagless, E., Karmagioli, M., Philips, S., Seal, S., Turner, R., and Foster, G.D., 1996. A carlavirus-specific PCR primer and partial nucleotide sequence provides further evidence for the recognition of cowpea mild mottle virus as a whitefly-transmitted carlavirus. *European Journal of Plant Pathology, 102:* 305–310.

Bagga, H.S. and Laster, M.L. 1968. Relation of insects to the initation and development of boll rot of cotton. *Journal of Economic Entomology, 61:* 1141.

Bahrani, Z., Sherwood, J.L., Sanborn, M.R., and Keyser, G.C. 1988. The use of monoclonal antibodies to detect wheat soil-borne mosaic virus. *Journal of General Virology, 69:* 1317.

Baharuddin, B., Rudolph, K., and Niepold, F. 1994. Production of monospecefic antiserum against the blood disease bacterium affecting banana and plantain. *Phytopathology, 84:* 570–575.

Bajet, N.B., Daquioag, R.D., and Hibnio, H. 1985. Enzyme-linked immunosorbent assay to diagnose rice tungro. *Journal of Plant Protection in the Tropics, 2:* 125–129.

Baker, C.A. and Purcifull, D.E. 1989. Antigenic diversity of papaya ring spot virus (PRSV) isolate detected by monoclonal antibodies to PRSV-w. *Phytopathology, 79:* 214.

Bakker, W. 1975. Rice yellow mottle virus. *CMI/AAB Descriptions of Plant Viruses,* No. 149.

Balesdent, M.H., Jedryczka, M., Jain, L. Mendes-Pereira, E., Betrandy, J., and Rouxel, T., 1998. Conidia as a substrate for internal transcribed spacer based PCR identification of components of *Leptosphaeria maculans*-species complex. *Phytopathology, 88:* 1210–1217.

Ball, E.M. 1971. Leaf dip serology. *Methods in Virology,* Vol. 5, pp. 445–450. Academic Press, New York.

Ball, E.M. and Brakke, M.K. 1968. Leaf dip serology for electron microscopic identification of plant viruses. *Virology, 36:* 152–155.

Bancroft, J.B. 1962. Purification and properties of bean pod mottle virus and associated centrifugal and electrophoretic components. *Virology, 16:* 419–427.

Bancroft, J.B., Bracker, C.E., and Wagner, G.W. 1969. Structures derived from cowpea chlorotic mottle and brome mosaic virus protein. *Virology, 38:* 324–335.

Bancroft, J.B., Hills, G.J., and Markham, R. 1967. A study of the self assembly process in a small spherical virus. *Virology, 31:* 354–379.

Bandla, M.D., Westcot, D.M., Chenault, K.D., Ulman, D.E., German, T.L., and Sherwood, J.L. 1994. Use of monoclonal antibody to the nonstructural protein encoded by the small RNA of the tomato spotted wilt tospovirus to identify the viruliferous thrips. *Phytopathology, 84:* 1427–1431.

Banks, J.N. and Cox, S.J. 1992. The solid phase attachment of fungal hyphae in an ELISA to screen for antifungal antibodies. *Mycopathologia, 120:* 79–85.

Banks, J.N., Cox, S.J., Clarke, J.H., Shamsi, R.H., and Northway, B.J. 1992. Towards the immunological detection of field and storage fungi. In: *Modern Methods in Food Mycology* (Eds.) R.A. Samson, A.D. Hocking, J.I. Pitt, and A.D. King, pp. 247–252. Elsevier, Holland.

Banks, J.N., Cox, S.J., Northway, B.J., and Rizvi, R.H. 1994. Monoclonal antibodies to fungi of significance to the quality of foods and feeds. *Food and Agricultural Immunology, 6:* 321–327.

Banttari, E.E. and Goodwin, P.H. 1985. Detection of potato viruses, S, X and Y by enzyme-linked immunosorbent assay on nitrocellulose membranes (Dot-ELISA). *Plant Disease, 69:* 202.

Barbara, D.J., Clark, M.F., and Thresh, J.M. 1978. Rapid detection and serotyping of *Prunus* necrotic ring spot virus in perennial crops by enzyme-linked immunosorbent assay. *Annals of Applied Biology, 90:* 395–399.

Barbara, D.J., Morton, A. Spence, N.J., and Miller, A. 1995. Rapid differentiation of closely related isolates of two plant viruses by polymerase chain reaction and restriction fragment length polymorphisms analysis. *Journal of Virological Methods, 55:* 121–131.

Bar-Joseph, M., Garnsey, S.M., Gonsalves, D., Moscovitz, M., Purcifull, D.E., Clark, M.F., and Loebenstein, G. 1979. The use of enzyme-linked immunosorbent assay for the detection of citrus tristeza virus. *Phytopathology, 69:* 190–194.

Bar-Joseph, M., Lobenstein, G., and Cohen, J. 1970. Partial purification of virus-like particles associated with citrus tristeza disease. *Phytopathology, 60:* 75–78.

Bar-Joseph, M., Rosner, A., Moscovitz, M., and Hull, R. 1983. A simple procedure for the extraction of double stranded RNA from virus-infected plants. *Journal of Virological Methods, 6:* 1–8.

Barker, H., Reavy, B., Kumar, A., Webster, K.D., and Mayo, M.A. 1992. Restricted virus multiplication in potatoes transformed with the coat protein gene of potato host gene-mediated resistance. *Annals of Applied Biology, 120:* 55–64.

Barker, H., Webster, K.D., Jolly, C.A., Reavy, B., Kumar, A. and Mayo, M.A. 1994. Enchancement of resistance to potato leafroll virus multiplication in potato by combining the effects of host genes and transgenes. *Molecular Plant-Microbe Interactions, 7:* 528–530.

Barnes, L.W. 1994. The role of plant clinics in disease diagnosis and education: A North American perspective. *Annual Review of Phytopathology, 32:* 601–609.

Barnett, D.W. and Murrant, A.F. 1970. Host range, properties and purification of raspberry bushy dwarf virus. *Annals of Applied Biology 65:* 433–449.

Barta, T.M., Kinscherf, T.G. and Willis, D.K. 1992. Regulation of tabtoxin production by the *lem A* gene in *Pseudomonas syringae. Journal of Bacteriology, 174:* 3021–3029.

Baudry, C. 1999. *Phytophthora fragariae* of raspberry and strawberry: detection using molecular biology. *Phytoma, No. 513,* 48–52.

Bawden, F.C. 1951. *Plant Pathology Department Report of Rothamstead Experiment Station 1950,* pp. 69–78.

Bawden, F.C. and Pirie, N.W. 1937. Liquid crystalline preparations of cucumber viruses 3 and 4. *Nature, 139:* 546–547.

Beal, J.M., Preston, W.H., and Mitchell, J.W. 1955. Use of 2,3,5-triphenyl tetrazolium chloride to detect the presence of viruses in plants. *Plant Disease Reporter, 39:* 558–560.

Beck, J.J., Beebe, J.R., Stewart, S.T., Bassin, C. and Etienne, L., 1996. Colorimetric PCR and ELISA diagnostics for the detection of *Pseudocercospora herpotrichoides* in field samples. *Brighton Crop Protection Conference: Pests and Diseases, 1:* 221–226.

Beck, J.J. and Ligon, J.M. 1995. Polymerase chain reaction assays for the detection of *Stagonospora nodorum* and *Septoria tritici* in wheat. *Phytopathology, 85:* 319–324.

Behjatnia, S.A.A. Dry, I.B. Krake, L.R., Conde, B.D., Connelly, M.I., Randles, J.W. and Rezaian, M.A. 1996. New potato spindle tuber viroid and tomato leafcurl geminivirus strains from wild *Solanum* sp. *Phytopathology, 86:* 880–886.

Beijersbergen, A. 1993. Trans-kingdom promiscuity: Similarities between T-DNA transfer by *Agrobacterium tumefaciens* and bacterial conjugation. Ph.D. Thesis, University of Leiden, Netherlands.

Belgrader, P., Benett, W., Hadley, D., Richards, J., Stratton, P., Mariella, R, Jr. and Milanovich, F. 1999. PCR detection of bacteria in seven minutes. *Science (Washington), 284:* 449–450.

Bell, K.S., Roberts, J., Verrall, S., Cullen, D.W., Williams, N.A., Harrison, J.G., Toth, I.K., Cooke, D.E.L., Duncan, J.M. and Claxton, J.R. 1999. Detection and quantification of *Spongospora subterranea* f.sp. *subterranea* in soils and on tubers using specific PCR primers. *European Journal of Plant Pathology, 105:* 905–915.

Bellaire, L. de, Chillet, M. and Mourichon, S. 2000. Elaboration of an early quantification method of quiescent infections of *Colletotrichum musae* on bananas. *Plant Disease, 84:* 128–133.

Beltz, G.A., Jacobs, J.A., Eickbush, T.H., Cherbas, P.J., and Kafatos, F.C. 1983. Isolation of multigene families and determination of homologies by filter hybridization methods. *Methods in Enzymology, 100:* 266–285.

Bem, F. and Murrant, A.F. 1980. Heracleum latent virus. *CMI/AAB Descriptions of Plant Viruses,* No. 228.

Benedict, A., Alvarez, A.M., Berestecky, J., Imanaka, W., Mizumoto, C., Pollard, L., Mew, T., and Gonzalez, C. 1989. Pathovar specific monoclonal antibodies for *Xanthomonas campestris* pv. *oryzae. Phytopathology, 79:* 322–328.

Bennett, C.W. 1949. Some unreported host plants of sugarbeet mosaic virus. *Phytopathology, 39:* 669–672.

Bennett, C.W. 1955. Recovery of water pimpernel from curly top and the reaction of recovered plants to reinoculation with different virus strains. *Phytopathology, 45:* 531–536.

Bennett, C.W. 1959. Lychnis ringspot. *Phytopathology, 49:* 706–713.

Bennett, C.W. 1963. Highly virulent strains of curly top virus in sugarbeet in western United States. *Journal of American Society of Sugarbeet Technologists, 12:* 515–520.

Bennett, C.W. and Costa, A.S. 1961. Sowbane mosaic caused by a seed-transmitted virus. *Phytopathology, 51:* 546–550.

Benson, A.P., Raymer, W.B., Smith, W., Jones, E., and Munro, J. 1965. Potato diseases and their control. *Potato Handbook, 10:* 32–36.

Benson, A.P., and Singh, R.P. 1964. Seed transmission of potato spindle tuber virus in tomato. *American Potato Journal, 41:* 294.

Benson, D.M. 1992. Detection by enzyme-linked immunosorbent assay of *Rhizoctonia* species on poinsettia cuttings. *Plant Disease, 76:* 578–581.

Benton, R.J., Bowman, F.T., Fraser, L., and Kebby, R.G. 1950. Stunting and scaly butt of citrus associated with *Poncirus trifoliata* root stock. *Agricultural Gazette, N.S. Wales, 61:* 20–22, 40.

Bercks, R. 1971. White clover mosaic virus. *CMI/AAB Descriptions of Plant Viruses No. 41.*

Bercks, R., and Brandes, J. 1963. Elektronenmikroskopische und serologische Untersuchungen zur klnosifizierung des clover yellow mosaic virus. *Phytopathologische Zeitschrift, 47:* 381–390.

Bercks, R. Koenig, R., and Querfurth, G. 1972. Plant virus serology. In: *Principles and Techniques in Plant Virology* (Eds.). C.I. Kado and H.O. Agrawal, pp. 466–490. Van Nostrand Reinhold Co., New York.

Bereswill, S. and Geider, K. 1996. Identification of *Erwinia amylovora* by PCR analyses. *Acta Horticulturae, No. 411,* 57–62.

Beretta, M.J.G., Barthe, G.A., Ceccardi, T.L., Lee, R.F., and Devick, K.S. 1997. A survey for strains of *Xylella fastidiosa* in citrus affected by citrus variegated chlorosis and citrus blight in Brazil. *Plant Disease, 81:* 1196–1198.

Berger, P.H., Thornbury, D.W., and Pirone, T.P. 1985. Detection of picogram quantities of potyviruses using a dot-blot immunobinding assay. *Journal of Virological Methods, 12:* 31.

Bergeson, G.B., Athow, K.L., Laviolette, F.A., and Thomasine, M. 1964. Transmission, movement and vector relationships of tobacco ringspot virus in soybean. *Phytopathology, 54:* 723–726.

Berisha, B., Chen, Y.D., Zhang, G.Y., Xu, B.Y. and Chen, T.A. 1998. Isolation of Pierce's disease bacteria from grapevines in Europe. *European Journal of Plant Pathology, 104:* 427–433.

Berlyn, G.P. and Miksche, J.P. 1976. Botanical microtechnique and cytochemistry. The Iowa State University Press, Ames, 326 pp.

Bermpohl, A., Dreier, J., Bahro, R. and Eichenlaub, R. 1996. Exopolysaccharides in the pathogenic interaction of *Clavibacter michiganensis* subsp. *michiganensis* with tomato plants. *Microbiological Research, 151:* 391–399.

Bernhard, F., Coplin, D.L. and Geider, K. 1993. A gene cluster for synthesis in *Erwinia amylovora:* characterization and relationship to *cps* genes in *Erwinia stewartii. Molecular and General Genetics, 239:* 158–168.

Bertaccini, A., Davis, R.E., Hammond, R.W., Vibio, M., Bellardi, M.G., and Lee, I.M. 1992. Sensitive detection of mycoplasma-like organisms in field-collected and in vitro propagated plants of *Brassica hydrangea* and *Chrysanthemum* by polymerase chain reaction. *Annals of Applied Biology, 121:* 593–599.

Bertaccini, A., Davis, R.E. and Lee, I.M. 1992. In vitro micropropagation for maintenance and mycoplasma-like organisms in infected plant tissues. *Horticultural Science, 27:* 1041–1043.

Bertaccini, A., Davis, R.E., Lee, I.M., Conti, M., Dally, E.L. and Douglas, S.M. 1990. Detection of chrysanthemum yellows mycoplasma-like organism by dot-hybridization. *Plant Disease, 74:* 40.

Best, R.J. 1968. Tomato spotted wilt virus. *Advances in Virus Research, 13:* 60–146.

Bezerra, I.C., Resende, R.O. de, Pozzer, L., Nagata, T., Kormelink, R. and De Avila, A.C. 1999. Increase of tospoviral diversity in Brazil with the identification of two new tospovirus species, one from chrysanthemum and one from zucchini. *Phytopathology, 89:* 823–830.

Bijaisoradat, M. and Kuhn, C.W. 1988. Detection of two viruses in peanut seeds by complementary DNA hybridization tests. *Plant Disease, 72:* 956–959.

Bils, R.F. and Hall, C.E. 1962. Electron microscopy of wound humour virus. *Virology, 17:* 123–130.

Bily, K.L.G. and Legunkova, R.M. 1966. Electron microscope studies on the soybean mosaic virus. *Mykrobiol. Zh., 28:* 39–42.

Bitancourt, A.A. 1944. A test for the early identification of citrus tristeza. *Biologico, 10:* 169–175.

Bittóová, M., Hrouda, M. and Kominek, P. 1997. Molecular detection of plum pox virus. *Ochrana Rostlin, 33:* 1–7.

Black, L.M. 1938. Properties of the potato yellow dwarf virus. *Phytopathology, 28:* 863–874.

Black, L.M. 1941. Specific transmission of varieties of potato yellow dwarf virus by related insects. *American Potato Journal, 18:* 231–233.

Black, L.M. 1944. Some viruses transmitted by agallian leaf hoppers. *Proceedings of American Philosophical Society, 88:* 132–144.

Black, L.M. 1953. Occasional transmission of some plant viruses through the eggs of their insect vectors. *Phytopathology, 43:* 9–10.

Black, L.M. 1965. Physiology of virus-induced tumours in plants. In: *Handbuck der Pflanzenphysiologie* (Ed.) A. Lang, pp. 236–266. Springer-Verlag, New York.

Black, L.M. and Brakke, M.H. 1952. Multiplication of wound tumour virus in an insect vector. *Phytopathology, 42:* 269–273.

Blas, D.De, Borja, M.J., Saiz M., and Romero J. 1994. Broad spectrum detection of cucumber mosaic virus (CMV) using polymerase chain reaction. *Journal of Phytopathology, 141:* 323–329.

Blok, V.C., Ziegler, A., Scott, K., Dangora, D.B., Robinson, D.J. and Murant, A.F. 1995. Detection of groundnut rosette umbravirus infections with radioactive and nonradioactive probes to its satellite RNA. *Annals of Applied Biology, 127:* 321–328.

Bock, K.R. 1967. Hop mosaic virus. *Report of East Malling Research Station, 1966*, pp. 163–165.

Bold, H.C., Alexopoulos, C.J., and Delevorges, T. 1980. *Morphology of Plants and Fungi.* Harper & Row Publishers, New York.

Bonants, P., Hagennar-de Weerdt, M., Gent - Pelzer, M. van, Lacourt, I., Cooke, D. and Duncan, J. 1997. Detection and identification of *Phytophthora fragariae* Hickman by polymerase chain reaction. *European Journal of Plant Pathology, 103:* 345–355.

Bonde, M.R., Peterson, G.L., Dowler, W.M., and May, B. 1984. Isozyme analysis to differentiate species of *Peronosclerospora* causing downy mildews of maize. *Phytopathology, 74:* 1278–1283.

Bonde, M.R., Peterson, G.L., and Matsumoto, T.T. 1989. The use of isozymes to identify teliospores of *Tilletia indica. Phytopathology, 79:* 596–599.

Bonnekamp, P.M., Pomp, H., and Gussenhoven, G.C. 1990. Production and characterization of monoclonal antibodies to potato virus A. *Journal of Phytopathology, 128:* 112.

Bonner, F., Saillard, C., Kollar, A., Seemuller, E., and Bove, J.M. 1990. Detection and differentiation of the mycoplasma-like organism associated with apple proliferation disease using cloned DNA probes. *Molecular Plant-Microbe Interaction, 3:* 438.

Boonham, and Barker, I. 1998. Strain-specific recombinant antibodies to potato virus Y potyvirus. *Journal of Virological Methods, 74:* 193–199.

Borja, M.J. and Ponz, F. 1992. An appraisal of different methods for the detection of the walnut strain of cherry leafroll virus. *Journal of Virological Methods, 36:* 73–83.

Börjesson, T. and Johnson, L. 1998. Detection of common bunt (*Tilletia caries*) infestation in wheat with an electronic nose and human panel. *Zeitschrift für Planzenkrankheiten und Pflanzenschutz, 105:* 306–313.

Bos, L. 1970a. *Symptoms of Virus Diseases in Plants.* Oxford & IBH Publishing Co., New Delhi.

Bos, L. 1970b. The identification of three new viruses isolated from *Wisteria* and *Pisum* in the Netherlands, and the problem of variation within the potato virus Y group. *Netherlands Journal Plant Pathology, 76:* 8–46.

Bos, L. and Van der Want, J.P.H. 1962. Early-browning of pea, a disease caused by a soil- and seed-borne virus. *T. Pl. Ziekten, 68:* 368–390.

Bosland, P.W. and Williams, P.H. 1987. An evaluation of *Fusarium oxysporum* from crucifers based on pathogenicity, isozyme polymorphism, vegetative compatibility, and geographic origin. *Canadian Journal of Botany, 65:* 2067–2073.

Bossi, R. and Dewey, F.M. 1992. Development of a monoclonal antibody-based immunodetection assay for *Botrytis cinerea. Plant Pathology, 41:* 472–482.

Boulton, M.I. and Markham, G. 1986. The use of squash-blotting to detect plant pathogens in insect vectors. In: *Developments and Applications in Virus Testing* (Eds.) R.A.C. Jones and L. Torrance, Dev. Appl. Biol. I., pp. 55–69. Association of Applied Biologists, Wellsbourne, U.K.

Bounou, S., Jabaji - Hare, S.H., Hogue, R., and Charest, P.M. 1999. Polymerase chain reaction-based assay for specific detection of *Rhizoctonia solani* AG-3 isolates. *Mycological Research, 103:* 1–8.

Bouwen, I. and Maat, D.Z. 1992. Pelargonium flower-break and pelargonium line pattern viruses in Netherlands: purification, antiserum preparation, serological identification and detection in pelargonium by ELISA. *Netherlands Journal of Plant Pathology, 98:* 141–156.

Bouzar, H., Jones, J.B., Minsavage, G.V., Stall, R.E., and Scott, J.E. 1994. Proteins unique to phenotypically distinct groups of *Xanthomonas campestris* pv. *vesicatoria* revealed by silver staining. *Phytopathology, 84:* 39–44.

Bradsley, E.S., Burgess, J., Daniels, A., and Nicholson, P. 1998. The use of a polymerase chain reaction diagnostic test to detect and estimate the severity of stem base diseases of winter wheat. *Brighton Crop Protection Conference: Pests and Diseases, 3:* 1041–1046.

Bragard, C. and Verhoyen, M. 1993. Monoclonal antibodies specific for *Xanthomonas campestris* bacteria pathogenic on wheat and other small grains in comparison with polyclonal antisera. *Journal of Phytopathology, 139:* 217–228.

Branch, A.D., Benenfeld, B.J., Franck, E.R., Shaw, J.F., Varban, M.L., Willis, K.K., Rosen, D.L., and Robertson, H.D. 1988. Interference between coinoculated viroids. *Virology, 163:* 538–546.

Brandes, J. 1957. Eine elektronen mikroskopische Schnell-methode zum nachweisfalen und stabchenformiger viren, insbesondere in Kartoffeldunkel-keimen. *Nachrbl. Deut. Pflanzenschutzdienst (Brawnschweig), 9:* 151–152.

Brandes, J. and Wetter, C. 1959. Classification of elongated plant viruses on the basis of particle morphology. *Virology 8:* 99–115.

Braun, H., Levivier, S., Eber, F., Renard, M., and Chevre, A.M., 1997. Electrophoretic analysis of natural populations of *Leptosphaeria maculans* directly from leaf lesions. *Plant Pathology, 46:* 147–154.

Breed, R.S., Murray, E.G.D., and Smith, N.R. 1957. *Bergey's Manual of Determinative Bacteriology,* 7th Edition. The Williams and Wilkins Company, Baltimore.

Breese, S.S. Jr. and Hsu, K.C. 1971. *Methods in Virology, 5:* 399–422.

Bremer, K. 1964. *Agropyron* mosaic virus in Finland. *Ann. Agricult. Fenniae, 3:* 324–333.

Breyel, E., Casper, R., Ansa, O.A., Kuhn, C.W., Misari, S.M. and Demski, S.W. 1988. A simple procedure to detect a ds RNA associated with groundnut rosette. *Journal of Phytopathology, 121:*118–124.

Briand, J.P., Al Moudallal, Z., and Van Regenmortel, M.H.V. 1982. Serological differentiation of tobamoviruses by means of monoclonal antibodies. *Journal of Virological Methods, 5:* 293.

Bridge, P.D., Hopkinson, L.A., and Rutherford, M.A., 1995. Rapid mitochondrial probes for analysis of polymorphisms in *Fusarium oxysporum* special forms. *Letters in Applied Microbiology, 21:* 198–201.

Brierley, P. 1951. Value of index plants for detecting dahlia viruses. *Plant Disease Reporter, 35:* 405–407.

Brierley, P. and Smith, F.F. 1950. Some vectors, hosts and properties of dahlia mosaic virus. *Plant Disease Reporter, 34:* 363–370.

Brill, L.M., McClary R.D., and Sinclair, J.B. 1994. Analysis of two ELISA formats and antigen preparations using polyclonal antibodies against *Phomopsis longicolla. Phytopathology, 84:* 173–179.

Brinkerhoff, L.A. 1960. Variability for pathogenicity of *Xanthomonas malvacearum. Empire Cotton Growing Review, 37:* 235.

Brlansky, R.H. and Derrick, K.S. 1979. Detection of seed-borne plant viruses using serologically specific electron microscopy. *Phytopathology, 69:* 96–100.

Brown, A.E., Muthumeenakshi, S., Sreenivasaprasad, S., Mills, P.R. and Swinburne, T.R. 1993. A PCR primer specific to *Cylindrocarpon heteronema* for detection of the pathogen in apple wood. *FEMS Microbiology Letters, 108:* 117–120.

Brown, A.E., Muthumeenakshi, S., Swinburne, T.R., and Li, R. 1994. Detection of the source of infection of apple trees by *Cylindrocarpon heteronema* using DNA polymorphism. *Plant Pathology, 43:* 338–343.

Brunt, A.A. 1966a. Narcissus mosaic virus. *Annals of Applied Biology 58:* 13–23.

Brunt, A.A. 1966b. Partial purification, morphology, and serology of dahlia mosaic virus. *Virology, 28:* 778–779.

Brunt, A.A. 1968. Some hosts and properties of bulbous iris mosaic virus. *Annals of Applied Biology, 61:* 187–194.

Brunt, A.A. 1971. Some hosts and properties of dahlia mosaic virus. *Annals of Applied Biology, 67:* 357–368.

Brunt, A.A. and Atkey, P.T. 1967. Rapid detection of narcissus yellow stripe and two other filamentous viruses in crude negatively stained narcissus sap. *Report of Glasshouse Crops Research Institute (1966),* pp. 155–159.

Brunt, A.A., Kenten, R.H., Gibbs, A.J., and Nixon, H.L. 1965. Further studies on cocoa yellow mosaic virus. *Journal of General Microbiology 38:* 81–90.

Brunt, A.A. and Richards, K.O. 1989. Biology and molecular biology of furoviruses. *Advances in Virus Research, 36:* 1–32.

Brunt, A.A. and Shikata, E. 1986. Fungus-transmitted and similar labile rod shaped viruses. In: *The Plant Viruses* (Eds.) M.H.V. Van Regenmortel and H. Fraenkel-Conrat, Vol. 2, pp. 305–335, Plenum Press, New York.

Bryan, G.T., Labourdette, E., Melton, R.E., Nicholson, P., Daniels, M.J., and Osbourn, A.E. 1999. DNA polymorphic and host range in take-all fungus *Gaeumannomyces graminis*. *Mycological Research, 103:* 319–327.

Buchanan, D. 1932. A bacterial disease of beans transmitted by *Heliothrips femoralis* Rent. *Journal of Economic Entomology, 25:* 49–53.

Buchanan, R.E. and Gibbons, N.E. (Eds.) 1974. *Bergey's Manual of Determinative Bacteriology,* 8th Edition. Williams & Wilkins, Baltimore.

Buhariwalla, H., Greaves, S., Magrath, R., and Mithen, R. 1995. Development of specific PCR primers for the amplification of polymorphic DNA from the obligate root pathogen *Plasmodiophora brassicae*. *Physiological and Molecular Plant Pathology, 47:* 83–94.

Bulman, S.R. and Marshall, J.W. 1998. Detection of *Spongspora subterranea* in potato tuber lesions using polymerase chain reaction (PCR). *Plant Pathology, 47:* 759–766.

Burger, P.H., Thornbury, D.W., and Pirone, T.P. 1985. Detection of picogram quantities of polyviruses using a dot-blot immunobinding assay. *Journal of Virological Methods,* 12: 31.

Burr, T.J., Bazzi, C., Süle, S., and Otten. L. 1998. Crown gall of grape: biology of *Agrobacterium vitis* and the development of disease control strategies. *Plant Disease, 82:* 1288–1297.

Burr, T.J., Norelli, J.I., Katz, B.H., and Bishop, A.L. 1990. Use of Ti-plasmid DNA probes for determining tumourigenicity of *Agrobacterium* strains. *Applied Environmental Microbiology, 56:* 1782.

Cabauatan, P.Q., Arboleda, M. and Azzam, O. 1998. Differentiation of rice tungro bacilliform virus strains by restriction analysis and DNA hybridization. *Journal of Virological Methods, 76:* 121–126.

Cabauatan, P.Q., Koganezawa, H., Tsuda, S., and Hibino, H. 1994. Applying rapid immunofilter paper assay to detect rice viruses. *International Rice Research Notes, 19(2):* 34.

Caciagli, P. and Bosco, D., 1996. Quantitative determination of tomato yellow leafcurl geminivirus DNA by chemiluminescent assay using digoxigenin-labeled probes. *Journal of Virological Methods, 57:* 19–29.

Caciagli, P. and Bosco, D. 1997. Quantitation of over time of tomato yellow leafcurl geminivirus DNA in its whitefly vector. *Phytopathology, 87:* 610–613.

Cadman, C.H., Dias, H.P., and Harrison, B.D. 1960. Sap transmissible viruses associated with diseases of grapevines in Europe and North America. *Nature, 187:* 577.

Cadman, C.H. and Harrison, B.D. 1959. Studies on the properties of soil-borne viruses of the tobacco rattle type occurring in Scotland. *Annals of Applied Biology, 47:* 542–556.

Cahill, D.M., and Hardham, A.R. 1994a. Exploitation of zoospore taxis in the development of a novel dipstick immunoassay for the specific detection of *Phytophthora cinnamomi. Phytopathology, 84:* 193–200.

Cahill, D.M. and Hardham, A.R. 1994b. A dipstick immunoassay for the specific detection of *Phytophthora cinnamomi* in soils. *Phytopathology, 84:* 1284–1292.

Cai, W.Q., Wang, R., Qin, B.Y., Ma, S.F., Tian, B., Zhang, C.L., Chen, J., and Li, L. 1989. Production of CMV specific monoclonal antibodies by electrofusion. *Acta Microbiologica Sinica, 29:* 444.

Cairoli, S., Arnoldi, A. and Pagani, S. 1996. Enzyme-linked immunosorbent assay for the quantitation of the fungicide in fruits and fruit juices. *Journal of Agricultural and Food Chemistry, 44:* 3849–3854.

Calavan, E.C., Frolich, E.F., Carpenter, J.B., Roistacher, C.N. and Christiansen, D.W. 1964. Rapid indexing for exocortis of citrus. *Phytopathology, 54:* 1359–1362.

Caldis, P.D. 1930. Souring of the figs by yeasts and transmission of the disease by insects. *Journal of Agricultural Research, 40:* 1031–1051.

Cammarco, R.F.E.A., Lima, J.A.A. and Pio-Ribeiro, G. 1998. Transmission and presence in the soil of papaya lethal yellowing virus. *Fitopatologia Brasileira, 23:* 453–458.

Cambra, M, Camarasa, E., Gorris, M.T., Roman, M.P., Asensio, M., Perez, E., Serra, J. and Cambra, M.A. 1994. Detection of structural proteins of virus by immunoprinting: ELISA and its use for diagnosis. *Investigacion Agraria, Produccion y Protection Vegetales, Fuera de Serie, No. 2,* 221–230.

Cambra, M., Olnos, A., Asensio, M., Esteban, O., Candresse, T., Gorris, M.T. and Boscia, D., 1998. Detection and typing of *Prunus* viruses in plant tissues and in vectors by print- and spot-capture PCR, heminested PCR and PCR-ELISA. *Acta Horticulturae, No. 472:* 257–263.

Campbell, R.N., Wipf-Scheibel, C. and Lecoq, H. 1996. Vector-assisted seed transmission of melon necrotic spot virus in melon. *Phytopathology, 86:* 1294–1298.

Cancino, M., Abouzid, A.M., Morales, F.J., Purcifull, D.E., Polston, J.E., and Hiebert, E. 1995. Generation and characterization of three monoclonal antibodies useful in detecting and distinguishing bean golden mosaic virus isolates. *Phytopathology, 85:* 484–490.

Candresse, T., Cambra, M., Dallot, S., Lanneau, M., Asensio, M., Gorris, M.T., Revers, F., Macquaire, G., Olmos, A., Boscia, D., Quiot, J.B. and Dunez, J. 1998. Comparison of monoclonal antibodies and polymerase chain reaction assays for the typing of isolates belonging to the D and M serotypes of plum pox potyvirus. *Phytopathology, 88:* 198–204.

Candresse, T., Kofalvi, S.A., Lanneau. M. and Dunez. J. 1998. A PCR-ELISA procedure for the simultaneous detection and identification of prunus necrotic ringspot (PNRSV) and apple mosaic (ApMV) ilarviruses. *Acta Horticulturae, No. 472,* 219–225.

Candresse, T., Macquaire, G., Brault, V., Monsion, M., and Dumez, J. 1990. ^{32}P and biotin-labeled in vitro transcribed cRNA probes for the detection of potato spindle tuber viroid and chrysanthemum stunt viroid. *Research in Virology, 141:* 97–107.

Canto, T., Ellis, P., Bowler, G., and Lopez-Abella, D. 1995. Production of monoclonal antibodies to potato virus Y helper component-protease and their use for strain differentiation. *Plant Disease, 79:* 234–237.

Carlier, J., Mourichon, X., Gonzalez-de-Leon, D., Zapter, M.F., and Lebrun, M.H. 1994. DNA restriction fragment length polymorphism in *Mycosphaerella* species that cause banana leafspot diseases. *Phytopathology, 84:* 751–756.

Carmichael, J.W. 1955. Lacto-fuchsin: a new medium for mounting fungi. *Mycologia, 47:* 611.

Carroll, T.W. 1970. Relation of barley stripe mosaic virus to plastids. *Virology, 42:* 1015–1022.

Carter, W. 1973. *Insects in Relation to Plant Disease.* John Wiley & Sons, New York.

Castello, J.D., Lakshman, D.K., Tavantzis, S.M., Rogers, S.O., Bachand, G.D., Jagels, R., Carlisle, R.J.J. and Liu, Y. 1995. Detection of infectious tomato tobamovirus in fog and clouds. *Phytopathology, 85:* 1409–1412.

Castilla, N.P., Elazegui, F.A., Lanip, W.M. and Savary, S. 1997. Method for detecting rice sheath blight pathogen in soil samples using mungbean. *International Rice Research Notes, 22 (2):* 48.

Castro, S., Carazo, G., Saiz, M., Romero, J., and Blas, C. 1993. Use of enzymatic cDNA amplification as a method of detection of bean yellow mosaic virus. *Netherlands Journal of Plant Pathology, 99:* 97–100.

Caudwell, A., Meigno, S.R., Kuszala, C., Schneider, C., Larrue, J., Fleury, A., and Boudon, E. 1982. Serological purification and visualization in the electron microscope of grapevine Flavescence Doreé pathogen (MLO) in diseased plants and infectious vector extracts. *Yale Journal of Biology, Mediterranea, 56:* 936–937.

Cavileer, T.D., Clarke, R.C., Corsini, D.L. and Berger, P.H. 1998. A new strain of potato carlavirus. *Plant Disease, 81:* 98–102.

Cayley, D.M. 1932. Breaking in tulips II. *Annals of Applied Biology 19:* 153–172.

Cazelles, O., Ruchta, A. and Schwarzel, R. 1995. Bacterial soft rot of potato tubers: progress in detection of latent infection of *Erwinia chrysanthemi* and forecasting field outbreaks. *Revue Suisse d' Agriculture, 27:* 17–22.

Celio, G.J. and Hausbeck, M.K. 1998. Conidial germination, infection structure formation and early colony development of powdery mildew on poinsettia. *Phytopathology, 88:* 105–113.

Centis, S., Guillas, I., Séjalon, N., Esquerre-Tugaye, M.T., and Dumas, B. 1997. En-
    dopolygalacturonase genes from *Colletotrichum lindemuthianum:* cloning of
    CLPG2 and comparison of its expression to that of *CLPG1* during saprophytic and
    parasitic growth of the fungus. *Molecular Plant-Microbe Interactions, 10:*
    769–775.
Ceramic-Zagorac, P. and Hiruki, C. 1996. Comparative molecular studies on aster yellows
    phytoplasmas. *Acta Horticulturae, No. 377,* 266–276.
Cervera, M., López, M.M., Navarro, L. and Peña. L., 1998. Virulence and supervirulence
    of *Agrobacterium tumefaciens* in woody fruit plants. *Physiological and Molecular
    Plant Pathology, 52:* 67–78.
Chahal, A. and Nassuth, A. 1992. Tomato spotted wilt virus: detection and location in leaf
    and stem tissues by press-blotting. *Plant Disease Research, 7:* 171–179.
Chakraborty, V., Basu, P. Das, R., Saha, A., and Chakraborty, B.N. 1996. Evaluation of an-
    tiserum raised against *Pestalotiopsis theae* for the detection of grey blight of tea by
    ELISA. *Folia Microbiologica, 41:* 413–418.
Chalfant, R.B. and Chapman, R.K. 1962. Transmission of cabbage viruses A and B by the
    cabbage aphid and the green aphid. *Journal of Economic Entomology, 55:* 584–
    590.
Chamberland, H., Charest, P.M., Ouellette, G.A., and Pauze, F.J. 1985. Chitinase-gold
    complex used to localize chitin ultrastructurally in tomato root cells infected by
    Fusarium oxysporum f.sp. *radicis lycopersici* compared with a chitin specific gold-
    conjugated lectin. *Histochemical Journal, 17:* 313–321.
Chamberlain, J.R., Culbreath, A.K., Todd, J.W. and Demski, J.W. 1993. Detection of
    tomato spotted wilt virus in tobacco thrips (Thysanoptera: Thripidae) over wintering
    in harvested peanut fields. *Journal of Economic Entomology, 86:* 40–45.
Chang, G.H. and Yu, R.C. 1997. Rapid immunoassay of fungal mycelia in rice and corn.
    Journal of Chinse Agricultural Chemical Society, 35: 533–539.
Chang, M.U., Doi, Y., and Yora, K. 1976. A rod-shaped virus found in the peony ringspot.
    Annals of Phytopathological Society, Japan, 42: 325–328.
Chastain, T.G. and King, B. 1990. A biochemical method for estimating viability of
    teliospores of *Tilletia controversa. Phytopathology, 80:* 474–476.
Chen ChunXian and Wan ShuYuan. 1995. Detection of citrus exocortis viroid with a pho-
    tobiotin-labeled F_2O probe. *Chinese Journal of Virology, 11:* 276–278.
Chen, J., Sako, N., Ohshima, K. and Watanabe, Y. 1996. Specific detection of the rakkyo
    strain of tobacco mosaic virus by reverse transcription and polymerase chain reac-
    tion. *Annals of Phytopathological Society, Japan, 62:* 513–516.
Chen, J., Torrance, L., Cowan, G.H., Mac Farlane, S.A., Stubbs, G. and Wilson, T.M.A.
    1997. Monoclonal antibodies detect a single amino acid difference between the coat
    proteins of soil-borne wheat mosaic virus isolates: implications for virus structure.
    Phytopathology, 87: 295–301.
Chen, J.C., Banks, D., Jarret, R.L., Chang, C.J. and Smith, B.J. 2000. Use of 16 S rDNA
    sequences as signature characters to identify *Xylella fastidiosa. Current Microbiol-
    ogy, 40:* 29–33.
Chen Jing, Hu WeiZhen, Yu. JiaLing and Han ChengGui, 1996. Rapid detection of tomato
    ringspot virus by the reverse transcription polymerase chain reaction. *Chinese Jour-
    nal of Virology, 12:* 190–192.

Chen, K.H., Credi, R., Loi, N., Maixner, M., and Chen, T.A. 1994. Identification and grouping of mycoplasma-like organisms associated with grapevine yellows and clover phyllody disease based on immunological and molecular analyses. *Applied and Environmental Microbiology, 60:* 1905–1913.

Chen, L.C., Durand, P., and Hill, J.H. 1982. Detection of pathogenic strains of soybean mosaic virus by enzyme-linked immunosorbent assay with polystyrene plates and beads as solid phase. *Phytopathology, 72:* 1177–1181.

Chen LanMing and Chen YongXuan. 1998. Direct competitive ELISA screening method for aflatoxin B1. *Journal of Nanjing Agricultural University, 21:* 62–65.

Chen, M., and Shikata, E. 1971. Morphology and intracellular localization of rice transitory yellowing virus. *Virology, 46:* 786–796.

Chen MiawFan and Lin ChanPin 1997. DNA probes and PCR primers for the detection of phytoplasma associated with peanut witches' broom. *European Journal of Plant Pathology, 130:* 137–147.

Chen, N.H., Yang, L., and Qiu, W.F. 1990. Combined immunosorbent and double decoration method for detecting plant viruses by electron microscopy. *Chinese Journal of Virology, 6:* 378–379.

Chen, T.A. and Liao, C.H. 1975. Corn stunt spiroplasma: isolation, cultivation and proof of pathogenicity. *Science, 188:* 1015–1017.

Chen, W., Gray, L.E. and Grau, C.R. 1996. Molecular differentiation of fungi associated with brown stem rot and detection of *Phialophora gregata* in resistant and susceptible soybean cultivars. *Phytopathology, 86:* 1140–1148.

Chen, W., Tien, P., Zhu, Y.X., and Liu, Y. 1983. Viroid-like RNAs associated with burdock stunt disease. *Journal of General Virology, 64:* 409–414.

Chen WeiDony, Gray, L.E., Kurle, J.E. and Grau, C.R. 1999. Specific detection of *Phialophora gregatum* and *Plectosporium tabacinum* in inoculated soybean plants using polymerase chain reaction. *Molecular Ecology, 8:* 871–877.

Cheng, C.P., Chen, C.T., Deng, T.C., and Su, H.J. 1993. Monoclonal antibodies against sugarcane mosaic virus. *Plant Pathology Bulletin, 2:* 227–231.

Cheng Ye, Chen JianPing and Hong Jian 1999. Ultrastructure of cystosori of *Polymyxa graminis* in barley roots. *Mycosystema, 18:* 30–34.

Chia, T.F., Chan, Y.S. and Chua, N.H. 1992. Detection and localization of viruses in orchids by tissue-print hybridization. *Plant Pathology, 4:* 355.

Chia, T.F., Chan, Y.S. and Chua, N.H. 1995. Tissue-print hybridization for the detection and localization of plant viruses. In: *Molecular Methods in Plant Pathology* (Eds.) R.P. Singh and U.S. Singh, pp. 145–149. CRC Press, Boca Raton, FL.

Childs, J.F.L., Norman, G.G., and Eichhorn, J.L. 1958. A colour test for exocortis infection in *Poncirus trifoliata*. *Phytopathology, 48:* 426–432.

Chittaranjan, S. and Boer, S.H. de. 1997. Detection of *Xanthomonas campestris* pv. *pelargonii* in geranium and greenhouse nutrient solution by serological and PCR techniques. *European Journal of Plant Pathology, 103:* 555–563.

Chiu, R.J. and Black, L.M. 1969. Assay of wound tumour virus by the fluorescent cell counting technique. *Virology, 37:* 667–677.

Chiu, R.J., Jean, J.H., Chen, M.H., and Lo, T.C. 1968. Transmission of transitory yellowing viruses of rice by two leafhoppers. *Phytopathology, 58:* 740–745.

Chiu, R.J., Liu, H.Y., MacLeold, R., and Black, L.M. 1970. Potato yellow dwarf virus in leafhopper cell culture. *Virology, 40:* 381–396.

Cho, J.J., Mau, R.F.L., German, T.L., Hartmann, R.W., Yudin, L.S., Gonsalves, D., and Providenti, R. 1989. A multidisciplinary approach to management of tomato spotted wilt virus in Hawaii. *Plant Disease, 73:* 375–383.

Cho, J.J., Mau, R.F.L., Hamsaki, R.T., and Gonsalves, D. 1988. Detection of tomato spotted wilt virus in individual thrips by enzyme-linked immunosorbent assay. *Phytopathology, 78:* 1348–1352.

Choi GugSeoum, Choi YongMun, Won SenYoung and Yiem MyoungSoon, 1998. Detection of cucumber mosaic virus using rapid immuno filter paper assay. *RDA Journal of Crop Protection, 40:* 115–119.

Choi, J., Matsuyama, N., and Wakimoto, S. 1980. Serovars of *Xanthomonas camprestris* pv. *oryzae* collected from Asian countries. *Annals of Phytopathological Society, Japan, 46:* 209–215.

Choudhury, M.M., and Rosenkranz, E. 1983. Vector relationship of *Graminella nigrifrons* to maize chlorotic dwarf virus. *Phytopathology, 73:* 685–690.

Christie, R.G. 1967. Rapid staining procedures for differentiating plant virus inclusions in epidermal strips. *Virology, 31:* 268–271.

Christie, R.G., Baltensperger, D.D., and Edwardson, J.R. 1988. Use of light microscope for the detection and diagnosis of viral diseases of crops. *Proceedings of Soil and Crop Science Society of Florida, 47:* 175–197.

Christie, R.G. and Edwardson, J.R. 1977. *Light and Electron Microscopy of Plant Virus Inclusions.* Florida Agricultural Experiment Station, Monograph Series No. 9, Institute of Food and Agricultural Sciences, University of Florida, Gainesville.

Christie, R.G., Edwardson, J.R. and Simone, G.W. 1995. Diagnosing plant virus disease by light microscopy. In: *Molecular Methods in Plant Pathology* (Eds.) R.P. Singh and U.S. Singh, pp. 31–51. CRC Press, Boca Raton, FL.

Chu, M.H., Desvoyes, B., Turina, M., Noad, R. and Scholthof, H.B. 2000. Genetic dissection of tomato bushy stunt virus p19-protein mediated host-dependent symptom induction and systemic invasion. *Virology, 266:* 79–87.

Chu, P.W.G., Waterhouse, P.M., Martin, R.R., and W.L. Gerlach, 1989. New approaches to the detection of microbial plant pathogens. *Biotechnology and Genetic Engineering Reviews, 7:* 45–111.

Chung, B.N., Choi, G.S. and Choi, Y.M. 1999. Identification of tomato aspermy virus (TAV) and chrysanthemum virus B (CVB) from *Dendranthema indicum* in Korea. *Plant Pathology Journal, 15:* 119–123.

Cilliers, A.J., Swart, W.J. and Wingfield, M.J. 1994. Selective medium for isolating *Lasiodiplodia theobromae. Plant Disease, 78:* 1052–1055.

Clark, M.F. 1992. Immunodiagnostic techniques for plant mycoplasma-like organisms. In: *Techniques for the Rapid Detection of Plant Pathogens* (Eds.) J.M. Duncan and L. Torrance, pp. 34–45. British Society for Plant Pathology/Blackwell Scientific, Oxford.

Clark, M.F., and Adams, A.N. 1977. Characteristics of the microplate method of enzyme-linked immunosorbent assay for the detection of plant viruses. *Journal of General Virology, 34:* 475–483.

Clark, M.F., Adams, A.N. and Barbara, D.J. 1976. The detection of plant viruses by enzyme-linked immunosorbent assay (ELISA). *Acta Horticulturae, 67:* 43–49.

Clark, M.F., Morton, A., and Buss, S.L. 1989. Preparation of mycoplasma immunogens from plants and a comparison of polyclonal and monoclonal antibodies made against primula yellows MLO-associated antigens. *Annals of Applied Biology, 114:* 161.

Clarke, R.G. and Wilde, G.E. 1971. Association of the green stink bug and the yeast spot organism of soybeans. III. Effect on soybean quality. *Journal of Economic Entomology, 64:* 222–224.

Clear, R.M. and Patrick, S.K. 1992. A simple medium to aid the identification of *Fusarium moniliforme, F. proliferatum* and *F. subglutinans. Journal of Food Protection, 55:* 120–122.

Clinch, P., Loughnane, J.B., and Murphy, P.A. 1936. A study of the acuba or yellow mosaics of the potato. *Scientific Proceedings of Royal Dublin Society, 21:* 431–448.

Clough, G.H. and Hamm, P.B. 1995. Coat protein transgenic resistance to watermelon mosaic and zucchini yellow mosaic virus in squash and cantaloupe. *Plant Disease, 79:* 1107–1109.

Clover, G. and Henry, C. 1999. Detection and discrimination of wheat spindle streak mosaic virus and wheat yellow mosaic virus using multiplex RT-PCR. *European Journal of Plant Pathology, 105:* 891–896.

Coddington, A., Matthews, P.M., Cultis, C., and Smith, K.M. 1987. Restriction digest patterns of total DNA from different races of *Fusarium oxysporum* f.sp. *pisi,* an improved method for race classification. *Journal of Phytopathology, 118:* 9–20.

Coff, C., Poupard, P., Xiao, Q., Graner, A. and Lund, V. 1998. Sequence of a plastocyanin cDNA from wheat and the use of the gene product to determine serologically tissue degradation after infection with *Pseudocercosporella herpotrichoides. Journal of Phytopathology, 146:* 11–17.

Coffin, R.S. and Coutts, R.H.A. 1992. Ds-RNA cloning and diagnosis of beet pseudo yellows virus by PCR and nucleic acid hybridization. *Intervirology, 33:* 197–203.

Cohen, T., Loebenstein, G., and Milne, R.G. 1982. Effect of pH and other condition on immunosorbent electron microscopy of several plant viruses. *Journal of Virological Methods, 4:* 323–330.

Colas, V., Lacourt, I., Ricci, P., Vanlerberghe-Masulli, F., Venard, P., Poupet, A. and Panabieres, F. 1998. Diversity of virulence in *Phytophthora parasitica* in tobacco as reflected by nuclear RFLPs. *Phytopathology, 88:* 205–212.

Colinet, D., Kummert, J., Lepoivre, P., and Semal, J. 1994. Identification of distinct potyvirus in mixedly infected sweet potato by polymerase chain reaction with degenerate primers. *Phytopathology, 84:* 65–69.

Colinet, D., Nguyen, M., Kummert, J., Lepoivre, P., and Xia, F.Z. 1998. Differentiation among potyviruses infecting sweet potato based on genus- and virus-specific reverse transcription-polymerase chain reaction. *Plant Disease, 82:* 223–229.

Comstock, J.C., Irey, M.S., Lockhart B.E.L. and Wang, Z.K. 1998. Incidence of yellow leaf syndrome in CP cultivars based on polymerase chain reaction and serological reactions. *Sugar Cane, No. 4,* 21–24.

Comstock, J.C., Wang, Z.K. and Perdomo, R. 1997. The incidence of leaf scald and its effect on yield components: detection by serological and isolation techniques with particular reference to CP80-1743 in Florida. *Sugar Cane, No. 4,* 18–21.

Condemine, G., Castillo, A., Passeri, F. and Enard, C. 1999. The PecT repressor coregulates synthesis of exopolysaccharides and virulence factors in *Erwinia chrysanthemi. Molecular Plant-Microbe Interactions, 12:* 45–52.

Conejero, V., Picazo, I., and Segado, P. 1979. Citrus exocortis viroid (CEV): protein alterations in different hosts following viroid infection. *Virology, 97:* 454–456.

Conti, M. 1984. Epidemiology and vectors of plant reo-like viruses. *Current Topics in Vector Research, 2:* 112–139.

Cook, D. and Sequira, I. 1991. Isolation and characterization of DNA clones specific for race 3 of *Pseudomonas solanacearum. Phytopathology, 81:* 696.

Cornuet, P. 1953. Sur une methode de diagnostic sur tubercle de la maladie del' enroulement de la pomme de terre. *C. R. Acad. Sci. Paris, 257:* 1364–1366.

Costa, A.S. and Muller, G.W. 1980. Control by cross-protection: a U.S.-Brazil cooperative sucess. *Plant Disease, 64:* 538–541.

Costa, A.S., and Penteado, M.P. 1951. Corn seedlings as test plants for the sugarcane mosaic virus. *Phytopathology, 41:* 758–763.

Cottyn, B., Bautista, A.T., Nelson, R.J., Leach, J.E., Swings, J., and Mew, T.W. 1994. Polymerase chain reaction amplification of DNA from bacterial pathogens of rice using specific oligonucleotide primers. *International Rice Research Notes, 19(1):* 30–32.

Couch, H.B. and Gold, A.H. 1954. Rod shaped particles associated with lettuce mosaic. *Phytopathology, 44:* 715–717.

Cousin, M.T., Roux, J., Boudon-Padieu, E., Berges, R., Seemüller, E. and Hiruki, C. 1998. Use of heteroduplex mobility analysis (HMA) for differentiating phytoplasma isolates causing witches' broom disease on *Populus nigra* cv. Italica and stolbur or bigbud symptoms on tomato. *Journal of Phytopathology, 146:* 97–102.

Cousin, R., Maillet, P.L., Allard, C., and Staron, T. 1969. Mosaique commone du pois. *Ann. Phytopathol. I:* 195–200.

Coyne, C.J., Mehlenbacher, S.A., Hampton, R.O., Pinkerton, J.N. and Johnson, K.B. 1996. Use of ELISA to rapidly screen hazelnut for resistance to eastern filbert blight. *Plant Disease, 80:* 1327–1330.

Credi, R. 1997. Indexing tests on a grapevine rugose wood disease and mechanical transmission of two associated viruses. *Phytopathologia Mediterranea, 36:* 1–7.

Cronjé, C.P.R., Bailey, R.A., Jones, P., and Suma, S. 1999. The phytoplasma associated with Ramu stunt disease of sugarcane is closely related to the white leaf phytoplasma group. *Plant Disease, 83:* 588.

Cronjé, C.P.R., Tymon, A.M., Jones, P. and Bailey, R.A. 1998. Association of phytoplasma with a yellow leaf syndrome of sugarcane in Africa. *Annals of Applied Biology, 133:* 177–186.

Cropley, R. 1961. Cherry leaf roll virus. *Annals of Applied Biology, 49:* 524–529.

Cropley, R. 1964. Transmission of apple chlorotic leafspot virus from *Chenopodium* to apple. *Plant Disease Reporter, 48:* 678.

Cropley, R., Gilmer, R.M., and Posnette, A.F. 1964. Necrotic ringspot and prune dwarf viruses in *Prunus* and in herbaceous indicators. *Annals of Applied Biology, 53:* 325–332.

Crossley, S.J., Jacobi, V., and Adams, A.N. 1998. IC-PCR amplification of apple stem grooving virus isolates and comparison of polymerase and coat protein sequences. *Acta Horticulturae, No. 472,* 113–117.

Crowdy, S.H. 1947. Observations on the pathogenicity of *Calonectria rigiduscula* (Berk. and Br.) Sacc. on *Theobroma cacao* L. *Annals of Applied Biology, 34:* 45–59.

Crowhurst, R.N., Binnie, S.J., Bowen, J.K., Hawthorne, B.T., Plummer, K.M. Rees-George, J., Rikkerink, E.H.A., and Templeton, M.D., 1997. Effect of disruption of a cutinase gene (*cut A*) on virulence and tissue specificity of *Fusarium solani* f.sp. *cucurbita* race 2 toward *Cucurbita maxima* and *C. moschata. Molecular Plant-Microbe Interactions, 10:* 355–368.

Croxall, H.E., Collingwood, C.A., and Jenkins, J.E.E. 1951. Observations on brown rot (*Sclerotinia fructigena*) of apples in relation to injury caused by earwig (*Forficula auricularia*). *Annals of Applied Biology, 38:* 833–843.

Cubero, J., Martinez, M.C., Llop, P., and Lopez, M. 1999. A simple and efficient PCR method for the detection of *Agrobacterium tumefaciens* in plant tumors. *Journal of Applied Microbiology, 86:* 591–602.

Culver, J.N. 1996. Viral avirulence genes In: *Plant-Microbe Interactions* (Eds.) G. Stacey and N.T. Keen, pp. 196–219. Chapman & Hall, New York.

Culver, J.N., Lindbeck, A.G.C., Desjardins, P.R., Dawson, W.O., Herrmann, R.G. and Larkins, B.A. 1991. Analysis of tobacco mosaic virus-host interactions by directed genome modification. In: *Plant Molecular Biology 2, NATO ASI Series A. Life Sciences 212:* 23–33.

Culver, J.N. and Sherwood, J.L. 1988. Detection of peanut stripe virus in peanut seed by an indirect ELISA using a monoclonal antibody. *Plant Disease, 72:* 676.

Culver, J.N., Sherwood, J.L., and Melonk, H.A. 1987. Resistance to peanut stripe virus in *Arachis* germplasm. *Plant Disease, 71:* 1080–1082.

Cuppels, D.A., Moore, R.A., and Morris, V.L. 1990. Construction and use of a nonradioactive DNA hybridization probe for detection of *Pseudomonas syringae* pv. *tomato* on tomato plants. *Applied Environmental Microbiology, 56:* 1743.

Dahal, G., Dasgupta, I., Lee, G., and Hall, R. 1992. Comparative transmission of and varietal reaction to three isolates of rice tungro virus disease. *Annals of Applied Biology, 120:* 287–300.

Dai, Q, He, F.T. and Lue, P.Y. 1997. Elimination of phytoplasma by stem culture from mulberry plants (*Morus alba*) with dwarf disease. *Plant Pathology, 46:* 56–61.

Dale, J.L. 1988. Rapid compression technique for detecting mycoplasma-like organisms in leaf midrib sieve tubes by fluorescence microscopy. *Phytopathology, 78:* 118–120.

Dale, J.L. and Allen, R.N. 1979. Avocado affected by sunblotch disease contains low molecular weight ribonucleic acid. *Australian Plant Pathology, 8:* 3–4.

Dale, W.T. 1949. Observations on a virus disease of cowpea in Trinidad. *Annals of Applied Biology, 36:* 327–333.

Damsteegt, V.D., Stone, A.L. Russo, A.J., Luster, D.G., Gildow, F.E. and Smith, O.P. 1999. Identification, characterization and relatedness of luteovirus isolates from forage legumes. *Phytopathology, 89:* 374–379.

Damsteegt, V.D., Stone, A.L., Waterworth, H.E., Mink, G.I., Howell, W.E. and Levy, I. 1998. The versatility of *Prunus tomentosa* as a bioindicator of viruses. *Acta Horticulturae, No. 472,* 143–146.

Damsteegt, V.D., Waterworth, H.E., Mink, G.I., Howell, W.E. and Levy, L. 1997. *Prunus tomentosa* as a diagnostic host for detection of plum pox virus and other *Prunus* viruses. *Plant Disease, 81:* 329–332.

Daniels, J., Marinho, V.L.A., Kummert, J., and Lepoivre, P. 1998. Developing a colorimetric RT-PCR test for apple stem grooving virus detection in apple trees. *Acta Horticulturae, No. 472,* 105–111.

D'Arcy, C.J., Murphy, J.C., and Miklatz, S.D. 1990. Murine monoclonal antibodies produced against two Illinois strains of barley yellow dwarf virus: production and use for virus detection. *Phytopathology, 80:* 377.

D'Arcy, C.J., Torrance, L., and Martin, R.R. 1989. Discrimination among luteoviruses and their strains by monoclonal antibodies and identification of common epitopes. *Phytopathology, 79:* 869.

Darby, J.F., Larson, R.H., and Walker, J.C. 1951. *Variation in Virulence and Properties of Potato Virus Y Strains.* University of Wisconsin, Madison, Research Bulletin 177.

Darda, G. 1998. A new test-combination: DAS-ELISA with subsequent amplified ELISA on the same microtitre plate for detection of potato virus X in potato tubers. *Zeitschrift für Pflanzenkrankheiten und Pflanzenschultz, 105:* 105–113.

Das, M. and Bora, K.N. 1998. Ultrastructural studies on infection processes by *Colletotrichum actuatum* on guava fruit. *Indian Phytopathology, 51:* 353–356.

Dasgupta, I., Das, B.K., Nath, P.S., Mukhopadhyay, S., Niazi, F.R. and Varma, A. 1996. Detection of rice tungro bacilliform virus in field and glasshouse samples by polymerase chain reaction. *Journal of Virological Methods, 58:* 53–58.

Daugherty, D.M. and Foster, J.E. 1966. Organism of yeast spot disease isolated from rice damaged by rice stink bug. *Journal of Economic Entomology, 59:* 1282–1283.

Davies, D.L. 1998. The occurrence of two distinct phytoplasma isolates associated with *Rubus* species in the UK. *Acta Horticulturae, No. 471,* 63–65.

Davies, J.W., Kaesberg, P., and Diener, T.O. 1974. Potato spindle tuber viroid XII: an investigation of viroid RNA as messenger for protein synthesis. *Virology, 61:* 281–286.

Davis, M.J., Kramer, J.B., Ferwerda, F.H. and Brunner, B.R. 1996. Association of a bacterium and not a phytoplasma with papaya bunchy top disease. *Phytopathology, 86:* 102–109.

Davis, R.E. and Fletcher, J. 1983. *Spiroplasma citri* in Maryland: isolation from field-grown plants of horseradish with brittle root symptoms. *Plant Disease, 67:* 900–903.

Davis, R.E. and Lee, L.M. 1982. Pathogenicity of spiroplasma, mycoplasma-like organisms and vascular-limited fastidious walled bacteria. In: *Phytopathogenic Prokaryotes* (Eds.) M.S. Mount and G.H. Lacy, Vol. 1, pp. 492–513. Academic Press, New York.

Davis, R.E. and Sinclair, W.A. 1998. Phytoplasma identity and disease etiology. *Phytopathology, 88:* 1372–1376.

Davis, R.E., Lee, I.M., Douglas S.M., and Dally, E.L. 1990. Molecular cloning and detection of chromosomal and extachromosomal DNA of the mycoplasma-like organisms associated with little leaf disease in periwinkle *(Cartharanthus roseus)*. *Phytopathology, 80:* 789–793.

Davis, R.E., Lee, I., Douglas, S.M., Dally, E.L., and Dewitt, N. 1988. Cloned nucleic acid hybridization probes in detection and classification of mycoplasma-like organisms. *Acta Horticulturae, 234:* 115–122.

Davis, R.E., Lee, I.M., Douglas, S.M., Dally, E.L., and Dewitt, N. 1990a. Development and use of cloned nucleic acid hybridization probes for disease diagnosis and detection of sequence homologies among uncultured mycoplasma-like organisms (MLOs). *Zbl. Suppl. 20:* 303–307. Gustav Fischer Verlag, New York.

Davis, R.E., Lee, I.M., Douglas, S.M., Dally, E.L., and Dewilt, N. 1990b. Cloned nucleic acid hybridization probes in detection and classification of mycoplasma-like organisms. *Acta Horticulturae, 34:* 115–122.

Davis, R.E., Sinclair, W.A., Lee, I.M., and Dally, E. 1992. Cloned DNA probes specific for detection for a mycoplasma-like organism associated with ash yellows. *Molecular Plant-Microbe Interaction, 5:* 163–169.

Davis, M.J., Tsai, J.H., Cox, R.L., McDaniel, L.L., and Harrison, N.A. 1988. Cloning of chromosomal and extra-chromosomal DNA of the mycoplasma-like organism that causes maize bushy stunt disease. *Molecular Plant-Microbe Interaction, 1:* 295.

deAvila, A.C., Huguenot, C., Resenda, R., de O, Kitajima, E.W., Goldbach, R.W., and Peters, W. 1990. Serological differentiation of 20 isolates of tomato spotted wilt virus. *Journal of General Virology, 71:* 2801–2807.

De Boer, S.H. and Hall, J.W. 2000. Proficiency testing in a laboratory accreditation program for the bacterial ring rot pathogen of potato. *Plant Disease, 84:* 649–653.

De Boer, S.H., and McNaughton, M.E. 1986. Evaluation of immunofluorescence with monoclonal antibodies for detecting latent bacterial ring rot infections. *American Potato Journal, 63:* 533–543.

De Boer, S.H. and McNaughton, M.E. 1987. Monoclonal antibodies to the lipopolysaccharides of *Erwinia carotovora* subsp. *atroseptica* sero group I. *Phytopathology, 77:* 828–832.

De Boer, S.H. and Ward, L.J. 1995. PCR detection of *Erwinia carotovora* subsp. *atroseptica* associated with potato tissue. *Phytopathology, 85:* 854–858.

De Boer, S.H. and Wieczorek, A. 1984. Production of monoclonal antibodies to *Corynebacterium sepedonicum*. *Phytopathology, 74:* 1431–1434.

De Bokx, J.A., Piron, P.G.M., and Maat, D.Z. 1980. Detection of potato virus X in tubers with enzyme-linked immunosorbent assay (ELISA). *Potato Research, 23:* 129–131.

Deeley, J., Stevens, W.A., and Fox, R.T.V. 1979. Use of Dienes' stain to detect plant diseases induced by mycoplasma-like organisms. *Phytopathology, 69:* 1169–1171.

Dekker, E.L., Dore, L., Porta, C., and Van Regenmortel M.H.V. 1987. Conformational specificity of monoclonal antibodies used in the diagnosis of tomato mosaic virus. *Archives of Virology, 94:* 191.

Dekker, E.L., Pinner, M.S., Markham, P.G., and Van Regenmortel, M.H.V. 1988. Characterization of maize streak virus isolates from different plant species by polyclonal and monoclonal antibodies. *Journal of General Virology, 69:* 983.

Delfosse, P., Reddy, A.S., Legréve, A., Devi, P.S., Devi, K.T., Maraite, H. and Reddy, D.V.R. 1999. Indian peanut clump virus (ICPV) infection on wheat and barley: symptoms, yield loss, and transmission through seeds. *Plant Pathology, 48:* 273–282.

Delgado-Sanchez, S. and Grogan, R.G. 1966. *Chenopodium quinoa,* a local lesion assay host for potato virus Y: Purification and properties of potato virus Y. *Phytopathology, 56:* 1394–1396.

Dellogi, A., Brisset, M.N., Paulin, J.P. and Expert, D. 1998. Dual role of desferrioxamine in *Erwinia amylovora* pathogenicity. *Molecular Plant-Microbe Interactions, 11:* 734–742.

Delwart, E.L.H., Sheppard, H.W., Walker, B.C., Goudsmit, J. and Mullins, I. 1994. Human immuno-deficiency virus type 1 evolution in vivo tracked by DNA heteroduplex mobility assays. *Journal of Virology, 68:* 6672–6683.

Délye, C., Laigret, F., and Corio-Costet, M.F. 1997. RAPD analysis provides insight into biology and epidemiology of *Uncinula nector. Phytopathology, 87:* 670–677.

Demski, J.W. 1968. Local lesion reactions of *Chenopodium* species to watermelon mosaic virus 2. *Phytopathology, 58:* 1196–1197.

Denes, A.S. and Sinha, R.C. 1991. Extra chromosomal DNA elements of plant pathogenic mycoplasma-like organisms. *Canadian Journal of Plant Pathology, 13:* 26–32.

Deng, S.J. and Hiruki, C. 1990a. Molecular cloning and detection of DNA of the mycoplasma-like organism associated with clover proliferation. *Canadian Journal of Plant Pathology, 12:* 383–388.

Deng, S.J. and Hiruki, C. 1990b. Enhanced detection of a plant pathogenic mycoplasma-like organism by polymerase chain reaction. *Proceedings of the Japan Academy of Sciences, B., Physical and Biological Sciences, 66:* 140–144.

Deng, S.J. and Hiruki, C. 1991. Genetic relatedness between two nonculturable mycoplasma-like organism revealed by nucleic acid hybridization and polymerase chain reaction. *Phytopathology, 81:* 1475–1479.

Deng, W.Y., Chen, L.S., Peng, W.T., Liang, X.Y., Sekiguchi, S., Gordon, M.P., Conai, L. and Nester, E.W. 1999. Vir E1 is a specific molecular chaperone for the exported single-stranded DNA binding protein Vir E2 in *Agrobacterium. Molecular Biology, 31:* 1795–1807.

Denny, T.P. 1995. Involvement of bacterial polysaccharides in plant pathogenesis. *Annual Review of Phytopathology, 33:* 173–197.

De Parasis, J. and Roth, D.A. 1990. Nucleic acid probes for identification of genus specific 16Sr RNA sequences. *Phytopathology, 80:* 618.

Derks, A.F.L.M., Lemmers, M.E.C. and Hollinger, T.C. 1997. Detectability of viruses in lily bulbs depends on virus, host and storage conditions. *Acta Horticulturae, No. 430,* 633–640.

Derrick, K.S. 1972. Immuno-specific grids for electron microscopy of plant viruses. *Phytopathology, 62:* 753 (Abstr.).

Derrick, K.S. 1973. Quantitative assay for plant viruses using serologically specific electron microscopy. *Virology, 56:* 652–653.

Derrick, K.S. and Brlansky, R.H. 1976. Assay for viruses and mycoplasma using serologically specific electron microscopy. *Phytopathology, 66:* 815–820.

Desjardins, P.R. and Wallace, J.M. 1962. Cucumber an additional herbaceous host of the infectious variegation strains of citrus psorosis virus. *Plant Disease Reporter. 46:* 416–418.

Des Scenzo, R.A. and Harrington, T.C. 1994. Use of (CAT) 5 as a DNA fingerprinting probe for fungi. *Phytopathology, 84:* 534–540.

Desvignes, J.C., Cornaggia, D., Grasseau, N., Ambrós, S and Flores, R. 1999. Pear blister canker viriod: host range and improved bioassay with two new pear indicators Fieud 37 and Fieud 110. *Plant Disease, 83:* 419–422.

Dewey, P., Macdonald, M., and Philips, S. 1989. Development of monoclonal antibody ELISA, dot-blot and dip-stick immunoassays for *Humicola lanuginosa* in rice. *Journal of General Microbiology, 135:* 361–374.

Dewey, F., Macdonald, M., Philips, S., and Priestley, R. 1990. Development of monoclonal antibody ELISA and dip-stick immunoassays for *Penicillium islandicum* in rice grains. *Journal of General Microbiology, 136:* 753–760.

Dewey, R.A., Semorile, L.C., and Grau, O. 1996. Detection of *Tospovirus* species by RT-PCR of the N-gene and restriction enzyme digestion of the products. *Journal of Virological Methods, 56:* 19–27.

De Zoeten, G.A., Gaard, G., and Diez. F.B. 1972. Nuclear vesiculation associated with pea enation mosaic virus-infected plants. *Virology, 48:* 638–647.

Dhanraj, K.S. and Raychaudhuri, S.P. 1969. A note on barley mosaic in India. *Plant Disease Reporter, 53:* 766–767.

Dhingra, K.L. and Nariani, J.K. 1963. A virus disease of onion. *Indian Phytopathology, 16:* 311–312.

Diacehun, S. and Valleau, W.D. 1950. Tobacco streak virus in sweet clover. *Phytopathology, 40:* 516–518.

Diaco, R.J., Hill, H., Hill, E.K., Tachibana, H., and Durand, D.P. 1985. Monoclonal antibody-based biotin-avidin ELISA for the detection of soybean mosaic virus in soybean seeds. *Journal of General Virology, 66:* 2089–2094.

Diamond, H., Cooke, B.M. and Cregg, B. 1998. Morphology of sporodochia of *Microdochium nivale* var. *nivale* on cereal hosts using cryoscanning electron microscope. *Mycologist, 12:* 118–120.

Dias, H.F. 1963. Host range and properties of grapevine yelow mosaic viruses. *Annals of Applied Biology, 51:* 85–95.

Dias, H.F. and Harrison, B.D. 1963. The relationship between grapevine fan leaf, grapevine yellow mosaic and arabis mosaic viruses. *Annals of Applied Biology, 51:* 97–105.

Dickman, M.B., Podila, G.K., and Kolattukudy, P.E. 1989. Insertion of cutinase gene into a wound pathogen enables it to infect intact host. *Nature, 342:* 446–448.

Diener, T.O. 1963. Physiology of virus-infected plants. *Annual Review of Phytopathology, 1:* 197–218.

Diener, T.O. 1971a. Potato spindle tuber "virus" IV: a replicating, low molecular weight RNA. *Virology, 45:* 411–428.

Diener, T.O. 1971b. A plant virus with properties of a free ribonucleic acid: potato spindle tuber virus. In: *Comparative Virology* (Eds.) K. Maramorosch and E. Kurstak, pp. 433–478. Academic Press, New York.

Diener, T.O. 1979. *Viroids and Viroid Diseases.* John Wiley & Sons, New York.

Diener, T.O. and Jenifer, F.G. 1964. A dependable local lesion assay for turnip yellow mosaic virus. *Phytopathology, 54:* 1258–1260.

Diener, T.O. and Lawson, R.H. 1973. Chrysanthemum stunt—a viroid disease. *Virology, 51:* 94–101.

Diener, T.O. and Schneider, I.R. 1966. The two components of tobacco ringspot virus nucleic acid: origin and properties. *Virology, 29:* 100–105.

Diener, T.O., Smith, D.R., and Galindo, A.J. 1985. PM antigen: A disease-associated host protein in viroid-infected tomato. In: *Subviral Pathogens of Plants and Animals: Viroids and Prions* (Eds.) K. Maramorosch and J.J. McKelvey, Jr., pp. 299–313. Academic Press, New York.

Dietzgen, R.G. 1983. Monoklonale Antikorper gegen PflanzenViren, Herstellung, Reinigung und Charakterisierung. Ph.d. Thesis, University of Tübingen, Germany.

Dietzgen, R.G. and Sander, E. 1982. Monoclonal antibodies against a plant virus. *Archives of Virology, 74:* 197.

Dimock, A.W. 1947. Chrysanthemum stunt. *New York State Flower Growers Bulletin 26:* 2.

Dimock, A.W., Geissinger, C.M., and Horst, R.K. 1971. Chlorotic mottle: a newly recognized disease of chrysanthemum. *Phytopathology, 61:* 415–419.

Dobrowolski, M.P. and O'Brien, P.A. 1993. Use of RAPD-PCR to isolate a species specific DNA probe for *Phytophthora cinnamomi. FEMS Microbiology Letters, 113:* 43–47.

Doi, Y., Ternaka, M., Yora, K., and Asuyama, H. 1967. Mycoplasma and PLT group-like microorganisms infected with mulberry dwarf, potato witches' broom, aster yellows and paulownia witches' broom. *Annals of Phytopathological Society, Japan, 33:* 259–266.

Dong, L.C., Sun, C.W., Thies, K.L., Luthe, D.S., and Graves, C.H., Jr., 1992. Use of polymerase chain reaction to detect pathogenic strains of *Agrobacterium. Phytopathology, 82:* 434–439.

D'Onghia, A.M., Djelouah, K., Alioto, D., Castellano, M.A. and Savino, V. 1998. ELISA correlates with biological indexing for detection of citrus psorosis-associated virus. *Journal of Plant Pathology, 80:* 157–163.

Doohan, F.M., Parry, D.W. and Nicholson. P. 1999. *Fusarium* ear blight of wheat: the use of quantitative PCR and visual disease assessment in studies of disease control. *Plant Pathology, 48:* 209–217.

Dore, I., Dekker, E.L., Porta, C., and Van Regenmortel, M.H.V. 1987. Detection by ELISA of two tobamoviruses in orchids using monoclonal antibodies. *Journal of Phytopathology, 20:* 317.

Dorsey, C.K. and Leach, J.G. 1956. The bionomics of certain insects associated with oak wilt with particular reference to the Nitidulidae. *Journal of Economic Entomology, 49:* 219–230.

Doster, M.A., and Michailides, T.J. 1998. Production of bright greenish yellow fluorescence in figs infected by *Aspergillus* species in California orchards. *Plant Disease, 82:* 669–673.

Dougherty, W.G., Willis, L., and Johnston, R.E. 1985. Topographic analysis of tobacco etch virus capsid protein epitopes. *Virology, 144:* 66.

Dow, J.M., Feng, J.X., Barber, C.E., Tang, J.L. and Daniels, M.J. 2000. Novel genes involved in the regulation of pathogenicity factor production within the *rpf* gene cluster of *Xanthomonas campestris*. *Microbiology (Reading), 146:* 885–891.

Dreier, J., Bermpohl, A., and Eichenlaub, R. 1995. Southern hybridization and PCR for specific detection of phytopathogenic *Clavibacter michiganensis* subsp. *michiganensis, Phytopathology, 85:* 462–468.

Dreier, J., Meletzus, D. and Eichenlaub, R. 1997. Characterization of the plasmid encoded virulence region *pat-1* of phytopathogenic *Clavibacter michiganensis* subsp. *michiganensis. Molecular Plant-Microbe Interactions, 10:* 195–206.

Drygin, Y.F., Korotaeva, S.G., and Dorokhov, Y.L. 1992. Direct RNA polymerase chain reaction for TMV detection in crude cell extracts. *FEBS Letters, 309:* 350–352.

Duban, M.E., Lee, K. and Lynn, D.G. 1993. Strategies in pathogenesis: mechanistic specificity in the detection of generic signals. *Molecular Microbiology, 7:* 637–645.

Duffus, J.E. 1985. Whitefly-borne viruses. *Phytoparasitica, 13:* 274.

Duffy, B.K. and Weller, D.M. 1994. A semiselective and diagnostic medium for *Gaeumannomyces graminis* var. *tritici. Phytopathology, 84:* 1407–1415.

Dunbar, K.B., Pinnow, D.L., Morris, J.B., and Pittman, R. 1993. Virus elimination from interspecific *Arachis* hybrids. *Plant Disease, 77:* 517–520.

Dushnicky, L.G., Ballance, G.M., Sumner, M.J. and MacGregor, A.W. 1998. Detection of infection and host responses in susceptible and resistant wheat cultivars to a toxin-producing isolate of *Pyrenophora tritici-ripentis. Canadiam Journal of Plant Pathology, 20:* 19–27.

Dye, D.W. 1962. The inadequacy of the usual determinative tests for the identification of *Xanthomonas* spp. *New Zealand Journal of Science, 5:* 393–416.

Ebisugi, H., Ooishi, S., Goda, T., Kubo, H. and Sakiyama, K. 1998. Effectiveness of membrane filtration for phage technique for detection of *Xanthomonas campestris* pv. *citri. Research Bulletin of Plant Protection Service, No. 34,* 113–115.

Edwards, G.J. and Ducharme, E.P. 1974. Attempt at previsual diagnosis of citrus young tree decline by use of remote sensing infrared thermometer. *Plant Disease Reporter, 58:* 793–796.

Edwards, S.G., Hetherington, R., Glynn, N.C., Hare, M.C., West, S.J.E., and Parry, D.W. 1998. Evaluation of fungicide seed treatments against *Fusarium* diseases of wheat using PCR diagnostic tests. *Brighton Crop Protection Conference: Pests and Diseases, 3:* 1017–1022.

Edwardson, J.R. 1966. Electron microscopy of cytoplasmic inclusions in cells infected with rod-shaped viruses. *American Journal of Botany, 53:* 359–363.

Edwardson, J.R. 1981. Fifth International Congress of Virology, Strasbourg, p. 244

El-Dougdoug, Kh. A., 1998. Occurrence of peach latent mosaic viroid in apple *(Malus domestica). Annals of Agricultural Science (Cairo), 43:* 21–30.

Elliott, C. and Poos, F.W. 1934. Overwintering of *Aplanobacter stewartii. Science, 80:* 289–290.

Elliott, M.L., DesJardin, E.A., and Henson, J.M. 1933. Use of a polymerase chain reaction assay to aid in identification of *Gaeumannomyces graminis* var. *graminis* from different grass hosts. *Phytopathology, 83:* 414–418.

Ellis, P.J. and Wieczorek, A. 1992. Production of monoclonal antibodies to beet western yellows virus and potato leafroll virus and their use in luteovirus detection. *Plant Disease, 76:* 75–78.

El-Nashaar, H.M., Moore, L.W., and George, R.A. 1986. Enzyme-linked immunosorbent assay quantification of wheat by *Gaeumannomyces graminis* var. *tritici* as moderated by biocontrol agents. *Phytopathology, 76:* 1319–1322.

Engvall, E. and Perlmann, P. 1971. Enzyme-linked immunosorbent assay (ELISA): quantitative assay of immunoglobulin G. *Immunochemistry, 8:* 871–874.

Enkerli, J., Bhatt, G. and Covert, S.F. 1998. Maackiain detoxification contributes to the virulence of *Nectria haematococca MPVI* on chickpea. *Molecular Plant-Microbe Interactions, 11:* 317–326.

Eriksson, A.R., Andersson, R.A., Pirhonen, M. and Palva, E.T. 1998. Two component regulation involved in the global control of virulence in *Erwinia carotovora* subsp. *carotovora*. *Molecular Plant-Microbe Interactions, 11:* 743–752.

Erlich, H.A. and Arnhein, N. 1992. Genetic analysis using polymerase chain reaction. *Annual Review of Genetics, 26:* 479–506.

Erlich, H.A., Gelfand, D., and Sninsky, J.J. 1991. Recent advances in the polymerase chain reaction. *Science, 252:* 1643–1651.

Erokhina, T.N., Zinovkin, R.A., Vitushkina, M.V., Jelkmann, W. and Agranovsky, A.A. 2000. Detection of beet yellows closterovirus methyl transferase-like and helicase-like proteins *in vivo* using monoclonal antibodies. *Journal of General Virology, 81:* 597–603.

Errampalli, D. and Fletcher, J. 1993. Production and monospecific polyclonal antibodies against aster yellows mycoplasma-like organisms associated antigen. *Phytopathology, 83:* 1279–1282.

Esau, K. 1968. *Viruses in Plant Hosts: Form, Distribution and Pathogenic Effect.* University of Wisconsin Press, Madison.

Esau, K. and Hoefert, L.L. 1971. Cytology of beet yellows virus infection in *Tetragonia* I. Parenchyma cells in infected leaf. *Protoplasma 72:* 255–273.

Esau, K. and Hoefert, L.L. 1972a. Development of infection with beet western yellows virus in the sugarbeet. *Virology, 48:* 724–738.

Esau, K. and Hoefert, L.L. 1972b. Ultrastructure of sugarbeet leaves infected with beet western yellows virus. *Journal of Ultrastructure Research, 40:* 556–571.

Esau, K. and Hoefert, L.L. 1972c. Conformations of virus-infected cells. *Protoplasma, 73:* 51–65.

Esau, K. and Hoefert, L.L. 1981. Beet yellow stunt virus in the phloem of *Sonchus oleraceus* L. *Journal of Ultrastructure Research, 75:* 326–338.

Esmenjaud, D., Abad, P., Pinck, L. and Walter, B. 1994. Detection of a region of the coat protein gene of grapevine fan leaf virus by RT-PCR in the nematode vector *Xiphinema index*. *Plant Disease, 78:* 1087–1090.

Eun, A.J.C. and Wong, S.M. 1999. Detection of cymbidium mosaic potexvirus and odontoglossum ringspot tobamovirus using immuno-capillary zone electrophoresis. *Phytopathology, 89:* 522–528.

Faggian, R., Bulman, S.R. and Porter, I.J. 1999. Specific polymerase chain reaction primers for the detection of *Plasmodiophora brassicae* in soil and water. *Phytopathology, 89:* 392–397.

Falk, B.W., Tsai, J.H., and Lommel, S.A. 1987. Differences in levels of detection for the maize stripe virus capsid and major noncapsid proteins in plant and insect hosts. *Journal of General Virology, 68:* 1801–1811.

Fang, C.T., Allen, O.N., Riker, A.J., and Dickson, J.G. 1950. The pathogenic, physiological, serological reactions of the form species *Xanthomonas translucens. Phytopathology, 40:* 44–64.

Faria, J.C. and Maxwell, D.P. 1999. Variability in geminivirus isolates associated with *Phaseolus* spp. in Brazil. *Phytopathology, 89:* 262–268.

Fegan, M., Croft, B.J., Teakle, D.S., Hayward, A.C. and Smith, G.R. 1998. Sensitive and specific detection of *Clavibacter xyli* subsp. *xyli,* causal agent of ratoon stunting disease of sugarcane with a polymerase chain reaction-based assay. *Plant Pathology, 47:* 495–504.

Fenby, N.S., Scott, N.W., Slater, A., and Elliott, M.C. 1995. PCR and nonisotopic labeling techniques for plant virus detection. *Cellular and Molecular Biology, 41:* 639–652.

Feng LanXiang, Lu AiLan and Li MaoLin. 1998. Application of the virobacterial agglutination for detecting vegetable viruses. *China Vegetables, No. 2,* 1–5.

Feodorova, R.N. 1997. New and improved methods of detecting smuts in wheat and barley seeds. *Biuletyn Instytutu Hodowli i Aklimatyzacji Ros'lin, No. 201,* 253–256.

Fernow, K.H. 1967. Tomato as a test plant for detecting mild strains of potato spindle tuber virus. *Phytopathology, 57:* 1347–1352.

Fernow, K.H., Peterson, L.C., and Plasited, R.L. 1970. Spindle tuber virus in seeds and pollen of infected potato plants. *American Potato Journal, 47:* 75–80.

Fessehaie, A., Wydra, K. and Rudolph, K. 1999. Development of a new semiselective medium for isolating *Xanthomonas campestris* pv. *manihotis* from plant material and soil. *Phytopathology, 89:* 591–597.

Field, T.K., Patterson, C.A., Gergerich, R.C., and Kim, K.S. 1994. Fate of viruses in bean leaves after deposition by *Epilachna varivestis,* a beetle vector of plant viruses. *Phytopathology, 84:* 1346–1350.

Figueira, A.R., Domier, L.L. and D'Arcy, C.J. 1997. Comparison of techniques for detection of barley yellow dwarf virus-PAV-IL. *Plant Disease, 81:* 1236–1240.

Finch, J.T. and Klug, A. 1967. Structure of broad bean mottle virus I. *Journal of Molecular Biology, 24:* 289–302.

Finch, J.T., Klug, A., and Leberman, R. 1970. The structure of turnip crinkle and tomato bushy stunt viruses II. *Journal of Molecular Biology, 50:* 215–222.

Firrao, G., Gobbi E., and Locel, R. 1994. Rapid diagnosis of apple proliferation mycoplasma-like organism using a polymerase chain reaction procedure. *Plant Pathology, 43:* 669–674.

Fitzell, R.F., Fahy, P.C., and Evans, G. 1980. Serological studies on some Australian isolates of *Verticillium* spp. *Australian Journal of Biological Sciences, 33:* 115–124.

Foster, H., Kinscherf, T.G., Leong, S.A., and Maxwell, D.P. 1987. Molecular analysis of the mitochondrial genome of *Phytophthora. Current Genetics, 12:* 215–218.

Fouly, H.M., Wilkinson, H.T. and Domier, L.L. 1996. Use of random amplified polymorphic DNA (RAPD) for identification of *Gaeumannomyces* species. *Soil Biology and Biochemistry, 28:* 703–710.

Fourie, J.F. and Holz, G., 1995. Initial infection processes by *Botrytis cinerea* on nectarine and plum fruit and the development of decay. *Phytopathology, 85:* 82–87.

Fraaije, B.A., Birnbaum, Y., Franken, A.A.J.M. and Bulk, R.W. van den. 1996. The development of a conductimetric assay for the automated detection of metabolically active soft rot *Erwinia* spp. in potato tuber peel extracts. *Journal of Applied Bacteriology, 81:* 375–382.

Fraaije, B.A., Lovell, D.J., Rohel, E.A. and Hollomon, D.W. 1999. Rapid detection and diagnosis of *Septoria tritici* epidemics in wheat using a polymerase chain reaction/PicoGreen assay. *Journal of Applied Bacteriology, 86:* 701–708.

Franck, A., Gera, A., Antignus, Y. and Loebenstein, G. 1996. Detection of pelargonium flower break carmovirus using ELISA and a transcribed RNA probe. *Acta Horticulturae, No. 432,* 338–344.

Francki, R.I.B. 1980. Limited value of the thermal inactivation point, longevity in vitro and dilution end point as criteria for the characterization, identification and classification of plant viruses. *Intervirology, 13:* 91–98.

Francki, R.I.B. and Hatta, T. 1981. Tomato spotted wilt virus. In: *Hand Book of Plant Virus Infections and Comparative Diagnosis* (Ed.) E. Kurstak, pp. 491–512. Elsevier/North-Holland, Amsterdam.

Francki, R.I.B., Kitajima, E.W., and Peters, D. 1981. Rhabdoviruses. In: *Hand Book of Plant Virus Infections and Comparative Diagnosis* (Ed.) E. Kurstak, pp. 455–489. Elsevier/North-Holland, Amsterdam.

Franconi, R., Roggero, P., Pirazzi, P., Arias, F.J., Desiderio, A., Bitti, O., Pashkonlo, V.D., Mattei, B., Bracci, L., Masenga, V., Milne, R.G. and Benvenuto, E. 1999. Functional expression in bacteria and plants of scFv antibody fragment against tospoviruses. *Immunotechnology, 4:* 180–201.

Franken, A.A.J.M. 1992. Application of polyclonal and monoclonal antibodies for the detection of *Xanthomonas campestris* pv. *campestris* in crucifer seeds using immunofluorescence microscopy. *Netherlands Journal of Plant Pathology, 98:* 95–106.

Franken, A.A.J.M., Zilverentant, J.F., BooneKamp, P.M., and Schots, A. 1992. Specificity of polyclonal and monoclonal antibodies for the identification of *Xanthomonas campestris* pv. *campestris*. *Netherlands Journal of Plant Pathology, 98:* 81–94.

Fránová, J., Karešová, R., Navratil, M., Šimková, M., Válová, P. and Nebesárová, J. 2000. A carrot proliferation disease associated with rickettsiae-like organisms in the Czech Republic. *Journal of Phytopathology, 148:* 53–55.

Franz, A., Makkouk, K.M. and Vetten, H.J. 1998. Acquisition, retention and transmission of faba bean necrotic yellows virus by two of its aphid vectors, *Aphis craccivora* (Koch) and *Acyrthosiphon pisum* (Harris). *Journal of Phytopathology, 146:* 347–355.

Franza, T., Sauvage, C., and Expert, D. 1999. Iron regulation and pathogenicity in *Erwinia chrysanthemi* 3937: a role of the Fur repressor protein. *Molecular Plant-Microbe Interactions, 12:* 119–128.

Fraser, R.S.S. 1990. The genetics of resistance to plant viruses. *Annual Review of Phytopathology, 28:* 179–200.

Fraser, D.E., Shoemaker, P.B. and Ristaino, J.B. 1999. Characterization of isolates of *Phytophthora infestans* from tomato and potato in North Carolina from 1993 to 1995. *Plant Disease, 83:* 633–638.

Frazer, H.L. 1944. Observations on the methods of transmission of internal boll disease of cotton by cotton stainer bug. *Annals of Applied Biology, 31:* 271–290.

Frazier, N.W. 1968a. Transmission of strawberry mottle virus by juice and aphids to herbaceous hosts. *Plant Disease Reporter, 52:* 64–67.

Frazier, N.W. 1968b. Transmission of strawberry crinkle virus by the dark strawberry aphid *Chaetosiphon jacobi. Phytopathology, 58:* 165–172.

Frechon, D., Exbrayat, P., Helias, V., Hyman, L.J. Jouan, B., Llop, P., Lopez, M.M., Payet, N., Perombelon, M.C.M., Toth, L.K., Beckhoven, J.R.C. M. van, Wolf, J.M. van der and Bertheaw, Y. 1998. Evaluation of a PCR kit for the detection of *Erwinia carotovora* subsp *carotovora* on potato tubers. *Potato Research, 41:* 163–173.

Freeman, S., Maimon, M., and Pinkas, Y. 1999. Use of GUS transformants of *Fusarium subglutinans* for determining etiology of mango malformation disease. *Phytopathology, 89:* 456–461.

Frei, U. and Wenzel, G. 1993. Differentiation and diagnosis of *Pseudocercosporella herpotrichoides* (Fon.) Delighton with genomic DNA probes. *Journal of Phytopathology, 139:* 229–237.

Freigoun, S.O., El Faki, H.I., Gelie, B., Schirmer, M., and Lemattre, M. 1994. Phage sensitivity in relation to pathogenicity and virulence of the cotton bacterial blight pathogen of Sudan. *Plant Pathology, 43:* 493–497.

Freitag, J.H. 1941. Insect transmission, host range and properties of squash mosaic virus. *Phytopathology, 31:* 8.

Freitag, J.H. 1956. Beetle transmission, host range and properties of squash mosaic virus. *Phytopathology, 46:* 73–81.

Freitag, J.H. and Milner, K.S. 1970. Host range, aphid transmission and properties of muskmelon vein necrosis virus. *Phytopathology, 60:* 166–170.

Frommel, M.I. and Pazos, G. 1994. Detection of *Xanthomonas campestris* pv. *undulosa* infested wheat seeds by combined liquid medium enrichment and ELISA. *Plant Pathology, 43:* 589–596.

Fuchs, E., Gruntzig, M., Kegler, H., Krczal, G., and Avenarius, V. 1995. A biological test for characterizing of plum pox virus strains. *Rasteniev dni Nauki, 32:* 33–35.

Fuchs, E., Schlufter, C. and Kegler, H. 1996. Occurrence of a plant virus in the northern sea. *Archives of Phytopathology and Plant Protection, 32:* 365–366.

Fuchs, J.G., Mönne-Loccoz, Y. and Défago, G. 1999. Ability of nonpathogenic *Fusarium oxysporum* Fo47 to protect tomato against *Fusarium* wilt. *Biological Control, 14:* 105–110.

Fuerst, J.A. and Perry, J.W. 1988. Demonstration of lipopolysaccharide on sheathed flagella of *Vibrio cholerae* 0: 1 by protein A-gold immunoelectron microscopy. *Journal of Bacteriology, 170:* 1488–1494.

Fuji, S.L. and Nakamal, H. 1999. Detection of Japanese yam mosaic virus by ELISA and RT-PCR. *Annals of Phytopathological Society, Japan, 65:* 207–210.

Fujiwara, A., Takamura, H., Majumder, P., Yoshida, H., and Kojima, M. 1998. Functional analysis of the protein encoded by the chromosomal virulence gene (acvB) of *Agrobacterium tumefaciens. Annals of Phytopathological Society, Japan, 64:* 191–193.

Fukushi, T. 1933. Transmission of the virus through the eggs of an insect. *Proceedings of Imperial Academy (Tokyo), 8:* 457–460.

Fukushi, T. 1940. Further studies of the dwarf disease of rice plant. *Journal of Faculty of Agriculture, Hokkaido University, 45:* 83–154.

Fukushi, T. and Kimura, I. 1959. Mechanical transmission of rice dwarf virus to *Nephotettix cincticeps*. *Annals of Phytopathological Society, Japan, 23:* 54.

Fulton, C.E., and Brown, A.E. 1997. Use of SSU rDNA group I intron to distinguish *Monilia fruticola* from *M. laxa* and *M. fructigena*, *FEMS Microbiology Letters, 157:* 307–312.

Fulton, J.P., Gergerich, R.C., and Scott, H.A. 1987. Beetle transmission of plant viruses. *Annual Review of Phytopathology, 25:* 111–123.

Fulton, R.W. 1948. Hosts of the tobacco streak virus. *Phytopathology, 38:* 421–428.

Fulton, R.W. 1952. Mechanical transmission and properties of rose mosaic virus. *Phytopathology, 42:* 413–416.

Fulton, R.W. 1967. Purification and serology of rose mosaic virus. *Phytopathology, 57:* 1197–1201.

Fulton, R.W. 1978. Superinfection by strains of tobacco streak virus. *Virology, 85:* 1–8.

Gabor, B.K., O'Gara, E.T., Philip, B.A., Horan, D.P., and Hardham, A.R. 1993. Specificities of monoclonal antibodies to *Phytophthora cinnamomi* in two rapid diagnostic assays. *Plant Disease, 77:* 1189–1197.

Gaitaitis, R., Sumner, D., Gay, D., Smittle, D., Mc Donald, G., Maw, B., Johnson, W.C., III, Tollner, B. and Hung, Y. 1997. Bacterial streak and bulb rot of onion. I. A diagnostic medium for semiselective isolation and enumeration of *Pseudomonas viridiflava*. *Plant Disease, 87:* 897–900.

Galindo, J.A., Smith, D.R., and Diener, T.O. 1982. Etiology of Planta Macho, a viroid disease of tomato. *Phytopathology, 72:* 49–54.

Gamez, R. and Leon, P. 1988. Maize raydofino and related viruses. In: *The Plant Viruses* (Ed.) R. Koenig, Vol. 3, pp. 213–233. Plenum Press, New York.

Gara, I.W., Kondo, H., and Maeda, T. 1997. Evaluation of dot immunobinding assay and rapid immunofilter paper assay for detection of cymbidium mosaic virus in orchids. *Bulletin of Research Institute for Bioresources, Okayama University, 5:* 39–46.

Garber, R.C. and Yoder, O.C. 1984. Mitochondrial DNA of the filamentous ascomycete *Cochliobolus heterostrophus*. *Current Genetics, 8:* 621–628.

Garcia, M.L., Sanchez, de la, Torre, M.I. Bo, E.D., Dielouah, K., Rouag, N., Luisoni, E., Milne, R.G. and Grau, O. 1997. Detection of citrus psorosis ringspot virus using RT-PCR and DAS-ELISA. *Plant Pathology, 46:* 830–836.

Garcia Pedrajas, M.D., Bainbridge, B.W., Heale, J.B., Perez Artés, E., and Jimenez Diaz, R.M. 1999. A simple PCR-based method for the detection of the chickpea wilt pathogen *Fusarium oxysporum* f.sp. *ciceris* in artificial and natural soils. *European Journal of Plant Pathology, 105:* 251–259.

Garde, S. and Bender, C.L. 1991. DNA probes for detection of copper resistance genes in *Xanthomonas campestris* pv. *vesicatoria*. *Applied Environmental Microbiology, 57:* 2435.

Gardiner, T.J., Pearson, M.N., Hopcroft, D.H., and Forster, R.L.S. 1995. Characterization of a labile RNA virus-like agent from white clover. *Annals of Applied Biology, 126:* 91–104.

Gardner, M.W. and Kendrick, J.B. 1921. Soybean mosaic. *Journal of Agricultural Research, 22:* 111–114.

Garnsey, S.M. 1968. Distribution of citrange stunt (tatter-leaf) virus in leaves of citrus hosts. *Phytopathology, 58:* 1041–1074.

Garnsey, S.M. and Whidden, R. 1973. Efficiency of mechanical inoculation procedures for citrus exocortis virus. *Plant Disease Reporter, 57:* 886–890.

Garret, A., Kerlan, C., and Thomas, D. 1993. The intestine is a site of passage for potato leaf roll virus from the gut lumen into the hemocoel in the aphid vector *Myzus persicae* Sulz. *Archives of Virology, 31:* 377–392.

George, M.L.C., Bustaman, M., Cruz, W.T., Leach, S.E. and Nelson, R.J. 1997. Movement of *Xanthomonas oryzae* pv. *oryzae* in Southeast Asia detected using PCR-based DNA fingerprinting. *Phytopathology, 87:* 302–309.

George, M.L.C., Nelson, R.J., Zeigler, R.S. and Leung, H. 1998. Rapid population analysis of *Magnaporthe grisea* by using rep-PCR and endogeneous repetitive DNA sequences. *Phytopathology, 88:* 223–229.

Gera, A., Lawson, R.H., and Hsu, H.T., 1995. Identification and assay. In: *Viruses and Virus-like Diseases of Bulb and Flower Crops* (Eds.) G. Loebenstein, R.H. Lawson and A.A. Brunt, pp. 165–180. John Wiley & Sons, Chichester, U.K.

Gerber, M. and Sarkar, S. 1988. Comparison between urease and phosphatase in virus diagnosis with ELISA technique. *Journal of Plant Diseases and Protection, 95:* 544–550.

Gerik, J.S., Lommel, S.A., and Huisman, O.C. 1987. A specific serological staining procedure for *Verticillium dahliae* in cotton root tissue. *Phytopathology, 77:* 261–265.

Gerola, F.M., Bassi, M., and Giussani, G. 1966. Some observations on the shape and localization of different viruses in experimentally infected plants and on the fine structure of the host cells. III. Turnip yellow mosaic virus in *Brassica chinensis* L. *Caryologia, 19:* 457–479.

Ghabrial, S.A. and Shepherd, R.J. 1967. A sensitive radioimmunosorbent assay for the detection of plant viruses. *Journal of General Virology, 48:* 311–317.

Ghanekar, A.M., Reddy, D.V.R., Izuka, N., Amin, P.W., and Gibbon, R.W. 1979. Bud necrosis in groundnut (*Arachis hypogaea*) in India caused by tomato spotted wilt virus. *Annals of Applied Biology, 93:* 173–179.

Gibb, K.S., Padovan, A.C., and Mogen, B.D. 1995. Studies on sweet potato little leaf phytoplasma detected in sweet potato and other plant species growing in Northern Australia. *Phytopathology, 85:* 169–174.

Gibb, K.S., Persley, D.M., Schneider, B. and Thomas, J.E. 1996. Phytoplasmas associated with papaya diseases in Australia. *Plant Disease, 80:* 174–178.

Gibb, K.S., Schneider, B., and Padovan, A.C. 1998. Differential detection and genetic relatedness of phytoplasmas in papya. *Plant Pathology, 47:* 325–332.

Gibbs, A., Armstrong, J., MacKenzie, A.M. and Weiller, G. 1998. The GPRIME package: Computer programs for identifying the best regions of aligned gencs to target in nucleic acid hybridization-based diagnostic tests and their use with plant viruses. *Journal of Virological Methods, 74:* 67–76.

Gibbs, A., Giussani-Belli, G., and Smith, H.C. 1968. Broad bean stain and true broad bean mosaic viruses. *Annals of Applied Biology, 61:* 99–107.

Gibbs, A.J. and Harrison, B.D. 1964. Nematode-transmitted viruses in sugarbeet in East Anglia. *Plant Pathology, 13:* 144–150.

Gibbs, A.J. and Harrison, B.D. 1969. Eggplant mosaic virus and its relationship to Andean potato latent virus. *Annals of Applied Biology, 64:* 225–231.

Gibbs, A.J., Kassanis, B., Nixon, H.L., and Woods, R.D. 1963. The relationship between barley stripe mosaic and *Lychnis* ringspot viruses. *Virology, 20:* 194–198.

Gibbs, A.J., Nixon, H.L., and Woods, R.D. 1963. Properties of purified preparations of lucerne mosaic virus. *Virology, 19:* 441–449.

Gilbertson, R.L., Maxwell, D.P., Hagedorn, D.J., and Leong, S.A. 1989. Development and application of a plasmid DNA probe for detection of bacteria causing common bacterial blight of bean. *Phytopathology, 79:* 518.

Gill, C.C. and Chong, J. 1979a. Cytopathological evidence for the division of barley yellow dwarf virus isolates into two subgroups. *Virology, 95:* 59–69.

Gill, C.C. and Chong, J. 1979b. Cytological alterations in cells infected with corn leaf aphid-specific isolates of barley yellow dwarf virus. *Phytopathology, 69:* 363–368.

Gillaspie, A.G. Jr., Mitchell, S.E., Stuart, G.W. and Bozarth, R.F. 1999. RT-PCR method for detecting cowpea mottle carmovirus in *Vigna* germplasm. *Plant Disease, 83:* 639–643.

Gillaspie, A.G. Jr., Pittman, R.N., Pinnow, D.L. and Cassidy, B.G. 2000. Sensitive method for testing peanut seed lots for peanut stripe and peanut mottle viruses by immunocapture–reverse transcription–polymerase chain reaction. *Plant Disease, 84:* 559–561.

Gillings, M.R., Fahy, P.C., Broadbent, P., and Barnes, D. 1995. Rapid identification of a second outbreak of Asiatic citrus canker in the Northern Territory using polymerase chain reaction and genomic fingerprinting. *Australasian Plant Pathology, 24:* 104–111.

Gindrat, D., and Pezet, R., 1994. Paraquat, a tool for rapid detection of latent fungal infections and of endophytic fungi. *Journal of Phytopathology, 141:* 86–98.

Gingery, R.E. 1988. The rice stripe virus group. In: *The Plant Viruses* (Ed.) R.G. Milne, Vol. 4, pp. 297–329. Plenum Press, New York.

Gispert, C., Perring, T.M. and Creamer, R. 1998. Purification and characterization of peach mosaic virus. *Plant Disease, 82:* 905–908.

Giunchedi, L. and Pollini, C.P. 1992. Cytopathological negative staining and serological electron microscopy of clostero-like virus associated with pear vine yellows disease. *Journal of Phytopathology, 134:* 329–335.

Giunchedi, L., Pollini, C.P., Biondi, S., and Babini, A.R. 1993. PCR detection of MLOs in quick decline affected trees in Italy. *Annals of Applied Biology, 124:* 399–403.

Gleason, M.L., Ghabrial, S.A., and Ferriss, R.S. 1987. Serological detection of *Phomopsis longicola* in soybean seeds. *Phytopathology, 77:* 371–375.

Gnanamanickam, S.S., Shigaki, T., Medalla, E.S., Mew, T., and Alvarez, A.M. 1994. Problems in detection of *Xanthomonas oryzae* pv. *oryzae* in rice seeds and potential for improvement using monoclonal antibodies. *Plant Disease, 78:* 173–178.

Goldstein, B. 1924. Cytological study of living cells of tobacco plants affected with mosaic diseases. *Bulletin of Torrey Botanical Club, 51:* 261–272.

Gonsalves, D. and Ishii, M. 1980. Purification and serology of papaya ringspot virus. *Phytopathology, 70:* 1028–1032.

Goodman, R.M. 1981. Geminiviruses. In: *Handbook of Plant Virus Infections and Comparative Diagnosis* (Ed.) E. Kurstak, pp. 883–910. Elsevier/North-Holland, Amsterdam.

Goodman, P.H. and Bantari, E.E. 1984. Increased sensitivity of ELISA for potato viruses S, X and Y by polystyrene pretreatments, additives and a modified assay procedure. *Plant Disease, 68:* 944–947.

Goodwin, P.H., English, J.T., Neber, D.A., Duniway, J.M., and Kirkpatrick, B.C. 1990. Detection of *Phytophthora parasitica* from soil and host tissue with a species-specific DNA probe. *Phytopathology, 80:* 277.

Goodwin, P.H., Kirkpatrick, B.C., and Duniway, J.M. 1989. Cloned DNA probes for identification of *Phytophthora parasitica. Phytopathology, 79:* 716–721.

Goodwin, P.H., Kirkpatrick, B.C., and Duniway, J.M. 1990. Identification of *Phytophthora citrophthora* with cloned DNA probes. *Applied Environmental Microbiology, 56:* 669.

Goodwin, P.H., Mahuku, G.S., Liu HongWei and Xue, B.G. 1999. Monitoring phytoplasma in population of aster leafhoppers from lettuce fields using the polymerase chain reaction. *Crop Protection, 18:* 98–99.

Goodwin, P.H. and Nassuth, A. 1993. Detection and characterization of plant pathogens. In: *Methods in Plant Molecular Biology and Biotechnology* (Eds.) B.R. Glick and J.C. Thompson, pp. 303–319. CRC Press, Inc., Boca Raton, Fla.

Goodwin, P.H., Xue, B.G., Kuske, C.R., and Sears, M.K. 1994. Amplification of plasmid DNA to detect plant pathogenic mycoplasma-like organisms. *Annals of Applied Biology, 124:* 27–36.

Goodwin, S.B., Schneider, R.E. and Fry, W.E. 1995. Use of cellulose acetate electrophoresis for rapid identification of allozyme genotypes of *Phytophthora infestans. Plant Disease, 79:* 1181–1185.

Gopalan, S., and He, S.Y. 1996. Bacterial genes involved in the elicitation of hypersensitive response and pathogenesis. *Plant Disease, 80:* 604–610.

Gough, K.C., Cockburn, W. and Whitelam, G.C. 1999. Selection of phage displayed peptides that bind to cucumber mosaic virus coat protein. *Journal of Virological Methods, 79:* 169–180.

Gough, K.H. and Shukla, D.D. 1980. Further studies on the use of protein A in immune electron microscopy for detecting virus particles. *Journal of General Virology, 51:* 415–419.

Gould, G.E. 1944. The biology and control of the striped cucumber beetle. *Bulletin of Indiana Agricultural Experiment Station, 490:* 28.

Goulden, M.G. and Baulcombe, D.C. 1993. Functionally homologous host components recognize potato virus X in *Gomphrena globosa* and potato. *The Plant Cell, 5:* 921–930.

Green, M.J., Thompson, D.A. and Mac Kenzie, D.J. 1999. Easy and efficient DNA extraction from woody plants for the detection of phytoplasmas by polymerase chain reaction. *Plant Disease, 83:* 482–485.

Griep, R.A., Twisk, C., van, and Schots, A. 1999. Selection of beet necrotic yellow vein virus specific single chain Fv antibodies from a semisynthetic combinatorial antibody library. *European Journal of Plant Pathology, 105:* 147–156.

Griep, R.A., van Twisk, C., van Beckhoven, J.R.C.M., van der Wolf, J.M. and Schots, A. 1998. Development of specific recombinant monoclonal antibodies against lipopolysaccharide of *Ralstonia solanacearum* race 3. *Phytopathology, 88:* 795–803.

Griffiths, H.M., Sinclair, W.A., Davis, R.E., Lee, I.M., Dally, E.L., Guo, Y.K., Chen, T.A., and Hibben, C.R. 1994. Characterization of mycoplasma-like organisms from *Fraxinus, Syringa* and associated plants from geographically diverse sites. *Phytopathology, 84:* 119–126.

Grogan, R.C. 1981. The science and art of plant disease diagnosis. *Annual Review of Phytopathology, 19:* 333–351.

Gross, H.J., Domdey, H., Lossow, C., Janak, P., Raba, M., Alberty, H., and Sanger, H.C. 1978. Nucleotide sequence and secondary structure of potato spindle tuber viroid. *Nature (Lond.) 273:* 203–208.

Gross, H.J., Domdey, H., Lossow, C., Janak, P., Raba, M., Alberty, H., Losson, C.H., Ramon, K., and Sanger, H.C. 1982. Nucleotide sequence and secondary structure of citrus exocortis and chrysanthemum stunt viroid. *Journal of Biochemistry, 121:* 249–257.

Gugerli, P. 1983. Use of enzyme immunoassay in phytopathology. In: *Immunoenzymatic Techniques* (Eds.) S. Avrameas, P. Duret, R. Masseyeff, and G. Feldmann, p. 369. Elsevier, Amsterdam.

Gugerli, P. 1990. Advanced immunological techniques for virus detection. In: *Control of Virus and Virus-like Diseases of Potato and Sweet Potato.* International Potato Center, Lima, Peru.

Gugerli, P., Brugger, J.J. and Ramel, M.E. 1997. Immuno-chemical identification of the sixth grapevine leafroll-associated virus (GLRaV-6) and improvement of the diagnostic features for the sanitary selection in viticulture. *Revue Swisse de Viticulture d' Arboriculture et d' Horticulture, 29:* 137–141.

Gugerli, P. and Frie, P. 1983. Characterization of monoclonal antibodies to potato virus Y and their use for virus detection. *Journal of General Virology, 64:* 2471.

Guitierrez, W.A. and Shew, H.D. 1998. Identification and quantification of ascospores as the primary inoculum for collar rot of greenhouse-produced tobacco seedlings. *Plant Disease, 82:* 485–490.

Guo, Y.H., Cheng, Z.M., Walla, J.A. and Zhang, Z. 1998. Diagnosis of X-disease phytoplasma in stone fruits by a monoclonal antibody developed directly from a woody plant. *Journal of Environmental Horticulture, 16:* 33–57.

Guthrie, J.N., White, D.T., Walsh, K.B. and Scott, P.T. 1998. Epidemiology of phytoplasma associated papaya disease in Queensland, Australia. *Plant Disease, 82:* 1107–1111.

Guzmán, P., Gepts, P., Temple, S., Mkandawire, A.B.C., and Gilbertson, R.L. 1999. Detection and differentiation of *Phaeoisariopsis griseola* isolation with polymerase chain reaction and group specific primers. *Plant Disease, 83:* 37–42.

Haase, A., Richter, J., and Rabenstein, F. 1989. Monoclonal antibodies for detection and serotyping of cucumber mosaic virus. *Journal of Phytopathology, 127:* 129.

Haas, J.H., Moore, L.W., Ream, W. and Manulis, S. 1995. PCR primers for detection of phytopathogenic *Agrobacterium* strains. *Applied and Environmental Microbiology, 61:* 2879–2884.

Haber, S. and Knapen, H. 1989. Filter paper sero-assay (FiPSA): a rapid, sensitive technique for sero-diagnosis of plant viruses. *Canadian Journal of Plant Pathology, 11:* 109–113.

Habili, N. 1993. Detection of Australian field isolates of citrus tristeza virus by double stranded RNA analysis. *Journal of Phytopathology, 138:* 308–316.

Habili, N., Fazeli, C.F. and Rezain, M.A. 1997. Identification of a cDNA clone specific to grapevine leafroll-associated virus 1 and occurrence of the virus in Australia. *Plant Pathology, 46:* 516–522.

Habili, N., McInnes, J.L., and Symons, R.H. 1987. Nonradioactive photobiotin-labeled DNA probes for the routine diagnosis of barley yellow dwarf virus. *Journal of Virological Methods, 16:* 225–237.

Haddow, W.R., and Newman, F.S. 1942. A disease of the Scots pine (*Pinus sylvestris* L.) caused by the fungus *Diplodia pinea* Kick associated with the pine spittle bug (*Aphrophora parallela* Say.). I. Symptoms and etiology. *Transactions of Royal Canadian Institute, 24:* 1–17.

Hadidi, A., Huang, C., Hammong, R.W., and Hashimoto, J. 1990. Homology of the agent associated with dapple disease to apple scar skin viroid and molecular detection of these viroids. *Phytopathology, 80:* 263–268.

Hadidi, A., Levy, L. and Podleckis, E.V. 1995. Polymerase chain reaction technology in Plant Pathology. In: *Molecular Methods in Plant Pathology* (Eds.) R.P. Singh and U.S. Singh, pp. 167–187. CRC Press, Boca Raton, FL.

Hadidi, A., Montasser, M.S., Levy, L., Goth, R.W., Converse, R.H., Madkour, M.H., and Skrzeckowski, L.S. 1993. Detection of potato leaf roll and strawberry mild yellow edge luteoviruses by reverse transcription polymerase chain reaction amplification. *Plant Disease, 77:* 595–601.

Hadidi, A. and Yang, X. 1990. Detection of pome fruit viroids by enzymatic cDNA amplification. *Journal of Virological Methods, 30:* 261–270.

Hafner, G.J., Harding, R.M. and Dale, J.L. 1995. Movement and transmission of banana bunchy top virus DNA component one in bananas. *Journal of General Virology, 76:* 2279–2285.

Hagedorn, D.J. and Walker, J.C. 1949. Wisconsin pea streak. *Phytopathology, 39:* 837–847.

Hahn, M. and Mendgen, K. 1997. Characterization of in planta induced rust genes isolated from a haustorium-specific cDNA library. *Molecular Plant-Microbe Interactions, 10:* 427–434.

Hahn, M., Neef, U., Struck, C., Göttfert, M. and Mendgen, K. 1997. A putative amino acid transporter is specifically expressed in haustoria of the rust fungus *Uromyces fabae*. *Molecular Plant-Microbe Interactions, 10:* 435–438.

Hajimorad, M.R., Dietzgen, R.G., and Francki, R.I.B. 1990. Differentiation and antigenic characterization of closely related alfalfa mosaic virus strains with monoclonal antibodies. *Journal of General Virology, 71:* 2809.

Halk, E.L., Hsu, H.T., Aebig, J., and Franke, J. 1984. Production of monoclonal antibodies against three ilarviruses and alfalfa mosaic virus and their use in serotyping. *Phytopathology, 74:* 367.

Halonen, P., Meurman, O., Lovgren, T., Hemmila, I., and Soini, E. 1986. Detection of viral antigens by time-resolved fluoroimmunoassay. In: *Rapid Methods and Automation in Microbiology and Immunology* (Eds.) K.O. Habermehl, pp. 133–146. Springer-Verlag, New York.

Halpern, B.T. and Hillman, B.I. 1996. Detection of blueberry scorch virus strain NJ2 by reverse transcriptase–polymerase chain reaction amplification. *Plant Disease, 80:* 219–222.

Ham, B.K., Lee, T.H., You, J.S., Nam, Y.W., Kim, J.K. and Paek, K.H. 1999. Isolation of a putative host factor interacting with cucumber mosaic virus–encoded 2b protein by yeast two-hybrid screening. *Molecules and Cells, 9:* 548–555.

Hamer, J., Farrell, L., Orbach, M., Valent, A., and Chumley, F. 1989. Host species specific conservation of a family of repeated DNA sequences in the genome of a fungal plant pathogen. *Proceedings of National Academy of Sciences, USA, 86:* 9981–9985.

Hammond, J. 1998. Serological relationships between cylindrical inclusion proteins of potyviruses. *Phytopathology, 88:* 965–971.

Han MuSeok, Noh Eun Woon and Yun JeongKoo, 1997. Occurrence of some phyllody disease in Korea and detection of its phytoplasma. *Korean Journal of Plant Pathology, 13:* 239–243.

Han, S.S., Ra, D.S., and Nelson, R.J. 1995. Relationship between DNA finger prints and virulence of *Pyricularia grisea* from rice and new hosts in Korea. *International Rice Research Notes, 20(1):* 26–27.

Han SangSop, Hiruki, ChuJi and Kim SongMun, 1998. Rapid analysis of genetic relationship of phytoplasma isolates by a DNA heteroduplex mobility assay. *Korean Journal of Plant Pathology, 14:* 382–385.

Hara, H., Hoga, K. and Tanaka, H. 1995. An improved selective medium for quantitative isolation of *Pseudomonas solanacearum. Annals of Phytopathological Society, Japan, 61:* 255.

Hardham, A.R., Gubler, F., Duniec, J., and Elliott, J. 1991. A review of methods for the production and use of monoclonal antibodies to study zoosporic plant pathogens. *Journal of Microscopy, 162:* 305–318.

Hardham, A.R., Suzaki, E., and Perkin, J.L. 1986. Monoclonal antibodies to isolate-species and genus-specific components on the surface of zoospores and cysts of the fungus. *Phytophthora cinnamomi. Canadian Journal of Botany, 64:* 311–321.

Hariri, D., Delarenay, T., Gomes, L., Filleur, S., Plovie, C. and Lapierre, H. 1996. Comparison and differentiation of wheat yellow mosaic virus (WYMV), wheat spindle mosaic virus (WSSMV) and barley yellow mosaic virus (BaYMV) isolates using WYMV monoclonal antibodies. *European Journal of Plant Pathology, 102:* 283–292.

Harjosudarmo, J., Ohshima, K., Uyeda, I., and Shikata, E. 1990. Relationship between hybridoma screening for use in direct double antibody sandwich ELISA for the detection of plant viruses. *Annals of Phytopathological Society, Japan, 56:* 569.

Harper, G., Dahal, G., Thottappilly, G. and Hull, R. 1999. Detection of episomal banana streak badnavirus by IC-PCR. *Journal of Virological Methods, 79:* 1–8.

Harper, K. and Creamer, R. 1995. Hybridization detection of insect-transmitted plant viruses with digoxigenin-labeled probes. *Plant Disease, 79:* 563–567.

Harper, K., Kerschbaumer, R.J., Ziegler, A., MacIntosh, S.M., Cowan, G.H., Himmler, G., Mayo, M.A. and Torrance, L. 1997. A scFv–alkaline phosphatase fusion protein which detects potato leafroll luteovirus in plant extracts by ELISA. *Journal of Virological Methods, 63:* 237–242.

Harris, K.F. 1979. Leafhoppers and aphids as biological vectors: vector-virus relationships. In: *Leafhopper Vectors and Plant Disease Agents* (Eds.) K. Maramorosch and K.F. Harris, pp. 217–308. Academic Press, New York.

Harris, J.R. and Horne, R.E. 1986. *Electron Microscopy of Proteins, Vol. 5. Viral Structure.* Academic Press, London.

Harris, K.F. and Maramorosch, K. 1977. *Aphids as Virus Vectors.* Academic Press, New York.

Harrison, B.D. 1958. Ability of single aphids to transmit both avirulent and virulent strains of potato leaf roll virus. *Virology, 6:* 278–286.

Harrison, B.D. 1985. Advances in geminivirus research. *Annual Review of Phytopathology, 23:* 55–82.

Harrison, B.D. and Crowley, N.C. 1965. Properties and structure of lettuce necrotic yellows virus. *Virology, 26:* 297–310.

Harrison, B.D., Finch, J.T., Gibbs, A.J., Hollings, M., Shepherd, R.J., Valenta, V., and Wetter, C. 1971. Sixteen groups of plant viruses. *Virology, 45:* 356–363.

Harrison, B.D., and Jones, R.A.C. 1970. Host range and some properties of potato moptop virus. *Annals of Applied Biology, 65:* 393–402.

Harrison, B.D., Muniyappa, V., Swanson, M.M., Roberts, I.M., and Robinson, D.J. 1991. Recognition and differentiation of seven whitefly-transmitted geminivirus from India and their relationships to African cassava mosaic and Thailand mungbean yellow mosaic viruses. *Annals of Applied Biology, 118:* 299–308.

Harrison, B.D. and Nixon, H.L. 1959. Separation and properties of particles of tobacco rattle virus with different lengths. *Journal of General Microbiology, 2:* 569–581.

Harrison, B.D. and Nixon, H.L. 1960. Purification and electron microscopy of three soil-borne plant viruses. *Virology, 12:* 104–117.

Harrison, B.D., Nixon, H.L. and Woods, R.D. 1965. Lengths and structure of particles of barley stripe mosaic virus. *Virology, 26:* 284–289.

Harrison, B.D. and Roberts, I.M. 1971. Pinwheel and crystalline structure induced by Atropa mild mosaic virus, a plant virus with particles 925 nm long. *Journal of General Virology, 10:* 71–78.

Harrison, B.D., Stefanac, Z., and Roberts, I.M. 1970. Role of mitochondria in the formation of X bodies in cells of *Nicoriana clevelandii* infected with tobacco rattle virus. *Journal of General Virology, 6:* 127–140.

Harrison, N.A., Bourne, C.M., Cox, R.L., Tsai, J.H., and Richardson, P.A. 1992. DNA probes for detection of mycoplasma-like organisms associated with lethal yellowing disease of palms in Florida. *Phytopathology, 82:* 216–224.

Harrison, N.A., Richardson, P.A., Kramer, J.B., and Tsai, J.H. 1994. Detection of the mycoplasma-like organism associated with lethal yellowing disease of palms in Florida by polymerase chain reaction. *Plant Pathology, 43:* 998–1008.

Harrison, N.A., Richardson, P.A., Tsai, J.H., Ebbert, M.A. and Kramer, J.B. 1996. PCR assay for detection of the phytoplasma associated with maize bushy stunt disease. *Plant Disease, 80:* 263–269.

Harrison, T.G., Rees, E.A., Barker, H., and Lowe, R. 1993. Detection of spore balls of *Spongospora subterranea* on potato tubers by enzyme-linked immunosorbent assay. *Plant Pathology, 42:* 181–186.

Hartung, F., Werner, R., Mühlbach, H.P., and Büttner, C. 1998. Highly specific PCR-diagnosis to determine *Pseudomonas solanacearum* strains of different geographic origins. *Theoretical and Applied Genetics, 96:* 797–802.

Hartung, J.S. and Civerolo, E.L. 1989. Restriction fragment length polymorphisms distinguish *Xanthomonas campestris* strains isolated from Florida citrus nurseries from *X. campestris* pv. *citri. Phtyopathology, 79:* 793–799.

Hartung, J.S. and Pooler, M.R. 1997. Immunocapture and multiplexed PCR assay for *Xanthomonas fragariae,* causal agent of angular leaf spot disease. *Acta Horticulturae, No. 439 II,* 821–828.

Hartung, J.S., Pruvost, O.P., Villermot, I. and Alvarez, A. 1996: Rapid and sensitive colorimetric detection of *Xanthomonas axonopodis* pv. *citri* by immunocapture and nested–polymerase chain reaction assay. *Phytopathology, 86:* 95–101.

Harvey, H.P. and Ophel-Keller, K. 1996. Quantification of *Gaeumannomyces graminis* var. *tritici* in infected roots and arid soil using slot-blot hybridization. *Mycological Research, 100:* 962–970.

Haseloff, B.J., Freeman, T.L., Valmeekam, V., Melkus, M.W., Oner, F., Valachovic, M.S. and San Francisco, M.J.D. 1998. The *exuT* gene of *Erwinia chrysanthemi* EC 16: nucleotide sequence, expression, localization and relevance of the gene product. *Molecular Plant-Microbe Interactions, 11:* 270–276.

Haseloff, J., Mohamed, N.A., and Symons, R.H. 1982. Viroid RNAs of the cadang-cadang disease of coconuts. *Nature (Lond.) 299:* 316–322.

Haseloff, J. and Symons, R.H. 1981. Chrysanthemum stunt viroid—primary sequence and secondary structure. *Nucleic Acids Research, 9:* 2741–2752.

Hataya, T., Inoue, A.K., and Shikata, E. 1994. A PCR-microplate hybridization method for plant virus detection. *Journal of Virological Methods, 46:* 223–236.

Hatta, T. and Francki, R.I.B. 1981. Cytopathic structures associated with tonoplasts of plant cells infected with cucumber mosaic and tomato aspermy viruses. *Journal of General Virology, 53:* 343–346.

Hatta, T. and Francki, R.I.B. 1989. Identification of small polyhedral virus particles in thin sections of plant cells by an enzyme cytochemical technique. *Journal of Ultrastructure Research, 74:* 116–129.

Have, A. ten, Mulder, W., Visser, J. and Van Kan, J.A.L. 1998. The endopolygalacturonase gene *Bcpg1* is required for full virulence of *Botrytis cinerea. Molecular Plant-Microbe Interactions, 11:* 1009–1016.

He, C.X., Li, W.B., Ayres, A.J., Hartung, J.S., Miranda, V.S. and Teixeira, D.C. 2000. Distribution of *Xylella fastidiosa* in citrus root stocks and transmission of citrus variegated chlorosis between sweet orange plants through natural root grafts. *Plant Disease, 84:* 622–626.

He, S.Y. 1998. Type III protein secretion systems in plant and animal pathogenic bacteria. *Annual Review of Phytopathology, 36:* 363–392.

He XiaoHua, Liu SiJun and Perry, K.C. 1998. Identification of epitopes in cucumber mosaic virus using a phage-displayed random peptide library. *Journal of General Virology, 79:* 3145–3153.

Helguera, M., Bravo-Almonacid, F., Kobayashi, K., Rabinowicz, P.D., Conci, V. and Mentaberry, A. 1997. Immunological detection of a Gar V–type virus in Argentine garlic cultivars. *Plant Disease, 81:* 1005–1010.

Helias, V., Roux, A.C. le, Bertheau, Y. Andrivon, D., Gauthier, J.P. and Jouan, B. 1998. Characterization of *Erwinia carotovora* subspecies and detection of *Erwinia caro-*

tovora subsp. *atroseptica* in potato plants, soil or water extracts with PCR-based methods. *European Journal of Plant Pathology, 104:* 685–699.

Hellmann, G.M., Thornbury, D.W., Hiebert, E., Shaw, J.G., Pirone, T.P., and Rhoads, R.E. 1983. Cell-free translation of tobacco vein mottling virus RNA. II. Immunoprecipitation of products by antisera to cylindrical inclusion, near inclusion and helper component proteins. *Virology, 124:* 434–444.

Hellqvist, S., Koponen, H., Lindqvist, H. and Valkonen, J. 1998. The downy mildew fungus *Peronospora sparsa* in wild arctic bramble and in cultivated hybrid arctic bramble-sampling in northern Sweden. *Växtskyddnotiser, 62:* 41–45.

Hemmila, I., Dakubu, S., Mukkala, V.M., Siitari, H., and Lovgren, T. 1984. Europium as a label in time-resolved fluoroimmunometric assays. *Analytical Biochemistry, 137:* 335.

Henco, K., Sanger, H.L., and Riesner, D. 1979. Fine structure melting of viroids as studied by kinetic methods. *Nucleic Acids Research, 6:* 3041–3059.

Henderson, R.G. 1931. Transmission of tobacco ringspot by seed of *Petunia. Phytopathology, 21:* 225–229.

Henderson, W.J. 1935. Yellow dwarf, a virus disease of onions and its control. *Iowa State College of Agriculture Research Bulletin, 188.*

Henriquez, N.P., Kenyone, L. and Quiroz, L. 1999. Corn stunt complex mollicutes in Belize. *Plant Disease, 83:* 77.

Henson, J.M. and French, R. 1993. The polymerase chain reaction and plant disease diagnosis. *Annual Review of Phytopathology, 31:* 81–109.

Henson, J.M., Goins, T., Grey, W., Mathre, D.E., and Elliott, M.L., 1993. Use of polymerase chain reaction to detect *Gaeumannomyces graminis* DNA in plants grown in artificially and naturally infested soil. *Phytopathology, 83:* 283–287.

Hernandez, J.F., Posada, M.A., Portillo, P. del and Arbelaez, G. 1999. Identification of molecular markers of *Fusarium oxysporum* f. sp. *dianthi* by RAPD. *Acta Horticulturae, No. 482,* 123–131.

Herold, F. and Munz, K. 1965. Electron microscopic demonstration of virus-like particles in *Peregrinus maidis* following acquisition of maize mosaic virus. *Virology, 34:* 583–589.

Herold, F. and Munz, K. 1969. Peanut mottle virus. *Phytopathology, 59:* 663–666.

Herold, F. and Weibel, J. 1963a. Electron microscopic demonstration and sugarcane mosaic virus particles in cells of *Saccharum officinarum* and *Zea mays. Phytopathology, 53:* 469–471.

Herold, F. and Weibel, J. 1963b. Electron microscopic demonstration of papaya ringspot virus. *Virology, 18:* 302–311.

Heuss, K., Liu, Q., Hammerschlag, F.A. and Hammond, R.W. 1999. A cRNA probe detects prunus necrotic ringspot virus in three peach cultivars after micrografting and in peach shoots following long term culture at 4°C. *Hort Science, 34:*346–347.

Hewish, D.R., Shukla, D.D., Johnstone, G.R., and Sward, R.J. 1983. Monoclonal antibodies to lueoviruses. *Proceedings of 4th International Congress of Plant Pathology, Melbourne, Australia,* 116p.

Hibben, C.R., Sinclair, W.A., Davis, R.E., and Alexander, J.H. 1991. Relatedness of mycoplama-like organisms associated with ash yellows and lilac witches' broom. *Plant Disease, 75:* 1227–1230.

Hibino, H. and Cabunagan, R.C. 1986. Rice tungro-associated viruses and their relations to the host plants and vector leafhoppers. *Tropical Agricultural Research Series* No. 19: 173–192, Tropical Agriculture Center, Japan.

Hibino, H., Daquioag, R., Cabauatan, P., and Dahal, G. 1988. Resistance to rice tungro spherical virus in rice. *Plant Disease, 72:* 843–847.

Hibino, H., Ishikawa, K., Omura, T., Cabauatan, P.Q., and Koganezawa, H. 1991. Characterization of rice tungro bacilliform and rice tungro spherical viruses. *Phytopathology, 81,* 1130–1132.

Hibino, H., Tsuchizaki, T., and Saito, Y. 1974a. Comparative electron microscopy of cytoplasmic inclusions induced by 9 isolates of soil-borne wheat mosaic virus. *Virology, 57:* 510–521.

Hibino, H., Tsuchizaki, T., and Saito, Y. 1974b. Electron microscopy of inclusion development in rye leaf cells infected with soil-borne wheat mosaic virus. *Virology, 57:* 522–530.

Hibrand, L., Gall, O. Le, Candresse, T., and Dunez, J. 1992. Immunodetection of the proteins encoded by grapevine chrome-mosaic nepovirus RNA2. *Journal of General Virology, 73:* 2093–2098.

Hidaka, Z. 1954. Morphology of the tobacco stunt virus. *Proceedings of the International Conference on Electron Microscopy. London,* pp. 54–55.

Hiebert, E. and McDonald, J.G. 1973. Characterization of some proteins associated with viruses in the potato Y group. *Virology, 56:* 349–361.

Hiebert, E., Purcifull, D.E., Christie, R.G., and Christie, S.R. 1971. Partial purification of inclusions induced by tobacco etch virus and potato virus Y. *Virology, 43:* 638–646.

Hiebert, E., Purcifull, D.E., and Christie, R.G. 1984. Purification and immunological analyses of plant viral inclusion bodies. In: *Methods in Virology* (Eds.) K. Maramorosch and H. Koprowski, Vol. 8, pp. 225–280. Academic Press, New York.

Hilborn, M.I., Hyland, F., and McCrum, R.E. 1965. Pathological anatomy of apple trees affected by stem pitting virus. *Phytopathology, 55:* 34–39.

Hildebrand, M., Dickler, E. and Geider, K. 2000. Occurrence of *Erwinia amylovora* on insects in a fire blight orchard. *Journal of Phytopathology, 148:* 251–256.

Hilf, M.E., Karasev, A.V. Albiach-Marti, M.R. Dawson, W.O. and Garnsey, S.M. 1999. Two paths of sequence divergence in the citrus tristeza virus complex. *Phytopathology, 89:* 332–336.

Hill, E.K., Hill, J.H., and Durand, D.P. 1984. Production of monoclonal antibodies to viruses in the potyvirus group: use in radioimmunoassay. *Journal of General Virology, 65:* 525.

Hill, I.H. and Durand, D.P. 1986. Soybean mosaic virrus. In: *Methods of Enzymatic Analysis.* (Eds.) K.V. Bergmeyer, J. Wergmeyer and M. Grabi, Vol. XI, pp. 455–474. VCH Publishers, Weinheim.

Hill, J.H., Benner, H.I., and Duesen, R.A. 1994. Rapid differentiation of soybean mosaic virus isolates by antigenic signature analysis. *Journal of Phytopathology, 142:* 152–162.

Himmler, G., Brix, U., Steinkellner, M., Laimer, M., Mattanovick, D., and Katinger, H.W.D. 1988. Early screening for anti-plum pox virus monoclonal antibodies with different epitope specificities by means of gold labeled immunosorbent electron microscopy. *Journal of Virological Methods, 22:* 351.

Hiruki, C. 1988. Rapid and specific detection methods for plant mycoplasmas. In: *Mycoplasma Diseases of Crops* (Eds.) K. Maramoroach and S.P. Raychaudhuri, pp. 77–101. Springer-Verlag, New York.

Hiruki, C. and da Rocha, A. 1986. Histochemical diagnosis of mycoplasma infections in *Catharanthus roseus* by means of a fluorescent DNA binding agent, 4,6-diamidino-2-phenyl indole-dihydrochloride. *Canadian Journal of Plant Pathology, 8:* 185–188.

Hiruki, C. and Deng, S. 1992. Distribution of plant pathogenic mollicutes in *Cathananthus roseus* determined by in situ molecular hybridization and DNA staining with DAPI. *Proceedings of Japan Academy Series B, Physical and Biological Sciences, 68(10):* 187–190.

Hiruki, C., Figueiredo, G., Inoue, M., and Furuya, Y. 1984. Production and characterization of monoclonal antibodies specific to sweet clover necrotic mosaic virus. *Journal of Virological Methods, 8:* 301.

Hirumi, H., Granados, R.R., and Maramorosch, K. 1967. Electron microscopy of a plant pathogenic virus in the nervous system of its insect vector. *Journal of Virology, 1:* 430–444.

Hitchborn, J.G., and Hills, G.J. 1965. The use of negative staining in the electron microscopic examination of plant viruses in crude extracts. *Virology, 27:* 528–540.

Hobbs, H.A., Reddy, D.V.R., Rajeshwari, R., and Reddy, A.S. 1987. Use of direct antigen coating and protein A coating ELISA procedures for detection of three peanut viruses. *Plant Disease, 71:* 747–749.

Hobbs, H.A., Reddy, D.V.R., and Reddy, A.S. 1989. Detection of a mycoplasma-like organism in peanut plants with witches' broom using indirect enzyme-linked immunosorbent assay (ELISA). *Plant Pathology, 36:* 164–167.

Hodgson, R.A.J., Wall, G.C. and Randles, J.W. 1998. Specific identification of coconut tinangaja viroid for differential field diagnosis of viroids in coconut palm. *Phytopathology, 88:* 774–781.

Hoefert, L.L., Mc Creight, J.D. and Christie, R.D. 1992. Microwave enhanced staining for plant virus inclusions. *Biotechnic and Histochemistry, 67:* 40–44.

Hoffman, C.H. and Moses, C.S. 1940. Mating habits of *Scolytus multistriatus* and the dissemination of *Ceratostomella ulmi. Journal of Economic Entomology, 33:* 818–819.

Hoffmann, K., Sackey, S.T., Maiss, E., Adomako, D. and Vetten, H.J. 1997. Immunocapture polymerase chain reaction for the detection and characterization of cacao swollen shoot virus 1A isolates. *Journal of Phytopathology, 145:* 205–213.

Hogenhout, S.A., Verbeek, M., Hans, F., Houterman, P.M., Fortass, M., Wilk, F. van der, Huttinga, H. and Henvel, J.F.J.M. van den, 1996. Molecular basis of the interactions between luteoviruses and aphids. *Agronomie, 16:* 167–173.

Holland, A.A. and Choo, Y.S. 1970. Immunoelectrophoretic characteristics of *Ophiobolus graminis* Sacc. as an aid in classification and determination. *Antonie van Leeuwenhoek, 36:* 541–548.

Hollings, M. 1955. Investigation of chrysanthemum viruses. I Aspermy flower distortion. *Annals of Applied Biology, 43:* 83–102.

Hollings, M. 1956. *Chenopodium amaranticolor* Coste and Reyn, as a test plant for plant viruses. *Plant Pathology, 5:* 57.

Hollings, M. 1959. Host range studies with fifty two plant viruses. *Annals of Applied Biology, 98:* 108.

Hollings, M. 1962. Studies on pelargonium leafcurl virus. I. Host range, transmission and properties in vitro. *Annals of Applied Biology, 50:* 189–202.

Hollings, M. 1965. Some properties of celery yellow vein—a virus serologically related to tomato black-ring virus. *Annals of Applied Biology, 55:* 459–470.

Hollings, M. 1966. Local lesion and other test plants for the identification and culture of viruses. In: *Viruses of Plants,* pp. 230–248.

Hollings, M. and Nariani, T.K. 1965. Some properties of clover yellow vein, a virus from *Trifolium repens* L. *Annals of Applied Biology, 56:* 99–109.

Hollings, M. and Stone, O.M. 1963. Turnip crinkle virus isolated from an ornamental variegated cabbage (*Brassica oleracea* var. *Capitata* L.) *Report of Glasshouse Crops Research Institute,* 1962, pp. 118–125.

Hollings, M. and Stone, O.M. 1964. Investigation of carnation viruses. I. Carnation mottle. *Annals of Applied Biology, 53:* 103–118.

Hollings, M. and Stone, O.M. 1965a. Studies of pelargonium leafcurl virus. II. Relationships to tomato bushy stunt and other viruses. *Annals of Applied Biology, 56:* 87–98.

Hollings, M. and Stone, O.M. 1965b. *Chenopodium quinoa* Willd, as an indicator plant for carnation latent virus. *Plant Pathology, 14:* 66–68.

Hollings, M. and Stone, O.M. 1965c. *Torenia fournieri* Lind as a test plant for plant viruses. *Plant Pathology, 14:* 165–168.

Hollings, M. and Stone, O.M. 1967. *Report on Glasshouse Crops Research Institute,* 1966, pp. 1–13.

Holmes, F.O. 1931. Local lesions of mosaic in *Nicotiana tabacum* L. *Contributions of Boyce Thompson Institute, 2:* 163–172.

Holmes, F.O. 1946. A comparison of the experimental host ranges of tobacco etch and tobacco mosaic viruses. *Phytopathology, 36:* 643–659.

Holy, S. and Abou Haidar, M.G. 1993. Production of infectious *in vitro* transcripts from a full length clover yellow mosaic virus cDNA clone. *Journal of General Virology, 74:* 781–784.

Holtz, B.A., Karu, A.E., and Weinhold, A.R. 1994. Enzyme-linked immunosorbent assay for detection of *Thielaviopsis basicola. Phytopathology, 84:* 977–983.

Honda, Y., Kameya-Iwaki, M., Hanada, K., Tochihara, K., and Tokashiki, I. 1989. Occurrence of tomato spotted wilt virus on watermelon in Japan. *Food and Fertilizer Technology Center Technical Bulletin, No. 114,* pp. 15–19.

Hood, M.E. and Shew, H.D. 1996. Applications of KOH–aniline blue fluorescence in the study of plant fungal interactions. *Phytopathology, 86:* 704–708.

Hooftman, R. Arts, M.J., Shamloul, A.M., Zaayen, A. van, and Hadidi, A. 1996. Detection of chrysanthemum stunt viroid by reverse transcription–polymerase chain reaction and by tissue blot hybridization. *Acta Horticulturae, No. 432,* 120–128.

Hooker, M.E., Lee, R.F., Civerolo, E.L., and Wang, S.Y. 1993. Reliability of gentisic acid, a fluorescent marker for diagnosis of citrus greening disease. *Plant Disease, 77:* 174–180.

Hooykaas, P.J.J. and Beijersbergen, A.G.M. 1994. The virulence system of *Agrobacterium tumefaciens. Annual Review of Phytopathology, 32:* 157–179.

Horn, N.M., Reddy, S.V., Roberts, I.M., and Reddy, D.V.R. 1993. Chickpea chlorotic dwarf virus, a new leafhopper-transmitted geminivirus of chickpea in India. *Annals of Applied Biology, 122:* 467–479.

Horn, N.M., Reddy, S.V., and Reddy, D.V.R. 1994. Virus vector relationships of chickpea chlorotic dwarf geminivirus and the leafhopper *Orosius orientalis* (Hemiptera: Cicadellidae). *Annals of Applied Biology, 124:* 441–450.

Horst, R.K. 1975. Detection of latent infectious agent that protects against infection by chrysanthemum chlorotic mottle viroid. *Phytopathology, 65:* 1000–1003.

Howell, W.E., Mink, G.I. Hurt, S.S., Foster, J.A. and Postman, J.D. 1996. Select *Malus* clones for rapid detection of apple stem grooving virus. *Plant Disease, 80:* 1200–1202.

Hoyer, V., Maiss, E., Jelkmann, W. Lesemann, D.E. and Vetten, H.J. 1996. Identification of the coat protein gene of a sweet potato sunken vein closterovirus isolate from Kenya and evidence for serological relationship among geographically diverse closterovirus isolates from sweet potato. *Phytopathology, 86:* 744–750.

Hrabak, E.M. and Willis, D.K. 1993. Involvement of the *lemA* gene in production of syringomycin and protease by *Pseudomonas syringae* pv. *syringae*. *Molecular Plant-Microbe Interactions, 6:* 368–375.

Hsieh, S.P.Y., Huang, R.Z. and Wang, T.C., 1996. Application of tannic acid in qualitative and quantitative growth assay of *Rhizoctonia* spp. *Plant Pathology Bulletin, 5:* 100–106.

Hsu, H.T. 1978. Cell fusion induced by a plant virus. *Virology, 84:* 9–18.

Hsu, H.T., Aebig, J., and Rochow, W.F. 1984. Differences among monoclonal antibodies to barley yellow dwarf viruses. *Phytopathology, 74:* 600.

Hsu, H.T. Halk, E.L., and Lawson, R.H. 1983. Monoclonal antibodies in plant virology. *Proceedings of 4th International Congress of Plant Pathology, Melbourne, Australia,* p. 25.

Hsu, H.T. and Lawson, R.H. 1985a. Comparison of mouse monoclonal antibodies and polyclonal antibodies of chicken egg yolk and rabbit for assay of carnation etched ring virus. *Phythopathology, 75:* 778.

Hsu, H.T. and Lawson, R.H. 1985b. Detection of carnation etched ring virus using mouse monoclonal antibodies and polyclonal chicken and rabbit antisera. *Acta Horticulturae, 164:* 199.

Hsu, H.T. and Lawson, R.H. 1991. Direct tissue blotting for detection of tomato spotted wilt virus in *Impatiens*. *Plant Disease, 75:* 292–295.

Hsu, H.T., Vongsasitron, D., and Lawson, R.H. 1992. An improved method for serological detection of cymbidium mosaic potex virus infection in orchids. *Phythopathology, 82:* 491–495.

Hsu, H.T., Wang, Y.C., Lawson, R.H., Wang, M., and Gonsalves, D. 1990. Splenocytes of mice with induced immunological tolerance to plant antigens for construction of hybridomas secreting tomato spotted wilt virus-specific antibodies. *Phythopathology, 80:* 158.

Hsu, Y.H., Annamalai, P., Lin, C.S., Chen, Y.Y., Chang, W.C. and Lin, N.S. 2000. A sensitive method for detecting bamboo mosaic virus (BaMV) and establishment of BaMV-free meristem-tip cultures. *Plant Pathology, 49:* 101–107.

Hu, C.C., Aboul-Ala, A.E., Naidu, R.A., and Ghabriel, S.A. 1997. Evidence for the occurrence of two distinct subgroups of peanut stunt cucumovirus strains: molecular characterization of RNA 3. *Journal of General Virology, 78:* 929–939.

Hu, C.C., Sanger, M.P. and Ghabriel, S.A. 1998. Production of infectious peanut stunt virus RNA from full length cDNA clones representing two subgroups of strains: mapping satellite RNA support to RNA1. *Journal of General Virology, 79:* 2013–2021.

Hu, J.S., Boscia, D., and Gonsalves, D. 1989. Use of monoclonal antibodies in the study of closteroviruses associated with grape leafroll disease. *Phytopathology, 79:* 1189.

Hu, J.S., Sether, D.M., Liu, X.P., Wang, M., Zee, F. and Ullman, D.E. 1997. Use of a tissue blotting immunoassay to examine the distribution of pineapple closterovirus in Hawaii. *Plant Disease, 81:* 1150–1154.

Hu, J.S., Sether, D.M. and Ullman, D.E. 1996. Detection of pineapple *Closterovirus* in pineapple plants and mealybugs using monoclonal antibodies. *Plant Pathology, 45:* 829–836.

Hu, J.S., Wang, M., Sether, D., Xie, W. and Leonhardt, K.W. 1996. Use of polymerase chain reaction (PCR) to study transmission of banana bunchy top virus by banana aphid *(Pentalonia nigronervosa)*. *Annals of Applied Biology, 128:* 55–64.

Hu QinXue, Zhang ChunLi, Chen Jie, Wu Deki and Lin Mulan, 1997. Biotin hydrazide–labeled and DIG-labeled cDNA probes for the diagnosis of citrus exocortis viroid in plants. *Chinese Journal of Virology, 13:* 159–163.

Hu WeiWen and Wong SekMan. 1998. The use of DIG-labeled cRNA probes for the detection of cymbidium mosaic potexvirus (CyMV) and odontoglossum ringspot tobamovirus (ORSV) in orchids. *Journal of Virological Methods, 70:* 193–199.

Hu, X., Nazar, R.N., and Robb, J. 1993. Quantification of *Verticillium* biomass in wilt disease development. *Physiological and Molecular Plans Pathology, 42:* 23–36.

Huang, B.C., Zhu, H., Hu, G.G., Li, Q.X., Gao, J.L., and Goto M. 1993. Serological studies of *Xanthomonas campestris* pv. *oryzae* with polyclonal and monoclonal antibodies. *Annals of Phytopathological Society, Japan, 59:* 123–127.

Huang, J. and Schell, M.A. 1995. Molecular characterization of the *eps* gene cluster of *Pseudomonas solanacearum* and its transcriptional regulation at a single promoter. *Molecular Microbiology, 16:* 977–989.

Huckett, B.I. and Botha, F.C. 1996. Progress towards a definitive diagnostic test for sugarcane mosaic virus infection. *Proceedings of Annual Congress of South African Sugar Technologist Association, No. 70,* 11–13.

Huertas-González, M.D., Ruiz-Roldan, M.C., Di Petro, A. and Roncero, M.I.G. 1999. Cross-protection provides evidence for race-specific avirulence factors in *Fusarium oxysproum*. *Physiological and Molecular Plant Pathology, 54:* 63–72.

Huet, H., Mahendra, S., Wang, J., Sivamani, E., Ong, C.A., Chen, L., de Kochko, A., Beachy, R.N. and Fauquet, C. 1999. Near immunity to rice tungro spherical virus achieved in rice by replicase-mediated resistance strategy. *Phytopathology, 89:* 1022–1027.

Huff, D.R., Bunting, T.E., and Plumley, K.A. 1994. Use of random amplified polymorphic DNA markers for the detection of genetic variation in *Magnaporthe poae*. *Phytopathology, 84:* 1312–1316.

Hughes, J. d'A., Adomako, D. and Ollennu, L.A.A. 1995. Evidence from the virobacterial agglutination test for the existence of eight serogroups of cocoa swollen shoot virus. *Annals of Applied Biology, 127:* 297–307.

Hughes, J.d'A. and Ollennu, L.A. 1993. The virobacterial agglutination test as a rapid means of detecting cocoa swollen shoot virus. *Annals of Applied Biology, 122:* 299–310.

Hughes, K.T.D., Inman, A.J., Beales, P.A., Cook, R.T.A., Fulton, C.E. and Mc Reynolds, A.D.K. 1998. PCR-based detection of *Phytophthora fragariae* in raspberry and strawberry roots. *Brighton Crop Protection Conference: Pests and Diseases, 2:* 687–692.

Huguenot, C., Dobbelsteen, G., De Van Den, Haen, P., Wagemakers, C.A.M., Dorst, G.A., Osterbaus, A.D.M.E., and Peters, D. 1990. Detection of tomato spotted wilt virus using monoclonal antibodies and riboprobes. *Archives of Virology, 110:* 47–62.

Huguenot, C., Givord, L., Sommermeyer, G., and Van Regenmortel, M.H.V. 1989. Differentiation of peanut clumpvirus serotypes by monoclonal antibodies. *Research in Virology, 140:* 87.

Huguenot, C., Van den Dobbelsteen, G., De Haan, P., Wagemakers, C.A.M., Osterhaus, A.D.M.E., and Peters, D. 1990. Detection of tomato spotted wilt virus using monoclonal antibodies and riboprobes. *Archives of Virology, 110:* 47.

Hull, R., and Adams, A.N. 1968. Groundnut rosette and its assistor virus. *Annals of Applied Biology, 62:* 139–145.

Hunter, W.B., Hiebert, E., Webb, S.E., Tsai, J.H. and Polston, J.E. 1998. Location of geminiviruses in the whitefly *Bemisia tabaci* (Homoptera: Aleyrodidae). *Plant Disease, 82:* 1147–1151.

Hung, T.H., Wu, M.L. and Su, H.J. 1999. Detection of fastidious bacteria causing citrus greening disease by non-radioactive DNA probes. *Annals of Phytopathological Society, Japan, 65:* 140–146.

Huss, B., Muller, S., Sommer Meyer, G., Walter, B., and Van Regenmortel, M.H.V. 1987. Grapevine fan leaf virus monoclonal antibodies: their use to distinguish different isolates. *Journal of Phytopathology, 119:* 358.

Iannelli, D., Barba, M., D'Apice, L., Pasquini, G., Capparelli, R., Monti, Z., Parrella, G., Scala, F. and Noviello, C. 1996. Cytofluorimetric method for the detection of the cucumber mosaic virus. *Phytopathology, 86:* 959–965.

Iannelli, D., D'Apice, L., Cottone, C., Viscardi, M., Scala, F., Zoina, A., Sorbo, G. del, Spigno, P. and Capparelli, R. 1997. Simultaneous detection of cucumber mosaic virus, tomato mosaic virus and potato virus Y by flow cytometry. *Journal of Virological Methods, 69:* 137–145.

Ieki, H., Yamaguchi, A., Kano, T., Koizumi, M. and Iwanami, T. 1997. Control of stem pitting disease caused by citrus tristeza virus using protective mild strains in Naval orange. *Annals of Phytopathological Society, Japan, 63:* 170–175.

IRRI (International Rice Research Institute) 1983. *Annual Report for 1982,* Los Banos, Philippines.

Ishiie, T., Doi, Y., Yora, K., and Asuyama, H. 1967. Suppressive effects of antibiotics of tetracycline group on symptom development of mulberry dwarf disease. *Annals of Phytopathological Society, Japan, 33:* 267–275.

Ito, S., Maehara, T., Maruno, E., Tanaka, S., Kameya-Iwaki, M. and Kishi, F. 1999. Development of a PCR-based assay for the detection of *Plasmodiophora brassicae* in soil. *Journal of Phytopathology, 147:* 83–88.

Ito, S., Ushijima, Y., Fujii, T., Tanaka, M., Kameya-Iwaki, S., Yoshiwara, S. and Kishi, F. 1998. Detection of viable cells of *Ralstonia solanacearum* in soil using a semis-elective medium and a PCR technique. *Journal of Phytopathology, 146:* 379–384.

Ito, T., Kanematsu, S., Koganezawa, H., Tsuchizaki, T., and Yoshida, K. 1993. Detection of viroid-associated with apple fruit crinkle disease. *Annals of Phytopathological Society, Japan, 59:* 520–527.

Iwanowski, D. 1903. Uber die Mosaikkrankheit der Tabakspflanze Z. *Pflanzenkhrankh, 13:* 1–41.

Iwasaki, M., Nakano, M., and Shinkai, A. 1985. Detection of rice grassy stunt virus in plan-thopper vectors and rice plants by ELISA. *Annals of Phytopathological Society, Japan, 51:* 450–458.

Jain, R.K., Pappu, S.S., Pappu, H.R., Culbreath, A.K. and Todd, J.W. 1998. Molecular di-agnosis of tomato spotted wilt tospovirus infection of peanut and other field and greenhouse crops. *Plant Disease, 82:* 900–904.

Jamaux, I.D. and Spire, D. 1999. Comparison of responses of ascospores and mycelium by ELISA with antimycelium and anti-ascospore antisera for the development of a method to detect *Sclerotinia sclerotiorum* on petals of oilseed rape. *Annals of Ap-plied Biology, 134:* 171–179.

Jamaux, I. and Spire, D. 1994. Development of a polyclonal antibody-based immunoassay for the early detection of *Sclerotinia sclerotiorum* in rapeseed petals. *Plant Pathol-ogy, 43:* 847–852.

James, D., Godkin, S.E., Rickson, F.R., Thompson, D.A., Eastwell, K.C. and Hansen, A.J. 1999. Electron microscopic detection of novel, coiled, virus-like particles associated with graft inoculation of *Prunus* spp. *Plant Disease, 83:* 949–953.

James, D. and Howell, W.E. 1998. Identification of a flexuous virus associated with peach mosaic disease. *Acta Horticulturae, No. 472,* 285–290.

James, D., Jelkmann, W. and Upton, C. 1999. Specific detection of cherry mottle leaf virus using digoxigenin-labeled cDNA probes and RT-PCR. *Plant Disease, 83,* 235–239.

James, D. and Mukerji, S. 1996. Comparison of ELISA and immunoblotting techniques for the detection of cherry mottle leaf virus. *Annals of Applied Biology, 129:* 13–23.

Jansing, H. and Rudolph, K. 1990. A sensitive and quick test for determination of bean seed infestation *Pseudomonas syringae* pv. *phaseolicola. Zeitschrift fur Pflanzenkrankheiten und Pflanzenschutz, 97:* 42–55.

Jarausch, W., Lansac, M., Saillard, C., Broquaire, J.M. and Dosba, F. 1998. PCR assay for specific detection of European stone fruit yellows phytoplasmas and its use for epi-demiology. *European Journal of Plant Pathology, 104:* 17–27.

Jarvekulg, L., Sober, J., Sinijarv, R., Toots, I., and Saarma, M. 1989. Time-resolved fluo-roimmuno-assay of potato virus M with monoclonal antibodies. *Annals of Applied Biology, 114:* 279.

Jayasinghe, C.K. and Fernando, T.H.P.S. 1998. Growth at different temperatures and on fungicide-amended media: two characteristics to distinguish *Colletotrichum* species pathogenic to rubber. *Mycopathologia, 143:* 93–95.

Jelkmann, W. and Keim-Konard, R. 1997. Immunocapture-polymerase chain reaction and plate-trapped ELISA for the detection of apple stem pitting virus. *Journal of Phy-topathology, 145:* 499–503.

Jelkmann, W., Kein-Konrad, R., Vitushkina, M. and Fechtner, B. 1998. Complete nucleotide sequence of little cherry closterovirus and virus detection by polymerase chain reaction. *Acta Horticulturae, No. 472,* 315–319.

Jenner, C.E., Keane, G.J., Jones, J.E. and Walsh, J.A. 1999. Serotypic variations in turnip mosaic virus. *Plant Pathology, 48:* 101–108.

Jennings, J.M., Newton, A.C. and Buck, K.W. 1997. Detection of polymorphism in *Puccinia hordei* using RFLP and RAPD markers, differential cultivars and analysis of the intergenic spacer region of rDNA. *Journal of Phytopathology, 145:* 511–519.

Jensen, D.D. and Gold, A.H. 1955. Hosts, transmission and electron microscopy of *Cymbidium* mosaic virus with special reference to cattleyea leaf necrosis. *Phytopathology 45:* 327–334.

Jiang, Y.P., Lei, S.D., and Chen, T.A. 1988. Purification of aster yellows agent from diseased lettuce using affinity chromoatography. *Phytopathology, 78:* 828–831.

Jianping, C. and Adams, M.J. 1991. Serological relationships between five fungally transmitted cereal viruses and other elongated viruses. *Plant Pathology, 40:* 226–231.

Jianping, C., Swaby, A.G., Adams, M.J., and Yili, R. 1991. Barley mild mosaic virus inside its fungal vector, *Polymyxa graminis. Annals of Applied Biology, 118:* 615–621.

Johansen, D.A. 1940. *Plant Microtechnique.* McGraw-Hill Book Company, Inc., New York.

Johanson, A., Crowhurst, R.N., Rikkerink, E.H.A., Fullerton, R.A., and Templeton, M.D. 1994. The use of species specific DNA probes for the identification of *Mycosphaerella fijiensis* and *M. musicola,* the causal agents of Sigatoka disease of banana. *Plant Pathology, 43:* 701–707.

Johanson, A. and Jeger, M.S. 1993. Use of PCR for detection of *Mycosphaerella fijiensis* and *M. musicola,* the causal agents of sigatoka leafspots in banana and plantain. *Mycological Research, 67:* 670–674.

John, V.T. 1963. Physiology of virus-infected plants. *Bulletin of National Institute of Science, India, 24:* 103–114.

Johnson, I. and Fulton, R.W. 1942. The broad bean ringspot virus. *Phytopathology, 32:* 605–612.

Joisson, C., Dubs, M.C., and Van Regenmortel, M.H.V. 1992. Cross-reactive potential of monoclonal antibodies raised against proteolysed tobacco etch virus. *Research in Virology, 143:* 155–166.

Joisson, C. and van Regenmortel, M.H.V. 1991. Influence of the C-terminus of the small protein subunit of bean pod mottle virus on the antigenicity of the virus determined using monoclonal antibodies and antipeptide antiserum. *Journal of General Virology, 72:* 2225.

Jones, J.B. and Vuurde, J.W.L. van. 1996. Immunomagnetic isolation of *Xanthomonas campestris* pv. *pelargonii. Journal of Applied Bacteriology, 81:* 78–82.

Jones, J.D., Buck, K.W., and Plumb, R.T. 1991. The detection of beet western yellows virus and beet mild yellowing virus in crop plants using the polymerase chain reaction. *Journal of Virological Methods, 35:* 287–296.

Jones, R., Barnes, L., Gonzalez, C., Leach, J., Alvarez, A., and Benedict, A. 1989. Identification of low virulence strains of *Xanthomonas campestris* pv. *oryzae* from rice in the United States. *Phytopathology, 79:* 984–990.

Jones, R.A.C. and Harrison, B.D. 1969. The behaviour of potato moptop virus in soil and evidence for its transmission by *Spongospora subreranea* (Wallr). *Lagerh. Annals of Applied Biology, 63:* 1–17.

Jordan, R. 1984. Evaluating the relative specific activities of monoclonal antibodies to plant viruses. In: *Hybridoma Technology in Agricultural and Veterinary Research* (Eds.) N. Stein and H.R. Gamble, p. 259. Rowan and Allanheld, Totowa, N.J.

Jordan, R. 1986. Diagnosis of plant viruses using double stranded RNA. In: *Tissue Culture as a Plant Production System for Horticultural Crops.* (Eds.) R.H. Zimmerman, R.T. Griesbach, F.A. Hammerschlag, and R.H. Lawson, pp. 125–134, Martinus Nijhoff, Dordrecht.

Jordan, R. 1989. Successful production of monoclonal antibodies to three carnation viruses using an admixture of only partially purified virus preparations as immunogen. *Phytopathology, 79:* 1213.

Judelson, H.S. and Messenger - Routh, B. 1996. Quantitation of *Phytophthora cinnamomi* in avocado roots using a species specific DNA probe. *Phytopathology, 86:* 763–768.

Juretic, N., Millicic, D., and Schmelzer, K. 1970. Zur Kennmis des Ringmosaikvirus der Kapuzinerkresse (Nasturtium Ringspot Virus) und seiner Zell-Einschlusskorper. *Acta Botanica, Croatia, 29:* 17–26.

Kado, C.I. 1971. *Methods in Plant Bacteriology—Laboratory Manual.* Department of Plant Pathology, University of California, Davis.

Kageyama, K., Ohyama, A. and Hyakumachi, M. 1997. Detection of *Pythium ultimum* using polymerase chain reaction with species-specific primers. *Plant Disease, 81:* 1155–1160.

Kaiser, W.J., Wyatt, S.D., and Pesbo, G.R. 1982. Natural hosts and vectors of tobacco streak virus in eastern Washington. *Phytopathology, 72:* 1508–1512.

Kalamkoff, J., and Tremaine, J.H. 1969. Some physical and chemical properties of carnation ringspot virus. *Virology, 33:* 10–16.

Kalmar, G.B. and Eastwell, K.C. 1989. Reaction of coat proteins of two comoviruses in different aggregation states with monoclonal antibodies. *Journal of General Virology, 70:* 3451.

Kanamori, H., Sugimoto, H., Ochiar, K., Kaku, H. and Tsuyumu, S. 1999. Isolation of *hrp* cluster from *Xanthomonas campestris* pv. *citri* and its application for RFLP analyses of xanthomonads. *Annals of Phytopathological Society, Japan, 65:* 110–115.

Karešová, R. and Paprštein, P. 1998. Occurrence of plum pseudopox (apple chlorotic leafspot virus) in plum germplasm. *Acta Horticulturae, No. 478,* 283–286.

Karjalainen, R., Kangasniemi, A., Hamalainen, J. and Tegel, J. 1995. Evaluation of PCR methods for the diagnosis of bacterial ringrot infections in potato. *Bulletin, OEPP, 25:* 169–175.

Karpovich-Tate, N., Spanu, P. and Dewey, F.M. 1998. Use of monoclonal antibodies to determine biomass of *Cladosporium fulvum* in infected tomato leaves. *Molecular Plant-Microbe Interactions, 11:* 710–716.

Kassanis, B. 1939. Intranuclear inclusions in virus-infected plants. *Annals of Applied Biology, 26:* 705–709.

Kassanis, B. 1955. Some properties of four viruses isolated from carnation plants. *Annals of Applied Biology, 43:* 103.

Kassanis, B. 1961. Potato para crinkle virus. *European Potato Journal, 4:* 13–24.

Kassanis, B. 1964. Properties of tobacco necrosis virus and its association with satelite virus. *Ann. Inst. Phytopath. Benati, 6:* 7–26.

Kassanis, B. 1970. Tobacco necrosis virus. *CMI/AAB descriptions of Plant Viruses, No. 14.*

Kastirr, R. 1990. Problems in the detection of plant pathogenic viruses in aphids by means of ELISA and interpretation of the results. *Nachrichtenblatt Pflanzenschutz, 44:* 201–204.

Kaufmann, P.J. and Weidemann, G.J. 1996. Isozyme analysis of *Collectorichum gloeosporioides* from five host genera. *Plant Disease, 80:* 1289–1293.

Kawano, T. and Takahashi, Y. 1997. Simplified detection of plant viruses using high density latex. *Annals of Phytopathological Society, Japan, 63:* 403–405.

Kawaradani, M., Kusakari, S., Kimura, M., Takizawa, H. and Nishihashi, H. 1998. A new method for measuing β-1,3-glucanase activity using *p*-nitrophenyl β-D-laminaritetraoside as a substrate to diagnose *Verticillium* wilt of eggplant. *Annals of Phytopathological Society, Japan, 64:* 489–493.

Kawaradani, M., Kusakari, S. Morita, S. and Tanaka, Y. 1994. The enzyme activities in eggplant infected with soil-borne diseases and application to diagnosis for diseases. *Annals of Phytopathological Society, Japan, 60:* 507–513.

Kawchuk, L.M., Lynch, D.A., Leggett, F.L., Howard, R.J. and McDonald, J.G. 1997. Detection and characterization of a Canadian tobacco rattle virus isolate using a PCR-based assay. *Canadian Journal of Plant Pathology, 19:* 101–105.

Kay, D.H. 1965. *Techniques for Electron Microscopy,* Second Edition, Blackwell, Oxford, U.K.

Kazinczi, G. and Horváth, J. 1998. Transmission of sowbane mosaic sobemovirus by seeds of *Chenopodium* species and viability of seeds. *Acta Phytopathologica et Entomologica, Hungarica, 33:* 21–26.

Kegler, H., Richter, J., and Schmidt, H.B. 1966. Untersuchungen zur Identifizierung und Differenzierung des Blattrolvirus der Kirsch (cherry leaf-roll virus). *Phytopathologische Zeitschrift, 56:* 313–330.

Keifer, M.C., Owens, R.A. and Diener, T.O. 1983. Structural similarities between viroids and transposable genetic elements. *Proceedings of National Academy of Sciences, USA, 80:* 6234–6238.

Kaller, J.R. 1953. Investigations on chrysanthemum stunt virus and chrysanthemum virus Q. *Cornell University Agricultural Experiment Station Memoir,* 324.

Keller, K.E., Johansen, E., Martin, R.R. and Hampton, R.O. 1998. Potyvirus genome–linked protein (VPg) determines pea seed-borne mosaic virus pathotype–specific virulence in *Pisum sativum. Molecular Plant-Microbe Interactions, 11:* 124–130.

Kelley, A., Alcala-Jimenez, A.R., Bainbridge, B.W., Heale, J.B., Perez-Artes, E., and Jimenez-Diaz, R.M. 1994. Use of genetic fingerprinting and random amplified polymorphic DNA to characterize pathotypes of *Fusarium oxysporum* f. sp. *ciceris. Phytopathology, 84:* 1293–1298.

Kelley, A.G., Bainbridge, B.W., Heale, J.B., Peréz Artes, E. and Jiménez Diaz, R.M. 1998. In planta polymerase chain reaction detection of the wilt-inducing pathotype of *Fusarium oxysporum* f.sp. *ciceris* in chickpea (*Cicer arietinium* L.). *Physiological and Molecular Plant Pathology, 52:* 397–409.

Kelman, A. 1954. The relationship of pathogenicity of *Pseudomonas solanacearum* to colony appearance in a tetrazolium medium. *Phytopathology, 44:* 693–695.

Kendall, J.J., Hollomon, D.W. and Selley, A. 1998. Immunodiagnosis as an aid to timing of fungicide sprays for the control of *Mycosphaerella graminicola* on winter wheat in the U.K. *Brighton Crop Protection Conference: Pests and Diseases, 2:* 701–706.

Kennedy, R., Wakeham, A.J. and Cullington, J.E. 1999. Production and immunodetection of ascospores of *Mycosphaerella brassicola:* ringspot of vegetable crucifers. *Plant Pathology, 48:* 297–307.

Kenten, R.H. and Legg, J.T. 1967. Some properties of cocoa mottle leaf virus. *Journal of General Virology, 1:* 465–470.

Khadhair, A.H., Kawchuk, L.M., Taillon, R.C. and Botar, G. 1998. Detection and molecular characterization of an aster yellows phytoplasma in parsley. *Canadian Journal of Plant Pathology, 20:* 55–61.

Khan, A.A., Furuya, N., Matsumoto, M. and Matsuyama, N. 1999. Trial for rapid identification of pathogens from blasted pear blossoms and rotted radish leaves by direct colony TLC and whole cellular fatty acid analysis. *Journal of Faculty of Agriculture, Kyushu University, 43:* 327–335.

Khan, A.A. and Matsuyama, N. 1998. Rapid identification of phytopathogenic bacteria by TLC. *Journal of Faculty of Agriculture, Kyushu University, 42:* 281–287.

Khan, I.A., Turpin, F.T., Lister, R.M., Scott, D.H. and Whitford, F. 1997. Using enzyme-linked immunosorbent assay (ELISA) for the detection of *Erwinia stewartii* (Smith) Dye in dent corn plants and corn flea beetles (*Chaetocnema pulicaria* Melsheimer). *Sarhad Journal of Agriculture, 13:* 391–397.

Khan, M.Q., Maxwell, D.P. and Maxwell, M.D. 1977. Cytochemistry and ultrastructure of red clover vein mosaic virus induced inclusions, *Virology, 78:* 173.

Khanzada, A.K. and Mathur, S.A. 1988. Influence of extraction rate and concentration of stain on loose smut infection detection in wheat seed. *Pakistan Journal of Agricultural Research, 9:* 218–222.

Kietreiber, M. 1962. Der *Septoria*-Befall von Weizenkornern (Zur Mthodik der Erkenung). *Proceedings of the International Seed Testing Association, 27:* 843–855.

Kietreiber, M. 1980. *Seproia*-Befall der Saatgerste festgestellt mit dem filter-papier-Fluoreszenz test. *Jahrbuch 1979 der Bundensanstalt für Polanzenbau Samen-prufung in Wien,* 103–109.

Kikkert, M., Meurs, C., van de Wetering, F., Dorfmüller, S., Peters, D., Kormelink, R. and Goldbach, R. 1998. Binding of tomato spotted wilt virus to a 94 kDa thrips protein. *Phytopathology, 83:* 63–69.

Kim, D.H., Martyn, R.D., and Magill, C.W. 1993. Mitochondrial DNA (mtDNA)-relatedness among formae speciales of *Fusarium oxysporum* in the Cucurbitaceae, *Phytopathology, 83:* 91–97.

Kim, K.S., and Lee, K.W. 1992. Gerninivirus-induced macrotubules and their suggested roll in cell-to-cell movement. *Phytopathology, 82:* 664–669.

Kimura, I., and Black, L.M. 1972. The cell infecting unit of wound tumor virus. *Virology, 49:* 549–561.

Kinard, G.R., Scott, S.W. and Barnett, O.W. 1996. Detection of apple chlorotic leafspot and apple stem grooving viruses using RT-PCR. *Plant Disease, 80:* 616–621.

Kiranmai, G., Satyanarayana, T. and Sreenivasulu, P. 1998. Molecular cloning and detection of cucumber mosaic cucumovirus causing infectious chlorosis disease of banana using DNA probes. *Current Science, 74:* 356–359.

Kirkpatrick, B.C. 1989. Strategies for characterizing plant pathogenic mycoplasma-like organisms and their effects on plants. In: *Plant Microbe Interactions*, (Eds.) T. Kosuge and E.W. Nester, p. 241. McGraw Hill, New York.

Kirkpatrick, B.C., Stenger, D.C., Morris, T.J., and Purcell, H. 1987. Cloning and detection of DNA from a nonculturable plant pathogenic mycoplasma-like organism. *Science, 238:* 197–200.

Kistler, H.C., Bosland, P.W., Benny, U., Leong, S., and Williams, P.H. 1987. Relatedness of strains of *Fusarium oxysporum* from crucifers measured by examination of mitochondrial and ribosomal DNA. *Phytopathology, 77:* 1289–1293.

Kitagawa, T., Sakamoto, Y., Furumi, K., and Ogura, H. 1989. Novel enzyme immunoassays for specific detection of *Fusarium oxysporum* f. sp. *cucumerianum* and for general detection of various *Fusarium* species. *Phytopathology, 79:* 162–165.

Kitajima, E.W., Camargo, I.J.B., and Costa, A.S. 1968. Morfologia do virus do mosaico da cenoura. *Bragantia, 27:* 13–14.

Klassen, G.R., Balcerzak, M. and deCock, A.W.A.M. 1996. 5S ribosomal RNA gene spacers as species-specific probes for eight species of *Pythium*. *Phytopathology, 86:* 581–587.

Klausner, A. 1987. Immunoassays flourish in new markets. *Biotechnology, 5:* 551–556.

Klein, R.E., Wyatt, S.D., Keiser, W.J., and Mink, G.I. 1992. Comparative immunoassays of bean common mosaic virus in individual bean (*Phaseolus vulgaris*) seed and bulked bean seed samples. *Plant Disease, 76:* 57–59.

Klich, M.A., and Mullaney, E. 1987. DNA restriction enzyme fragment polymorphism as a tool for rapid differentiation of *Aspergillus flavus* from *Aspergillus oryzae*. *Experimental Mycology, 11:* 170–175.

Klich, M.A., Mullaney E.J., and Daly, C.B. 1993. Analysis of interspecific variability of three common species of *Aspergillus* section versicolores using DNA restriction fragment length polymorphisms. *Mycologia, 85:* 852–855.

Klinskowski, M. and Schmelzer, K. 1960. A necrotic type of potato virus Y. *American Potato Journal, 37:* 221–228.

Klug, A. and Finch, J.T. 1966. The symmetries of the protein and nucleic acid in turnip yellow mosaic virus: X-ray diffraction studies. *Journal of Molecular Biology, 2:* 201–215.

Ko, H.J., Hwang, B.K., and Hwang, B.G. 1993. Restriction fragment length polymorphisms of mitochondrial and nuclear DNAs among Korean races of *Magnaporthe grisea*. *Journal of Phytopathology, 138:* 41–54.

Ko, H.C. and Lin, C.P. 1994. Development and application of cloned DNA probes for a mycoplasma-like organism associated with sweet potato witches' broom. *Phytopathology, 84:* 468–473.

Koenig, R., Lüddecke, P., and Haeberle, A.M. 1995. Detection of beet necrotic yellow vein virus strains, variants and mixed infections by examining single-strand conformation polymorphisms of immunocapture RT-PCR products. *Journal of General Virology, 76:* 2051–2055.

Kohnen, P.D., Dougherty, W.G., and Hampton, R.O. 1992. Detection of pea seed-borne mosaic potyvirus by sequence specific enzymatic amplification. *Journal of Virological Methods, 37:* 253–258.

Kohnen, P.D., Johansen, I.E., and Hampton, R.O. 1995. Characterization and molecular detection of the P4 pathotype of pea seed borne mosaic potyvirus. *Phytopathology, 85:* 789–793.

Koike, M., Fujita, M., Nagao, H. and Ohshima, S. 1996. Random amplified polymorphic DNA analysis of Japanese isolates of *Verticillium dahliae* and *V. albo-atrum. Plant Diseases, 80:* 1224–1227.

Koike, S.T., Barak, J.D., Handerson, D.M., and Gilbertson, R.L. 1999. Bacterial blight of leek: a new disease in California caused by *Pseudomonas syringae. Plant Disease, 83:* 165–170.

Kokko, H.I., Kivineva, M. and Karenlampi, S.O. 1996. Single-step immunocapture RT-PCR in the detection of raspberry bushy dwarf virus. *Biotechniques, No. 18,* 47–50.

Köklü, G. 1999. Production of antisera to grapevine leafroll associated virus-2 and evaluation of the serological diagnosis of infected plants. *Journal of Turkish Phytopathology, 28:* 119–131.

Kölber, M., Németh, M., Krizbal, L., Szemes, M., Kiss-Toth, E., Dorgal, L. and Kálmán, M. 1998. Detectability of prunus necrotic ringspot and plum pox viruses by RT-PCR, multiplex RT-PCR, ELISA and indexing on woody indicators. *Acta Horticulturae, No. 472:* 243–247.

Kolmer, J.A., Lim, J.Q., and Sies, M. 1995. Virulence and molecular polymorphisms in *Puccinia recondita* f.sp. *tritici* in Canada. *Phytopathology, 85:* 276–285.

Konate, G. and Barro, N. 1993. Dissemination and detection of peanut clump virus in groundnut seed. *Annals of Applied Biology, 123:* 623–629.

Konate, G., Barro, N., Fargette, D., Swanson, M.M., and Harrison, B.D. 1995. Occurrence of whitefly-transmitted geminiviruses in crops in Burkino Fasa and their serological detection and differentiation. *Annals of Applied Biology, 126:* 121–129.

Konate, G., Traore, O. and Coulibaly, M.M. 1997. Characterization of rice yellow mottle virus isolates in Sudano-Sahelian areas. *Archives of Virology, 142:* 1117–1124.

Korschineck, I., Himmler, G., Sagl, R., Stein Kellner, H., and Katinger, H.W.D. 1991. A PCR membrane spot assay for the detection of plum pox virus RNA in bark of infected trees. *Journal of Virological Methods, 31:* 139–146.

Koshimizu, Y. and Iizuka, N. 1957. Origins and formation of intracellular inclusions associated with two leguminous virus diseases. *Protoplasma, 48:* 113–133.

Kovocs, N. 1956. Identification of *Pseudomonas pyocyanese* by the oxidase reaction. *Nature (Lond.), 178:* 703.

Kraft, J.M. and Boge, W.L. 1994. Development of an antiserum to quantify *Aphanomyces euteiches* in resistant pea lines. *Plant Disease, 78:* 179–183.

Kraft, J.M. and Boge, W.L. 1996. Identification of characteristics associated with resistance to root rot caused by *Aphanomyces euteiches* in pea. *Plant Disease, 80:* 1383–1386.

Kralikova, K.P. and Kegler, 1967. Investigations on the properties of a virus isolate from prune dwarf-diseased plum trees. *Biologia (Bratislava), 22:* 673–678.

Krieg, N.R., and Holt, J.G. (Eds.) 1984. *Bergey's Manual of Systematic Bacteriology,* Vol. I. Williams & Wilkins, Baltimore.

Kristek, J. and Polak, J. 1990. Diagnosis of strawberry latent ringspot virus by an immunoenzymatic test (ELISA). *Sbornik UYTIZ, Zahradnickvi, 17:* 299–303.

Kronstad, J.W. 1997. Virulence and cAMP in smuts, blasts and blights. *Trends in Plant Science, 2:* 193–199.

Kruse, M., Koenig, R., Hoffmann, A., Kaufmann, A., Commandeur, U., Solovyev, A.G., Saverekov, I., and Burgemeister, W. 1994. Restriction fragment length polymorphism analysis of reverse transcription PCR products reveals the existence of two major strain groups of beet necrotic yellow vein virus. *Journal of General Virology, 75:* 1835–1842.

Kryczynski, S., Paduch-Cichal, E., and Skrzeczkowski, 1988. Transmission of three viroids through seed and pollen of tomato plants. *Journal of Phytopathology, 121:* 51–57.

Kuflu, K.M. and Cuppels, D.A. 1997. Development of a diagnostic cDNA probe for xanthomonads causing bacterial leaf spot of peppers and tomatoes. *Applied and Environmental Microbiology, 63:* 4462–4470.

Kuhn, C.W. 1964. Purification, serology and properties of a new cowpea virus. *Phytopathology, 54:* 853–857.

Kuhn, C.W. 1965. Symptomatology, host range, and effects on yield of a seed transmitted peanut virus. *Phytopathology, 55:* 880–884.

Kulske, C.R. and Kirkpatrick, B.C. 1992. Distribution and multiplication of western aster yellows mycoplasma-like organism in *Catharanthus roseus* as determined by DNA hybridization analysis. *Phytopathology, 82:* 457–462.

Kulske, C.R., Kirkpatrick, B.C., Davis, M.S., and Seemuller, E. 1991a. DNA hybridization between westen aster yellows mycoplasma-like organism plasmids and extra-chromosomal DNA from other plant pathogenic mycoplasma-like organisms. *Molecular Plant-Microbe Interaction, 4:* 75–80.

Kulske, C.R., Kirkpatrick, B.C., and Seemuller, E. 1991b. Differentiation of virescence MLOs using western aster yellows mycoplasma-like organism chromosomal DNA probes and restriction fragment length polymorphism analysis. *Journal of General Microbiology, 137:* 153.

Kummert, J., Marinho, V.L.A., Rufflard, G., Colinet, D. and Lepoivre, P. 1998. Sensitive detection of apple stem grooving and apple stem pitting viruses from infected apple trees by RT-PCR. *Acta Horticulturae, No. 472:* 97–104.

Kuniyuki, H., Betti, J.A. and Yuki, V.A. 1998. Detection of the virus and virus-like diseases of grapevine through leaf grafting. *Summa Phytopathologica, 24:* 105–107.

Kunkel, L.O. 1934. Studies on acquired immunity with tobacco and aucuba mosaics. *Phytopathology, 24:* 437–466.

Kuo Kerchung, 1999. In corporation of cotton blue and Calcofluor White M2R into conventional paraffin section. *Plant Protection Bulletin, (Taipei), 41:* 79–82.

Kuo, T.T., Huang, T.C., Wu, R.Y., and Yang, C.M. 1967. Characterization of three bacteriophages of *Xanthomonas oryzae* (Uyeda et Ishiyama) Dowson. *Botanical Bulletin of Academia Sinica, 8:* 246–254.

Kutcher, H.R., and Mortensen, K. 1999. Genotypic and pathogenic variation of *Colletotrichum gloeosporides* f.sp. *malvae. Canadian Journal of Plant Pathology, 21:* 37–41.

Lacourt, I. and Duncan, J.M. 1997. Specific detection of *Phytophthora nicotianae* using polymerase chain reaction and primers based on the cDNA sequence of its elicitin gene *ParA1. European Journal of Plant Pathology, 103:* 73–83.

Laemonlen, F.F. 1969. The association of the mite *Siteroptes reniformis* and *Nigrospora oryzae* in Nigrospora lint rot of cotton. *Phytopathology, 59:* 1036.

Lakshman, D.K., Hiruki, C., Wu, X.N., and Leung, W.C. 1986. Use of the [^{32}P] RNA probes for the dot hybridization detection of potato spindle tuber viroid. *Journal of Virological Methods, 14:* 309–319.

Lange, L., and Heide, M. 1986. Dot immunobinding (DIB) for detection of virus in seed. *Canadian Journal of Plant Pathology, 8:* 373–379.

Lange, L., Jomanlor, A., and Heide, M. 1991. Testing seeds for viruses by dot immunobinding (DIB) directly on plain paper. *Tidiskrift for Planteaul; 93:* 93–96.

Langenberg, W.G. 1974. Leafdip serology for the determination of strain relationship of elongated plant viruses. *Phytopathology, 64:* 128–131.

Langeveld, S.A., Derks, A.F.L.M., Konicheva, V., Munoz, D., Zhinnan, C., Denkova, S.T., Lemmers, M.E.C. and Boonekamp, P.M. 1997. Molecular identification of potyviruses in Dutch stocks of *Narcissus. Acta Horticulturae II. No. 430,* 641–648.

Langeveld, S.A., Dore, J.M., Memelink, J., Derks, A.F.L.M., Vulgt, C.I.M., Van der, Asjes, C.J., and Bol, J.F. 1991. Identification of potyviruses using the polymerase chain reaction with degenerate primers. *Journal of General Virology, 72:* 1531–1541.

Lapidot, M., Friedmann, M., Lachman, O., Yehezkel, A., Nahon, S., Cohen, S. and Pilowsky, M. 1997. Comparison of resistance level to tomato yellow leafcurl virus among commercial cultivars and breeding lines. *Plant Disease, 81:* 1425–1428.

Larson, R.A., 1944. The identity of the virus causing punctate necrosis and mottle in potatoes. *Phytopathology, 34:* 1006.

Laugé, R., Joosten, M.H.A.J., Haanstra, J.P.W., Goodwin, P.H., Lindhout, P. and Wit, P.J.G.M. de. 1998. Successful search for a resistance gene in tomato targeted against a virulence factor of a fungal pathogen. *Proceedings of National Academy of Sciences, U.S.A. 95:* 9014–9018.

Leach, J.E. and White, F.E. 1991. Molecular probes for disease diagnosis and monitoring. In: *Rice Biotechnology* (Eds.) G.S. Khush and G.H. Toenniessen, pp. 281–307. C.A.B. International, U.K., and International Rice Research Institute, Philippines.

Leach, J.E., White, F., Rhoads, M., and Leung, H. 1990. A repetitive DNA sequence differentiates *Xanthomonas campestris* pv. *oryzae* from other pathovars of *Xanthomonas campestris. Molecular Plant-Microbe Interactions, 3:* 238–246.

Leach, J.G. 1993. The method of survival of bacteria in the puparia of the seed corn maggot *(Hyelmyia cilicrura). Z. Angew. Entomol. 20:* 150–161.

Lee, I.M., Bartoszy, K., Gundersen, D.E. Mogen, B. and Davis, R.E. 1997. Nested PCR for ultrasensitive detection of the potato ring rot bacterium, *Clavibacter michiganensis* subsp. *sepedonicus. Applied and Environmental Microbiology, 63:* 2625–2630.

Lee, C.L. and Black, L.M. 1955. Anatomical studies of *Trifolium incarnatum* infected by wound tumour virus. *American Journal of Botany, 42:* 160–168.

Lee, I.M. and Davis, R.E. 1983. Phloem-limited prokaryotes in sieve elements isolated by enzyme treatment of diseased plant tissues. *Phytopathology, 73:* 1540–1543.

Lee, I.M. and Davis, R.E. 1988. Detection and investigation of genetic relatedness among aster yellows and other mycoplasma-like organisms by using cloned DNA and RNA probes. *Molecular Plant-Microbe Interactions, 1:* 303–310.

Lee, I.M., Davis, R.E., Chen, T.A., Chlykowkski, L.N., and Fletcher, J. 1992a. A genotypes based system for identification and classification of mycoplasma-like organisms (MLOs) in the aster yellows MLO cluster. *Phytopathology, 82:* 977–986.

Lee, I.M., Gundersen, D.E., Hammond, R.W., and Davis, R.E. 1994. Use of mycoplasma-like organism (MLO) group specific oligonucleotide primers for nested-PCR assays to detect mixed MLO infections in a single host plant. *Phytopathology, 84:* 559–566.

Lee, I.M., Hammond, R.W., Davis, R.E., and Gunderson, D.E. 1992b. Phylogenetic relationships among plant and pathogenic mycoplasma-like organisms (MLOs) based on 165 rRNA sequence analysis. *Phytopathology, 82:* 1094.

Lee, I.M., Hammond, R.W., Davis, R.E., and Gunderson, D.E. 1993. Universal amplification and analysis of pathogen 16S rDNA for classification and identification of mycoplasma-like organisms. *Phytopathology, 83:* 834–842.

Lee, JimSang, Nou HeeSun, Hong DaeKi, Leekyeong Koog, Lee Yoon Su and Choi JangKyung, 1998. Detection of viral disease in *Lilium oriental* plants using RT-PCR techniques. *RDA Journal of Crop Protection, 40:* 50–56.

Lee, KuiJae, 1998. Analysis and detection of coat protein gene of barley yellow mosaic virus and barley mild mosaic virus by RT-PCR. *Korean Journal of Plant Pathology, 14:* 314–318.

Lee, P.E. 1967. Morphology of wheat striate mosaic virus and its localization in infected cells. *Virology, 33:* 84–94.

Lee, P.E. 1968. Partial purification of wheat striate mosaic virus and fine structural studies of the virus. *Virology, 34:* 583–589.

Lee, R.P., Garnsey, S.M., Briansky, R.H. and Calvert, L.A. 1982. Purification of inclusion bodies of citrus tristeza virus. *Phytopathology, 72:* 953.

Lee, S.B., White, T.J., and Taylor, J.W. 1993. Detection of *Phytophthora* species by oligonucleotide hybridization to amplified DNA spacers. *Phytopathology, 83:* 177–181.

Lejour, C. and Kummert, J. 1986. Comparative study of the detection of double stranded RNA in different viral systems. *Annales de Gambloux, 92:* 177–188.

Lemmetty, A. and Lehto, K. 1999. Successful back-inoculation confirms the role of black currant reversion associated virus as the causal agent of reversion disease. *European Journal of Plant Pathology, 105:* 297–301.

Lemmetty, A., Susi, P., Latvala, S. and Lehto, K. 1998. Detection of the putative causal agent of black currant reversion disease. *Acta Horticulturae, No. 471,* 93–98.

Lenardon, S.L., Gordan, D.T., and Gregory, R.E. 1993. Serological differentiation of maize dwarf mosaic potyvirus strains A, D, E, and F by electroblot immunoassay. *Phytopathology, 83:* 86–91.

Leone, G., Schijndel, H.B. van, Gemen, B. van and Schoen, C.D. 1997. Direct detection of potato leafroll virus in potato tubers by immunocapture and the isothermal nucleic acid amplification method - NASBA. *Journal of Virological Methods, 66:* 19–27.

Leone, G. and Schoen, C.D. 1997. Novel methods based on genomic amplification for routine detection of potato leafroll virus in potato tubers. In: *Diagnosis and Identification of Plant Pathogens* (Eds.) H.W. Dehne, G. Adam, M. Diekmann, J. Frahm, A.

Mauler-Machnik and P. van Halteren, pp. 161–167. Kluwer Academic Publishers, Netherlands.

Le Pelley, R.H. 1942. The food and feeding habits of Anstistia in Kenya. *Bulletin of Entomological Research, 33:* 71–89.

Lesemann, D.E. 1977. Virus group specific and virus specific cytological alterations induced by members of the tymovirus group. *Phytopathologische Zeitschrift, 90:* 315–336.

Lesemann, D.E. and Vetten, J. 1984 as quoted by Milne and Lesemann, 1984.

Leu, H.H., Leu, L.S. and Lin, C.P. 1998. Development and application of monoclonal antibodies against *Xylella fastidiosa*, the causal bacterium of pear leaf scorch. *Journal of Phytopathology, 146:* 31–37.

Lévesque, C.A., Harlton, C.E. and de Cock, A.W.A.M. 1998. Identification of some oomycetes by reverse dot-blot hybridization. *Phytopathology, 88:* 213–222.

Levesque, C.A., Vrain, T.C., and De Boer, S.M. 1994. Development of a species specific probe for *Pythium ultimum* using amplified ribosomal DNA. *Phytopathology, 84:* 474–478.

Levy, L., Hadidi, A., Kolber, M., Tokes, G. and Nemeth, M. 1995. 3′ Non-coding region RT-PCR detection and molecular hybridization of plum pox virus in anthers of infected stone fruit. *Acta Horticulturae, No. 386,* 331–339.

Levy, L., Lee, I.M., and Hadidi, A. 1994. Simple and rapid preparation of infected plant tissue extracts for PCR-amplification of virus, viroid and MLO nucleic acids. *Journal of Virological Methods, 49:* 295–304.

Levy, M., Romao, J., Marchetti, M.A., and Hamer, J.E. 1991. DNA finger printing with a dispersed repeated sequence resolves pathotype diversity in the rice blast fungus. *The Plant Cell, 3:* 95–102.

Lewis, F.A., Griffiths, S., Dunnicliff, R., Wells, M., Dudding, N., and Bird, C.C. 1987. Sensitive in situ hybridization technique using biotin-streptavidin-polyalkaline phospharase complex. *Journal of Clinical Pathology, 40(2):* 163–166.

Lherminier, J., Prensier, G., Boudon-Padieu, E., and Caudwell, A. 1990. Immunolabeling of grapevine flavescence dorée MLO in salivary glands of *Euscelidices variegatus:* a light and electron microscopy study. *Journal of Histochemistry and Cytochemistry, 38:* 79–85.

Li, R.H., Wisler, G.C., Liu, H.Y. and Duffus, J.E. 1998. Comparison of diagnostic techniques for detecting tomato infectious chlorosis virus. *Plant Disease, 82:* 84–88.

Li, S.F., Hataya, T., Furuya, K., Horita, H., Sano, T. and Shikata, E. 1997. Occurrence of chrysanthemum stunt disease in Kokkaido and detection of chrysanthemum stunt viroid by electrophoresis and hybridization. *Annual Report of Society of Plant Protection of North Japan, No. 48,* 113–117.

Li, X., Boer, S.H. de, and Ward, L.J. 1997. Improved microscopic identification of *Clavibacter michiganensis* subsp. *sepedonicus* by combining in situ hybridization with immunofluorescence. *Letters in Applied Microbiology, 24:* 431–434.

Li, X. and De Boer, S.H. 1995. Selection of polymerase chain reaction primers from an RNA intergenic spacer region for specific detection of *Clavibacter michiganensis* subsp. *sepedonicus. Phytopathology, 85:* 837–842.

Li, X., Wong, W.C., and Hayward, A.C. 1993. Production and use of monoclonal antibodies to *Pseudomonas andropogonis. Journal of Phytopathology, 138:* 21–30.

References453

Li, Y., Wei, C., Tien, P., Pan, N. and Chen, Z. 1996. Immunodetection of beet necrotic yellow vein RNA3-encoded protein in different host plants and tissues. *Acta Virologica, 40:* 67–72.

Libby, R.R. and Ellis, D.E. 1954. Transmission of *Colletotrichum lagenarium* by the spotted cucumber beetle. *Plant Disease Reporter, 38:* 200.

Liew, E.C.Y., Maclean, D.J. and Irwin, J.A.G. 1998. Specific PCR based detection of *Phytophthora medicaginis* using the intergenic spacer region of the ribosomal DNA. *Mycological Research, 102:* 73–80.

Lim, S.T., Wong, S.M., Yeong, C.Y., Lee, S.C., and Goh, C.J. 1993. Rapid detection of cymbidium mosaic virus by the polymerase chain reaction (PCR). *Journal of Virological Methods, 41:* 37–46.

Lim, W.L. and Hagedorn, D.J. 1977. Bimodal transmission of plant viruses. In: *Aphids as Virus Vectors* (Eds.) K.F. Harris and K. Maramorosch, pp. 237–251. Academic Press, New York.

Lima, J.E.O.de, Miranda, V.S., Hartung, J.S., Brlansky, R.H., Coutinho, A., Roberto, S.R. and Carlos, E.F. 1998. Coffee leaf scorch bacterium: axenic culture, pathogenicity and comparison with *Xylella fastidiosa* of citrus. *Plant Disease, 82:* 94–97.

Lin, C.P. and Chen, T.A. 1985. Production of monoclonal antibodies against *Spiroplasma citri. Phytopathology, 75:* 845–851.

Lin, C.P. and Chen, T.A. 1986. Comparison of monoclonal antibodies and polyclonal antibodies in detection of aster yellows mycoplasma-like organism. *Phytopathology, 76:* 45–50.

Lin, C.P., Shen, W.C., Ko, H.C., Chang, F.L., Yu, Y.L., Chen, M.F., and Wu, F.Y. 1993. Application of monoclonal antibodies and cloned DNA probes for the detection and differentiation of phytopathogenic mycoplasma-like organisms. *Plant Pathology Bulletin, 2:* 161–168.

Lin, R.F., Yu, J.B., Jin, D.D., and Ye, L. 1991. Studies on methods of detection of two solanaceous vegetable viruses. *Acta Agriculture Zheijiangensis, 3:* 127–132.

Lindberg, M. and Collmer, A. 1992. Analysis of eight *out* genes in a cluster required for pectic enzyme secretion by *Erwinia chrysanthemi:* sequence comparison with secretion genes from other gram negative bacteria. *Journal of Bacteriology, 174:* 7385–7397.

Lindgren, P.B. 1997. The role of *hrp* genes during plant-bacterial interactions. *Annual Review of Phytopathology, 35:* 129–152.

Lindner, R.C. 1961. Chemical tests in the diagnosis of plant virus diseases. *Botanical Review, 27:* 501–521.

Lindner, R.C., Kirkpatrick, H.C., and Weeks, T.E. 1950. A simple staining technique for detecting virus diseases in some woody plants. *Science, 112:* 119–120.

Lindqvist, H., Koponea, H. and Valkonen, J.P.T. 1998. *Peronospora sparsa* on cultivated *Rubus articus* and its detection by PCR-based on ITS sequences. *Plant Disease, 82:* 1304–1311.

Ling, K.C. 1972. *Rice virus diseases.* International Rice Research Institute, Los Banos, Philippines, 141p.

Lipp, R.L., Alvarez, A.M., Benedict, A.A., and Berestecky, J. 1992. Use of monoclonal antibodies and pathogenicity tests to characterize strains of *Xanthomonas campestris* pv. *diffenbachiae* from aroids. *Phytopathology, 82:* 677–682.

List, G.M. and Kreutzer, W.A. 1942. Transmission of the causal agent of the ringrot disease of potatoes by insects. *Journal of Economic Entomology, 35:* 455–465.

Lister, R.M., Bancroft, J.B., and Nadakavukaren, M.J. 1965. Some sap-transmissible viruses from apple. *Phytopathology, 55:* 859.

Lister, R.M., Bancroft, J.B., and Shay, J.R. 1964. Chlorotic leafspot from a mechanically transmissible virus from apple. *Phytopathology, 54:* 1300.

Liu HsingLung, Lin ChienYih and Chen LungChung, 1995. A baiting technique for detection of *Rhizoctonia solani* by angle spinach seed. *Bulletin of Taichung District Agricultural Improvement Station, No. 49,* 1–7.

Liu, S. and Dean, R.A. 1997. G protein α-subunit genes control growth, development and pathogenicity of *Magnaporthe grisea. Molecular Plant-Microbe Interactions, 10:* 1075–1086.

Liu SiJun, Briddon, R.W., Bedford, I.D. Pinner, M.S., and Markham, P.G. 1999. Identification of genes directly involved in insect transmission of African cassava mosaic geminivirus by *Bemisia tabaci. Virus Genes, 18:* 5–11.

Livieratos, I.C., Avgelis, A.D. and Coutts, R.H.A. 1999. Molecular characterization of the cucurbit yellow stunting disorder virus coat protein gene. *Phytopathology, 89:* 1050–1055.

Lizarraya, C. and Fernandez-Northcote, E.M. 1989. Detection of potato viruses K and Y in sap extracts by a modified indirect enzyme-linked immunosorbent assay on nitrocellulose membranes (NCM-ELISA). *Plant Disease, 73:* 11.

Loebenstein, G. 1972. Inhibition, interference and acquired resistance during infection. In *Principles and Techniques in Plant Virology* (Eds.) C.I. Kado and H.O. Agrawal, pp. 33–61. Van Nostrand Reinhold Co., New York.

Loebenstein, G., Akad, F., Filatov, V., Sadvakasova, G., Manadilova, A., Bakelman, H., Teverovsky, E., Lachmann, O. and David, A. 1997. Improved detection of potato leafroll luteovirus in leaves and tubers with a digoxigenin-labeled cRNA probe. *Plant Disease, 81:* 489–491.

Loi, N., Ermacora, P., Chen, T.A., Carraro, L. and Osler, R. 1998. Monoclonal antibodies for the detection of tagetes witches' broom agent. *Journal of Plant Pathology, 80:* 171–174.

Lopes, P.C., Matsumura, A., Porto, M.D.M. Simonetti, A., Sand, S. van der, and Guimares, A. 1998. Production of polyclonal antisera against *Bipolaris sorokiniana. Fitopatologia Brasileira, 23:* 155–157.

Lopez-Moya, J.J., Cubero, J., Lopez-Abella, D., and Diaz-Ruiz, J.R. 1992. Detection of cauliflower mosaic virus (CaMV) in single aphid by polymerase chain reaction (PCR). *Journal of Virological Methods, 37:* 129–138.

Lorenz, K.H., Schneider B., Ahrens, U., and Seemuller, E. 1995. Detection of the apple proliferation and pear decline phytoplasmas by PCR amplification of ribosomal and nonribosomal DNA. *Phytopathology, 85:* 771–776.

Lotrakul, P., Valverde, R.A., Clark, C.A. Sim, J. and De La Torre, R. 1998. Detection of a geminivirus infecting sweet potato in the United States. *Plant Disease, 82:* 1253–1257.

Louro, D. 1995. Use of tissue-print immunoassay for the practical diagnosis of tomato spotted wilt tospovirus. *Bulletin OEPP, 25:* 277–281.

Louro, D., and Lesemann, D.E. 1984. Use of protein A-gold complex for specific labeling of antibodies bound to plant viruses. I. Viral antigens in suspensions. *Journal of Virological Methods, 9:* 107.

Louws, F.J., Fulbright, D.W., Stephens, C.T. and de Brujin, F.J. 1995. Differentiation of genomic structure of rep-PCR fingerprinting to rapidly classify *Xanthomonas campestris* pv. *vesicatoria. Phytopathology, 85:* 528–536.

Lovic, B.R., Martyn, R.D., and Miller, M.E. 1995. Sequence analysis of the ITS regions of rDNA in *Monosporascus* spp. to evaluate its potential for PCR-mediated detection. *Phytopathology, 85:* 655–661.

Lovisolo, O. 1960. *Ocimum basilicum* L., a new test plant for lucerne mosaic virus. *Proceedings of 4th Congress Potato Virus Diseases, Braunschweig,* pp. 138–140.

Lovisolo, O. 1966. Observations on *Ocimum basilicum* as a test plant for plant viruses. In: *Viruses of Plants,* pp. 242–246. North-Holland Publishing Company, Amsterdam.

Lovisolo, M. and Conti, M. 1966. Identification of an aphid-transmitted cowpea mosaic virus. *Netherlands Journal of Plant Pathology, 72:* 265–269.

Lu, Y.J., Zhou, M.G., Ye, Z.Y., Hollomon, D.W. and Butters, J.A. 2000. Cloning and characterization of β-tubulin gene fragment from carbendazim resistance strain of *Fusarium graminearum. Acta Phytopathologica Sinica, 30:* 30–34.

Lübeck, P.S., Alekhina, I.A., Lübeck, M. and Bulat, S.A. 1998. UP-PCR genotyping and rDNA analysis of *Ascochyta pisi* Lib. *Journal of Phytopathology, 146:* 51–55.

Lucy, P.A., Boulton, M.I., Davies, J.W. and Maule, A.J. 1996. Tissue specificity of *Zea mays* infection by maize streak virus. *Molecular Plant-Microbe Interactions, 9:* 22–31.

Lunsgaard, T. 1992. N protein of *Festuca* leaf streak virus (Rhabdoviridae) detected in cytoplasmic viroplasms by immunogold labeling. *Journal of Phytopathology, 134:* 27–32.

Lynch, M. 1988. Estimation of relatedness by DNA finger printing. *Molecular Biological Evolution, 5:* 584–599.

Lyons, N.F. and Taylor, J.D. 1990. Serological detection and identification of bacteria from plants by the conjugated *Staphylococcus aureus* slide agglutination test. *Plant Pathology, 39:* 584–590.

Lyons, N.F. and White, J.G. 1992. Detection of *Pythium violae* and *Pythium sulcatum* in carrots with cavity spot using competition ELISA. *Annals of Applied Biology, 120:* 235–244.

Maat, D.Z. and De Bokx, J.A. 1978a. Potato leafroll virus: antiserum preparation and detection in potato leaves and sprouts by the enzyme-linked immunosorbent assay (ELISA). *Netherlands Journal of Plant Pathology, 84:* 149–156.

Maat, D.Z. and De Bokx, J.A. 1978b. Enzyme-linked immunosorbent assay (ELISA) for the detection of potato virus A and Y in potato leaves and sprouts. *Netherlands Journal of Plant Pathology, 84:* 167–173.

Macfarlane, L., Jenkins, J.E.E., and Melville, S.C. 1968. A soil-borne virus of winter oats. *Plant Pathology, 17:* 167–170.

MacIntosh, S., Robinson, D.I., and Harrison, B.D. 1992. Detection of three whitefly-transmissible geminiviruses occurring in Europe by tests with heterologous monoclonal antibodies. *Annals of Applied Biology, 121:* 297–303.

MacKenzie, D. and Ellis, P.J. 1992. Resistance to tomato spotted wilt virus infection in transgenic tobacco expressing the viral nucleocapsid gene. *Molecular Plant-Microbe Interaction, 5:* 34–40.

Maclachian, D.S., Larson, R.H., and Walker, J.C. 1953. Strain inter-relationships in potato virus A. *University of Wisconsin Research Bulletin, No. 180.*

MacLeold, R., Black, L.M., and Moyer, F.H. 1966. The fine structure and intracellular localization of potato yellow dwarf virus. *Virology, 29:* 540–552.

Maeda, T. Sako, N. and Inouye, N. 1997. Rapid and sensitive detection of cucumber mosaic virus by a simplified ELISA using two monoclonal antibodies. *Bulletin of Research Institute for Bioresources, Okayama University, 5:* 23–30.

Maeda, T., Sako, N., and Inouye, N. 1988. Production of monoclonal antibodies to cucumber mosaic virus and their use in ELISA. *Annals of Phytopathological Society, Japan, 54:* 600.

Maes, M., Garbeva, P. and Kamoen, O. 1996. Recognition and detection in seed of the *Xanthomonas* pathogens that cause cereal leaf streak using rDNA spacer sequences and polymerase chain reaction. *Phytopathology, 86:* 63–69.

Maeso, D., Pagani, C., Mirabelle, I. and Conci, V.C. 1997. Studies on viruses affecting garlic in Uruguay. *Acta Horticulturae, No. 433,* 617–622.

Magee, W.I., Beck, C.F., and Ristown, S.S. 1986. Monoclonal antibodies specific for *Corynebacterium sepedonicum,* the causative agent of potato ringrot. *Hybridoma, 5:* 231–235.

Magome, H., Terauchi, H., Yoshikawa, N. and Takahashi, T. 1997. Analysis of double-stranded RNA in tissues infected with apple stem grooving capillovirus. *Annals of Phytopathological Society, Japan, 63:* 450–454.

Magome, H., Yoshikawa, N. and Takahashi, T. 1999. Single-strand conformation analysis of apple stem grooving capillovirus sequence variants. *Phytopathology, 89:* 136–140.

Mahmood, T., Hein, G.L. and French, R.C. 1997. Development of serological procedures for rapid and reliable detection of wheat streak mosaic virus in a single wheat curl mite. *Plant Disease, 81:* 250–253.

Mahmood, T. and Rush, C.M. 1999. Evidence of cross-protection between beet soil-borne mosaic virus and beet necrotic yellow vein virus in sugarbeet. *Plant Disease, 83:* 521–526.

MaHong, Zhao XiaoLi, Wong XingHua, Cheng Dan, Xu Abing and Huang Chun Nong, 1997. The detection of barley yellow mosaic virus with monoclonal antibody to BaYMV. *Acta Phytophylacica, Sinica, 24:* 309–312.

Mahuku, G.S. and Goodwin, P.H. 1997. Presence of *Xanthomonas fragariae* in symptomless crowns in oat and detection using a nested-polymerase chain reaction (PCR). *Canadian Journal of Plant Pathology, 19:* 366–370.

Mahuku, G.S., Platt (Bud), H.W. and Maxwell, P. 1999. Comparison of polymerase chain reaction-based method with plating on media to detect and identify *Verticillium* wilt pathogen of potato. *Canadian Journal of Plant Pathology, 21:* 125–131.

Mäki-Valkama, T. and Karjalainen, R. 1994. Differentiation of *Erwinia carotovora* subsp. *citroseptica* and *carotovora* by RAPD-PCR. *Annals of Applied Biology, 125:* 301–305.

Mäki-Valkama, T., Pehu, T., Santala, A., Valkonen, J.P.T., Koivu, K., Lehto, K. and Pehu, E. 2000a. High level of resistance to potato virus Y by expressing P1 sequence in antisense orientation in transgenic potato. *Molecular Breeding, 6:* 95–104.

Mäki-Valkama, T., Valkonen, J.P.T., Kreuze, J.F. and Pehu, E. 2000b. Transgenic resistance to PVY associated with post-transcriptional silencing of P1 transgene is overcome by PVY^N strains that carry homologous P1 sequences and recover transgene expression at infection. *Molecular Plant-Microbe Interactions, 13:* 366–373.

Makkouk, K.M., Hsu, H.T., and Kumari, S.G. 1993. Detection of three plant viruses by dot-blot and tissue blot immunoassays using chemiluminiscent and chromogenic substrates. *Journal of Phytopathology, 139:* 97–112.

Malandrin, L., Huard, A. and Samson, R. 1996. Discriminant envelope profiles between *Pseudomonas syringae* pv. *pisi* and *P. syringae* pv. *syringae*. *FEMS Microbiology Letters, 41:* 11–17.

Malandrin, L. and Samson, R. 1998. Isozyme analysis for the identification of *Pseudomonas syringae* pv. strains. *Journal of Applied Microbiology, 84:* 895–902.

Malcult, I., Marano, M.R., Kavanagh, T.A., Jong, W.de., Forsyth, A. and Baulcombe, D.C. 1999. The 25-kDa movement protein of PVX elicits *Nb*-mediated hypersensitive cell death in potato. *Molecular Plant-Microbe Interactions, 12:* 536–543.

Malinowski, T., Gieślińska, M., Zawadzka, B., Interwicz, B. and Porebska, A. 1997. Characterization of monoclonal antibodies against apple chlorotic leaf spot virus (ACLSV) and their application for detection of ACLSV and identification of its strain. *Phytopathologia Polonica, No. 14,* 35–40.

Malinowski, T., Komorowska, B., Golis, T. and Zawadzki, B. 1998. Detection of apple stem pitting virus and pear vein yellows virus using reverse transcription-polymerase chain reaction. *Acta Horticulturae, No. 472,* 87–95.

Malisano, G., Firrao, G. and Locci, R. 1996. 16S rDNA-derived oligonucleotide probes for the differential diagnosis of plum leptonecrosis and apple proliferation phytoplasmas. *Bulletin, OEPP, 26:* 421–428.

Manandhar, H.R., Jørgenden, H.J.L., Mathur, S.B. and Snedegaard-Petersen, V. 1998. Suppression of rice blast by preinoculation with avirulent *Pyricularia oryzae* and the nonrice pathogen *Bipolaris sorokiniana*. *Phytopathology, 88:* 735–739.

Manickam, K. 2000. Molecular detection of banana bunchy top nanavirus and production of disease-free banana plantlets. Doctoral Thesis, Tamil Nadu Agricultural University, Coimbatore, India.

Manicom, B.Q., Bar-Joseph, M., Rosner, A., Vigodsky-Haas, H., and Kotze, J.M. 1987. Potential applications of random DNA probes and restriction fragment length polymorphisms in the taxonomy of the *Fusaria*. *Phytopathology, 77:* 669–672.

Mansky, L.M., Andrews, R.E., Jr., Durand, D.P., and Hill, J.H. 1990. Plant virus location in leaf tissue by press blotting. *Plant Molecular Biology Reporter, 8:* 13–17.

Mansvelt, L. and Hattingh, M.J. 1987. Scanning electron microscopy of pear blossom invasion by *Pseudomonas syringae* pv. *syringae*. *Canadian Journal of Botany, 65:* 2523–2529.

Manulis, S., Kogan, N., Reuven, M., and Ben-Yephety, Y. 1994. Use of RAPD technique for identification of *Fusarium oxysporum* f.sp. *dianthi* from carnation. *Phytopathology, 84:* 98–101.

Manulis, S., Kogan, N., Valinksy, L., Dror, O. and Kleitman, F. 1998. Detection of *Erwinia herbicola* pv. *gysophilae* in gysophila plants by PCR. *European Journal of Plant Pathology, 104:* 85–91.

Maramorosch, K. 1952. Direct evidence for the multiplication of aster yellows virus in its vector. *Phytopathology, 42:* 59–64.

Maramorosch, K. 1955. Seedlings of *Solanum tuberosum* as indicator plants for potato leafroll virus. *American Potato Journal, 32:* 49–50.

Marchoux, G., Gebre-Selassie, K., and Villevieille, M. 1991. Detection of tomato spotted wilt virus and transmission by *Frankliniella occidentalls* in France. *Phytopathology, 40:* 347–351.

Marcone, C., Hergenhahn, F., Ragozzino, A. and Seemüller, R. 1999. Dodder transmission of pear decline, European stone fruit yellows, rubus stunt, *Pieris echioides* yellows and cotton phyllody phytoplasmas to periwinkle. *Journal of Phytopathology, 147:* 187–192.

Marcone, C., Ragozzino, A. and Seemüller, E. 1996. Detection of an elm yellows-related phytoplasma in eucalyptus trees affected by little leaf disease in Italy. *Plant Disease, 80:* 669–673.

Marcone, C., Ragozzino, A. and Seemüller, E. 1997. Detection of Bermuda grass white leaf disease in Italy and characterization of the associated phytoplasma by RFLP analysis. *Plant Disease, 81:* 862–866.

Marie-Jeanne, V., Ioos, R., Peyre, J., Alliot, B. and Signoret, P. 2000. Differentiation of Poaceae potyviruses by reverse transcription–polymerase chain reaction and restriction analysis. *Journal of Phytopathology, 148:* 141–151.

Maris, B. and Rozendaal, A. 1956. Enkele proeven met stammen van hetz-en bet aucubabontvirus van de aardappel. *Tijdschr. Plzieki, 62:* 12–18.

Markham, P.G., Townsend, R., Bar-Joseph, M., Daniels, M.J., Plaskitt, A., and Meddins, 1974. Spiroplasmas are the causal agents of citrus little leaf disease. *Annals of Applied Biology, 78:* 49–57.

Markham, R. and Smith, K.M. 1949. Studies on the virus of turnip yellow mosaic. *Parasitology, 39:* 330–342.

Martelli, G.P. 1981. Tombus viruses. In: *Plant Virus Infections and Comparative Diagnosis* (Ed.) E. Kurstak, pp. 61–90. Elsevier/North-Holland, Amsterdam.

Martelli, G.P. and Russo, M. 1969. *Annual Review of Phytopathology, 1* (hort. ser.): 339.

Martelli, G.P. and Russo, M. 1973. Electron microscopy of artichoke mottled crinkle virus in leaves of *Chenopodium quinoa* Wild. *Journal of Ultrastructure Research, 42:* 93–107.

Martelli, G.P. and Russo, M. 1976. Unusual cytoplasmic inclusions induced by watermelon mosaic virus. *Virology, 72:* 352–362.

Martelli, G.P. and Russo, M. 1977. Rhabdoviruses of plants. In: *The Atlas of Insect and Plant Viruses* (Ed.) K. Maramorosch, pp. 181–213. Academic Press, New York.

Martelli, G.P. and Russo, M. 1984. Use of thin sectioning for visualization and identification of plant viruses. *Methods in Virology, 8:* 143–224.

Martelli, G.P. and Russo, M. 1985. Virus-host relationships: symptomatological and ultrastructural aspects. In: *The Plant Viruses* (Ed.) R.I.B. Francki, Vol. I, pp. 163–205. Plenum Press, New York.

Martin, C. 1954. Sur la presence de sucres reducteurs chez les pommes de terre atteintes de virus de l'enroulement. *C.R. Acad. Sci., Paris, 238:* 724–726.

Martin, F.N. 1991. Selection of DNA probes useful for isolate identification of two *Phythium* spp. *Phytopathology, 81:* 742.

Martin, L.A., Fox, R.T.V., Baldwin, B.C., and Connerton, I.F. 1992. Use of polymerase chain reaction for the diagnosis of MBC resistance in *Botrytis cinerea.* In: *Brighton Crop Protection Conference, Pests and Diseases,* Vol. 1, Fernherm, U.K.

Martin, R.R., and Stace-Smith, R. 1984. Production and characterization of monoclonal antibodies specific to potato leaf roll virus. *Canadian Journal of Plant Pathology, 6:* 206.

Martinez-Soriano, J.P., Galino-Arlonso, J. Maroon, C.J.M., Yucel, I., Smith, D.R. and Diener, T.O. 1996. Mexican papita viroid: putative ancester of crop viroids. *Proceedings of National Academy of Sciences, USA, 93:* 9397–9401.

Marzachi, C., Veratti, F. and Bosco, D. 1998. Direct PCR detection of phytoplasmas in experimentally infected insects. *Annals of Applied Biology, 133:* 45–54.

Masuta, C., Nishimura, M., Marishita, H. and Hataya, T. 1999. A single amino acid change in viral genome–associated protein of potato virus Y correlates with resistance breaking in "Virgin A Mutant" tobacco. *Phytopathology, 89:* 118–123.

Mathews, D.M., Heick, J.A. and Dodds, J.A. 1997. Detection of avocado sunblotch viroid by polymerase chain reaction (PCR). *California Avocado Society, Yearbook, 81:* 91–96.

Mathews, D.M., Riley, K. and Dodds, J.A. 1997. Comparisons of detection methods for citrus virus in field trees during months of nonoptimal titer. *Plant Disease, 81:* 525–529.

Matsui, C. and Yamaguchi, A. 1964. Electron microscopy of host cells infected with tobacco etch virus. I. Fine structures of leaf cells at later stages of infection. *Virology, 22:* 40–47.

Matsumoto, M. and Matsuyama, N. 1998. Trials of identification of *Rhizoctonia solani* AG1-1A, the causal agent of rice sheath rot disease using specifically primed PCR analysis in diseased plant tissues. *Bulletin of the Institute of Tropical Agriculture, Kyushu University, 21:* 27–32.

Matsuo, K., Ando, T., Ohshima, K. and Sako, N. 1998. Detection by reverse transcription and polymerase chain reaction of melon necrotic spot virus strains distributed in Japan. *Annals of Phytopathological Society, Japan, 64:* 208–212.

Matsuyama, N. 1995. Trials for rapid identification of phytopathogenic bacteria by HPLC and direct colony TLC. *Journal of Faculty of Agriculture, Kyushu University, 40:* 87–91.

Matsuyama, N. 1998. Presumptive identification of several phytopathogenic bacteria by novel diagnostic tests. *Journal of Faculty of Agriculture, Kyushu University, 43:* 337–343.

Matsuyama, N. and Furuya, N. 1993a. Application of the direct colony TLC method for identification of phytopathogenic bacteria. *Journal of Faculty of Agriculture, Kyushu University, 37(3–4):* 283–287.

Matsuyama, N. and Furuya, N. 1993b. Application of the direct colony TLC for identification of phytopathogenic bacteria (II): chromatographic profile of *Erwinia* and *Pseu-*

domonas spp. *Journal of Faculty of Agriculture, Kyushu University, 38(1–2):* 89–95.

Matsuyama, N., Mian, I.H., Akanda, A.M., and Furuya, N. 1993a. Comparative studies on thin layer chromatography of lipids from various phytopathogenic bacteria. *Annals of Phytopathological Society, Japan, 59:* 528–534.

Matsuyama, N., Mian, I.H., Akanda, A.M., and Furuya, N. 1993b. On a rapid identification of phytopathogenic bacteria by direct colony thin-layer chromatography. *Proceedings of Association for Plant Protection of Kyushu, 39:* 60–63.

Matsuyama, N., Ueda, Y., Iiyama, K., Furuya, N., Ura, H., Khan, A.A. and Matsumoto, M. 1998. Rapid extraction-HPLC, as a tool for presumptive identification of *Burkholderia gladioli, B. glumae* and *B. plantarii,* causal agents of various rice diseases. *Journal of Faculty of Agriculture, Kyushu University, 42:* 265–272.

Matthews, R.E.F. 1957. *Plant Virus Serology.* Cambridge University Press, Cambridge.

Matthews, R.E.F. 1977. Tymovirus (turnip yellow mosaic virus) group. In: *The Atlas of Insect and Plant Viruses* (Ed.) K. Maramorosch, pp. 347–361. Academic Press, New York.

Matthews, R.E.F. 1978. Are viroids negative stranded viruses? *Nature (Lond), 276:* 850.

Matthews, R.E.F. 1991. *Plant Virology,* 3rd Edition. Academic Press, New York.

Matthysee, A.G. 1994. Conditioned medium promotes the attachment of *Agrobacterium tumefaciens* strains NT 1 to carrot cells. *Protoplasma, 183:* 131–136.

Matthysse, A.G. and McMahan, S. 1998. Root colonization by *Agrobacterium tumefaciens* is reduced in *cel, attB, attD* and *attR* mutants. *Applied and Environmental Microbiology, 64:* 2341–2345.

Mayers, C., N., Palukaitis, P. and Carr, J.P. 2000. Subcellular distribution analysis of the cucumber mosaic virus 2b protein. *Journal of General Virology, 81:* 219–226.

Mazarei, M. and Kerr, A. 1990. Distinguishing pathovars of *Pseudomonas syringae* on peas: nutritional pathogenicity and serological tests. *Plant Pathology, 39:* 278–285.

Mazzola, M., Wong, OiTak. and Cook, R.J. 1996. Virulence of *Rhizoctonia oryzae* and *R. solani AG-8* on wheat and detection of *R. oryzae* in plant tissue by PCR. *Phytopathology, 86:* 354–360.

McClean, A.P.D. 1931. Bunchy top diseases of tomato. South Africa Department of Science, *Bulletin,* 100.

McClean, A.P.D. 1935a. Bunchy top disease of tomato. South Africa Department of Agriculture, *Science Bulletin,* 100.

McClean, A.P.D. 1935b. Further investigations on the bunchy top disease of tomatoes. Union of South Africa Department of Agriculture, *Science Bulletin,* 139.

McClean, A.P.D. 1948. Bunchy top disease of tomato: additional host plants and the transmission of the virus through the seed of infected plants. South Africa Department of Agriculture, *Science Bulletin,* 256.

McDermott, J.M., Brandle, U., Dutty, F., Haemmerli, U.A., Keller, S., Muller, K.E., and Wolfe, M.S. 1994. Genetic variation in powdery mildew of barley: development of RAPD-SCAR and VNTR markers. *Phytopathology, 84:* 1316–1321.

McDonald, J.G., Wong, E., Kristjansson, G.T. and White, G.P. 1999. Direct amplification by PCR of DNA from ungerminated teliospores of *Tilletia* species. *Canadian Journal of Plant Pathology, 21:* 78–80.

McInnes, J.L., Habili, N., and Symons, R.H. 1989. Nonradioactive photobiotin-labeled DNA probes for routine diagnosis of viroids in plant extracts. *Journal of Virological Methods, 23:* 299–312.

McInnes, J.L. and Symons, R.H. 1989. Nucleic acid probes in the diagnosis of plant viruses and viroids. In: *Nucleic Acid Probes* (Ed.) R.H. Symons. CRC Press, Boca Raton, FL.

McIntosh, A.H., Skowronski, B.S., and Maramorosch, K. 1974. Rapid identification of *Spiroplasma citri* and its relation to other yellows agent. *Phytopathologische Zeitschrift, 80:* 153–156.

McIntyre, J.L. and Sands, D.C. 1977. How disease is diagnosed. In: *Plant Disease—an Advanced Treatise* (Eds.) J.G. Horsfall and E.B. Cowling, Vol. I, pp. 35–53. Academic Press, New York.

McKay, G.J., Brown, A.E., Bjourson, A.J. and Mercer, P.C. 1999. Molecular characterization of *Alternaria linicola* and its detection in linseed. *European Journal of Plant Pathology, 105:* 157–166.

McKay, G.J. and Cooke, L.R. 1997. A PCR-based method to characterize and identify benzimidazole resistance in *Helminthosporium solani. FEMS Microbiology Letters, 152:* 371–378.

McKinney, H.H. 1929. Mosaic diseases in the Canary Islands, West Africa and Gibraltar. *Journal of Agricultural Research, 37:* 557–558.

McKinney, H.H. 1953. Soil-borne wheat mosaic viruses in the Great Plains. *Plant Disease Reporter, 37:* 24–26.

McLaughlin, R.J., Chen, T.A., and Wells, J.M. 1989. Monoclonal antibodies against *Erwinia amylovora*—characterization and evaluation of a mixture for detection of enzyme-linked immunosorbent assay. *Phytopathology, 79:* 610–613.

McLean, A.P.D. 1931. Bunchy top disease of tomato. *South African Department of Science, Bulletin, 100:* 36 pp.

McManus, P.S. and Jones, A.C. 1996. Detection of *Erwinia amylovora* by nested PCR and PCR-dot blot and reverse blot hybridization. *Acta Horticulturae, No. 411,* 87–90.

McMullen, C.R., Gardner, W.S., and Myers, G.A. 1977. Ultrastructure of cell-wall thickenings and paramural bodies induced by barley stripe mosaic virus. *Phytopathology, 67:* 462–467.

McMullen, C.R., Gardner, W.S., and Myers, G.A. 1978. Aberrant plastids in barley leaf tissue infected with barley stripe mosaic virus. *Phytopathology, 68:* 317–325.

McWhorter, F.P. 1941a. Isometric crystals produced by *Pisum* virus 2 and *Phaseolus* virus 2. *Phytopathology, 31:* 760–761.

McWhorter, F.P. 1941b. Plant-virus differentiation by trypan blue reactions within infected tissue. *Stain Technology, 16:* 143–149.

McWhorter, F.P. 1965. Plant virus inclusions. *Annual Review of Phytopathology, 3:* 287–312.

Medeiros, R.B., Rasochova, L. and German, T.L. 2000. Simplified, rapid method for cloning of virus-binding polypeptides (putative receptors) via the far-Western screening of a cDNA expression library using purified virus particles. *Journal of Virological Methods, 86:* 155–166.

Mehta, P., Brlansky, R.A., Gowda, S. and Yokomi, R.A. 1997. Reverse transcription–polymerase chain reaction detection of citrus tristeza virus in aphids. *Plant Disease, 81:* 1066–1069.

Melton, D.A., Krieg, P.A., Rebaghiati, M.R., Maniatus, T., Zinn, K., and Green, M.R. 1984. Efficient in vitro synthesis of biologically active RNA and RNA hybridization probes from plasmids containing bacteriophage SP6 promoter. *Nucleic Acids Research, 12:* 7033–7056.

Melton, R.E., Flegg, L.M., Brown, J.K.M. Oliver, R.P., Daniels, M.J. and Osburn, A.E. 1998. Heterologous expression of *Septoria lycopersici* tomatinase in *Cladosporium fulvum:* effects of compatible and incompatible interactions with tomato seedlings. *Molecular Plant-Microbe Interactions, 11:* 228–236.

Mendes, L.O.T. 1956. Podridao interna dos capulhos do algodoeiro obtida por meio de insetos. *Braganna, 15:* 9–11.

Mendgen, K. and Lesemann, D.E. 1991. *Electron Microscopy of Plant Pathogens.* Springer-Verlag, Berlin.

Meng BaoZhong, Johnson, R., Peressini, S., Forsline, P.L. and Gonsalves, D. 1999. Rupestris stem pitting associated virus-1 is consistently detected in grapevines that are infected with rupestris stem pitting. *European Journal of Plant Pathology, 105:* 191–199.

Mercier, L. 1911. Sur le role des insects comma agents de propagation de l'ergot des graminees. *Compt. Rend. Soc. Biol., 70:* 300–302.

Merighi, M., Sandrini, A., Landini, S., Ghini, S., Girotti, S., Malaguti, S. and Bazzi, C. 2000. Chemiluminescent and colorimetric detection of *Erwinia amylovora* by immunoenzymatic determination of PCR amplicons from plasmid pEA29. *Plant Disease, 84:* 49–54.

Merten, D.W., Hebeile-Bors, E., Himmler, G., Reiter, S., Messner, P., and Katinger, H. 1985. Monoclonal ELISA for the determination of BNYV-virus. *Dev. Biol. Stand., 60:* 451.

Mes, J.J., Van Dorn, J., Roebroeck, E.J.A., van Egmond, E., van Aartrijk, J., and Bonekamp, P.M. 1994. Restriction fragment length polymorphisms, races, and vegetative compatibility groups within a worldwide collection of *Fusarium oxysporum* f. sp. *gladioli. Plant Pathology, 43:* 362–370.

Meshi, T., Motoyishi, F., Maeda, T., Yoshiwoka, S., Watanabe, Y. and Okada, Y. 1989. Mutations in the tobacco mosaic 30 kDa protein gene overcome Tm-2 resistance in tomato. *The Plant Cell, 1:* 515–522.

Mesquita, A.G.G., Paula, T.J.Jr., Moreira, M.A. and de Barros, E.G. 1998. Identification of races of *Colletotrichum lindemuthianum* with the aid of PCR-based molecular markers. *Plant Disease, 82:* 1084–1087.

Milicic, D., Wrischer, M., and Juretic, N. 1954. Intracellular inclusion bodies of broad bean wilt virus. *Phytopathologische Zeitschrift, 80:* 127–135.

Miller, S.A., Bhat, R.G., and Schmitthenner, 1994. Detection of *Phytophthora capsici* in pepper and cucurbit crops in Ohio with two commercial immunoassay kits. *Plant Disease, 78:* 1042–1046.

Miller, S.A., Madden, L.V. and Schmutthenner, A.F. 1997. Distribution of *Phytophthora* spp. in field soils determined by immunoassay. *Phytopathology, 87:* 101–107.

Mills, D., Russell, B.W. and Hanus, J.W. 1997. Specific detection of *Clavibacter michiganensis* subsp. *sepedonicus* by amplification of three unique DNA sequences isolated by subtraction hybridization. *Phytopathology, 87:* 853–861.

Milne, R.G. 1967. Electron microscopy of leaves infected with sowbane mosaic virus and other small polyhedrosis viruses. *Virology, 32:* 589–600.

Milne, R.G. 1970. An electron microscope study of tomato spotted wilt virus in sections of infected cells and in negative strain preparations. *Journal of General Virology, 6:* 267–276.

Milne, R.G. 1972. Electron microscopy of viruses. In: *Principles and Techniques in Plant Virology* (Eds.) C.I. Kado and H.O. Agrawal, pp. 79–128. Van Nostrand Reinhold Company, New York.

Milne, R.G. 1981. Notes for the Course on Immunoelectron Microscopy of Plant Viruses. Association of Applied Biologists Workshop in Electron Microscope Serology. John Innes Institute, England, 1981.

Milne, R.G. 1984. Electron microscopy for the identification of plant viruses in in vitro preparations. *Methods in Virology, 7:* 87.

Milne, R.G. 1992. Immunoelectron microscopy of plant viruses and mycoplasmas. *Advances in Disease Vector Research, 9:* 283–312.

Milne, R.G. 1993a. Electron microscopy of in vitro preparations. In: *Diagnosis of Plant Virus Diseases* (Ed.) R.E.F. Matthews, pp. 216–251, CRC Press Inc., Boca Raton, USA.

Milne, R.G. 1993b. Solid phase immune electron microscopy of virus preparations. In: *Immune Electron Microscopy for Virus Diagnosis* (Eds.) A.D. Hyat and B.T. Eaton, pp. 27–70, CRC Press Inc., Boca Raton, USA.

Milne, R.G. and Lesemann, D.E. 1984. Immunosorbent electron microscopy in plant virus studies. *Methods in Virology, 7:* 85–101.

Milne, R.G. and Luisoni, E. 1975. Rapid high-resolution immunoelectron microscopy of plant viruses. *Virology, 68:* 270–274.

Milne, R.G. and Luisoni, E. 1977. Rapid immune electron microscopy of virus preparations. *Methods in Virology, 6:* 265–281.

Milne, R.G., Ramasso, E., Lenzi, R., Masenga, V., Sarindu, N., and Clark, M.F. 1995. Pre- and post-embedding immunogold labeling and electron microscopy in plant host tissues of three antigenically unrelated MLOs: primula yellows, tomato bigbud and bermuda grass whiteleaf. *European Journal of Plant Pathology, 101:* 57–67.

Mills, P.R., Sreenivasaprasad, J., and Brown, A.E. 1992. Detection and differentiation of *Colletotrichum gloeosporioides* isolates using PCR. *FEMS Microbiology Letters, 98(1–3):* 137–143.

Minafra, A., Hadidi, A. and Saldarelli, P. 1993. Sensitive immunocapture and multiplex reverse transcription polymerase chain reaction for the detection of grapevine leafroll associated virus III and grapevine virus B. *Proceedings of 11th Meeting of the International Council for the Study of Viruses and Virus Diseases of the Grapevine* (IIVG), Abst-137.

Mink, G.I., Howell, W.E., and Fridlund, P.R. 1985. Apple tip leaf antigens that cause spurious reactions with tomato ringspot virus antisera in ELISA. *Phytopathology, 75:* 325–329.

Minsavage, G.V., Thompson, C.M., Hopkins, D.L., Leite, R.M.V.B.C., and Stall, R.E. 1994. Development of a polymerase chain reaction protocol for detection of *Xylella fastidiosa* in plant tissue. *Phytopathology, 84:* 456–461.

Miyamoto, S. and Miyamoto, Y. 1966. Notes on aphid transmission of potato leaf roll virus. Scientific Report, Hyogo University, Agricultural Series. *Plant Protection, 7:* 51–66.

Miyoshi, T., Sawada, H., Tachibana, Y. and Matsuda, I. 1998. Detection of *Xanthomonas campestris citri* by PCR using primers from the spacer region between the 16S and 23S rRNA genes. *Annals of Phytopathological Society, Japan, 64:* 249–254.

Mizenina, O.A., Borisova, O.V., Novikov, V.K., Ertushenko, O.A., Baykov, A.A., and Atabekov, J.G. 1991. Inorganic pyrophosphatase from *E. coli* as a label for the detection of plant viruses by ELISA. *Journal of Phytopathology, 133:* 278–288.

Mohamed, N.A., Haseloff, J., Imperial, J.S. and Symons, R.H. 1982. Characterization of the different electrophoretic forms of the cadang-cadang viroid. *Journal of General Virology, 63:* 181–188.

Möller, M. and Harling, R. 1996. Randomly amplified polymorphic DNA (RAPD) profiling of *Plasmodiophora brassicae. Letters in Applied Microbiology, 22:* 70–75.

Momol, M.T., Momol, E.A., Lamboy, W.F. Norelli, J.L., Beer, S.V. and Aldwinckle, H.S. 1997. Characterization of *Erwinia amylovora* strains using random amplified polymorphic DNA fragments (RAPDs). *Journal of Applied Microbiology, 82:* 389–398.

Monis, J. and Bestwick, R.K. 1997. Serological detection of grapevine associated closteroviruses in infected grapevine cultivars. *Plant Disease, 81:* 802–808.

Montana, J.R., Hunger, R.M. and Sherwood, J.L. 1996. Serological characterization of wheat streak mosaic virus isolates. *Plant Disease, 80:* 1239–1244.

Montasser, M.S., Tousignant, M.E. and Kaper, J.M. 1998. Viral satellite RNAs for the prevention of cucumber mosaic virus (CMV) disease in field grown pepper and melon plants. *Plant Disease, 82:* 1298–1303.

Moon, J.S., Allen, R.G., Domier, L.L. and Hewings, A.D. 2000. Molecular and biological characterization of a trackable Illinois isolate of barley dwarf virus-PAV. *Plant Disease, 84:* 483–486.

Morgan, L.W., Bateman, G.L., Edwards, S.G., Marshall, J., Nicholson, P., Nuttall, M., Parry, D.W., Sckancher, M. and Turner, A.S. 1998. Fungicide evaluation and risk assessment of wheat stem base diseases using PCR. *Brighton Crop Protection Conference: Pests and Diseases, 3:* 1011–1016.

Morricca, S., Ragazzi, A., Kasuga, T. and Mitchelson, K.R. 1998. Detection of *Fusarium oxysporum* f.sp. *vasinfectum* in cotton tissue by polymerase chain reaction. *Plant Pathology, 47:* 486–494.

Mosqueda-Cano, G. and Herrera-Estrella, L. 1997. A simple and efficient PCR method for the specific detection of *Pseudomonas syringae* pv. *phaseolicola* in the bean seeds. *World Journal of Microbiology and Biotechnology, 13:* 463–467.

Morris, T.J. and Wright, N.S. 1975. Detection of polyacrylamide gel of a diagnostic nucleic acid from tissue infected with potato spindle tuber viroid. *American Potato Journal, 52:* 57–63.

Moukhamedov, R., Hu, X., Nazar, R.N., and Robb, J. 1994. Use of polymerase chain reaction amplified ribosomal intergenic sequences for the diagnosis of *Verticillium tricorpus. Phytopathology, 84:* 256–259.

Mowat, W.P. 1971. Stabilization, culture and some properties of tulip halo necrosis virus. *Annals of Applied Biology, 69:* 147–153.

Mráz, I., Petrzik, K., Fránová-Honetšlegrová. J. and Sip, M. 1997. Detection of strawberry vein banding virus by polymerase chain reaction and dot-blot hybridization. *Acta Virologica, 41:* 241–242.

Muchalski, T. 1997. Use of RT-PCR technique for detection of tobacco rattle virus in potato tubers. *Phytopathologia Polonica, No. 13,* 31–37.

Mueller, W.C. and Koeing, R. 1965. Nuclear inclusions produced by bean yellow mosaic virus as indicators of cross protection. *Phytopathology, 55:* 242–243.

Müller, R., Pasberg-Gauhl, C., Gauhl, F., Ramser, J. and Kahl, G. 1997. Oligonucleotide fingerprinting detects genetic variability at different levels in Nigerian *Mycosphaerella fijiensis. Journal of Phytopathology, 145:* 25–30.

Mumford, D.L., 1974. Purification of curly top virus. *Phytopathology, 64:* 136.

Mumford, R.A., Barker, I., and Wood, K.R. 1994. The detection of tomato spotted wilt virus using polymerase chain reaction. *Journal of Virological Methods, 46:* 303–311.

Mumford, R.A., Barker, I. and Wood, K.R. 1996. An improved method for the detection of *Tospoviruses* using the polymerase chain reaction. *Journal of Virological Methods, 57:* 109–115.

Mumford, R.A. and Seal, S.E. 1997. Rapid single-tube immunocapture RT-PCR for the detection of two yam potyviruses. *Journal of Virological Methods, 69:* 73–79.

Muniyappa V., Swanson, M.M., Duncan, G.H., and Harrison, B.D. 1991. Particle purification, properties and epitope variability of Indian tomato leafcurl geminivirus. *Annals of Applied Biology, 118:* 595–604.

Murillo, I., Cavallarin, L. and Segundo, B.S. 1998. The development of a rapid PCR assay for detection of *Fusarium moniliforme. European Journal of Plant Pathology, 104:* 301–311.

Murillo, I., Cavallarin, L. and San Segundo, B. 1999. Cytology of infection of maize seedlings by *Fusarium moniliforme* and immunolocalization of the pathogenesis-related PRms protein. *Phytopathology, 89:* 737–747.

Murphy, F.A., Fauquet, C.M. Bishop, D.H.L. Ghabrial, S.A., Jarvis, A.W., Martelli, G.P. Mayo, M.A. and Summers, M.D. (Eds.) 1996. Virus taxonomy: classification and nomenclature of viruses. *Sixth Report of the International Committee on Taxonomy of Viruses,* pp. 24–29. Scottish Crop Research Institute, Invergowrie, Dundee, Scotland.

Murphy, J.F., Sikora, E.J., Sammons, B. and Kaniewski, W.K. 1998. Performance of transgenic tomatoes expressing cucumber mosaic virus CP gene under epidemic conditions. *Hort Science, 33:* 1032–1035.

Murphy, D.M. and Pierce, W.H. 1937. Common mosaic of the garden pea, *Pisum sativum. Phytopathology, 27:* 710–726.

Murrant, A.F. and Goold, R.A. 1967. Anthriscus yellows virus. *13th Annual Report of Scottish Horticultural Research Institute,* 1966. p. 64.

Murrant, A.F., Goold, R.A., Roberts, I.M., and Cathro, J. 1969. Carrot mottle—a persistent aphid-borne virus with unusual properties and particles. *Journal of General Virology, 4:* 329–341.

Murrant, A.F., Taylor, C.E., and Chambers, J. 1968. Properties, relationship and transmission of a strain of raspberry ringspot virus infecting raspberry cultivars immune to the common Scottish strain. *Annals of Applied Biology, 61:* 175–186.

Mushin, R., Naylor, J., and Lahovary, N. 1959. Studies on plant pathogenic bacteria. I. Cultural and biochemical characters. II. Serology, *Australian Journal of Biological Sciences, 12:* 223–233.

Muthulakshmi, P. and Narayanasamy, P. 1996. Serological evidence for induction of resistance to rice tungro viruses in rice using antiviral principles. *International Rice Research Notes, 21*(2–3):77.

Myrta, A., Terlizzi, B. di, Boscia, D., Caglayan, K., Gavriel, I., Ghanen, G., Varveri, C. and Savino, V. 1998. Detection and serotyping of Mediterranean plum pox virus isolates by means strain-specific monoclonal antibodies. *Acta Virologica, 42:* 251–253.

Nachmias, A., Bar-Joseph, M., Solel, Z., and Barash, I. 1979. Diagnosis of mal seco disease in lemon by enzyme-linked immunosorbent assay. *Phytopathology, 69:* 559–561.

Nachmias, A., Buchner, V., and Krikun, J. 1982. Differential diagnosis of *Verticillium dahliae* in potato with antisera to partially purified pathogen-produced extracellular antigen. *Potato Research, 25:* 321–328.

Nagaraj, A.N. 1965. Immunofluorescence studies on synthesis and distribution of tobacco mosaic virus antigen in tobacco. *Virology, 25:* 133–142.

Nagaraj, A.N., Sinha, R.C., and Black, L.M. 1961. A smear technique for detecting virus antigen in individual vectors by the use of fluorescent antibodies. *Virology, 15:* 205–208.

Nagata, T. and Avila, A.C. de. 2000. Transmission of chrysanthemum stem necrosis virus, a recently discovered tospovirus by two thrips species. *Journal of Phytopathology, 148:* 123–125.

Nagel, J., Zettler, F.W., and Hiebert, E. 1983. Strains of bean yellow mosaic virus compared to clover yellow vein virus in relation to gladiolus production in Florida. *Phytopathology, 73:* 449–454.

Naidu, R.A., Robinson, D.J. and Kummins, F.M. 1998. Detection of each of the causal agents of groundnut rosette disease in plants and vector aphids by RT-PCR. *Journal of Virological Methods, 76:* 9–18.

Najar, A., Bouachem, S., Danet, J.L., Saillard, C., Garnier, M. and Bové, J.M. 1998. Presence of *Spiroplasma citri* the pathogen responsible for citrus stubborn disease and its vector leafhopper *Circulifer haematoceps* in Tunisia: contamination of both *C. haematoceps* and *C. opacipennis. Fruits (Paris), 53:* 391–396.

Nakahara, K., Hataya, T., Hayashi, Y., Sugimoto, T., Kimura, I. and Shikata, E. 1998. A mixture of synthetic oligonucleotide probes labeled with biotin for the sensitive detection of potato spindle tuber viroid. *Journal of Virological Methods, 71:* 219–227.

Nakahara, K., Hataya, T., Kimura, I. and Shikata, E. 1997. Reactions of potato cultivars in Japan to potato spindle tuber viroid and its gene diagnosis. *Annual Report of the Society of Plant Protection of North Japan, No. 48,* 69–74.

Nakahara, K., Hataya, T. and Uyeda, I. 1999. A simple, rapid method of nucleic acid extraction without tissue homogenization for detecting viroids for hybridization and RT-PCR. *Journal of Virological Methods, 77:* 47–58.

Nakahara, K., Hataya, T., Uyeda, I. and Ieki, H. 1998. An improved procedure for extracting nucleic acids from citrus tissues for diagnosis of citrus viroids. *Annals of Phytopathological Society, Japan, 64:* 532–538.

Nakamura, H., Kaneko, S., Yamaoka, Y. and Kakishima, M. 1998. PCR-SSCP analysis of the ribosomal DNA ITS regions of the willow rust fungi in Japan. *Annals of Phytopathological Society, Japan, 64:* 102–109.

Nakamura, H., Yoshikawa, N., Takahashi, T., Sahashi, N., Kubono, T. and Shoji, T. 1996. Evaluation of primer pairs for the reliable diagnosis of paulownia witches' broom disease using a polymerase chain reaction. *Plant Disease, 80:* 302–305.

Nakashima, K., Cabauatan, P.Q., and Koganezawa, H. 1993. Use of DNA probes to distinguish mycoplasma-like organisms (MLOs) of yellow dwarf (RYD) and orange leaf (ROL) in rice. *International Rice Research Notes, 18(4):* 29–30.

Nakashima, K. and Hayashi, T. 1995. Extrachromosomal DNAs of rice yellow dwarf and sugarcane white leaf phytoplasmas. *Annals of Phytopathological Society, Japan, 61:* 451–462.

Nakashima, K., Ohtsu, Y. and Prommintara, M. 1998. Detection of citrus greening organisms in citrus plants and psylla *Diaphorina citri* in Thailand. *Annals of Phytopathological Society, Japan, 64:* 153–159.

Nakashima, K., Prommintara, M., Ohtsu, Y., Kanto, T., Imada and Koizumi, M. 1996. Detection of 16 S rDNA of Thai isolates of bacterium-like organism associated with greening disease of citrus. *JIRCAS Journal, No. 3,* 1–8.

Namba, S., Kato, S., Iwanami, S., Oyaizu, H., Shiozawa, H., and Tsuchizaki, T. 1993. Detection and differentiation of plant pathogenic mycoplasma-like organism, using polymerase chain reaction. *Phytopathology, 83:* 791.

Nam KiWoong, Kim ChoongHoe, Huang Haiseong, 1996. Studies on the pear abnormal leafspot disease. 5. Selection of indicator plants. *Korean Journal of Plant Pathology, 12:* 214–216.

Naqvi, V.Z. and Futrell, M.C. 1970. Aphid transmission material produced by *Sclerospora sorghi* in corn and sorghum plants. *Phytopathology, 60:* 586.

Narayanasamy, P. 1989. Suitability of iodine test for detecting rice tungro virus infection. *International Rice Research Newsletter, 14(2):* 34.

Narayanasamy, P. and Doraiswamy, S. 1996. *Plant Viruses and Viral Diseases.* New Century Book House Pvt., Ltd., Madras, India.

Narayanasamy, P. and Natarajan, C. 1974. A rapid test for the identification of virus-infected groundnut (*Arachis hypogaea* L.). *Current Science, 43:* 700–701.

Narayanasamy, P. and Ramakrishnan, K. 1966. Studies on the sterility mosaic disease of pigeonpea. III. Nitrogen metabolism of infected plants. *Proceedings of Indian Academy of Sciences B, LX III 288:* 296.

Nariani, T.K. and Sastry, K.S.M. 1958. Two additional vectors of chili mosaic virus. *Indian Phytopathology, II:* 193–194.

Narváez, G., Slimane Skander, B., Ayllón, M.A., Rubio, L., Guerri, J. and Moreno, P. 2000. A new procedure to differentiate citrus tristeza virus isolate by hybridisation with digoxigenin-labeled-cDNA probes. *Journal of Virological Methods, 85:* 83–92.

Nasu, S. 1963. Studies on some leafhoppers and planthoppers which transmit virus diseases of rice plant in Japan. *Kyushu Agricultural Experiment Station Bulletin, 8:* 153–349.

Nault, L.R. 1911. Transmission biology, vector specificity and evolution of planthopper transmitted plant viruses. In: *Planthoppers, Their Ecology, Genetics and Management* (Eds.) R.F. Denno and T.J. Perfect. Chapman & Hall, New York.

Nault, L.R. and Ammar, E.D. 1989. Leafhopper and planthopper transmission of plant viruses. *Annual Review of Entomology, 34:* 503–529.

Nault, L.R. and Gondon, D.T. 1988. Multiplication of maize stripe virus in *Peregrinus maidis. Phytopathology, 78:* 991–995.

Naumann, K., Karl, H., Zielke, R., Schmidt, A., and Griesbach, E. 1988. Comparative analysis of bean seeds on incidence of the causal agent of haloblight *Pseudomonas syringae* pv. *phaseolicola* by different identification methods. I. Detection by isolation on bacterial nutrient media and serological methods. *Zentralblattfur Mikrobiologies, 143:* 487–498.

Navas-Castillo, J., Diaz, J.A., Sanchez Campos, S. and Moriones, E. 1998. Improvement of the print-capture polymerase chain reaction procedure for efficient amplification of DNA virus genomes from plants and insect vectors. *Journal of Virological Methods, 75:* 195–198.

Navot, N., Ber, R., and Czosnek, H. 1989. Rapid detection of tomato yellow leafcurl virus in squashes of plants and insect vectors. *Phytopathology, 79:* 562–568.

Navot, N., Zeidan, M., Picheraky, E., Zamir, D., and Czosnek, H. 1992. Use of the polymerase chain reaction to amplify tomato yellow leafcurl virus DNA from infected plants and viruliferous whiteflies. *Phytopathology, 82:* 1199–1202.

Nazar, R.N., Hu, X., Schmidt, J., Culham, D., and Robb, J. 1991. Potential use of a PCR-amplified detection and differentiation of *Verticillium* wilt pathogens. *Physiological and Molecular Plant Pathology, 39:* 1–11.

Neergaard, P. 1977. *Seed Pathology,* 2 Volumes. Macmillan, London.

Nei, M. and Li, W. 1979. Mathematical model for studying genetic variation in terms of restriction endonucleases. *Proccedings of National Academy of Sciences, USA, 76:* 5264–5273.

Nelson, R.J., Elias, K.S., Arévalo, G.E., Darlington, L.C. and Bailey, B.A. 1997. Genetic characterization by RAPD analysis of isolates of *Fusarium oxysporum* f.sp. *erythroxyli* associated with an emerging epidemic in Peru. *Phytopathology, 87:* 1220–1225.

Nemchinov, L., Hadidi, A. and Faggioli, F. 1998. PCR-detection of apple stem pitting virus from pome fruit hosts and sequence variability among viral isolates. *Acta Horticulturae, No. 472,* 67–73.

Nemec, S., Jabaji-Hare, S., and Charest, P.M. 1991. ELISA and immunocytochemical detection of *Fusarium solani*-produced naphthazarin toxins in citrus trees in Florida. *Phytopathology, 87:* 1497–1503.

Nene, Y.L. and Sheila, V.K. 1994. A potential substitute for agar in microbiological media. *International Chickpea and Pigeonpea Newsletter, No. 2,* 42–44.

Newhall, W.F. 1975. A chromatographic procedure for detecting citrus tree decline. *Plant Disease Reporter, 59:* 581–585.

Neustroeva, N.P., Dzantier, B.B., Markaryan, A.N., Bobkova, A.F., Igorov, A.M., and Atabekov, I.G. 1989. Enzyme-immunoassay of potato virus X using antibodies labeled by β-galactosidase of *E. coli. Brologiya, 1:* 115–118.

Newman, M.A., Conrads-Strauch, J., Scofield, G., Daniels, M.J. and Dow, J.H. 1994. Defense-related gene induction in *Brassica campestris* in response to defined mutants of *Xanthomonas campestris* with altered pathogenicity. *Molecular Plant-Microbe Interactions, 7:* 553–562.

Niblett, C.L., Dickson, E., Fernow, K.H., Horst, R.K., and Zaitlin, M. 1978. Cross protection among four viroids. *Virology, 91:* 198–203.

Nicholson, P. and Parry, D.W. 1996. Development and use of a PCR assay to detect *Rhizoctonia cerealis,* the cause of sharp eyespot in wheat. *Plant Pathology, 45:* 872–883.

Nicholson, P., Rezanoor, H.N., Simpson, D.R. and Joyce, D. 1997. Differentiation and quantification of cereal eyespot fungi *Tapesia yallundae* and *Tapesia acuformis* using PCR assay. *Plant Pathology, 46:* 842–856.

Nicholson, P., Rezanoor, H.N., and Hollins, T.N. 1994. The identification of a pathotype specific DNA probe for the R type of *Pseudocercosporella herpo-trichoides, Plant Pathology, 43:* 694–700.

Nicholson, P., Rezanoor, H.N., and Su, H. 1993. Use of random amplified polymorphic DNA (RAPD) analysis and genetic fingerprinting to differentiate isolates of race O, C and T of *Bipolaris maydis. Journal of Phytopathology, 139:* 261–267.

Nicholson, P., Simpson, D.R., Weston, G., Rezanoor, H.N., Lees, A.K., Parry, D.W. and Joyce, D. 1998. Detection and quantification of *Fusarium culmorum* and *Fusarium graminearum* in cereals using PCR assays. *Physiological and Molecular Plant Pathology, 53:* 17–37.

Nielsen, S.L., Husted, K. and Rasmussen, H.N. 1998. Detection of potato potyvirus Y in dormant tubers by a PCR based method DIAPOPS (detection and immobilized amplified product in a one phase system). *DJF Rapport, Markburg, No. 3,* 73–80.

Niepold, F. 1999. A simple and fast extraction procedure to obtain amplifiable DNA from *Ralstonia solanacearum* and *Clavibacter michiganensis* subsp. *michiganensis* inoculated potato tuber extracts and naturally infected tubers to conduct a polymerase chain reaction. (PCR). *Journal of Phytopathology, 147:* 249–256.

Niepold, F. and Schöber-Butin, G. 1995. Application of PCR technique to detect *Phytophthora infestans* in potato tubers and leaves. *Microbiological Research, 150:* 379–385.

Niepold, F. and Schöber-Butin, B. 1997. Application of the one-tube PCR technique in combination with a fast DNA extraction procedure for detecting *Phytophthora infestans* in infected potato tubers. *Microbiological Research, 152:* 345–351.

Nikolaeva, O.V., Karasev, A.V., Garnsey, S.M. and Lee, R.F. 1998. Serological differentiation of the citrus tristeza virus isolates causing stem pitting in sweet orange. *Plant Disease, 82:* 1276–1280.

Noel, M.C., Kerlan, C., Garnier, M., and Dunez, J. 1978. Possible use of immune electron microscopy (IEM) for the detection of plurn pox virus in fruit trees. *Ann. Phytopathol. 10:* 381–386.

Nolan, P.A. and Campbell, R.N. 1984. Squash mosaic virus detection in individual seeds and seed lots of cucurbits by enzyme-linked immunosorbent assay. *Plant Disease, 68:* 971–985.

Nolasco, G., Blas, C. De, Torres, V., and Ponz, F. 1993. A method of combining immunocapture and PCR amplification in a microtitre plate for the detection of plant viruses and subviral pathogens. *Journal of Virological Methods, 45:* 201–218.

Nolasco, G., Sequeira, Z., Bonacalza, B., Mendes, C., Torres, V., Sanchez, F., Urgoiti, B., Ponzi, F., Febres, V.J., Cevik, B., Lee, R.F. and Niblett, C.L. 1997. Sensitive CTV diagnosis using immunocapture reverse transcriptional polymerase chain reaction and an exonuclease fluorescent probe assay. *Fruits (Paris), 52:* 391–396.

Nolt, B.L., Rajeshwari, R., Bharathan, N., and Reddy, D.V.R. 1983. Improved serological techniques for the detection and identification of groundnut viruses. In: *Management of Diseases of Oilseed Crops* (Ed.) P. Narayanasamy, pp. 1–5. Tamil Nado Agricultural University, Madurai, India.

Nomura, K., Nasser, W., Kawagishi, H. and Tsuyumu, S. 1998. The *pir* gene of *Erwinia chrysanthemi* EC 16 regulates hyperinduction of pectate lyase virulence genes in response to plant signals. *Proceedins of National Academy of Sciences, USA, 95:* 14034–14039.

Norelli, J.L., Burr, T.J., Lolicero, A.M., Gilbert, M.T. and Katz, B.H. 1991. Homologous streptomycin resistance gene present among diverse gram-negative bacteria in New York state apples. *Applied Environmental Microbiology, 57:* 486.

Norman, D.J. and Yuen, J.M.F. 1998. A distinct pathotype of *Ralstonia (Pseudomonas) solanacearum* race 1, biovar 1, entering Florida in pothos *(Epipremnum aureum)* cuttings. *Canadian Journal of Plant Pathology, 20:* 171–175.

Noronha, Fonseca, M.E. de, Marcellino, H. and Gander, E. 1996. A rapid and sensitive dot-blot hybridization assay for the detection of citrus exocortis viroid in *Citrus medica* with digoxigenin labeled RNA probes. *Journal of Virological Methods, 57:* 203–207.

Northover, J. and Cerkauskas, R.F. 1994. Detection and significance of symptomless latent infections of *Monilinia fructicola* in plums. *Canadian Journal of Plant Pathology, 16:* 30–36.

Notte, P.la, Minafra, A. and Saldarelli, P. 1997. A spot-PCR technique for the detection of phloem-limited grapevine viruses. *Journal of Virological Methods, 66:* 103–108.

Nozu, Y., Usugi, T., and Nishimori, K. 1983. Production of monoclonal antibodies to a plant virus. *Seikagaku, 55:* 837.

Nozu, Y., Usugi, T., and Nishimori, K. 1986. Production of monoclonal antibodies to satsuma dwarf virus. *Annals of Phytopathological Society, Japan, 52:* 86.

Nümi, Y., Gondaira, T., Kutsuwada, Y. and Tsuji, H. 1999. Detection by ELISA and DIBA tests of lily symptomless virus (LSV), tulip breaking virus-lily (TBV-L) and cucumber mosaic virus (CMV). *Journal of the Japanese Society for Horticultural Science, 68:* 176–183.

Nutter, F.W. Jr., Schultz, P.M. and Hill, J.H. 1998. Quantification of within-field spread of soybean mosaic virus in soybean using strain-specific monoclonal antibodies. *Phytopathology, 88:* 895–901.

O'Brien, M.J., and Raymer, W.B. 1964. Symptomless hosts of the potato spindle tuber virus. *Phytopathology, 54:* 1045–1047.

Ochs, G. 1960. Paper chromatographic colorimetric method for expeditious diagnosis of leaf roll virus of grapevines. *Botanical Gazette, 121:* 198–200.

Odu, B.O., Hughes, J. d'A., Shoyinka, S.A. and Dongo, L.N. 1999. Isolation, characterization and identification of a potyvirus from *Dioscorea alata* L. (water yam) in Nigeria. *Annals of Applied Biology, 134:* 65–71.

Ogras, T.T., El-Fadly, G., Baloglu, S., Yilmaz, M.A., Cirakoglu, B., and Bermek, E. 1994. Development of a scientific detection technique for tobacco mosaic virus (TMV) based on dot-blot hybridization using an oligonucleotide probe. *Turkish Journal of Biology, 18:* 91–98.

Oh ChangSik, Heu SungGi, and Choi YongChul. 1999. Sensitive and pathovar-specific detection of *Xanthomonas campestris* pv. *glycines* by DNA hybridization and analysis. *Plant Pathology Journal, 15:* 57–61.

Ohmann-Kreuzberg, G. 1962. Das streifenmosaik virus der Gerste. *Phytopathologische Zeitschrift, 45:* 260–288.

Ohshima, K., Harjosudarma, J., Ishikawa, Y., and Shikata, E. 1990. Relationship between hybridoma screening procedures and the characteristics of monoclonal antibodies for use in direct double antibody sandwich ELISA for the detection of plant viruses. *Annals of Phytopathological Society, Japan, 56:* 569.

Ohshima, K. and Shikata, E. 1990. On the screening procedures of ELISA for monoclonal antibodies against three luteoviruses. *Annals of Phytopathological Society, Japan, 56:* 219.

Ohshima, K., Uyeda, I., and Shikata, E. 1988. Production and characteristics of monoclonal antibodies to potato leafroll virus. *Journal of Faculty of Agriculture, Hokkaido University, 68:* 373.

Ohshima, K., Uyeda, I., and Shikata, E. 1989. Characterization of monoclonal antibodies against tobacco necrotic dwarf virus. *Annals of Phytopathological Society, Japan, 55:* 420.

Okoli, C.A.N., Carder, J.H., and Barbara, D.J. 1994. Restriction fragment length polymorphisms (RFLPs) and the relationships of some host-adopted isolates of *Verticillium dahliae. Plant Pathology, 43:* 33–40.

O'Laughlin, G.T. and Chambers, T.C. 1967. The systemic infection of an aphid by a plant virus. *Virology, 33:* 262–271.

Old, K.M., Dudzinsky, M.J., and Bell, J.C. 1988. Isozyme variability in field populations of *Phytophthora cinnamomi* in Australia. *Australian Journal of Botany, 36:* 355–360.

Old, K.M., Moran, G.F. and Bell, J.C. 1984. Isozyme variability among isolates of *Phytophthora cinnamomi* from Australia and Papua New Guinea. *Canadian Journal of Botany, 62:* 2016–2022.

Oliver, R.M. 1973. Negative stain in electron microscopy of protein macromacromolecules. *Methods in Enzymology, 27:* 616–672.

Oliver, R.P., Farman, M.L., Jones, J.D.G., and Hammon-Kosack, K.E. 1993. Use of fungal transformants expressing β-glucuronidase activity to detect infection and measure hyphal biomass in infected plant tissues. *Molecular Plant-Microbe Interactions, 6:* 521–525.

Olivier, C. and Lorna, R. 1998. Detection of *Helminthosporium solani* from soil and plant tissue with species specific PCR primers. *FEMS Microbiology Letters, 168:* 235–241.

Olmos, A., Cambra, M., Dasi, M.A., Candressi, T., Esteban, O. Gorris, M.T. and Asensio, M. 1997. Simultaneous detection and typing of plum pox potyvirus (PPV) isolates by heminested-PCR and PCR-ELISA. *Journal of Virological Methods, 68:* 127–137.

Olmos, A., Dasi, M.A., Candresse, T. and Cambra, M. 1996. Print-capture PCR: a simple and highly sensitive method for the detection of plum pox virus (PPV) in plant tissues. *Nucleic Acids Research, 24:* 2192–2193.

Olson, E.O. 1968. Review of recent research on exocortis disease. In: *Proceedings of 4th Conference on International Organization of Citrus Virologists* (Ed.) J.F.L. Childs, pp. 92–96. University of Florida Press, Gainesville.

Olsson, C.H.B. and Heiberg, N. 1997. Sensitivity of the ELISA test to detect *Phytophthora fragariae* var. *rubi* in raspberry roots. *Journal of Phytopathology, 145:* 285–288.

Omura, T., Takahashi, Y., Shohara, K., Minobe, Y., Tsuchizaki, T., and Nozu, Y. 1986. Production of monoclonal antibodies against rice stripe virus for the detection of virus antigen in infected plants and viruliferous insects. *Annals of Phytopathological Society, Japan, 52:* 270.

Opel, H., Schmidt, H.B., and Kegler, H. 1963–64. Anreicherung und Darstellung von Ringfleckenviren der Kiroche. *Phytopathologische Zeitschrift, 49:* 105–113.

Opgenorth, D.C., Smart, C.D., Louws, P.J. Bruijn, F.J. de, and Kirkpatrick, B.C. 1996. Identification of *Xanthomonas fragariae* field isolates by rep-PCR genomic fingerprinting. *Plant Disease, 80:* 868–873.

Orihara, S. and Yamamoto, T. 1998. Detection of resting spores of *Plasmodiophora brassicae* from soil and plant tissue by enzyme immunoassay. *Annals of Phytopathological Society, Japan, 64:* 569–573.

Orita, M., Suzuki, Y., Sekiya, T., and Hayashi, K. 1989. Rapid and sensitive detection of point mutations and DNA polymorphisms using the polymerase chain reaction. *Genomics, 5:* 874–879.

Osaki, H., Kudo, A. and Ohtsu, Y. 1996. Japanese pear fruit dimple disease by apple scar skin viroid (ASSVd). *Annals of Phytopathological Society, Japan, 62:* 379–385.

Osborn, H.T. 1937. Vein mosaic of red clover. *Phytopathology, 27:* 1051.

Osbourn, A.E., Clarke, B.R., Dow, J.M. and Daniels, M.J. 1991. Partial characterization of avenacinase from *Gaeumannomyces graminis* var. *avenae. Physiological and Molecular Plant Pathology, 38:* 301–312.

Otsuki, Y. and Takebe, I. 1969. Fluorescent antibody straining of tobacco mosaic virus and antigen in tobacco mesophyll protoplasts. *Virology, 38:* 497–499.

Ou, S.H. 1985. *Rice Diseases,* 2nd Edition, Commonwealth Mycological Institute, Surrey, U.K.

Oudemans P. and Coffey, M.D. 1991a. Isozyme comparison within and among world wide sources of three morphologically distinct species of *Phytophthora. Mycological Research, 95:* 19–30.

Oudemans, P. and Coffey, M.D. 1991b. A revised systematics of twelve papillate *Phytophthora* species based on isozyme analysis. *Mycological Research, 95:* 1025–1046.

Overman, M.A., Ko, N.J., and Tsai, J.H. 1992. Identification of viruses and mycoplasmas in maize by use of light microscopy. *Plant Disease, 76:* 318–322.

Owen, H. 1956. Further observations on the pathogenicity of *Calonectria rigidiscula* (Berk. and Br.) Sacc. to *Theobroma cacao* L. *Annals of Applied Biology, 44:* 307–321.

Owens, R.A. 1990. Hybridization techniques for viroid and virus detection: recent refinements. In: *Control of Virus and Viruslike Diseases of Potato and Sweet Potato.* International Potato Center, Lima, Peru, pp. 35–40.

Owens, R.A. and Diener, T.O. 1981. Sensitive and rapid diagnosis of potato spindle tuber virold disease by nucleic acid hybridization. *Science, 213:* 670–672.

Owens, R.A., and Diener, T.O. 1984. Spot hybridization for detection of viroids and viruses. *Methods in Virology, 7:* 173–187.

Owens, R.A., Kiefer, M.C., and Cress, D.E. 1985. Construction of infectious potato spindle tuber viroid cDNA clones: Implications for investigations of viroid structure-function relationships. In: *Subviral Pathogens of Plants and Animals: Viroids and Pirions* (Eds.) K. Maramorosch and J.J. McKelvey Jr., pp. 315–334. Academic Press, New York.

Owens, R.A., Smith, D.R., and Diener, T.O. 1978. Measurement of viroid sequence homology by hybridization with complementary DNA prepared in vitro. *Virology, 89:* 388–394.

Paavanen-Huhtala, S., Hyvönen, J., Bulat, S.A. and Yli-Mattilia, T. 1999. RAPD-PCR, isozyme, rDNA, RFLP and rDNA sequence analysis in identification of Finnish *Fusarium oxysporum* isolates. *Mycological Research, 103:* 625–634.

Padmanabhan, P., Mohanraj, D., Alexander, K.C., and Jothl, R. 1995. Early and rapid detection of sugarcane smut by histological/immunological methods. In: *Detection of Plant Pathogens and Their Management* (Eds.) J.P. Verma, A. Varma, and Dinesh Kumar, pp. 349–356. Angkor Publishers, New Delhi.

Paduch-Cichal, E., Welnicki, M., Skrzeczkowski, S. and Slowinski, A. 1996. Occurrence of the apple scar-skin viroids group in Polish orchards. *Phytopathologia Polonica, No. 11,* 121–126.

Paduch-Cichal, E., Welnicki, M. and Slowinski, A. 1998. Detection of the apple scar-skin viroids group in the different parts of apple trees by dot-blot test. *Phytopathologia Polonica, No. 15,* 33–40.

Palacio, A. and Duran-Vila, N. 1999. Single strand conformation polymorphism (SSCP) analysis as a tool for viroid characterization. *Journal of Virological Methods, 77:* 27–36.

Palacio-Bielsa, A., Foissac, X. and Duran-Villa, N. 1999. Indexing citrus viroids by imprint hybridization. *European Journal of Plant Pathology, 105:* 897–903.

Paliwal, Y.C. 1980. Relationship of wheat streak mosaic and barley stripe mosaic viruses to vector and nonvector eriophyid mites. *Archives in Virology, 63:* 123–132.

Paludan, N. 1965. Carnation virus testing. *Saertryk Tidss. F. Planteavl., 69:* 38–46.

Palukaitis, P. 1984. Detection and characterization of subgenomic RNA in plant viruses. *Methods in Virology, 7:* 259–317.

Palukaitis, P., Rakowskii, A.G., Alexander, D.M., and Symons, R.H. 1981. Rapid indexing of the sunblotch disease of avocados using a complementary DNA probe to avocado sunbloch viroid. *Annals of Applied Biology, 98:* 439–449.

Pan, S.Q., Jin, S., Boulton, M.I., Hawes, M., Gordon, M.P. and Nester, E.W. 1995. An *Agrobacterium* virulence factor encoded by a Ti-plasmid gene or a chromosomal gene is required for T-DNA transfer into plants. *Molecular Microbiology, 17:* 259–269.

Pan, Y.B., Grisham, M.P. and Burner, D.M. 1997. A polymerase chain reaction protocol for the detection of *Xanthomonas albilineans,* the causal agent of sugarcane leaf scald disease. *Plant Disease, 81:* 189–194.

Pan, Y.B., Grisham, M.P., Burner, D.M. Damann, K.E. Jr., and Wei, Q. 1998. A polymerase chain reaction protocol for the detection of *Clavibacter xyli* subsp. *xyli,* the causal bacterium of sugarcane ratoon stunting disease. *Plant Disease, 82:* 285–290.

Pan, Y.B., Grisham, M.P., Burner, D.M., Legendre, B.L. and Wei, Q. 1999. Development of polymerase chain reaction primers highly specific for *Xanthomonas albilineans*, the causal bacterium of sugarcane leaf scald disease. *Plant Disease, 83:* 218–222.

Pan, Y.B., Grisham, M.P., Burner, D.M., Wei, Q. and Damann, K.E. Jr. 1998. Detecting *Clavibacter xyli* subsp. *xyli* by tissue blot DNA hybridization. *Sugar Cane, No. 3,* 3–8.

Panabieres, F., Marais, A., Trentin, F., Bonnet, P., and Ricci, P. 1989. Repetitive DNA polymorphism analysis as a tool for identifying *Phytophthora* species. *Phytopathology, 79:* 1105–1109.

Pappu, S.S., Pappu, H.R., Chang, C.A., Culbreath, A.K. and Todd, J.A. 1998. Differentiation of biologically distinct peanut stripe potyvirus strains by a nucleotide polymorphism—based assay. *Plant Disease, 82:* 1121–1125.

Pardue, M.L., 1985. In situ hybridization. In: *Nucleic Acid Hybridization—Practical Approach* (Eds.) B.D. Hames and S.J. Higgins, pp. 179–202, IRL Press, Oxford.

Parent, J.G., Lacroix, M., Page, D., Vézina, L. and Végicard, S. 1996. Identification of *Erwinia carotovora* from soft rot diseased plants by random amplified polymorphic DNA (RAPD) analysis. *Plant Disease, 80:* 494–499.

Pares, R.D., Gillings, M.R., and Gunn, L.V. 1992. Differentiation of biologically distinct cucumber mosaic virus isolates by PAGE of double stranded RNA. *Intervirology, 34:* 23–29.

Pares, R.D., and Whitecross, M.I. 1982. Gold-labeled antibody decoration (GLAD) in the diagnosis of plant viruses by immuno-electron microscopy. *Journal of Immunological Methods, 51:* 23.

Park WonMok, Choi SeulRan, Kim SuJoong, Choi SeungKook and Ryu KiHyun. 1998. Characterization and RT-PCR detection of turnip mosaic virus isolated from chinese cabbage in Korea. *Korean Journal of Plant Pathology, 14:* 223–228.

Park, WonMok, Shim KirlBo, KimSuJoong, and Ryu KiHyun, 1998. Detection of cymbidium mosaic virus and odontoglossum ringspot virus by ELISA and RT-PCR from cultivated orchids in Korea. *Korean Journal of Plant Pathology, 14:* 130–135.

Pasquini, C., Simeone, A.M., Conte, L. and Barba, M. 1998. Detection of plum pox virus in apricot seeds. *Acta Virologica, 42:* 260–263.

Patel, M.K., Dhande, G.M., and Kulkarmi, Y.S. 1951. Studies on some species of *Xanthomonas*. *Indian Phytopathology, 4:* 123–140.

Paulsen, A.Q., and Fulton, R.W. 1968. Hosts and properties of plum line pattern virus. *Phytopathology, 58:* 766–772.

Paulsen, A.Q. and Fulton, R.W. 1969. Purification, serological relationships and some characteristics of plum line-pattern virus. *Annals of Applied Biology, 63:* 233–240.

Peng RiHe, Han Cheng Gui, Yang Lili, Yu Jian-Lin and Liu Yi, 1998. Cytological localization of beet necrotic yellow vein virus transmitted by *Polymyxa betae. Acta Phytopathologica Sinica, 28:* 257–261.

Pennington, R.E., Sherwood, J.L., and Hunger, R.M. 1993. A PCR-based assay for wheat soil-borne mosaic virus in hard red winter wheat. *Plant Disease, 77:* 1202–1205.

Penrose, L.J. 1974. Micro-inclusions associated with sugarcane mosaic virus infection of sorghum and maize. *Phytopathologische Zeitschrift, 80:* 157–162.

Peralta, E.L., Diaz, C., Lima, H. and Martinez, Y. 1997. Diagnosis of citrus tristeza virus utilising the fluorogenic ultra micro ELISA system. *Fitopatologia, 32:* 112–115.

Peramul, K., Pillay, D. and Pillay, B. 1996. Random amplified polymorphic DNA (RAPD) analysis shows intraspecies differences among *Xanthomonas albilineans* strains. *Letters in Applied Microbiology, 23:* 307–311.

Permar, T.A. and Gottwald, T.R. 1989. Specific recognition of a *Xanthomonas campestris* Florida citrus nursery strain by a monoclonal antibody probe in a microfiltration enzyme immunoassay. *Phytopathology, 79:* 780–783.

Permar, T.A., Garnsey, S.M., Gumpf, D.J., and Lee, R.F. 1990. A monoclonal antibody that discriminates strains of citrus tristeza virus. *Phytopathology, 80:* 224.

Pesic, Z., Hiruki, C., and Chen, M.H. 1988. Detection of viral antigen by immunogold cytochemistry in ovules, pollen and anthers of alfalfa infected with alfalfa mosaic virus. *Phytopathology, 78:* 1027–1032.

Peters, D. 1967. The purification of potato leaf roll virus from its vector *Myzus persicae*. *Virology, 31:* 46–54.

Peters, D. and Black, L.M. 1970. Infection of primary cultures of aphid cells with a plant virus. *Virology, 40:* 847–853.

Peters, D. and Runia, W.T. 1985. The host range of viroids. In: *Subviral Pathogens of Plants and Animals: Viroids and Prions* (Eds.) K. Maramorosch and J.J. McKelvey Jr., pp. 21–31. Academic Press, New York.

Peters, R.D., Platt (Bud), H.W. and Hall, R. 1999. Use of allozyme markers to determine genotype to *Phytophthora infestans* in Canada. *Canadian Journal of Plant Pathology, 21:* 144–153.

Petri, L. 1910. Utersuchungen uber die Dambak—terein der Olivenfliege. *Zentr. Bakteriol. II. 26:* 357–367.

Phan, T.T.H., Khetarpal, P.K., Le, T.A.H. and Maury, Y. 1997. Comparison of immuno-capture-PCR and ELISA in quality control of pea seed for pea seed–borne mosaic potyvirus. In: *Seed Health Testing—Progress Towards the 21st Century* (Eds.) J.D. Hutchins and J.C. Reeves, pp. 193–199. CAB International, U.K.

Phillips, S., Briddon, R.W., Brunt, A.A. and Hull, R. 1999. The partial characterization of a badnavirus infecting the greater asiatic or water yam *(Dioscorea alata). Journal of Phytopathology, 147:* 265–269.

Piacitelli, J. and Santilli, V. 1961. Relationship of tobacco mosaic virus (TMV) lesion number and concentration to the rate of lesion production on pinto bean. *Nature, 191:* 624–625.

Pierce, W.H. 1934. Viroses of the bean. *Phytopathology, 24:* 87–115.

Pierce, W.H. 1935. Identification of certain viruses affecting leguminous plants. *Journal of Agricultural Research, 51:* 1017–1039.

Pietro, A.D. and Roncero, M.I. 1998. Cloning, expression and role in pathogenicity of *pg1* encoding the major extracellular endopolygalacturonase of the vascular wilt pathogens *Fusarium oxysporum. Molecular Plant-Microbe Interactions, 11:* 91–98.

Pio-Ribeiro, G., Winter, S., Jarret, R.L. Demski, J.W. and Hamilton, R.I. 1996. Detection of sweet potato virus disease–associated closterovirus in a sweet potato accession in the United States. *Plant Disease, 80:* 551–554.

Platiño Alvarez, B., Rodríguez Cámara, M.C., Rodriguez Fernández, T., González Jaen, M.T. and Vázquez Estévez, C. 1999. Immunodetection of an exopolygalacturonase in tomato plants infected with *Fusarium oxysporum* f.sp. *radicis lycopersici. Boleín de Sanidad Vegetal, Plagas, 25:* 529–536.

Plyler, T.R., Simone, G.W., Fernandez, D. and Kistler, H.C. 1999. Rapid detection of the *Fusarium oxysporum* lineage containing the Canary Island date palm wilt pathogen. *Phytopathology, 89:* 407–413.

Podleckis, E.V., Hammond, R.W., Hurtt, S.S., and Hadidi, A. 1993. Chemiluminescent detection of potato and porne fruit viroids by digoxigenin-labeled dot-blot and tissue blot hybridization. *Journal of Virological Methods, 43:* 147–158.

Polák, J. 1998. Enzyme-amplified ELISA for detection of beet mild yellowing virus in aphids. *Ochrana Rostlin, 34:* 45–48.

Polak, J. and Kristek, J. 1988. Use of horseradish peroxidase labeled antibodies in ELISA for plant virus detection. *Journal of Phytopathology, 122:* 200–207.

Pollini, C.P., Giunchedi, L. and Bissani, R. 1997. Immunoenzymatic detection of PCR products for the identification of phytoplasmas in plants. *Journal of Phytopathology, 145:* 371–374.

Pollini, C.P., Giunchedi, L. and Bissani, R. 1999. Specific detection of D- and M- isolates of plum pox virus by immunoenzymatic determination of PCR products. *Journal of Virological Methods, 67:* 127–133.

Pooler, M.R., Myung, I.S., Bentz, J., Sherald, J. and Hartung, J.S. 1997. Detection of *Xylella fastidiosa* in potential insect vectors by immuno-magnetic separation and nested polymerase chain reaction. *Letters in Applied Microbiology, 25:* 123–126.

Poplawsky, A.R. and Chun, W. 1998. *Xanthomonas campestris* pv. *campestris* requires a functional *pigB* for epiphytic survival and host infection. *Molecular Plant-Microbe Interactions, 11:* 466–475.

Poplawsky, A.R., Chun, W., Slater, H., Daniels, M.J. and Dow, J.M. 1998. Synthesis of extracellular polysaccharide extracellular enzymes and xanthomonadin in *Xanthomonas campestris:* evidence for the involvement of two intercellular regulatory signals. *Molecular Plant-Microbe Interactions, 11:* 68–70.

Porta, C., Devergne, J.C., Cardin, L., Briand, J.P., and Van Regenmortel, M.H.V. 1989. Serotype specificity of monoclonal antibodies to cucumber mosaic virus. *Archives of Virology, 104:* 271.

Porter, C.A., and McWhorter, F.P. 1952. Crystalline inclusions produced by red clover vein mosaic virus. *Phytopathology, 42:* 518.

Poul, F. and Dunez, J. 1989 Production and use of monoclonal antibodies for the detection of apple chlorotic leafspot virus. *Journal of Virological Methods, 25:* 153.

Powell, C.A. 1987. Detection of three plant viruses by dot-immunobinding assay. *Phytopathology, 77:* 306–309.

Powell, C.A. 1990. Detection of tomato ringspot virus with monoclonal antibodies. *Plant Disease, 7:* 904.

Powell, C.A. and Marquez, E.D. 1983. The preparation of monoclonal antibody to tomato ringspot virus. *Proceedings of 4th International Congress of Plant Pathology, Melbourne, Australia,* pp. 121.

Pradel, K.S., Ullrich, C.I., Santa Cruz, S. and Opar Ka, K.J. 1999. Symplastic continuity in *Agrobacterium tumefaciens*–induced tumours. *Journal of Experimental Botany, 50:* 183–192.

Pratt, M.J. 1961. Studies on clover yellow mosaic and white clover mosaic viruses. *Canadian Journal of Botany, 39:* 655–665.

Prentice, I.W. 1948. Resolution of strawberry virus complexes. II. Virus 2 (mild yellow edge virus). *Annals of Applied Biology, 35:* 279.

Prentice, I.W. and Harris, R.V. 1946. Resolution of strawberry virus complexes by means of the aphid vector *Capitophorus fragariae Theob. Annals of Applied Biology, 33:* 50–53.

Preston, J.F., Rice, J.D., Ingram, L.O. and Keen, N.T. 1992. Differential depolymerization mechanisms of pectate lyases secreted by *Erwinia chrysanthemi*-EC 16. *Journal of Bacteriology, 174:* 2039–2042.

Price, W.C. 1936. Virus concentration in relation to acquired immunity from tobacco ringspot. *Phytopathology, 26:* 503–529.

Price, W.C. 1971. Cadang-cadang of coconut—a review. *Plant Science, 3:* 1–13.

Price, W.C., Williams, R.C., and Wyckoff, R.W.G. 1946. Electron micrographs of crystalline plant viruses. *Archives of Biochemistry, 9:* 175–185.

Priestley, R.A. and Dewey, F.M. 1993. Development of a monoclonal antibody immunoassay for the eyespot pathogen *Pseudocercosporella herpotrichoides. Plant Pathology, 42:* 403–412.

Pring, R.J., Nash, C., Zakaria, M. and Bailey, J.A. 1995. Infection process and host range of *Colletotrichum capsici. Physiological and Molecular Plant Pathology, 46:* 137–152.

Priou, S. and French, E.R. 1997. A simple baiting technique to detect and quantify *Pythium aphanidermatum* in soil. *Fitopatologia, 32:* 187–193.

Prossen, D., Hatziloukas, E., Panopoulos, N.J., and Schaad, N.W. 1991. Direct detection of the halo blight pathogen *Pseudomonas syringae* pv. *phaseolicola* in bean seeds by DNA amplification. *Phytopathology, 81:* 1159.

Prüfer, D., Kawchuk, L., Monecke, M. Nowok, S., Fischer, R. and Rohde, W. 1999. Immunological analysis of potato leafroll luteovirus (PLRV) P1 expression identifies a 25 kDa RNA-binding protein derived via P1 processing. *Nucleic Acids Research, 27:* 421–425.

Pruvost, O., Hartung, J.S., Civeroto, E.L., Dubios, C., and Perrier, X. 1992. Plasmid DNA fingerprints distinguish pathotypes of *Xanthomonas campestris* pv. *citri*, the causal agent of citrus bacterial canker disease. *Phytopathology, 82:* 485–490.

Puchta, H. and Sanger, H.L. 1989. Sequence analysis of minute amounts of viroid RNA using the polymerase chain reaction. *Archives of Virology, 106:* 335–340.

Pugsley, A.P. 1993. The complete general protein secretory pathway in gram-negative bacteria. *Microbiological Reviews, 57:* 50–108.

Purcell, A.H., Saunders, S.R., Hendson, M. Grebus, M.E. and Henry, M.J. 1999. Causal role of *Xylella fastidiosa* in oleander leaf scorch disease. *Phytopathology, 89:* 53–58.

Purcifull, D.E., and Hiebert, E. 1971. Papaya mosaic virus. *CMI/AAB Descriptions of Plant Viruses, No. 56.*

Purcifull, D.E., Hiebert, E., and McDonald, J.G. 1973. Immunochemical specificity of cytoplasmic inclusions induced by viruses in the potato Y group. *Virology, 55:* 275–279.

Purcifull, D.E. and Shepard, J.F. 1967. Western celery mosaic virus in Florida celery. *Plant Disease Reporter. 51:* 502–505.

Qiu, W. and Moyer, J.W. 1999. Tomato spotted wilt tospovirus adapts to the TSWV N gene–derived resistance by genome reassortment. *Phytopathology, 89:* 575–582.

Querci, M., Owens, R.A., Vargas, C., and Salazar, L.F. 1995. Detection of potato spindle tuber viroid in avocado growing in Peru. *Plant Disease, 79:* 196–202.

Quino, A.J. 1989. Serology of *Xanthomonas campestris* pv. *oryzae. Proceedings of International Workshop on Bacterial Blight of Rice,* pp. 19–30. International Rice Research Institute, Philippines.

Rafay, S.A. 1935. Physical properties of sugarcane mosaic virus. *Indian Journal of Agricultural Science, 5:* 663–670.

Rafin, C., Nodet, P., and Tirilly, R. 1994. Immunoenzymatic staining procedure for *Pythium* species with filamentous noninflated sporangia in soilless cultures. *Mycological Research, 98:* 535–541.

Ragupathi, N. 1995. Studies on leafcurl virus disease of tomato (*Lycopersicon esculentum* Mill.). Doctoral Thesis, Tamil Nadu Agricultural University, Coimbatore, India.

Raj, S.K., Saxena, S., Hallan, V. and Singh, B.P. 1998. Reverse transcription–polymerase chain reaction (RT-PCR) for direct detection of cucumber mosaic virus (CMV) in gladiolus. *International Biochemistry and Molecular Biology, 44:* 89–95.

Rajeshwari, N., Shylaja, M.D., Krishnappa, M., Shetty, H.S., Mortensen, C.N. and Mathur, S.B. 1998. Development of ELISA for the detection of *Ralstonia solanacearum* in tomato: its application in seed health testing. *World Journal of Microbiology and Biotechnology, 14:* 697–704.

Rajeswari, S., Palaniswami, A. and Rajappan, K. 1997. A chemodiagnostic method for the detection of symptomless latent infection of *Colletotrichum musae* in banana fruits. *Plant Disease Research, 12:* 52–55.

Ramel, M.E., Gugerli, P., Saugy, R., Crausaz, P.H. and Brugger, J.J. 1998. Virus control of apple, pear and quince by rapid graft-indexing in the greenhouse and isolation of the agents of the major virus diseases. *Revue suisse de Viticulture d' Arboriculture et d' Horticulture, 30:* 13–21.

Ramsdell, D.C., Andrews, R.W., Gillett, J.M., and Morris, C.E. 1979. A comparison between enzyme-linked immunosorbent assay (ELISA) and *Chemopodium quinoa* for detection of peach rosette mosaic virus in 'Concord' grapevines. *Plant Disease Reporter, 63:* 74–78.

Rand, F.V. and Cash, L.C. 1920. *Bacterial Wilt of Cucurbits.* U.S. Department of Agriculture Technical Bulletin No. 828.

Randles, J.W. 1975. Association of two ribonucleic acid species with cadang-cadang disease of coconut palm. *Phytopathology, 65:* 163–167.

Randles, J.W. 1985. Coconut cadang-cadang viroid. In: *Subviral Pathogens of Plants and Animals: Viroids and Prions* (Eds.) K. Maramorosch and J.J. McKelvey Jr., pp. 39–74. Academic Press, New York.

Randles, J.W., Boccardo, G., Reuterma, M.C., and Rillo, E.P. 1977. Transmission of the RNA species associated with cadang-cadang of coconut palm, and the insensitivity of the disease to antibiotics. *Phytopathology, 67:* 1211–1216.

Randles, J.W., Rillo, E.P. and Diener, T.O. 1976. The viroid-like structure and cellular location of anomalous RNA associated with the cadang-cadang disease. *Virology, 74:* 128–139.

Rao, G.N. 1988. Studies on Rice Yellow Dwarf with Special Reference to Epidemilogy and Host Resistance. Doctoral Thesis. Tamil Nadu Agricultural University, Coimbatore, India.

Rassel, A. 1972. Demonstration of paracrystalline inclusions in the endoplasmic reticulum of *Pisum sativum* leaf excrescences induced by pea enation mosaic virus. *C. R. Hebd. Acad. Sci. Scannes Ser. D. 274:* 2871–2878.

Ravelonandro, M., Scorza, R., Bachelier, J.C. Laborme, G., Levy, L. Damsteegt, V., Callahan, A.M. and Dunez, J. 1997. Resistance of transgenic *Prunus domestica* to plum pox virus infection. *Plant Disease, 81:* 1231–1235.

Raymer, W.B., and O'Brien, M.J. 1962. Transmission of potato spindle tuber virus to tomato. *American Potato Journal, 39:* 401–408.

Reddy, A.S., Hobbs, H.A., Delfose, P., Murthy, A.K. and Reddy, D.V.R. 1998. Seed transmission of Indian peanut clump virus (IPCV) in peanut and millets. *Plant Disease, 82:* 343–346.

Reddy, A.V. 1986. Studies on Yellow Dwarf Disease of Rice. Doctoral Thesis, Tamil Nadu Agricultural University, Coimbatore, India.

Reddy, D.V.R. and Black, L.M. 1972. Increase of wound tumour virus in leafhoppers as assayed on vector cell monolayers. *Virology, 50:* 412–421.

Reddy, D.V.R., Ratna, A.S., Sudharsana, M.R., Poul, F., and Kiran Kumar, I. 1992. Serological relationships and purification of bud necrosis virus, a tospovirus occurring in peanut (*Arachis hypogaea* L.) in India. *Annals of Applied Biology, 120:* 279–286.

Reddy, D.V.R., Richins, R.D., Rajeshwari, R. Iizuka, N., Manohar, S.K. and Shepherd, R.J. 1993. Peanut chlorotic streak virus, a new caulimovirus infecting peanuts (*Arachis hypogaea*) in India. *Phytopathology, 83:* 129–133.

Reddy, M. and Reddy, A. 1984. Serotypes of *Xanthomonas campestris* pv. *oryzae. International Rice Research Newsletter, 14:* 17–18.

Reeves, P.J., Whitcombe, D. Wharam, S., Gibson, M., Allison, G. et al., 1993. Molecular cloning and characterization of 13 *out* genes from *Erwinia carotovora* subsp. *carotovora:* genes encoding members of a general secretion pathway (GSP) widespread in gram-negative bacteria. *Molecular Microbiology, 8:* 443–456.

Revers, F., Lot, H., Souche, S., LeGall, O. Candresse, T. and Dunez, J. 1997. Biological and molecular variability of lettuce mosaic virus isolates. *Phytopathology, 87:* 397–403.

Rey, M.E.C. D'Andrea, E., Calvert-Evers, J., Paximadis, M. and Boccardo, G. 1999. Evidence for a phytoreovirus associated with tobacco exhibiting tobacco leafcurl symptoms in South Africa. *Phytopathology, 89:* 303–307.

Rezaian, M.A., Krake, L.R., and L. Golino, D.A. 1992. Common identity of grapevine viroids from USA and Australia revealed by PCR analysis. *Intervirology. 34:* 38–43.

Rich, J.J., Hirano, S.S. and Willis, D.K. 1992. Pathovar specific requirement for *Pseudomonas syringae lemA* gene in disease lesion formation. *Applied and Environmental Microbiology, 58:* 1440–1446.

Richter, J., Augustin, W. and Kleinhempel, H. 1977. Nachweis des Kartofel-S-virus mit Hilfe des ELISA-Testes. *Arch. Phytopathologische Pflanzenschutz. 14:* 73–80.

Richter, J., Proll, E., Rabenstein, F., and Stanarius, A. 1994. Serological detection of members of the Potyviridae with polyclonal antisera. *Journal of Phytopathology. 142:* 11–18.

Ristains, J.B., Madritch, M., Trout, C.L. and Parra, G. 1998. PCR amplification of ribosomal DNA for species identification in the plant pathogenic genus *Phytophthora. Applied and Environmental Microbiology, 64:* 948–954.

Rittenburg, J.H., Petersen, F.P., Grothaus, G.D., and Miller, S.A. 1988. Development of a rapid, field usable immunoassay format, for detection and quantification of *Pythium, Rhizoctonia* and *Sclerotinia* spp. in plant tissue. *Phytopathology, 78:* 156.

Rizos, H., Gunn, L.V., Pares, R.D., and Gillings, M.R. 1992. Differentiation of cucumber mosaic virus isolates using the polymerase chain reaction. *Journal of General Virology, 73:* 2099–2103.

Robb, M.S. 1963. A method for the detection of dahlia mosaic virus in *Dahlia. Annals of Applied Biology, 52:* 145–148.

Robb, M.S. 1964. Location, structure and cytochemical staining reactions of the inclusion bodies found in *Dahlia variabilis* infected with dahlia mosaic virus. *Virology, 23:* 141–144.

Roberts, I.M. 1981. Notes for the course on immunoelectron microscopy of plant viruses. Association of Applied Biologists Workshop on Electron Microscopy Serology, John Innes Institute, England, 1981.

Roberts, I.M. and Harrison, B.D. 1970. Inclusion bodies and tubular structures in *Chemopodium amaranticolor* plants infected with strawberry latent ringspot virus. *Journal of General Virology, 7:* 47.

Roberts, I.M. and Harrison, B.D. 1979. Detection of potato leaf roll and potato mop top viruses by immunosorbent electron microscopy. *Annals of Applied Biology. 93:* 289–297.

Roberts, R.G., Reymond, S.T. and Anderson, B. 2000. RAPD fragment pattern analysis and morphological segregation of small-spored *Alternaria* species and species groups. *Mycological Research, 104:* 151–160.

Robertson, W.L., French, R., and Gray, S.M. 1991. Use of group specific primers and the polymerase chain reaction for the detection and identification of luteoviruses. *Journal of General Virology, 72:* 1473–1477.

Robinson, D.J. and Romero, J. 1991. Sensitivity and specificity of nucleic acid probes for potato leaf roll luteovirus detection. *Journal of Virological Methods, 34:* 209–219.

Rocha-Peña, M.A., Lee, R.F. Lastra, R.A., Niblett, C.L., Ochoa-Corona, F.M., Garsney, S.M. and Yokomi, R.K. 1995. Citrus tristeza virus and its aphid vector *Toxoptera citricida:* threats to citrus production in the Caribbean and Central and North America. *Plant Disease, 79:* 437–445.

Rodoni, B.C., Ahlawat, Y.S., Varma, A., Dale, J.L. and Harding, R.M. 1997. Identification and characterization of banana bract mosaic virus in India. *Plant Disease, 81:* 669–672.

Rogowsky, P.M., Powell, B.S., Shirasu, K. Lin, T.S., Morel, P. et al. 1990. Molecular characterization of the vir regulon of *Agrobacterium tumefaciens* complete nucleotide sequence and gene organization of the 28.63. kbp regulon cloned as single unit. *Plasmid, 23:* 85–106.

Rohloff, I. 1968. Virusprufung von Salatsamen nach dem *Chenopodium quinoa*-test von Marrou. *Saatgat Writ. 20:* 157–159.

Roland, G. 1962. Etude d'un virus isolé mecaniquement du cerisier. *Parasitica, 18:* 264–285.

Roland, G. 1969. Etude du Tomato black-ring disease. *Parasitica, 25:* 9–18.

Rollo, F., Amici, A., Foesi, F., and di Silvestro, L. 1987. Construction and characterization of a cloned probe for the detection of *Phoma tracheiphlia* in plant tissues. *Applied Microbiological Biotechnology, 26:* 352.

Rollo, F., Salvi, R., and Torchin, P. 1990. Highly sensitive and fast detection of *Phoma tracheiphila* by polymerase chain reaction. *Applied Microbiological Biotechnology, 32:* 572–576.

Romaine, C.P. and Horst, R.K. 1975. Suggested viroid etiology for chrysanthemum chlorotic mottle disease. *Virology, 64:* 86–95.

Romeiro, R.S., Silva, H.S.A., Beriam, L.O.S. Rodrigues Neto, J. and Carvalho, M.G. de. 1999. A bioassay for detection of *Agrobacterium tumefaciens* in soil and plant material. *Summa Phytopathologica, 25:* 359–362.

Romero-Durban, J., Cambra, M. and Duran-Vila, N. 1995. A simple imprint-hybridization method for detection of viroids. *Journal of Virological Methods, 55:* 37–47.

Rose, D.G., and Hubbard, A.L. 1986. Production of monoclonal antibodies for the detection of potato virus Y. *Annals of Applied Biology, 109:* 317.

Rosell, R.C., Torres-Jerez, I. and Brown, J.K. 1999. Tracing the geminivirus-whitefly transmission pathway by polymerase chain reaction in whitefly extracts, saliva, hemolymph and honey dew. *Phytopathology, 89:* 239–246.

Rosner, A. and Maslenin, L. 1999. Differentiation of PVY^{NTN} by unique single-restriction cleavage of PCR products. *Potato Research, 42:* 215–221.

Rosner, A., Maslenin, L. and Spiegel, S. 1997. The use of short and long PCR products for improved detection of prunus necrotic ringspot virus in woody plants. *Journal of Virological Methods, 67:* 135–141.

Rosner, A., Maslenin, L. and Spiegel, S. 1998. Differentiation among isolates of *Prunus* necrotic ringspot virus by transcript conformation polymorphism. *Journal of Virological Methods, 74:* 109–115.

Rosner, A., Maslenin, L. and Spiegel, S. 1999. Double-stranded conformation polymorphism of heterologous RNA transcripts and its use for virus strain differentiation. *Plant Pathology, 48:* 235–239.

Rosner, A., Shiboleth, Y., Spiegel, S., Krisbai, L. and Kölber, M. 1998. Evaluating the use of immunocapture and sap-dilution PCR for the detection of prunus necrotic ringspot virus. *Acta Horticlturae, No. 472,* 227–233.

Ross, A.F. 1948. Local lesions with potato virus Y. *Phytopathology, 38:* 930–932.

Rott, P., Davis, M.J., and Baudin, P. 1994. Serological variability in *Xanthomonas albilineans,* causal agent of leaf scald disease of sugarcane. *Plant Pathology, 43:* 344–349.

Rouhiainen, L., Laaksonen, M., Karjalainan, R., and Soderland, H. 1991. Rapid detection of a plant virus by solution hybridization using oligonucleotide probes. *Journal of Virological Methods, 34:* 81–90.

Rowhani, A., Biardi, L., Routh, G., Daubert, S.D. and Golino, D.A. 1998. Development of a sensitive colorimetric PCR assay for detection of viruses in woody plants. *Plant Disease, 82:* 880–884.

Rowhani, A., Maningas, M.A., Lile, L.S., Daubert, S.D., and Golino, D.A. 1995. Development of a detection system for viruses of woody plants based on PCR analysis of immobilized virions. *Phytopathology, 85* 347–352.

Roy, B.P., Abou Haider, M.G., Sit, J.L., and Alexander, A. 1988. Construction and use of cloned cDNA biotin and ^{32}P labeled probes for the detection of papaya mosaic potex virus RNA in plants. *Phytopathology, 78:* 1425–1429.

Rubinson, E., Galiakparov, N., Radian, S., Sela, I., Tanne, E. and Gafny, R. 1997. Serological detection of grapevine virus A using antiserum to a nonstructural protein, the putative movement protein. *Phytopathology, 87:* 1041–1045.

Rubio, L., Ayylón, M.A., Guerri, J., Pappu, H., Niblett, C. and Moreno, P. 1996. Differentiation of citrus tristeza closterovirus (CTV) isolates by single-stranded conformation polymorphism analysis of coat protein gene. *Annals of Applied Biology, 129:* 479–489.

Rubio-Huertos, M. 1950. Estudios sobre inclusiones intracellulares producids for virus en las plantas. *Microbiol. Espan. 3:* 207–232.

Rubio-Huertos, M. 1956. Origin and composition of cell inclusions associated with certain tobacco and crucifer viruses. *Phytopathology, 46:* 553–556.

Rubio-Huertos, M. 1972. Inclusion bodies. In: *Principles and Techniques in Plant Virology* (Eds.) C.I. Kado and H.O. Agrawal, pp. 62–75. Van Nostrand Reinhold Co., New York.

Rudolph, B.A. 1943. A possible relationship between the walnut erinose mite and walnut blight. *Science, 98:* 430–431.

Rush, C.M., French, R., and Heidel, G.B. 1994. Differentiation of two closely related furoviruses using the polymerase chain reaction. *Phytopathology, 84:* 1366–1369.

Russo, M., Martelli, G.P., and DiFranco, A. 1981. The fine structure of local lesions of beet necrotic yellow vein virus in *Chenopodium amaranticolor. Physiological Plant Pathology, 19(2):* 237–242.

Russo, M., Martelli, G.P., and Savino, V. 1982. Immunosorbent electron microscopy for detecting sap-transmissible viruses of grapevine. *Proceedings of the 7th Meeting of International Council for the Study of Viruses and Virus-Like Diseases of the Grapevine* (Ed.) A.J. McGinnis, Niagara Falls, 1980, pp. 251–257.

Ryba-White, M., Notteghem, J.L., and Leach, J.E. 1995. Comparison of *Xanthomonas oryzae* pv. *oryzae* strains from Africa, North America and Asia by restriction fragment length polymorphism analysis. *International Rice Research Notes, 20(1):* 25–26.

Rybicki, E.P. 1995. *Bromoviridae.* In: *Virus Taxonomy: Sixth Report of the International Committee on Taxonomy of Viruses* (Eds.) F.A. Murphy C.M. Fauquet, D.H.L. Bishop, S.A. Ghabriel, A.W. Jarvis, G.P. Martelli, M.A. Mayo and M.D. Summers, pp. 450–457. Springer-Verlag, New York.

Rybicki, E.P. and Hughes, F.L. 1990. Detection and typing of maize streak virus and other distinctly related geminiviruses of grasses by polymerase chain reaction amplification of a conserved viral sequence. *Journal of General Virology, 71:* 2519–2526.

Ryu, K.H., Kim, C.H. and Palukaitis, P. 1998. The coat proteins of cucumber mosaic virus is a host range determinant for infection of maize. *Molecular Plant-Microbe Interactions, 11:* 351–367.

Saarma, M., Jarvekulg, L., Hemmila, I., Sütari, H., and Sinijarv, R. 1989. Simultaneous quantification of two plant viruses of double-label time-resolved immuno-fluorometric assay. *Journal of Virological Methods, 23:* 47.

Sachadyn, P. and Kur, J. 1997. A new PCR system for *Agrobacterium tumefaciens* detection based on amplification of T-DNA fragment. *Acta Microbiologica Polonica, 46:* 145–156.

Saeed, E.M., Roux, J., and Cousin, M.T. 1993. Studies of polyclonal antibodies for the detection of MLOs associated with faba bean (*Vicia faba* L.) using different ELISA methods and dot blot. *Journal of Phytopathology, 137:* 33–43.

Saeed, E., Seemuller, E., Schneider, B., Saillard, C., Blanchard, B., Bertheau, Y., and Cousin, M.T. 1994. Molecular cloning, detection of chromosomal DNA of the mycoplasma-like organism (MLO) associated with faba bean (*Vicia faba* L.) phyllody by southern blot hybridization and polymerase chain reaction (PCR). *Journal of Phytopathology, 142:* 97–106.

Saito, M., Kiguchi, T. and Tamada, T. 1997. Non-radioactive digoxigenin—labeled DNA probes for the detection of five RNA species present in beet necrotic yellow vein virus. *Bulletin of Research Institute for Bioresources, Okayama University, 5:* 79–96.

Saito, N., Ohara, T., Sugimoto, T., Hayashi, Y., Hataya, T. and Shikata, E. 1995. Detection of citrus exocortis viroid by PCR-microplate hybridization. *Research Bulletin of Plant Protection Service, Japan, No. 31,* 47–55.

Saito, Y., Yamanaka, K., Watanabe, Y., Takamatsu, N., Meshi, T. and Okada, Y. 1989. Mutational analysis of the coat protein gene of tobacco mosaic virus in relation to hypersensitive response in tobacco plants with the N' gene. *Virology, 173:* 11–20.

Sakimura, K. 1962. The present status of thrips-bone viruses. In: *Biological Transmission of Disease Agents* (Ed.) K. Maramorosch, pp. 33–40. Academic Press, New York.

Saksena, K.N. and Mink, G.L. 1969. Purification and properties of apple chlorotic leafspot virus. *Phytopathology, 59:* 84–88.

Salaman, R.N. 1933. Protective inoculation against a plant virus. *Nature (Lond.) 131:* 468.

Salazar, L.F., Querci, M., Bartolini and Lazarte, V. 1995. Aphid transmission of potato spindle tuber viroid assisted by potato leafroll virus. *Fitopatologia, 30:* 56–58.

Salazar, O., Julian, M.C. and Rubio, V. 2000. Primers based on specific rDNA-ITS sequences for PCR detection of *Rhizoctonia solani, R. solani* AG2 subgroups and ecological types and binucleate *Rhizoctonia. Mycological Research, 104:* 281–285.

Saldarelli, P., Barbarossa, L., Griew, F. and Gallitelli, D. 1996. Digoxigenin-labeled riboprobes applied to phytosanitary certification of tomato in Italy. *Plant Disease, 80:* 1343–1346.

Saldarelli, P., Minafra, A., Martelli, G.P., and Walter, B. 1994. Detection of grapevine leaf roll associated closterovirus III by molecular hybridization. *Plant Pathology, 43:* 91–96.

Salinas, J. and Schots, A. 1994. Monoclonal antibodies based immunofluorescence test for detection of conidia of *Botrytis cinerea* on cut flowers. *Phytopathology, 84:* 351–356.

Salmond, G.P.C. 1994. Secretion of extracellular virulence factors by plant pathogenic bacteria. *Annual Review of Phytopathology, 32:* 181–200.

Samson, R.W. and Imle, E.P. 1942. A ringspot type of virus disease of tomato. *Phytopathology, 32:* 1037–1047.

Sanchez-Navarro, J.A., Aparicio, F. Rowhani, A. and Pallas, V. 1998. Comparative analysis of ELISA, nonradioactive molecular hybridization and PCR for the detection of prunus necrotic ringspot virus in herbaceous and *Prunus* hosts. *Plant Pathology, 47:* 780–786.

Sandrock, R.W. and VanEtten, H.D. 1998. Fungal sensitivity to and enzymatic degradation of the phytoanticipin α-tomatine. *Phytopathology, 88:* 137–143.

Sänger, H.L., Klotz, G., Riesner, D., Gross, H.J., and Kleinachmidt, A.K. 1976. Viroids are single-stranded covalently closed circular RNA molecules existing as highly base-paired rodlike structures. *Proceedings of National Academy of Sciences, USA, 73:* 3852–3856.

Sanger, M., Järlfors, U.E. and Ghabriel, S.A. 1998. Unusual cytoplasmic inclusions induced in tobacco by peanut stunt virus subgroup II strains map to RNA3. *Phytopathology, 88:* 1192–1199.

Sano, T. and Ishiguro, A. 1996. A simple and sensitive nonradioactive microplate hybridization for the detection and quantification of picograms of viroid and viral RNA. *Archives of Phytopathology and Plant Protection, 30:* 303–312.

Sano, T., Smith, C.L., and Cantor, C. 1992. Immuno-PCR very sensitive antigen detection by means of specific antibody DNA complexes. *Science, 258:* 120–122.

Sano, T., Uyeda, I., and Shikata, E. 1983. Comparative studies of hop stunt viroid and cucumber pale fruit viroid by polyacrylamide gel electrophoresis. Rept. Grant-in-Aid for Cooperative Research, B. 1982. Ministry of Education, Japan, pp. 26–36.

Santa Cruz, S. and Baulcombe, D.C. 1993. Molecular analysis of potato virus X isolates in relation to the potato hypersensitivity gene *Nx. Molecular Plant-Microbe Interactions, 6:* 707–714.

Sasaki, M., Fukamizu, T., Yamamoto, K., Ozawa, T., Kagami, Y., and Shikata, E. 1981. Detection of HSV in the latent-infected hops by mass diagnosis and its application. *Annals of Phytopathological Society, Japan, 47:* 416.

Sasaki, M. and Shikata, E. 1977a. Studies on the host range of hop stunt disease in Japan. *Proceedings of Japan Academy, 53B:* 103–108.

Sasaki, M. and Shikata, E. 1977b. On some properties of hop stunt disease agent, a viroid. *Proceedings of Japan Academy of Sciences, 53B:* 109–112.

Sasaki, M. and Shikata, E. 1980. Hop stunt disease, a new viroid disease occurring in Japan. *Review of Plant Protection Research, 13:* 97–113.

Sasaya, T., Nou, Y. and Koganezawa, H. 1998. Biological and serological comparisons of bean yellow mosaic virus (BYMV) isolates in Japan. *Annals of Phytopathological Society, Japan, 64:* 24–33.

Satyanarayana, T., Sreenivasulu, P., Naidu, M.V. and Padmanabhan, G. 1997. Nonradioactive DNA probe for detection of peanut chlorotic streak virus in groundnut. *Indian Journal of Experimental Biology, 35:* 1007–1020.

Sauvage, C. and Expert, D. 1994. Differential regulation by iron of *Erwinia chrysanthemi* pectate lyases: pathogenicity of iron transport regulatory (*cbr*) mutants. *Molecular Plant-Microbe Interactions, 7:* 71–77.

Savage, S.D. and Sall, M.A. 1981. Radioimmunoassay for *Botrytis cinerea. Phytopathology, 71:* 411–415.

Schaad, N.W. 1988. *Laboratory Guide for Identification of Plant Pathogenic Bacteria.* American Phytopathological Society, St. Paul, Minnesota.

Schaad, N.W., Azad, H., Peet, R.C., and Panopoulos, N.J. 1989. Identification of *Pseudomonas syringae* pv. *phaseolicola* by a DNA hybridization probe. *Phytopathology, 79:* 903.

Schaad, N.W., Cheong, S.S., Tamaki, S., Hatziloukas, E., and Panopoulos, N.J. 1995. A combined biological and enzymatic amplification (BIO-PCR) technique to detect *Pseudomonas syringae* pv. *phaseolicola* in bean seed extracts. *Phytopathology, 85:* 243–248.

Schaad, N.W. and Forster, R.L. 1985. A semi-selective agar medium for isolating *Xanthomonas campestris* pv. *translucens* from wheat seeds. *Phytopathology, 75:* 260–263.

Schadewijk, A.R. van. 1996. Detection of tomato spotted wilt virus in dahlia. *Acta Horticulturae, No. 432,* 384–391.

Schaff, D., Lee, I.M., and Davis, R.E. 1992. Sensitive detection and identification of mycoplasma-like organisms in plants by polymerase chain reaction. *Biochemical and Biophysical Research Communication, 186:* 1503–1509.

Scheffer, N.P., Nelson, R.R. and Ullstrup, A.J. 1967. Inheritance of toxin production and pathogenicity in *Cochliobolus carbonum* and *Cochlibolus victoriae. Phytopathology, 57:* 1288–1291.

Schenck, S. 1998. Evaluation of a PCR amplification method for detection of systemic smut infections in sugarcane. *SugarCane, No. 6,* 2–5.

Schesser, K., Luder, A., and Henson, J.M. 1991. Use of polymerase chain reaction to detect the take all fungus *Gaeumannomyces graminis* in infected wheat plants. *Applied Environmental Microbiology, 57:* 553.

Schieber, O., Seddas, A., Belin, C. and Walter, B. 1997. Monoclonal antibodies for detection, serological characterization and immunopurification of grapevine fleck virus. *European Journal of Plant Pathology, 103:* 767–774.

Schilling, A.G., Möller, E.M. and Geiger, H.H. 1996. Polymerase chain reaction-based assays for species specific detection of *Fusarium culmorum, F. graminearum* and *F. avenaceum. Phytopathology, 86:* 515–522.

Schlegel, D.E. and De Lisle, D.E. 1971. Viral protein in early stages of clover yellow mosaic virus infection of *Vicia faba. Virology, 45:* 747–754.

Schlenzig, A., Habermeyer, J. and Zinkernagel, V. 1999. Serological detection of latent infection with *Phytophthora infestans* in potato stems *Zeitschrift für Pflanzenkrankheiten und Pflanzenschutz, 106:* 221–230.

Schmechel, D., McCartney, H.A. and Magan, N. 1997. A novel spore trap for monitoring airborne *Alternaria brassicae* spores using monoclonal antibodies. In: *Diagnosis and Identification of Plant Pathogens* (Eds.) H.W. Dehne, G. Adam, M. Dickmann, F. Frahm, A. Mauler-Machnik and P. van Halteren, pp. 234–242. Kluwer Academic Publishers, Netherlands.

Schmelzer, K. 1958. Wirtspflanzen des Tomatenzwerbusch virus. *Z. Pflkrankh, 65:* 80–89.

Schmelzer, K. 1965. Das Wassermelonen mosaik-Virus tritt auch in Deutschland auf. *Z. Nachr. Pflanzensch, 3:* 69–71.

Schmelzer, K. 1969. Das Ulmenscheckungs-Virus. *Phytopathologische Zeitschrift, 64:* 39–67.

Schmelzer, K. 1970. Zur Differezierung von Herkunsten des Tomaten-schwarzring Virus (tomato black ring virus) durch Serologie und Pramunitat. *Arch. Pflanzenschutz, 6:* 273–287.

Schmidt, H.B. and Schmelzer, K. 1964. Elektronmikroskopische Darstellung und Vermessung des Zwiebelgelbstreifen-Virus. *Phytopathologische Zeitschrift, 50:* 191–195.

Schnabel, G., Schnabel, E.L. and Jones, A.L. 1999. Characterization of ribosomal DNA from *Venturia inaequalis* and its phylogenetic relationship to rDNA from other tree-fruit *Venturia* species. *Phytopathology, 89:* 100–108.

Schneider, B. and Gibb, K.S. 1997. Detection of phytoplasmas in declining pears in southern Australia. *Plant Disease, 81:* 254–258.

Schnell, R.J., Kuhn, D.N., Ronning, C.M. and Harkins, D. 1997. Application of RT-PCR for indexing avocado sunblotch viroid. *Plant Disease, 81:* 1023–1026.

Schonbeck, F. and Spengler, 1979. Nachweis von TMV in Mycorrhizen-haltengen zellen der Tomate mit Hilfe der Immunofluoreszenz. *Phytopathologische Zeitschrift, 94:* 84–86.

Schramm, G. and Röttger B. 1959. Untersuchungen Über das Tabakmosaikvirus mit fluoreszierenden Antikörpern. *Z. Naturforsch, 146:* 510–515.

Schumacher, J., Randles, J.W., and Riesner, D. 1983. Viroid and virusoid detection: an electrophoretic technique with the sensitivity of molecular hybridization. *Analytical Biochemistry, 135:* 288–295.

Schwarz, K. and Jelkmann, W. 1998. Detection and characterization of European apple stem pitting virus sources from apple and pear by PCR and partial sequence analysis. *Acta Horticulturae, No. 472,* 75–85.

Schwarz, R.E. 1968. Indexing of greening and exocortis through fluorescent marker substance. *Proceedings of 4th Conference of Citrus Virologists,* University of Florida, pp. 118–123.

Scoft, S.W., McLaughlin, M.R., and Ainsworth, A.J. 1989. Monoclonal antibodies produced to bean yellow mosaic virus and pea mosaic virus which cross-react among the three viruses. *Archives in Virology, 108:* 161.

Scortichini, M., Dettori, M.T., Marchesi, V., Palombi, M.A. and Rossi, M.P. 1998. Differentiation of *Pseudomonas avellanae* strains from Greece and Italy by rep-PCR genomic fingerprinting. *Journal of Phytopathology, 146:* 417–420.

Scortichini, M., Marchesi, U., Rossi, M.P., Angelucci, L. and Dettori, M.T. 2000. Rapid identification of *Pseudomonas avellanae* field isolates causing hazelnut decline in central Italy by repetitive PCR genomic fingerprinting. *Journal of Phytopathology, 148:* 153–159.

Scott, H.A. 1963. Purification of cucumber mosaic virus. *Virology, 20:* 103–106.

Scott-Craig, J.S., Panaccione, D.G., Cervone, F. and Walton, J.D. 1990. Endopolygalacturonase is not required for pathogenicity for *Cochliobolus carbonum* on maize. *Plant Cell, 2:* 1191–1200.

Seal, S.E., Taghavi, M., Fegan, N., Hayward, A.C. and Fegan, M. 1999. Determination of *Ralstonia (Pseudomonas) solanacearum* rDNA subgroups by PCR tests. *Plant Pathology, 48:* 115–120.

Seddas, A., Haidar, M.M., Greif, C., Jacquet, C., Cloquemin, G. and Walter, B. 2000. Establishment of a relationship between grapevine leafroll closteroviruses 1 and 3 by use of monoclonal antibodies. *Plant Pathology, 49:* 80–85.

Sefc, K.M., Leonhardt, W., and Steinkellner, H. 2000. Partial sequence identification of grapevine leafroll-associated virus-1 and development of a highly sensitive IC-RT-PCR detection method. *Journal of Virological Methods, 86:* 101–106.

Seifers, D.L., Harvey, T.L., Martin, T.J., and Jensen, S.G. 1997. Identification of the wheat curl mite as the vector of the High Plains virus of corn and wheat. *Plant Disease, 81:* 1161–1166.

Semancik, J.S., Conejero, V., and Gerhart, J. 1977. Citrus exocortis viroid: survey of protein synthesis in *Xenopus laevis* oocytes following addition of viroid RNA. *Virology, 80:* 218–221.

Semancik, J.S., Gumpf, D.J., and Bash, J.A. 1992. Interference between viroids inducing exocortis and cachexia diseases in citrus. *Annals of Applied Biology, 121:* 577–583.

Semancik, J.S. and Weathers, L.G. 1972. Exocortis virus: an infectious free nucleic acid plant virus with unusual properties. *Virology, 47:* 456–466.

Seoh M.-L., Wong S.-M. and Lee Z. 1998. Simultaneous TD/RT-PCR detection of cymbidium mosaic potexvirus and odontoglossum ringspot tobamovirus with a single pair of primers. *Journal of Virological Methods, 72:* 197–204.

Sequeira, O.A. De. 1967. Purification and serology of an apple mosaic virus. *Virology, 31:* 314–322.

Serio, F. Di, Aparicio, F., Alioto, D., Ragozzino, A. and Flores, R. 1996. Identification and molecular properties of a 306-nucleotide viroid associated with apple dimple fruit disease. *Journal of General Virology, 77:* 2833–2837.

Serjeant, E.P. 1967. Some properties of cocks foot mottle virus. *Annals of Applied Biology, 59:* 31–38.

Severin, H.H.P. 1929. Additional host plants of curly top. *Hilgardia, 3:* 595–636.

Shaarwy, M.A., Hu, J.S., Xie, W.S., and Sether, D. 1997. Reactivity of some Hawaiian banana bunchy top diseased samples using DAS-ELISA and a digoxigenin-labeled DNA probe. *Egyptian Journal of Agricultural Research, 75:* 515–527.

Shalitin, D., Mawassi, M., Gafry, R., Leitner, O., Cabilly, S., Eshhar, Z., and Bar-Joseph, M. 1994. Serological characterization of citrus tristeza virus isolates from Israel. *Annals of Applied Biology, 125:* 105–113.

Shalla, T.A. and Amici, A. 1967. The distribution of viral antigen in cells infected with tobacco mosaic virus as revealed by electron microscopy. *Virology, 31:* 78–91.

Shalla, T.A. and Shephard, J.F. 1972. The structure and antigenic analysis of amorphous inclusion bodies induced by potato virus. *Virology, 49:* 654–667.

Shalla, T.A., Shepherd, R.J., and Peterson, L.J. 1980. Comparative cytology of nine isolates of cauliflower mosaic virus. *Virology, 102:* 381–388.

Shamloul, A.M., Hadidi, A., Madkour, M.A. and Makkouk, K.M. 1999. Sensitive detection of banana bunchy top and faba bean necrotic yellows virus from infected leaves, in vitro tissue cultures and viruliferous aphids using polymerase chain reaction. *Canadian Journal of Plant Pathology, 21:* 326–337.

Shamloul, A.M., Hadidi, A., Zhu, S.F., Singh, R.P. and Sagredo, B. 1997. Sensitive detection of potato spindle tuber viroid using RT-PCR and identification of a viroid variant naturally infecting pepino plants. *Canadian Journal of Plant Pathology, 19:* 89–91.

Shamloul, A.M., Minafra, A., Hadidi, A., Waterworth, H.E., Giunchedi, L. and Allam, E.K. 1995. Peach latent mosaic viroid: nucleotide sequence of an Italian isolate, sensitive detection using RT-PCR and geographic distribution. *Acta Horticulturae, No. 386*, 522–530.

Shan Z.-H., Liao B.-S., Tan Y.J., LiDong, LeiYong and Shen M.Z. 1997. ELISA technique used to detect latent infection of groundnut by bacterial wilt *(Pseudomonas solanacearum)*. *Oilcrops of China, 19:* 45–47.

Shankar, M., Gregory, A., Kalkhoven, M.J., Cowling, W.A. and Sweetingham, M.W. 1998. A competitive ELISA for detecting resistance to latent stem infection by *Diaporthe toxica* in narrow-leafed lupins. *Australasian Plant Pathology, 27:* 251–258.

Sheffield, F.M. 1931. The formation of intracellular inclusions in solanaceous hosts infected with aucuba mosaic of tomato. *Annals of Applied Biology, 21:* 440–453.

Shelby, R.A. and Kalley, V.C. 1992. Detection of ergot alkaloids from *Claviceps* sp. in agricultural products by competitive ELISA, using a monoclonal antibody. *Journal of Agricultural and Food Chemistry, 40:* 1090–1092.

Shen, K.A., Meyers, B.C., Nurul Islam Faridi, M., Chin, D.B., Stelly, D.M. and Michelmore, R.W. 1998. Resistance gene candidates identified by PCR with degenerate oligonucleotide primers map to clusters of resistance genes in lettuce. *Molecular Plant-Microbe Interactions, 11:* 815–823.

Sheng, J. and Citrovsky, V. 1996. *Agrobacterium*-plant cell DNA transport have virulence proteins, will travel. *The Plant Cell, 8:* 1699–1710.

Shepherd, J.F., and Grogan, R.G. 1971. Celery mosaic virus. *CMI/AAB Descriptions of Plant Viruses, No. 50.*

Shepherd, R.J. 1970. Cauliflower mosaic virus. *CMI/AAB Descriptions of Plant Viruses, No. 24.*

Shepherd, R.J. 1971. Southern bean mosaic virus. *CMI/AAB Descriptions of Plant Viruses, No. 57.*

Shepherd, R.J. and Pound, G.S. 1960. Purification of turnip mosaic virus. *Phytopathology, 50:* 797–803.

Shepherd, R.J., Wakeman, R.J., and Romanko, R.R. 1968. DNA in cauliflower mosaic virus. *Virology, 36:* 150–152.

Shepherdson, S., Esau, K., and McCrum, R. 1980. Ultrastructure of potato leaf phloem infected with potato leaf roll virus. *Virology, 105:* 379–392.

Sherwood, J.L., Sanborn, M.R., and Keyser, G.C. 1985. Production of monoclonal antibodies to peanut mottle virus and wheat streak mosaic virus. *Phytopathology, 75:* 1358.

Sherwood, J.L., Sanborn, M.R., and Keyser, G.C., 1987. Production of monoclonal antibodies to peanut mottle virus and their use in enzyme-linked immunosorbent assay and dot immunobinding assay. *Phytopathology, 77:* 1158–1161.

Sherwood, J.L., Sanborn, M.R., Keyser, G.C., and Myers, L.D. 1989. Use of monoclonal antibodies in detection of tomato spotted wilt virus. *Phytopathology, 79:* 61–64.

Shikata, E. 1962. Observations of rice dwarf virus by means of electron microscope. *Plant Protection, 16:* 313–338.

Shikata, E. 1981. Reoviruses: In: *Hand Book of Plant Virus Infections and Comparative Diagnosis* (Ed.) E. Kurstak, pp. 423–451. Elsevier/North-Holland. Amsterdam.

Shikata, E. 1985. Hop stunt viroid and hop stunt disease. In: *Subviral Pathogens of Plants and Animals: Viroids and Prions* (Eds.) K. Maramorosch and J.J. McKelvey Jr., pp. 101–121, Academic Press, New York.

Shikata, E. and Maramorosch, K. 1967. Electron microscopy of wound tumor virus assembly sites in insect vectors and plants. *Virology, 32:* 363–377.

Shiwaku, K., Iwai, T. and Yamamoto, Y. 1996. Detection of chrysanthemum stunt viroid by dot-blot hybridization using DIG-labeled probe. *Proceedings of Kansai Plant Protection Society, No. 38,* 1–6.

Shoen, C.D., Knorr, D. and Leone, G. 1996. Detection of potato leafroll virus in dormant potato tubers by immunocapture and a flurogenic 5' nuclease RT-PCR assay. *Phytopathology, 86:* 993–999.

Sicard, D., Michalakis, Y., Dron, M. and Neema, C. 1997. Genetic diversity and pathogenic variation of *Colletotrichum lindemuthianum* in the three centers of diversity of its host *Phaseolus vulgaris. Phytopathology, 87:* 807–813.

Siddique, A.B.M., Guthrie, J.N., Walsh, K.B., White, D.T. and Scott, P.T. 1998. Histopathology and within-plant distribution of the phytoplasma associated with Australian papaya dieback. *Plant Disease, 82:* 1112–1120.

Siegel, B.M. 1964. *Modern Developments in Electron Microscopy.* Academic Press, New York, 432 p.

Siitari, H. and Kurppa, A. 1987. Time-resolved fluoroimmunoassay in the detection of plant viruses. *Journal of General Virology, 68:* 1423.

Siitari, H., Lovgren, T., and Halonen, P. 1986. Detection of viral antigen by direct one incubation time-resolved fluoroimmunoassay. In: *Developments and Applications in Virus Testing* (Eds.) R.A.C. Jones and L. Torrance, Vol. 1, p. 155. Association of Applied Biologists, Wellesbourne, U.K.

Silué, D., Tharreau, D., Talbot, N.J., Clergeot, P.H., Notteghem, J.L. and Lebrum, M.H. 1998. Identification and characterization of *apf1⁻* in a nonpathogenic mutant of the rice blast fungus *Magnaporthe grisea* which is unable to differentiate appressia. *Physiological and Molecular Plant Pathology, 53:* 239–251.

Silva, M.C., Nicole, M., Rijo, L., Geiger, J.P. and Rodriguez, C.J.Jr. 1999. Cytochemical aspects of the plant-rust fungus interface during compatible interaction *Coffea arabica* (cv. Catura)–*Hemileia vastatrix* (race III). *International Journal of Plant Sciences, 160:* 79–91.

Sinclair, W.A., Iuli, R.T., Dyer, A.T., and Larsen, A.D. 1989. Sampling and histological procedures for diagnosis of ash yellows. *Plant Disease, 73:* 432–435.

Sinclair, W.A., Griffiths, H.M. Davis, R.E., and Lee, I.M. 1992. Detection of ash yellows mycoplasma-like organisms in different tree organs and in chemically preserved specimens by cDNA probe vs DAPI. *Plant Disease, 76:* 154–158.

Singh, M., Singh, R.P. and Moore, L. 1999. Evaluation of NASH and RT-PCR for the detection of PVY in dormant tubers and its comparison with visual symptoms and ELISA in plants. *American Journal of Potato Research, 76:* 61–66.

Singh, R.P. 1970. Occurrence, symptoms and diagnostic hosts of strains of potato spindle tuber virus. *Phytopathology, 60:* 1314.

Singh, R.P. 1971. A local lesion host for potato spindle tuber virus. *Phytopathology, 61:* 1034–1035.

Singh, R.P. and Bagnall, R.H. 1968. *Solanum rostratum* Dunal., a new test plant for the potato spindle tuber virus. *American Potato Journal, 45:* 335–336.

Singh, R.P., Boucher, A., and Seabrook, J.E.A. 1988. Detection of the mild strains of potato spindle tuber viroid from single true potato seed by return electrophoresis. *Phytopathology, 78:* 663–667.

Singh, R.P., Kurz, J., Boiteau, G. and Moore, L.M. 1997. Potato leafroll virus detection by RT-PCR in field-collected aphids. *American Potato Journal, 74:* 305.

Singh, R.P. and Singh, M. 1998. Specific detection of potato virus A in dormant tubers by reverse transcription-polymerase chain reaction. *Plant Disease, 82:* 230–234.

Singh, R.P. Singh, M. and McDonald, J.G. 1998. Screening by a 3-primer PCR of North American PVY^N isolates for European type member of the tuber necrosis-inducing PVY^{NTN} subgroup. *Canadian Journal of Plant Pathology, 20:* 227–233.

Sinha, O.K., Singh Kishan and Misra, R. 1982. Stain technique for detection of smut hyphae in nodal buds of sugarcane. *Plant Disease, 66:* 932–933.

Sinha, R.C. 1960. Red clover mottle virus. *Annals of Applied Biology, 48:* 742–748.

Sinha, R.C. 1965. Sequential infection and distribution of wound tumor virus in the internal organs of a vector after ingestion of virus. *Virology, 26:* 673–686.

Sinha, R.C. 1988. Serological detection of mycoplasma-like organism from plants affected with yellows diseases. In: *Tree Mycoplasmas and Mycoplasma Diseases* (Ed.) C. Huruki, pp. 143–156. University of Alberta Press, Edmonton.

Sinha, R.C. and Benhamou, N. 1983. Detection of mycoplasma-like organism antigens from aster yellows-diseased plants by two serological procedures. *Phytopathology, 73:* 1199–1202.

Sinijarv, R., Jarvekulg, L., Andreeva, E., and Saarma, M. 1988. Detection of potato viruses X by one incubation europium time-resolved fluoro-immunoassay and ELISA. *Journal of General Virology, 69:* 991.

Sippell, D.N. and Hall, R. 1995. Glucose phosphate isomerase polymorphisms distinguish weakly virulent from highly virulent strains of *Leptosphaeria maculans. Canadian Journal of Plant Pathology, 17:* 1–6.

Sjöstrand, F.S. 1963. *The Electron Microscopy of Cells and Tissues.* Vol. I. Instrumentation and Techniques. Academic Press, New York.

Skaf, J.S., Schultz, M.H., Hirata, H. and Zoeten, G.A. de. 2000. Mutational evidence that the VPg is involved in the replication and not the movement of pea enation mosaic virus 1. *Journal of General Virology, 81:* 1103–1109.

Skrzeczkowski, L.J., Howell, W.E. and Mink, G.I. 1996. Occurrence of peach latent mosaic viroid in commercial peach and nectarine cultivars in the US. *Plant Disease, 80:* 823.

Slack, S.A., Drennan, J.L., Westra, A.A.G., Gudmestad, N.C. and Oleson, A.E. 1996. Comparison of PCR, ELISA, and DNA hybridization for the detection of *Clavibacter michiganensis* subsp. *sepedonicus* in field-grown potatoes. *Plant Disease, 80:* 519–524.

Slykhuis, J.T. 1970. Factors determining the development of wheat spindle streak mosaic caused by a soil-borne virus in Ontario. *Phytopathology, 60:* 319–331.

Slykhuis, J.T. 1972. Transmission of plant viruses by eriophyid mites. In: *Principles and Techniques in Plant Virology* (Eds.) C.I. Kado and H.O. Agrawal, pp. 204–225. Van Nostrand Reinhold Co., New York.

Smith, F.F. and Weiss, F. 1942. *Relationship of insects to the spread of azalea flower spot.* U.S. Department of Agriculture Technical Bulletin, No. 788, pp. 1–44.

Smith, G.R., van de Velde, R., and Dale, J.L. 1992. PCR amplification of a specific double stranded RNA region of Fiji disease virus from diseased sugarcane. *Journal of Virological Methods, 39:* 237–246.

Smith, H.G., Barker, I., Brewer, G., Stevens, M. and Hallsworth, P.B. 1996. Production and evaluation of monoclonal antibodies for the detection of beet mild yellowing luteovirus and related strains. *European Journal of Plant Pathology, 102:* 163–169.

Smith, K.M. 1933. The present status of plant virus research. *Biological Reviews, 8:* 136–179.

Smith, K.M. 1935. A new virus disease of the tomato. *Annals of Applied Biology, 22:* 731–741.

Smith, K.M. 1950. Some new virus diseases of ornamental plants *Journal of Royal Horticultural Society, 75:* 350–353.

Smith, K.M. 1972. *A Text-Book of Plant Virus Diseases.* 3rd Edition. Longman, London.

Smith, K.M. and Bald, J.G. 1935. A necrotic virus disease affecting tobacco and other plants. *Parasitology, 27:* 231–245.

Smith, O.P., Damsteegt, V.D., Keller, C.J., and Beck, R.J. 1993. Detection of potato leafroll virus in leaf and aphid extracts by dot blot hybridization. *Plant Disease, 77:* 1098–1102.

Smith, O.P., Peterson, S.L., Beck, R.J., Schaad, N.W. and Bonde, M.R. 1996. Development of a PCR-based method for identification of *Tilletia indica,* causal agent of Karnal bunt of wheat. *Phytopathology, 86:* 115–122.

Sobi, G.S., and Sandhu, M.S. 1968. Relationship between leaf miner (*Phyllocnistic citrella* Stainton) injury and citrus canker (*Xanthomonas citri* (Hasse) Dowson) incidence on citrus leaves. *Journal of Research, Punjab Agricultural University, 5:* 66.

Sokmen, M.A., Barker, H. and Torrance, L. 1998. Factors affecting the detection of potato mop-top virus in potato tubers and improvement of test procedures for more reliable assays. *Annals of Applied Biology, 133:* 55–63.

Solomon, J.J. and Govindan Kutty, M.P. 1991. Etiology—mycoplasma-like organisms. In: *Coconut Root (Wilt) Disease* (Eds.) M.K. Nair, K.K.N. Nambiar, P.K. Koshy, and N.P. Jayasankar, pp. 35–40. Central Plantation Crops Research Institute, Kasaragod, India.

Sommereyns, G.H. 1959. Note relative à des reactions necrotiques provoquees pon le virus A de la pomme de terre sur *Nicotiana glutinosa* L. *Parasitica, 15:* 29–34.

Somowiyarjo, S., Sako, N., and Nonaka, F. 1988. The use of monoclonal antibody for detecting zucchini yellow mosaic virus. *Annals of Phytopathological Society, Japan, 54:* 436.

Somowiyarjo, S., Sako, N., and Nonaka, F. 1990a. The use of monoclonal antibody for detecting zucchini mosaic virus. *Annals of Phytopathological Society, Japan, 54:* 436.

Somowiyarjo, S., Sako, N., and Nonaka, F. 1990b. Production and characterization of monoclonal antibodies to watermelon mosaic virus 2. *Annals of Phytopathological Society, Japan, 56:* 541.

Soria, S.L., Vega, R., Damsteegt, V.D., McDaniel, L.L., Kitto, S.L. and Evans, T.A. 1998. Occurrence and partial characterization of a new mechanically transmissible virus in mashua from the Ecuadorian Highlands. *Plant Disease, 82:* 69–73.

Sousa Santos, M., Cruz, L., Norskov, P. and Ramussen, O.F. 1997. A rapid and sensitive detection of *Clavibacter michiganensis* subsp. *michiganensis* in tomato seeds. *Seed Science and Technology, 25:* 581–584.

Spiegel, S. and Martin, R.R. 1993. Improved detection of potato leaf roll virus in dormant potato tubers and microtubers by the polymerase chain reaction and ELISA. *Annals of Applied Biology, 122:* 493–500.

Spiegel, S., Scott, S.N., Bowman-Vance, V., Tam, Y., Galiakparov, N.N. and Rosner, A. 1996. Improved detection of prunus necrotic ringspot virus by the polymerase chain reaction. *European Journal of Plant Pathology, 102:* 681–685.

Spence, N.J. 1997. The molecular genetics of plant virus interactions. In: *The Gene-for-Gene Relationship in Plant-Parasite Interactions* (Eds.) I.R. Crute, E.B. Holub and J.J. Burdon, pp. 347–357. CAB International, U.K.

Spieker, R.L., Marinkovic, S. and Sänger, H.L. 1996. A new sequence variant of *Coleus blumei* viroid 3 from *Coleus blumei* cultivar "Fairway Ruby." *Archives of Virology, 141:* 1377–1386.

Srinivasan, N. 1982. Simple diagnostic technique for plant diseases of mycoplasma etiology. *Current Science, 51:* 883–885.

Srinivasiu B. and Narayanasamy, P. 1995a. Serological detection of phyllody disease in sesamum and leafhopper *Orosius albicinctus. Journal of Mycology and Plant Pathology, 25:* 154–157.

Srinivasulu, B. and Narayanasamy, P. 1995b. Monitoring phyllody disease in sesamum. *Journal of Mycology and Plant Pathology, 25:* 165–167.

Stace-Smith, R. 1955. Studies on Rubus virus diseases in British Columbia II. Black raspberry necrosis. *Canadian Journal of Botany, 33:* 314–322.

Stace-Smith, R. 1962. Studies on *Rubus* virus diseases in British of Columbia. IX. Ringspot disease of red raspberry. *Canadian Journal of Botany, 40:* 905–912.

Stace-Smith, R. 1966. Purification and properties of tomato ringspot virus and an RNA-deficient component. *Virology, 29:* 240–247.

Stace-Smith, R., Bowler, D.J. Mackenzie, D.J., and Ellis, P. 1993. Monoclonal antibodies differentiate the weakly virulent from highly virulent strain of *Leptosphaeria maculans,* the organism causing blackleg of canola. *Canadian Journal of Botany, 15:* 127–133.

Stachel, S.E., Messens, E., Van Montagu, M. and Zambryski, P. 1985. Identification of the signal molecules produced by wounded plant cells that activate T-DNA transfer in *Agrobacterium tumefaciens. Nature (London) 318:* 624–629.

Stachel, S.E., Timmerman, B. and Zambryski, P. 1986. Generation of single-stranded T-DNA molecule during the initial stages of T-DNA transfer from *Agrobacterium tumefaciens* to plant cells. *Nature (London), 322:* 706–712.

Stachel, S.E. and Zambryski, P.C. 1986. *virA* and *virG* control the plant-induced activation of the T-DNA transfer process of *A. tumefaciens. Cell, 46:* 325–333.

Stack-Lorenzen, P., Guilton, M.C., Werner, R. and Muhlbach, H.P. 1997. Detection and tissue distribution of potato spindle tuber viroid in infected tomato plants by tissue imprint hybridization. *Archives of Virology, 142:* 1289–1296.

Sta. Cruz, F.C. and Hibino, H. 1987. Cellular inclusions in rice grassy stunt virus (GSV) infected rice. *International Rice Research Newsletter, 12(6):* 23.

Stahl, D.J. and Schäfer, W. 1992. Cutinase is not required for fungal pathogenicity on pea. *Plant Cell, 4:* 621–629.

Stanghellini, M.E., Rasmussen, S.L. and Kim, D.H. 1999. Aerial transmission of *Thielaviopsis basicola,* a pathogen of corn-salad by adult shore flies. *Phytopathology, 89:* 476–479.

Staples, R. and Brakke, M. 1963. Relation of *Agropyron repens* mosaic and wheat streak mosaic viruses. *Phytopathology, 53:* 969–972.

Stapp, C. 1966. *Bacterial Plant Pathogens.* Oxford University Press, London.

Starkbauerová, Z. and Šrobarová, A. 1993. *Agrobacterium tumefaciens* after in vivo inoculation of potato (*Solanum tuberosum* L) (scanning electron microscope). *Journal of Phytopathology, 139:* 238–246.

Stasz, T.E., Nixon, K., Harman, G.E., Weeden, N.F., and Kuter, G.A. 1989. Evaluation of phenetic species and phylogenetic relationships in the genus *Trichoderma* by cladistic analysis of isozyme polymorphism. *Mycologia, 81:* 391–403.

Stead, D.E., Ebbels, D.L. and Pemberton, A.W. 1996. The value of indexing for disease control strategies. *Advances in Botanical Research, 23:* 1–26.

Stevens, M., Hull, R. and Smith, H.G. 1997. Comparison of ELISA and RT-PCR for the detection of beet yellows closterovirus in plants and aphids. *Journal of Virological Methods, 68:* 9–16.

Stevens, M., Smith, H.G. and Hallsworth, P.A. 1995. Detection of the luteoviruses, beet mild yellowing virus and beet Western yellows virus in aphids caught in sugar beet and oilseed rape crops. *Annals of Applied Biology, 127:* 309–320.

Stevens, M., Smith, H.G., and Hallsworth, P.B. 1994. Identification of a second strain of beet mild yellowing luteovirus using monoclonal antibodies and transmission studies. *Annals of Applied Biology, 125:* 515–520.

Stobbs, L.W. and Barker, D. 1985. Rapid sample analysis with a simplified ELISA. *Phytopathology, 75:* 492–495.

Storey, H.H. 1932. The inheritance by an insect vector of the ability to transmit a plant virus. *Proceedings of Royal Society, B., 112:* 46–60.

Storey, H.H. 1933. Investigations on the mechanism of the transmission of plant viruses by insect vectors. *Proceedings of Royal Society, B, 113:* 463–485.

Streissle, G. and Granados, R.R. 1968. The fine structure of wound rumour virus and reovirus. *Arch. Ges. Virusforsch, 25:* 369–372.

Stubbs, L.L. and Grogan, R.G. 1963. Necrotic yellows: a newly recognized virus disease of lettuce. *Australian Journal of Agricultural Research, 14:* 439–459.

Su, H.J. and Wu, R.Y. 1989. *Characterization and monoclonal antibodies of the virus causing banana bunchy top.* Technical Bulletin No. 115, pp. 1–10, Food and Fertilizer Technology Center, Taipei, Taiwan.

Sudharsana, M.R. and Reddy, D.V.R. 1989. Penicillinase based enzyme-linked immunosorbent assay for the detection of plant viruses. *Journal of Virological Methods, 26:* 45–52.

Sugimura, Y. and Ushiyama, R. 1975. Cucumber green mottle mosaic virus infection and its bearing on cytological alterations in tobacco mesophyll protoplasts. *Journal of General Virology, 29:* 93–98.

Sukhacheva, E., Novikov, V., Plaksin, D., Pavlova, I. and Ambrosova, S. 1996. Highly sensitive immunoassays for detection of barley stripe mosaic virus and beet necrotic yellow vein virus. *Journal of Virological Methods, 56:* 199–207.

Sulzinski, M.A., Schlagnhaufer, B., Moorman, G.W. and Romaine, C.P. 1998. PCR-based detection of artificial latent infections of geranium by *Xanthomonas campestris* pv. *pelargoni. Journal of Phytopathology, 146:* 111–114.

Summanwar, A.S. and Mazama, T.S. 1982. Diagnostic technique for the detection of bunchy top and infectious chlorosis in banana suckers. *Current Science, 61:* 47–49.

Sundaram, S., Plasencia, S., and Banttari, E.E. 1991. Enzyme-linked immunosorbent assay for detection of *Verticillium* spp. using antisera produced to *V. dahliae* from potato. *Phytopathology, 81:* 1485–1489.

Surguchova, N., Chirkov, S. and Atabekova, J. 1998. Enzyme–linked immunosorbent assay (ELISA) of nepoviruses using an inorganic pyrophosphatase of *Escherichia coli* as an enzyme label. *Journal of Phytopathology and Plant Protection, 31:* 535–541.

Surico, G. 1993. Scanning electron microscopy of olive and oleander leaves colonized by *Pseudomonas syringae* subsp. *savastanoi. Journal of Phytopathology, 138:* 31–40.

Suzuki, F., Zhu, Y., Sawada, H. and Matsuda, I. 1998. Identification of proteins involved in toxin production by *Pseudomonas glumae. Annals of Phytopathological Society, Japan, 64:* 75–79.

Svircev, A.M., Gardiner, R.B., McKeen, W.E., Day, A.W., and Smith, R.J. 1986. Detection by protein A-gold of antigens to Botrytis cinerea in cytoplasm of infected *Vicia faba. Phytopathology, 76:* 622–626.

Swanson, M.M., Valand, G.B., Muniyappa, V. and Harrison, B.D. 1998. Serological detection and antigenic variation of two whitefly-transmitted geminiviruses: tobacco leafcurl and cotton yellow vein mosaic viruses. *Annals of Applied Biology, 132:* 427–435.

Sweigard, J.A., Carroll, A.M., Farrall, L., Chumley, F.G. and Valent, B. 1998. *Magnaporthe grisea* pathogenicity genes obtained through insertional mutagenesis. *Molecular Plant-Microbe Interactions, 11:* 404–412.

Sylvester, E.S. 1969. Evidence for transovarial passage of sowthistle yellow vein virus in the aphid *Hyperomyzus lactucae. Virology, 38:* 440–446.

Sylvester, E.S. 1973. Reduction of excretion, reproduction and survival in *Hyperomyzus lactucae* fed on plants infected with isolates of sowthistle yellow vein virus. *Virology, 42:* 1023–1042.

Sylvester, E.S. and Richardson, J. 1966. "Recharging" pea aphids with enation mosaic virus. *Virology, 30:* 592–597.

Sylvester, E.S., Richardson, J. and Frazier, N.W. 1974. Serial passage of strawberry crinkle virus in the aphid *Chaetosiphon jacobi. Virology, 59:* 301–306.

Symons, R.H. 1981. Avocado sun blotch viroid: primary sequence and proposed secondary structure. *Nucleic Acids Research, 9:* 6527–6537.

Szychowski, J.A., McKenry, M.V., Walker, M.A. Wolpert, J.A., Credi, R. and Semancik, J.S. 1995. The vein-banding disease syndrome: a synergistic reaction between grapevine viroids and fanleaf virus. *Vitis, 34:* 229–232.

Takahaski, Y., Kameya-Iwaki, M., and Shihara, K. 1989. Characteristics of epitopes on the tomato strain of tobacco mosaic virus detected by monoclonal antibodies. *Annals of Phytopathological Society, Japan, 55:* 179.

Takahashi, Y., Omura, T., Hibino, H. and Sato, M. 1996. Detection and identification of *Pseudomonas syringae* pv. *atropurpurea* by PCR amplification of specific fragments from an indigenous plasmid. *Plant Disease, 80:* 783–788.

Takahashi, Y., Tiongco, E.R., Cabauatan, P.Q., Koganezawa, H., Hibino, H., and Omura, T. 1993. Detection of rice tungro bacilliform virus by polymerase chain reaction for assessing mild infection of plants and viruliferous vector leafhoppers. *Phytopathology, 83:* 655–659.

Takaichi, M., Yamamoto, M., Naga Kubo, I. and Oeda, K. 1998. Four garlic viruses identified by reverse transcription–polymerase chain reaction and their regional distribution in Northern Japan. *Plant Disease, 82:* 694–698.

Takamatsu, S., Nakano, M., Yokota, H. and Kunoh, H. 1998. Detection of *Rhizoctonia solani* AG-2-2 IV, the causal agent of large patch of zoysia grass, using plasmid DNA as probe. *Annals of Phytopathological Society, Japan, 64:* 451–457.

Takenaka, S. and Kawasaki, S. 1994. Characterization of alanine-rich hydroxyproline-containing cell wall proteins and their application for identifying *Pythium* species. *Physiological and Molecular Plant Pathology, 45:* 249–261.

Takeshita, M., Uchiba, T. and Takanami, Y. 1999. Sensitive detection of cucumber mosaic cucumovirus RNA with digoxigenin-labeled cRNA probe. *Journal of Faculty of Agriculture, Kyushu University, 43:* 349–354.

Takeuchi, T., Sawada, H., Suguki, F. and Matsuda, I. 1997. Specific detection of *Burkholderia plantarii* and *B. glumae* by PCR using primers selected from the 16S–23S rDNA spacer regions. *Annals of Phytopathological Society, Japan, 63:* 455–462.

Tamada, T. 1975. Beet necrotic yellow vein virus. *CMI/AAB Description of Plant Viruses, No. 144.*

Talbot, N.J. 1998. Molecular variability of fungal pathogens: using the rice blast fungus as a case study. In: *Molecular Variability of Fungal Pathogens* (Eds.) P. Bridge, Y. Couteaudier and J. Clarkson, pp. 1–18. CAB International, U.K.

Tamada, T., Uchino, H., Kusume, T. and Saito, M. 1999. RNA3 deletion mutants of beet necrotic yellow vein virus do not cause rhizomania disease in sugar beets. *Phytopathology, 89:* 1000–1006.

Tan, M.K., Timmer, L.W., Broadbent, P., Priest, M. and Cain, P. 1996. Differentiation by molecular analysis of *Elsinoe* spp causing scab diseases of citrus and its epidemiological implications. *Phytopathology, 86:* 1039–1044.

Tanaka, S., Nishii, H., Ito, S., Kameya Iwaki, M. and Sommartya, P. 1997. Detection of cymbidium mosaic potexvirus and odontoglossum ringspot tobamovirus from Thai orchids by rapid immunofilter assay. *Plant Disease, 81:* 167–170.

Tanne, E., Spiegel-Roy, P. and Shalmovitz, N. 1996. Rapid in vitro indexing of grapevine viral diseases: the effect of stress inducing agents on the diagnosis of leafroll. *Plant Disease, 80:* 972–974.

Tarr, S.A.J. 1951. *Leaf Curl Disease of Cotton*. Commonwealth Mycological Institute, Kew.

Taylor, C.E. 1972. Transmission of viruses by nematodes. In: *Principles and Techniques in Plant Virology* (Eds.) C.I. Kado and H.O. Agrawal, pp. 226–247. Van Nostrand Reinhold Co., New York.

Taylor, C.E. 1980. Nematodes. In: *Vectors of Plant Pathogens* (Eds.) K.F. Harris and K. Maramorosch, pp. 375–416. Academic Press, New York.

Teakle, D.S. 1968. Sowbane mosaic virus infecting *Chenopodium trigonon* in Queensland. *Australian Journal of Biological Sciences, 21:* 649–653.

Teakle, D.S. 1972. Transmission of plant viruses by fungi. In: *Principles and Techniques in Plant Virology* (Eds.) C.I. Kado and H.O. Agrawal, pp. 248–266. Van Nostrand Reinhold Co., New York.

Teakle, D.S. 1980. Fungi. In: *Vectors of Plant Pathogens* (Eds.) K.F. Harris and K. Maramorosch, pp. 417–438. Academic Press, New York.

Tempel, A. 1959. Serological investigations in *Fusarium oxysporum. Meded. No. 138 Landbouwhogeschool. Wageningen, 59:* 1–60.

Terranova, G., Caruso, A. and Starrantino, A. 1995. Sanitary status and sanitation of 18 clones of "Tarocco" sweet orange. *Petria, 5:* 105–110.

Tharaud, M., Baudowin, E. and Paulin, J.P. 1993. Protection against fire blight by avirulent strains of *Erwinia amylovora:* modulation of the interaction by avirulent mutants. *Acta Horticulturae, No. 338,* 321–327.

Tharaud, M., Menggad, M., Paulin, J.P. and Laurent, J. 1994. Virulence, growth and surface characteristics of *Erwinia amylovora* mutants with altered pathogenicity. *Microbiology (UK), 140:* 659–669.

Themann, K. and Werres, S. 1997. In vitro comparison of different diagnostic methods for the detection of *Phytophthora* species in water. *Proceedings of 9th International Congress on Soilless Culture,* pp. 535–547. Wageningen, The Netherlands.

Thomas, B.T. 1980. The detection by serological methods of viruses infecting the rose. *Annals of Applied Biology, 94:* 91–101.

Thomas, J.E., Geering, A.D.W., Gambley, C.F. Kessling, A.P. and White, M. 1997. Purification, properties and diagnosis of banana bract mosaic potyvirus and its distinction from abaca mosaic potyvirus. *Phytopathology, 87:* 698–705.

Thomas, J.E., Massalski, P.R., and Harrison, B.D. 1986. Production of monoclonal antibodies to African cassava mosaic virus and differences in their reactivities with other whitefly-transmitted geminivirus. *Journal of General Virology, 67:* 2739.

Thomas, W. and Mohamed, M.A. 1979. Avocado sunblotch—a viroid disease. *Australian Plant Pathology, 8:* 1–3.

Thomas, W.D. Jr., Baker, R.R., and Zoril, J.G. 1951. The use of ultraviolet light as a means of diagnosing carnation mosaic. *Science, 113:* 576–577.

Thompson, E., Leary, J.V., and Chun, W.W.C. 1989. Specific identification of *Clavibacter michiganense* subsp. *michiganense* by a homologous DNA probe. *Phytopathology, 79:* 311.

Thompson, K.G., Dietzgen, R.G., Thomas, J.E. and Teakle, D.S. 1996. Detection of pineapple bacilliform virus using the polymerase chain reaction. *Annals of Applied Biology, 129:* 57–69.

Thornton, C.R., Dewey, F.M., and Gilligan, C.A. 1993. Development of monoclonal antibody-based immunological assays for the detection of live propagules of *Rhizoctonia solani* in soil. *Plant Pathology, 42:* 763–773.

Thornton, C.R., O'Neill, T.M. Hilton, G. and Gilligan, C.A. 1999. Detection and recovery of *Rhizoctonia solani* in naturally infested glasshouse soils using a combined baiting double monoclonal antibody ELISA. *Plant Pathology, 48:* 627–634.

Thottappilly, G., Dahal, G. and Lockhart, B.E.L. 1998. Studies on a Nigerian isolate of banana streak badnavirus. I. Purification and enzyme-linked immunosorbent assay. *Annals of Applied Biology, 132:* 253–261.

Thottappilly, G., Qiu, W.P., Batten, J.S., Hughes, J.N. and Scholthof, K.B.G. 1999. A new virus on maize in Nigeria: maize mild mottle virus. *Plant Disease, 83:* 302.

Thottappilly, G., and Schmutterer, H. 1968. Zur Kenntnis eines mechanisch samen-pilz- und insektenuber-tragbaren neuen Virus der Erbse. *Zeitschrift für Pflanzenkrankheiten und Pflanzenschutz, 75:* 1–8.

Tien, P.O. 1985. Viroids and viroid diseases in China. In: *Subviral Pathogens of Plants and Animals: Viroids and Prions* (Eds.) K. Maramorosch and J.J. McKelvy, Jr., pp. 123–136. Academic Press, Orlando, FL.

Timmer, L.W., Menge, J.A., Zitko, S.E., Pond, E., Miller, S.A., and Johnson, E.L. 1993. Comparison of ELISA techniques and standard isolation methods for *Phytophthora* detection in citrus orchards in Florida and California. *Plant Disease, 77:* 791–796.

Tiongco, E.R., Cabunagan, R.C., Flores, Z.M., Hibino, H., and Koganezawa, H. 1993. Serological monitoring of rice tungo disease development in the field: its implication in disease management. *Plant Disease, 77:* 877–882.

Tirry, L., Welvaert, W., and Samyin, G. 1988. Differentiation of the hop strain from other arabis mosaic virus isolates by means of monoclonal antibodies. *Meded Fac. Landbouwwet. Rijksuniv. Gent, 53:* 423.

Tisserat, N.A., Hulbert, S.H., and Nus, A. 1991. Identification of *Leptosphaeria korrae* by cloned DNA probes. *Phytopathology, 81:* 917–921.

Tomassoli, L., Lumia, V., Cerato, C. and Ghedini, R. 1998. Occurrence of potato tuber necrotic ringspot disease (PTNRD) in Italy. *Plant Disease, 82:* 350.

Tompkins, C.M. 1938. A mosaic disease of turnip. *Journal of Agricultural Research, 57:* 589–602.

Tompkins, C.M. 1939. A mosaic disease of radish in California. *Journal of Agricultural Research, 58:* 119–130.

Tooley, P.W., Bunyard, B.A., Carras, M.M. and Hatziloukas, E. 1997. Development of PCR primers from internal transcribed spacer region 2 for detection of *Phytophthora* species–infected potatoes. *Applied and Environmental Microbiology, 63:* 1467–1475.

Tooley, P.W., Carras, M.M. and Lambert, D.H. 1998. Application of a PCR-based test for detection of potato late blight and pink rot in tubers. *American Journal of Potato Research, 75:* 187–194.

Toriyama, S. 1983a. Rice stripe virus. *CMI/AAB Description of Plant Viruses, No. 269.*

Toriyama, S. 1983b. Characterization of rice stripe virus: a heavy component carrying infectivity. *Journal of General Virology, 61:* 187–195.

Torrance, L., Hutchins, A., Larkins, A.P., and Butcher, G.N. 1984. Characterization of monoclonal antibodies to potato virus X and assessment of their potential for use in large scale surveys of virus infection. *Proceedings of Sixth International Congress of Virology, Sendai, Japan,* p. 27.

Torrance, L., and Jones, R.A.C. 1982. Increased sensitivity of detection of plant viruses obtained by using a fluorogenic substrate in enzyme-linked immunosorbent assay. *Annals of Applied Biology, 101:* 501.

Torrance, L., Pead, M.T., and Buxton, G. 1988. Production and some characteristics of monoclonal antibodies against beet necrotic yellows vein virus. *Annals of Applied Biology, 113:* 519.

Toth, I.K., Hyman, L.J., Taylor, R. and Birch, P.R.J. 1998. PCR-based detection of *Xanthomonas campestris* pv. *phaseoli* var. *fuscans* in plant material and its differentiation from *X. campestris* pv. *phaseoli. Journal of Applied Microbiology, 85:* 327–336.

Toth, I.K., Hyman, L.J. and Wood, J.R. 1999. A one step PCR-based method for the detection of economically important soft rot *Erwinia* species in micropropagated potato plants. *Journal of Applied Microbiology, 87:* 158–166.

Toth, R.L., Harper, K., Mayo, M.A. and Torrance, L. 1999. Fusion proteins of single-chain variable fragments derived from phage display libraries are effective reagents for routine diagnosis of potato leafroll virus infection in potato. *Phytopathology, 89:* 1015–1021.

Towbin, H., Stachelin, T., and Gordon, J. 1979. *Proceedings of National Academy of Sciences, USA, 76:* 4350–4354.

Trans-Kersten, J., Guam, Y. and Allen, C. 1998. *Ralstonia solanacearum* pectin methyl esterase is required for growth on methylated pectin but not for bacterial vriulence. *Applied and Environmental Microbiology, 64:* 4918–4923.

Tremaine, J.H. and Goldsack, D. 1968. The structure of regular viruses in relation to their subunit amino acid composition. *Virology, 35:* 227–239.

Tremaine, J.H. and Ronald, W.P. 1983. Serological studies of sobemovirus antigens. *Phytopathology, 73:* 794.

Trout, C.L., Ristaino, J.B. Madritch, M. and Wangsomboonde, T. 1997. Rapid detection of *Phytophthora infestans* in late-blight infected potato and tomato using PCR. *Plant Disease, 81:* 1042–1048.

Truesdell, G.M., Yang, Z.H. and Dickman, M.B. 2000. A Gα subunit gene from the phytopathogenic fungus *Colletotrichum trifolii* is required for conidial germination. *Physiological and Molecular Plant Pathology, 56:* 131–140.

Truve, E., Naess, V., Blystad, D.R., Järvekülg, L., Mäkinen, K., Tamm, T. and Mumthe, T. 1997. Detection of cocksfoot mottle virus particles and RNA in oat plants by immunological, biotechnological and electron microscopical techniques. *Archives of Phytopathology and Plant Protection, 30:* 473–485.

Tsao, P.W. 1963. Intranuclear inclusion bodies in the leaves of cotton plants infected with leaf crumple virus. *Phytopathology, 53:* 243–244.

Tsuda, S., Fujisawa, I., Hanada, K., Hidaka, S., Higo, K., Kameya-Iwaki, M., and Tomaru, K. 1994. Detection of tomato spotted wilt virus SRNA in individual thrips by reverse transcription and polymerase. *Annals of Phytopathological Society, Japan, 60:* 99–103.

Tsuda, S., Fujisawa, I., Ohnishi, J., Hosokawa, D. and Tomaru, K. 1996. Localization of tomato spotted wilt tospovirus in larvae and pupae of the insect vector *Thrips setosus. Phytopathology, 86:* 1199–1203.

Tsuda, S., Kameya-Iwaki, M., Hanada, K., Kouda, Y., Hikata, M., and Tomaru, K. 1992. A novel detection and identification technique for plant viruses: rapid immunofilter paper assay (RIPA). *Plant Disease, 76:* 466–469.

Tsuda, S., Kameya-Iwaki, M., Hanada, K., Pujisawa, I., and Tomaru, K. 1993. Simultaneous diagnosis of plants infected with multiple viruses employing rapid immunofilter paper assay (RIPA) with two step method: Multi-RIPA. *Annals of Phytopathological Society, Japan, 59:* 200–203.

Tsuneyoshi, T. and Sumi, S. 1996. Differentiation among garlic viruses in mixed infections based on RT-PCR procedures and direct tissue blotting immunoassays. *Phytopathology, 86:* 253–259.

Tsushima, S., Narimatsu, C., Mizuno, A., and Kimura, R. 1994. Cloned DNA probes for detection of *Pseudomonas glumae* causing bacterial grain rot of rice. *Annals of Phytopathological Society, Japan, 60:* 576–584.

Tu, J.C. 1976. Lysosomal distribution and acid phosphatase activity in white clover infected with clover yellow mosaic virus. *Phytopathology, 66:* 588–593.

Turturo, C., Minafra, A., Ni, H., Wang, G., Terlizzi, B., di and Savino, V. 1998. Occurrence of peach latent mosaic viroid in China and development of an improved detection method. *Journal of Plant Pathology, 80:* 165–169.

Turturo, C., D'Onghia, A.M., Minafra, A. and Savino, V. 1998. PCR detection of citrus exocortis and citrus cachexia viroids. *Phytopathologica Mediterranea, 37:* 99–105.

Tymon, A.M., Jones, P. and Harrison, N.A. 1997. Detection and differentiation of African coconut phytoplasmas: RFLP analysis of PCR-amplified 16S rDNA and DNA hybridization. *Annals of Applied Biology, 131:* 91–102.

Uchiba, T., Takeshita, M. and Takanami, Y. 1999. Quantitative detection of cucumber mosaic cucumovirus RNA with microplate hybridization using digoxigenin-labeled oligodeoxy-ribonucleotide probe. *Journal of Faculty of Agriculture, Kyushu University, 43:* 343–348.

Uehara, T. and Hosokawa, D. 1994. Distribution of tobacco mosaic virus RNA molecules in infected tobacco protoplasts revealed by in situ hybridization. *Annals of Phytopathological Society, Japan, 60:* 649–657.

Ueng, P.P., Bergstrom, G.C., Slay, R.M., Geiger, F.A., Shaner, G., and Scharen, A.L. 1992. Restriction fragment length polymorphism in the wheat glume blotch fungus, *Phaeosphaeria nodorum. Phytopathology, 82:* 1302–1305.

Ueng, P.P. and Chen, W. 1994. Genetic differentiation between *Phaeosphaeria nodorum* and *P. avenaria* using restriction fragment length polymorphisms. *Phytopathology, 84:* 800–806.

Ueng, P.P., Cunfer, B.M., Alano, A.S., Youmans, J.D., and Chen, W. 1995. Correlation between molecular and biological characters in identifying the wheat and barley biotypes of *Stagonospora nodorum. Phytopathollogy, 85:* 44–52.

Uhde, K., Kerschbaumer, R.J., Koenig, R., Hirschl, S., Lemaire, O., Boonham, N., Roake, W. and Himmler, G. 2000. Improved detection of beet necrotic yellow vein virus in a DAS-ELISA by means of antibody single chain fragments (scFv) which were selected to protease-stable epitopes from phage display libraries. *Archives of Virology, 145:* 179–185.

Ullman, D.E., Cho, J.J., Mew, R.F.L., Westcot, D.M., and Custer, D.M. 1992. A midgut barrier to tomato spotted wilt virus acquisition by adult western flower thrips. *Phytopathology, 82:* 1333–1342.

Ullman, D.E., German, T.L., Sherwood, J.L., Westcot, D.M., and Cantone, F.A. 1993. Tospovirus replication in insect vector cells: immunochemical evidence that the nonstructural protein encoded by SRNA of tomato spotted wilt tospovirus is present in thrips vector cells. *Phytopathology, 83:* 456–463.

Ura, H., Matsumoto, M., Iiyama, K., Furuya, N. and Matsuyama, N. 1998. PCR-RFLP analysis for distinction of *Burkholderia* species, causal agents of various rice diseases. *Proceedings of Association for Plant Protection of Kyushu, 44:* 5–8.

Vaira, A.M., Roggero, P., Luisoni, E., Masenga, V., Milne, R.G., and Lisa, V. 1993. Characterization of two tospoviruses in Italy: tomato spotted wilt and impatients necrotic spot. *Plant Pathology, 42:* 530–542.

Vaira, A.M., Vecchiati, M., Masenga, V. and Accotto, G.P. 1996. A polyclonal antiserum against a recombinant viral protein combines specificity with versatility. *Journal of Virological Methods, 48:* 209–219.

Valverde, R.A., Dodds, J.A., and Heick, J.A. 1986. Double stranded RNA from plants infected with viruses having elongated particles and undivided genomes. *Phytopathology, 76:* 459–465.

Van der Vlugt, C.I.M., Derks, A.F.L.M., Bonekamp, P.M., and Goldbach, R.W. 1993. Improved detection of Iris severe mosaic virus in secondarily infected iris bulbs. *Annals of Applied Biology, 122:* 279–288.

Van der Vlugt, R.A.A., Steffens, P., Cuperus, C., Barg, E., Lesemann, D.E. Bos, L. and Vetten, H.J. 1999. Further evidence that shallot yellow stripe virus (SYSV) is a distinct potyvirus and reidentification of Welsh onion yellow stripe virus as a SYSV strain. *Phytopathology, 89:* 148–155.

Van Dorst, H.J.M. and Peters, D. 1974. Some biological observations on pale fruit, a viroid-incited disease of cucumber. *Netherlands Journal of Plant Pathology, 80:* 85–96.

Van Gijsegem, F., Genin, S. and Boucher, C. 1995. *hrp* and *avr* genes—key determinants controlling the interactions between plants and gram-negative phytopathogenic bacteria. In: *Pathogens and Host Specificity in Plant Diseases: Histochemical, Biochemical, Genetic and Molecular Bases* (Eds.) U.S. Singh and R.P. Singh, Vol. I, pp. 273–292. Pergamon Press, U.K.

Van Gijsegem, F., Gough, C., Zischek, C., Niquex, E., Arlat, M., Genin, S., Barberis, P., German, S., Catello, P. and Boucher, C. 1995. The *hrp* gene locus of *Pseudomonas solanacearum* which controls the production of a type III secretion system, encodes eight proteins related to components of the bacterial flagellar biogenesis complex. *Molecular Microbiology, 15:* 1095–1114.

Van Kammen, A. 1967. Purification and properties of the components of cowpea mosaic virus. *Virology, 31:* 633–642.

Van Kan, J.A.L., Van't Klooster, J.W., Wagemakers, A.M., Dees, D.C.T. Van der Vlugt and Bergmans, C.J.B. 1997. Cutinase A of *Botrytis cinerea* is expressed, but not essential during penetration of gerbera and tomato. *Molecular Plant-Microbe Interactions, 10:* 30–38.

Van Regenmortel, M.H.V. 1982. *Serology and Immunochemistry of Plant Viruses.* Academic Press, New York.

Van Regenmortel, M.H.V. and Burckard, J. 1980. Detection of a wide spectrum of tobacco mosaic virus strains by indirect enzyme-linked immunosorbent assay (ELISA). *Virology, 106:* 327–334.

Van Regenmotel, M.H.V. and Dubs, M.C. 1993. Serological procedures. In: *Diagnosis of Plant Virus Diseases* (Ed.) R.E.F. Matthews, pp. 160–214. CRC Press, Boca Raton, FL.

Varma, A., Gibbs, A.J., and Woods, R.D. 1970. A comparative study of red clover vein mosaic virus and some other plant viruses. *Journal of General Virology, 8:* 21–32.

Varma, A., Gibbs, A.J., Woods, R.D., and Finch, J.T. 1968. Some observations on the structure of the filamentous particles of several plant viruses. *Journal of General Virology, 2:* 107–114.

Varveri, C. and Boutsika, K. 1998. Application of immunocapture PCR technique for plum pox potyvirus detection under field conditions in Greece and assays to simplify standard technique. *Acta Horticulturae, No. 472,* 475–481.

Varveri, C., Candrease, T., Cugusi, M., Ravelonandro, M., and Dunez, J. 1988. Use of a ^{32}P labeled transcribed RNA probe for dot hybridization detection of plum pox virus. *Phytopathology, 78:* 1280–1283.

Vejaratpimol, R., Channuntapipat, C., Liewsaree, P., Pewnim, T., Ito, K., Izuka, M. and Minamiura, N. 1998. Evaluation of enzyme-linked immunosorbent assays for the detection of cymbidium mosaic virus in orchids. *Journal of Fermentation and Bioengineering, 86:* 65–71.

Vela, C., Cambra, M., Cortes, E., Morerro, P., Miguet, J.G., Perez de San Roman, C., and Sanz, A. 1986. Production and characterization of monoclonal antibodies specific for citrus tristeza virus and their use for diagnosis. *Journal of General Virology, 67:* 91.

Velichetl, R.K., Lamison, C., Brill, L.M., and Sinclair, J.B. 1993. Immunodetection of *Phomopsis* species in asymptomatic soybean plants. *Plant Disease, 77:* 70–77.

Venkitesh, S.R. and Koganezawa, H. 1995. Detection of rice tungo bacilliform virus in nonvector insect species following genomic amplification. *International Rice Research Notes, 20(1):* 35–36.

Vera, C. and Milne, R.G. 1994. Immunosorbent electron microscopy and gold label antibody decoration of MLOs from crude preparations of infected plants and insects. *Plant Pathology, 43:* 190–199.

Vera Cruz, C.M., Ardales, E.Y., Skinner, D.Z., Talag, J., Nelson, R.J. Louws, F.J., Leung, H., Mew, T.M. and Leach, J.E. 1996. Measurement of halotypic variation in *Xanthomonas oryzae* pv. *oryzae* within single field by rep-PCR and RFLP analysis. *Phytopathology, 86:* 1352–1359.

Verdier, V., Mosquera, G. and Assigbétsé, K. 1998. Detection of the cassava bacterial blight pathogen, *Xanthomonas axonopodis* pv. *manihotis* by polymerase chain reaction. *Plant Disease, 82:* 79–83.

Verreault, H., Lafond, M., Asselin, A., Banville, G., and Bellemare, G. 1988. Characterization of two DNA clones specific for identification of *Corynebacterium sepedonicum. Canadian Journal of Microbiology, 34:* 993.

Vidano, C. 1970. Phases of maize rough dwarf virus multiplication in the vector *Laodelphax striatellus* Fallen. *Virology, 41:* 218–232.

Vidano, C., Lovisolo, O., and Conti, M. 1966. Transmissione del virus del nanismo ruvido del mais (MRDV) a *Triticum vulgare* L. per mezzodi *Laodlephax striarella* Fall. *Accademia Scienza Torrino, 199:* 125–140.

Visvader, J.E., Gould, A.R., Bruening, G.E., and Symons, R.H. 1982. Citrus exocortis viroid nucleotide sequence and secondary structure of an Australian isolate. *FEBS Letters, 137:* 288–292.

Viswanathan, R. 1997a. Detection of phytoplasmas associated with grassy shoot disease by
ELISA techniques. *Zeitschrift fur Pflanzenkrankheiten und Pflanzenschutz, 104:*
9–16.

Viswanathan, R. 1997b. Detection of ratoon stunting disease (RSD) bacterium by ELISA.
Madras Agricultural Journal, 84: 374–377.

Viswanathan, R. and Alexander, K.C. 1995. Production of polyclonal antisera to grassy
shoot disease (MLOs) of sugarcane and their use in disease detection. In: *Detection
of Plant Pathogens and Their Management* (Eds.) J.P. Verma, A. Varma, and Dinesh Kumar, pp. 153–158. Angkor Publishers, New Delhi.

Viswanathan, R., Alexander, K.C., and Garg, I.D. 1996. Detection of sugarcane bacilliform
virus in sugarcane germplasm. *Acta Viologica, 40:* 5–8.

Viswanathan, R., Padmanaban, P., Mohanraj, D. and Nallathambi, P. 1998. Immunological techniques for the detection of leaf scald disease of sugarcane. *Indian Phytopathology, 51:* 187–189.

Viswanathan, R. and Premachandran, M.N. 1998. Occurrence and distribution of sugarcane bacilliform virus in the sugarcane germplasm collection in India. *Sugar Cane,
No. 6,* 9–18.

Viswanathan, R., Samiyappan, R. and Padmanaban, P. 1998. Specific detection of *Colletotrichum falcatum* in sugarcane by serological techniques. *Sugar Cane. No. 6,*
18–23.

Vitushkina, M., Fechtner, B. and Jelkmann, W. 1997. Development of an RT-PCR for the
detection of little cherry virus and characterization of some isolates occurring in Europe. *European Journal of Plant Pathology, 103:* 803–808.

Vivian-Nauer, A., Hoffmann-Boller, P. and Gafner, J. 1997. In vivo detection of folpet and
its metabolite phthalimide in grape must and wine. *American Journal of Enology
and Viticulture, 48:* 67–70.

Vlugt, R.A.A., Van der, Berendsen, M. and Koenraadt, 1997. Immunocapture reverse transcriptase PCR for detection of lettuce mosaic virus. In: *Seed Health Testing:
Progress towards the 21st Century* (Ed.) J.D. Hutchins and J.C. Reeves, pp.
185–198. CAB International, U.K.

Voigt, K., Schleier, S. and Wöstemeyer, J. 1998. RAPD-based molecular probes for black
leg fungus *Leptosphaeria maculans (Phoma lingam):* evidence for pathogenicity
group-specific sequences in the fungal genomes. *Journal of Phytopathology, 146:*
567–576.

Voller, A., Bartlett, A., Bidwell, D.E., Clark, M.F., and Adams, A.N. 1976. The detection
of viruses by enzyme-linked immunosorbent assay (ELISA). *Journal of General Virology, 33:* 165–167.

Vunsh, R., Rosner, A., and Stein, A. 1990. The use of the polymerase chain reaction (PCR)
for the detection of bean yellow mosaic virus. *Annals of Applied Biology, 17:*
561–569.

Vunsh, R., Rosmer, A., and Stein, A. 1991. Detection of bean yellow mosaic virus in gladioli corms by the polymerase chain reaction. *Annals of Applied Biology, 119:*
289–294.

Wah, Y.F.W.C. and Symons, R.H. 1997. A high sensitivity RT-PCR for the diagnosis of grapevine viroids in field and tissue culture samples. *Journal of Virological Methods, 63:* 57–69.

Wah, Y.F.W.C. and Symons, R.H. 1999. Transmission of viroids via grape seeds. *Journal of Phytopathology, 147:* 285–291.

Wakimoto, S. 1957. A simple method for comparisons of bacterial population in a large number of samples by phage technique. *Annals of Phytopathological Society, Japan, 22:* 159–163.

Wakimoto, S. 1960. Classification of strains of *Xanthomonas oryzae* on the basis of their susceptibility against bacteriophages. *Annals of Phytopathological Society, Japan, 25:* 193–198.

Walcott, R.R. and Gitaitis, R.D. 2000. Detection of *Acidovorax avenae* subsp. *citrulli* in watermelon seed using immunomagnetic separation and the polymerase chain reaction. *Plant Disease, 84:* 470–474.

Walkey, D.G.A., Lyons, N.F., and Taylor, J.D. 1992. An evaluation of a virobacterial aggulutination test for the detection of plant viruses. *Plant Pathology, 41:* 462–471.

Wallace, J.M. 1951. Recent developments in the study of quick decline and related diseases. *Phytopathology, 41:* 785–793.

Wallace, J.M. and Drake, R.J. 1961. Induction of woody galls by wounding of citrus infected with vein enation virus. *Plant Disease Reporter, 45:* 682–686.

Walter, B. 1981. Un viroide de la tomate en Afrique de l'ouest: identité avec le viroide du "potato spindle tuber." *C. R. Academy of Science, Paris, 292 III:* 537–542.

Wang, C., Chin, C.K. and Chen, A. 1998. Expression of the yeast δ-9 desaturase gene in tomato enhances its resistance to powdery mildew. *Physiological and Molecular Plant Pathology, 52:* 371–383.

Wang, H.L. and Gonsalves, D. 1999. Utilization of monoclonal and polyclonal antibodies to monitor the protecting and challenging strains of zucchini yellow mosaic virus in cross-protection. *Plant Pathology Bulletin, 8:* 111–116.

Wang, P. and Van Etten, H.D. 1992. Cloning and properties of a cyanide hydratase gene from the pathogenic fungus *Gloeocercospora sorghi. Biochemical and Biophysics Research Communications, 187:* 1048–1054.

Wang, R.V., Gergerich, R.C., and Kim, K.S. 1994. Entry of ingested plant viruses into hemocoel of the beetle vector *Diabrotica undecimpunctata howardii. Phytopathology, 84:* 147–153.

Wang, S. and Gergerich, R.C. 1998. Immuno-fluorescent localization of tobacco ringspot nepovirus in the vector nematode *Xiphinema americanum. Phytopathology, 88:* 885–889.

Wang, W.Y., Mink, G.I., and Silbernagel, M.J. 1985. A broad spectrum monoclonal antibody prepared against bean common mosaic virus. *Phytopathology, 75:* 1352.

Wang Yu Chi and Yu Roch Chui. 1998. Detection of toxigenic *Aspergillus* spp. in rice and corn by ELISA. *Journal of Chinese Agricultural Chemical Society, 36:* 512 –520.

Wang, Z.K., Comstock, J.C. Hatziloukas, E. and Schaad, N.W. 1999. Comparison of PCR, BIO-PCR, DIA, ELISA and isolation on semiselective medium for detection of *Xanthomonas albilineans*, the causal agent of leaf scald of sugarcane. *Plant Pathology, 48:* 245–252.

Wang, Zhong Kang, Luo Huai Hai and Shu Zheng Yi. 1997. Application of DIA (dot immunobinding assay) for rapid detection of *Xanthomonas axonopodis* pv. *citri. Journal of Southwest Agricultural University, 19:* 529–532.

Ward, E. and Gray, R.M. 1992. Generation of ribosomal DNA probe by PCR and its use in
identification of fungi with the *Gaeumannomyces-Phialophora* complex. *Plant
Pathology, 41:* 730–736.

Ward, L.J. and De Boer, S.H. 1984. Specific detection of *Erwinia carotovora* subsp.
atroseptica with digoxigenin-labeled DNA probe. *Phytopathology, 84:* 180–186.

Ward, L.J. and De Boer, S.H. 1990. A DNA probe specific for serologically diverse strains
of *Erwinia carotovora. Phytopathology, 80:* 665.

Warmke, H.E. and Edwardson, J.R. 1966a. Use of potassium permanganate as a fixative
for virus particles in plant tissues. *Virology, 28:* 693–700.

Warmke, H.E. and Edwardson, J.R. 1966b. Electron microscopy of crystalline inclusions
of tobacco mosaic virus in leaf tissue. *Virology, 30:* 45–57.

Watanabe, S., Ito, S., Omoda, N., Mimakata, H., Hayashi, M. and Yuasa, Y. 1998. Development of a competitive enzyme-linked immunosorbent assay based on a monoclonal antibody for a fungicide flutolanil. *Analytica Chimica Acta, 376:* 93–96.

Watson, D.H., Le Bouvier, G.L., Tomlinson, J.A., and Walkey, D.G. 1966. Electron microscopy and antigen precipitates extracted from gel diffusion plates. *Immunology,
10:* 305.

Watson, M.A. 1972. Transmission of plant viruses by aphids. In: *Principles and Techniques in Plant Virology* (Eds.) C.I. Kado and H.O. Agrawal, pp. 131–167. Van Nostrand Reinhold Co., New York.

Watson, M.A. and Okusanya, B.A.M. 1967. Studies on the transmission of groundnut
rosette virus by *Aphis craccivora* Koch. *Annals of Applied Biology, 60:* 199–208.

Watson, W.T., Kenerley, C.M. and Appel, D.N. 2000. Visual and infrared assessment of
root colonization of apple trees by *Phymatotrichopsis omnivora. Plant Disease, 84:*
539–543.

Weathers, L.G. 1965. Transmission of citrus exocortis virus of citrus by *Cuscuta subin-
clusa. Plant Disease Reporter, 49:* 189–190.

Weathers, L.G. and Greer, F.C. 1972. *Gynura* as a host for exocortis virus and citrus. *Proceedings of 5th Conference of International Organization of Citrus Virologists,*
(Ed.) W.C. Price, pp. 95–98. University of Florida Press, Gainesville.

Webb, R.E. and Buck, R.W. 1955. A diagnostic host for potato virus A. *American Potato
Journal, 32:* 248–253.

Weber, H. and Pfitzner, A.J.P. 1998. Tm-2^2 resistance in tomato requires recognition of the
carboxy terminus of the movement protein of tomato mosaic virus. *Molecular Plant-
Microbe Interactions, 11:* 498–503.

Weeds, P.L., Beever, R.E., Sharrock, K.R. and Long, P.G. 1999. A major gene controlling
pathogenicity in *Botryotinia fuckeliana (Botrytis cinerea). Physiological and
Molecular Plant Pathology, 54:* 13–35.

Wei, W.S., Plovanich-Jones, A., Deng, W.L., Jin, Q.L., Collmer, A., Huang, H.C. and He,
S.Y. 2000. The gene coding for the Hrp pilus structural protein is required for type
III secretion of HrP and Avr proteins in *Pseudomonas syringae* pv. *tomato. Pro-
ceedings of the National Academy of Sciences, U.S.A., 97:* 2247–2252.

Weidemann, H.L., and Buchta, U. 1998. A simple and rapid method for the detection of potato spindle tuber viroid (PSTVd) by RT-PCR. *Potato Research, 41:* 1–8.

Weier, H.U. and Gray, J.W. 1988. A programmable system to perform the polymerase chain reaction. *DNA, 7:* 441–447.

Weiland, J.J., Steffenson, B.J., Cartwright, R.D. and Webster, R.K. 1999. Identification of molecular genetic markers in *Pyrenophora teres* f. *teres* associated with low virulence on "Harbin" barley. *Phytopathology, 89:* 178–181.

Weintraub, M. and Ragetli, H.W.B. 1966. Fine structure of inclusions and organelles in *Vicia faba* infected with bean yellow mosaic virus. *Virology, 28:* 290–302.

Weintraub, M. and Ragetli, H.W.J. 1970. Electron microscopy of the bean and cowpea strains of southern bean mosaic virus within leaf cells. *Journal of Ultrastructure Research, 32:* 167–189.

Welliver, R.A. and Halbrendt, J.M. 1992. Dodder transmission of tomato ring spot virus. *Plant Disease, 76:* 642.

Welnicki, M. and Hiruki, C. 1992. Highly sensitive digoxigenin-labeled DNA probe for the detection of potato spindle tuber viroid. *Journal of Virological Methods, 39:* 91–99.

Wesley, S.V., Miller, J.S., Devi, P.S., Delfosse, P., Naidu, R.A., Mayo, M.A., Reddy, D.V.R. and Jana, M.K. 1996. Sensitive broad-spectrum detection of Indian peanut clump virus by nonradioactive nucleic acid probes. *Phytopathology, 86:* 1234–1237.

Wetter, C. 1971. Potato Virus S. *CMI/AAB Descriptions of Plant Viruses No. 60.*

Wetter, C. and Brandes, J. 1956. Untersuchungen uber das Kartoffel-S-virus. *Phytopathologische Zeitschrift, 26:* 81–92.

Wetter, C., Paul, H.L., Brandes, J., and Quantz, L. 1960. Vergleich zwischen Eigenschaften des Echten Acker bohnenmosaik Virus and des broad bean mottle Virus. *Zeitschr. f. Naturforsch, 15b, No. 7:* 444–447.

Wetter, C. and Quantz, L. 1958. Serologische Verwandtschaft zwvisehen Stein kleevirus, Stauchevirus der Erbe and Wisconsin pea streak virus. *Phytopathologische Zeitschrift. 33:* 430–432.

Wetzel, T., Candresse, T., Ravelonandro, M., and Dunez, J. 1991. A polymerase chain reaction assay adapted to plum pox potyvirus detection. *Journal of Virological Methods, 33:* 355.

Wetzel, T., Candresse, T., Macquaire, G., Ravelonandro, M., and Dunez, J. 1992. A highly sensitive immunocapture polymerase chain reaction method for plumpox potyvirus detection. *Journal of Virological Methods, 39:* 27–37.

Whisson, D.L., Herdina and Francis, L. 1995. Detection of *Rhizoctonia solani*-AG-8 in soil using a specific DNA probe. *Mycological Research, 99:* 1299–1302.

Whitcomb, R.F. 1972. Transmission of viruses and mycoplasma by the auchenorrhynchous Homoptera. In: *Principles and Techniques in Plant Virology* (Eds.) C.I. Kado and H.O. Agrawal, pp. 168–203. Van Nostrand Reinhold Co., New York.

Whitcomb, R.F. and Tully, J.G. (Eds.) 1989. *Mycoplasmas. Vol. V. Spiroplasmas. Acholeplasmas and Mycoplasmas of Plants and Arthropods.* Academic Press, San Diego.

Wiggerich, H.G. and Pühler, A. 2000. The exbD2 gene as well as the iron uptake genes *ton B, exbB* and *exbD1* of *Xanthomonas campestris* pv. *campestris* are essential for the induction of a hypersensitive response of pepper *(Capsicum annuum). Microbiology (Reading), 146:* 1053–1060.

Wiglesworth, M.D., Nesmith, W.C., Schardl, C.L., Li, D., and Siegel, M.R. 1994. Use of specific repetitive sequences in *Peronospora tabacina* for the early detection of the tobacco blue mold pathogen. *Phytopathology 84:* 425–430.

Wijkamp, I., Lent, J., Van Kormelin, K.R., Golbach, R., and Peters, D. 1993. Multiplication of tomato spotted wilt virus in its vector, *Frankliniella occidentalis*. *Journal of General Virology, 74:* 341–349.

Wilkinson, R.E., and Ghodgett, F.M. 1948. *Gomphrena globosa* a useful plant for qualitative and quantitative work with potato virus X. *Phytopathology, 38:* 28.

Williams, R.C. and Steere, R.L. 1951. Electron microscopic observations on the unit length of the particles of tobacco mosaic virus. *Journal of American Chemical Society, 73:* 2057.

Williamson, D.L. and Whitcomb, R.F. 1975. Plant mycoplasmas: a cultivable spiroplasma causes com stunt disease. *Science, 188:* 1018–1020.

Willits, D.A. and Sherwood, J.E. 1999. Polymerase chain reaction detection of *Ustilago hordei* in leaves of susceptible and resistant barley varieties. *Phytopathology, 89:* 212–217.

Wilson, D.M., Sydenham, E.W. Lombaert, G.A., Trucksess, M.W., Abramson, D. and Bennett, G.A. 1998. Mycotoxin analytical techniques. In: *Mycotoxins in Agriculture and Food Safety* (Eds.) K.K. Sinha and D. Bhatnagar, pp. 135–182. Marcel Dekker, New York.

Wirawan, I.G.P., Kang, H.W. and Kojima, M. 1993. Isolation and characterization of a new chromosomal virulence gene of *Agrobacterium tumefaciens*. *Journal of Bacteriology, 175:* 3208–3212.

Wisler, G.C., Baker, C.A., Purcifull, D.E., and Hiebert, E. 1989. Partial characterization of monoclonal antibodies to zucchini yellow mosaic virus ZYMV and water melon mosaic virus (WMV-2). *Phytopathology, 79:* 1213.

Wisler, G.C., Li, R.H., Liu, H.Y., Lowry, D.S. and Duffus, J.E. 1998. Tomato chlorosis virus: a new whitefly transmitted, phloem-limited bipartite closterovirus of tomato. *Phytopathology, 88:* 402–409.

Wisler, G.C., Liu, H.Y., Klassen, V.A., Duffus, J.E. and Falk, B.W. 1996. Tomato infectious chlorosis virus has a bipartite genome and induces phloem-limited inclusions characteristic of the closteroviruses. *Phytopathology, 86:* 622–626.

Wisniewski, L.A., Powell, P.A., Nelson, R.S. and Beachy, R.W. 1990. Local and systemic spread of tobacco mosaic virus in transgenic tobacco. *Plant Cell, 2:* 559.

Wolf, J.M. van der, Hyman, L.J., Jones, D.A.C., Grevesse, C., Beckhoven, J.R.C.M. van, Vuurde, J.W.L. van and Pérombelon, M.C.M. 1996 Immunomagnetic separation of *Erwinia carotovora* subsp. *atroseptica* from potato peel extracts to improve detection sensitivity on a crystal violet pectate medium or by PCR. *Journal of Applied Bacteriology, 80:* 487–495.

Wolf, P. 1969. *Chenopodium* Arten als Lokallasionswirte des Selleriemosaik Virus (western celery mosaic virus). *Monatsberichte, 11:* 293–298.

Woo, S.L., Zoina, A., Del Sorbo, G., Lorito, M., Nanni, B., Scala, F. and Noviello, C. 1996. Characterization of *Fusarium oxysporum* f.sp. *phaseoli* by pathogenic races, VCGs, RFLPs or RAPD. *Phytopathology, 86:* 966–973.

Wu, R.Y. and Ko, W.H. 1998. A simple method for preserving serological activity of banana bunchy top virus in diseased tissue. *Annals of Phytopathological Society, Japan, 64:* 34–37.

Wu, R.Y. and Su, H.J. 1990. Production of monoclonal antibodies against banana bunchy top virus and their use in enzyme-linked immunosorbent assay. *Journal of Phytopathology, 128:* 203.

Wu, W.C., Lee, S.T., Kuo, H.P. and Wang, L.Y. 1993. Use of phages for identifying the citrus canker bacterium *Xanthomonas campestris* pv. *citri* in Taiwan. *Plant Pathology, 42:* 389–395.

Wu Ya Qin, Chen Shuang Ying, Wang Wen Hui and Wang Xiao Feng, 1998. Comparison of three ELISA methods for the detection of apple chlorotic leafspot virus and apple stem grooving virus. *Acta Phytopathologica, Sinica, 25:* 245–248.

Wu, Z.C. and Hu, J.S. 1996. Comparison of ELISA, dot-blot and PCR assays for detection of whitefly-transmitted geminivirus. *International Journal of Tropical Plant Diseases, 13:* 205–211.

Wullings, B.A., Beuningen, A.R. van, Janse, J.D. and Akkermans, A.D.L. 1998. Detection of *Ralstonia solanacearum* which causes brown rot of potato by fluorescent in situ hybridization with 23S rRNA-targeted probes. *Applied and Environmental Microbiology, 64:* 4546–4554.

Wyatt, S.D. and Brown, J.K. 1996. Detection of subgroup III geminivirus isolates in leaf extracts by degenerate primers and polymerase chain reaction. *Phytopathology, 86:* 1288–1293.

Xiao, H.G. and Hu, J.S. 1999. Detection of banana bunchy top virus by polymerase chain reaction assays. *Journal of South China Agricultural University, 20:* 5–8.

Xie, G.L. and Mew, T.W. 1998. A leaf inoculation method for detection of *Xanthomonas oryzae* pv. *oryzicola. Plant Disease, 82:* 1007–1011.

Xie, W.S. and Hu, J.S. 1995. Molecular cloning, sequence analysis and detection of banana bunchy top virus in Hawaii. *Phytopathology, 85:* 339–347.

Xu, M.L., Melchinger, A.E. and Lübberstedt, T. 1999. Species-specific detection of the maize pathogens *Sporisorium reiliana* and *Ustilago maydis* by dot-blot hybridization and PCR-based assays. *Plant Disease, 83:* 390–395.

Xu, J.R. and Hamer, J.E. 1996. MAP kinase and cAMP signaling regulate infection structure formation and pathogenic growth in the rice blast fungus *Magnaporthe grisea. Gene and Development, 10:* 2696–2706.

Xu, J.R. Urban, M., Sweigard, J.A. and Hamer, J.E. 1997. The CPKA gene of *Magnaporthe grisea* is essential for appressorial penetration. *Molecular Plant-Microbe Interactions, 10:* 187–194.

Xu, L. and Hampton, R.O. 1996. Molecular detection of bean common mosaic and bean common mosaic necrosis potyviruses and pathogroups. *Archives of Virology, 141:* 1961–1977.

Xue, B., Goodwin, P.H., and Annis, S.L. 1992. Pathotype identification of *Leptosphaeria maculans* with PCR and oligonucleotide primers from ribosomal internal transcribed spacer sequences. *Physiological and Molecular Plant Pathology, 141:* 179–188.

Yamada, K. and Shikata, E. 1969. *Journal of Faculty of Agriculture, Hokkaido University*, *56:* 91.

Yamaji, Y., Horikoshi, K., Yamashita, H. and Matsumoto, T. 1998. Detection of tobacco rattle virus on gladiolus by RT-PCR. *Research Bulletin of Plant Protection Service, Japan, No. 34,* 107–111.

Yamashita, H., Holikoshi, K., Shimada, H. and Matsumoto, T. 1996. Analysis of double-stranded RNA for plant virus diagnosis. *Research Bulletin of Plant Protection Service, Japan, No. 32,* 111–114.

Yambao, M.L.M., Cabauatan, P.Q. and Azzam, O. 1998. Differentiation of rice tungro spherical virus variants by RT-PCR and RFLP. *International Rice Research Notes, 23(2):* 22–24.

Yang, S.S. and Kim, C.H. 1996. Studies on cross-protection of *Fusarium* wilt of cucumber. IV. Protective effect by a nonpathogenic isolate of *Fusarium oxysporum* in greenhouse and fields. *Korean Journal of Plant Pathology, 12:* 137–141.

Yang, X., Hadidi, A., and Garnsey, S.M. 1992. Enzymatic cDNA amplification of citrus exocortis and cachexia viroids from infected citrus hosts. *Phytopathology, 82:* 279–285.

Yang, Y., Kim, K.S. and Anderson, E.J. 1997. Seed transmission of cucumber mosaic virus in spinach. *Phytopathology, 87:* 924–931.

Yang, Z.N. and Mirkov, T.E. 1997. Sequence and relationships of sugarcane mosaic and sorghum virus strains and development of RT-PCR based RFLPs for strain discrimination. *Phytopathology, 87:* 932–939.

Yao, C.L., Magill, C.W., and Frederiksen, R.A. 1991. Use of an A-T rich DNA clone for identification and detection of *Peronosclerospora sorghi*. *Applied Environmental Microbiology, 57:* 2027.

Yashitola, J., Krishnaveni, D., Reddy, A.P.K. and Sonti, R.V. 1997. Genetic diversity within the population of *Xanthomonas oryzae* pv. *oryzae* in India. *Phytopathology, 87:* 760–765.

Yates, I.E., Hiett, K.L., Kapczynski, D.R. Smart, W. Glenn, A.E., Hinton, D.M., Bacon, C.W., Meinersmann, R., Liu, S. and Jaworski, A.J. 1999. GUS transformation of the maize endophyte *Fusarium moniliforme*. *Mycological Research, 103:* 129–136.

Ye, R., Xu, L., Gao, Z.Z., Yang, J.P., Chen, J., Chen, J.P., Adams, M.J. and Yu, S.Q. 2000. Use of monoclonal antibodies for the serological differentiation of wheat and oat furoviruses. *Journal of Phytopathology, 148:* 257–262.

Yoder, O.C. and Gracen, V.E. 1975. Segregation of pathogenicity types and host-specific toxin production in progenies of crosses between races T and O of *Helminthosporium maydis (Cochliobolus heterostrophus)*. *Phytopathology, 65:* 273–276.

Yokoyama, M., Ogawa, M., Nozu, Y., and Hashimoto, J. 1990. Detection of specific RNAs by in situ hybridization in plant protoplasts. *Plant and Cell Physiology, 31:* 403–406.

Yoneyama, K., Kono, Y., Yamaguchi, I., Horikoshi, M. and Hirooka, T. 1998. Toxoflavin is an essential factor for virulence of *Burkholderia glumae* causing rice seedling rot disease. *Annals of Phytopathological Society, Japan, 64:* 91–96.

Young, B.D. and Anderson, M.L.M. 1985. Quantitative analysis of solution hybridization. In: *Nucleic Acid Hybridization—Practical Approach* (Eds.) B.D. Hames and S.J. Higgins, pp. 47–71. IRL Press, Oxford.

Young, J.M., Takikawa, Y., Gardan, L. and Stead, D.E. 1992. Changing concepts in the taxonomy of plant pathogenic bacteria. *Annual Review of Phytopathology, 30:* 67–105.

Young, J.M. Saddler, G.S., Takikawa, Y., De Boer, S.H., Vauterin, I., Gardan, L., Gvozdyak, R.I. and Stead, D.E. 1996. Names of plant pathogenic bacteria 1864–1995. *Review of Plant Pathology, 75:* 721–762.

Yuen, G.Y., Craig, M.L., and Avila, F. 1993. Detection of *Pythium ultimum* with a species specific monoclonal antibody. *Plant Disease, 77:* 692–698.

Yuen, G.Y., Xia, J.Q. and Sutula, C.L. 1998. A sensitive ELISA for *Pythium ultimum* using polyclonal and species-specific monoclonal antibodies. *Plant Disease, 82:* 1029–1032.

Zaitlin, M. and Hariharasubramanian, V. 1972. A gel electrophoretic analysis of proteins from plants infected with tobacco mosaic and potato spindle tuber viruses. *Virology, 47:* 296–305.

Zambryski, P.C. 1992. Chronicles from the *Agrobacterium*-plant cell DNA transfer story. *Annual Review of Plant Physiology and Plant Molecular Biology, 43:* 465–490.

Zeidan, M. and Czosnek, H. 1991. Acquisition of yellow leafcurl virus by the whitefly *Bemisia tabaci*. *Journal of General Virology, 72:* 2607–2614.

Zeigler, R., Aricapa, G., and Hoyos, E. 1987. Distribution of fluorescent *Pseudomonas* spp. causing grain and sheath discolouration in rice in Latin America. *Plant Disease, 71:* 896–900.

Zeller, S.M. and Childs, L. 1925. Perennial canker of apple trees: a preliminary report. Oregon State College, *Agriculture Experiment Station Bulletin, No. 217.*

Zerbini, F.M., Koike, S.T., and Gilberton, R.L. 1995. Biological and molecular characterization of lettuce mosaic potyvirus isolates from the Salinas Valley of California. *Phytopathology, 85:* 746–752.

Zhang, A.P. 1999. Direct tissue blotting ELISA for detection of carnation mosaic virus in carnation. *Acta Agriculturae Shangai, 15:* 84–86.

Zhang, A.W. Hartman, G.L., Curio-Penny, G., Pedersen, W.L., and Becker, K.B. 1999. Molecular detection of *Diaporthe phaseolorum* and *Phomopsis longicolla* from soybean seeds. *Phytopathology, 89:* 796–804.

Zhang, A.W., Hartman, G.L. Riccioni, L. Chen, W.D. Ma, R.Z. and Pedersen, W.L. 1997. Using PCR to distinguish *Diaporthe phaseolorum* and *Phomopsis longicolla* from other soybean fungal pathogens and to detect them in soybean tissues. *Plant Disease, 81:* 1143–1149.

Zhang, S. and Goodwin, P.H. 1997. Rapid and sensitive detection of *Xanthomonas fragariae* by simple alkaline DNA extraction and polymerase chain reaction. *Journal of Phytopathology, 145:* 267–270.

Zhang, Y., Merighi, M., Bazzi, C. and Geider, K. 1998. Genomic analysis by pulse-field gel electrophoresis of *Erwinia amylovora* strains from the Mediterranean region including Italy. *Journal of Plant Pathology, 80:* 225–232.

Zhang, Y.P. Uyemoto, J.K., Golino, D.A. and Rowhani, A. 1998. Nucleotide sequence and RT-PCR detection of a virus associated with grapevine rupestris stem-pitting disease. *Phytopathology, 88:* 1231–1237.

Zheng, F.C. and Ward, E. 1998. Variation within and between *Phytophthora* species for rubber and citrus trees in China by polymerase chain reaction using RAPDs. *Journal of Phytopathology, 146:* 103–109.

Zhu, H., Gao, J.L., Li, Q.X., and Hu, G.X. 1988. Detection of *Xanthomonas oryzae* in rice by monoclonal antibodies. *Journal of Jiangsu Agricultural College, 9(2):* 41–43.

Zhu, S.F., Gillet, J.M., Ramsdell, D.C. and Hadidi, A. 1995a. Isolation and characterization of viroid-like RNAs from mosaic diseased blueberry tissue. *Acta Horticulturae, No. 385,* 132–140.

Zhu, S.F., Hadidi, A., Yang, X., Hammand, R.W. and Hansen, A.J. 1995b. Nucleotide sequence and secondary structure of pome fruit viroids from dapple apple diseased apples, pear rusty skin diseased pears and apple scar skin symptomless pears. *Acta Horticulturae, No. 386,* 554–559.

Ziegler, A., Mayo, M.A. and Torrance, L. 1998. Synthetic antigen from a peptide library can be an effective positive control in immunoassays for the detection and identification of two geminiviruses. *Phytopathology, 88:* 1302–1305.

Zimand, G., Valinksy, L., Elad, Y., Chet, I., and Manulis, S. 1994. Use of RAPD procedure for the identification of *Trichoderma* strains. *Mycological Research, 98:* 531–534.

Zimmermann, D., Sommermeyer, G., Walter, B., and Van Regenmortel, M.H.V. 1990. Production and characterization of monoclonal antibodies specific to closterovirus-like particles associated with grapevine leafroll disease. *Journal of Phytopathology, 130:* 279.

Zrein, M., Burckard, J., and Van Regenmortel, M.H.V. 1986. Use of the biotin-avidin system for detecting a broad range of serologically related plant viruses by ELISA. *Journal of Virological Methods, 13:* 121–128.

Index

Printed in the United States
by Baker & Taylor Publisher Services

Printed in the United States
by Baker & Taylor Publisher Services